Invading Nature - Springer Series in Invasion Ecology

Volume 10

Biological Invasions represent one of those rare themes that cut across the disciplines of academic biology, while having profound environmental, philosophical, socioeconomic, and legislative implications at a global scale. There can be no doubt that biological invasions represent the single greatest threat to biodiversity past the activities of humankind itself. The implications are far reaching. Novel ecological and evolutionary forces are now directing the future expression of life itself, as native species and the communities that they comprise contend with invading species. The rules of the game have been suddenly and irrevocably changed.

Invading Nature - Springer Series in Invasion Ecology is a new book series topically spanning the breadth of invasion biology. The series is of singular importance as an integrative venue focusing on the broader ecological and evolutionary issues arising from non-native species, the impacts such species have in particular environs, trends patterns and processes, as well as causes and correctives. The series seeks novel and synthetic approaches to invasions including experimental, theoretical, systematic and conceptual treatments

Prospective authors and/or editors should consult the **Series Editor Daniel Simberloff** for more details:

e-mail: tebo@utk.edu

More information about this series at http://www.springer.com/series/7228

Demetrio Boltovskoy
Editor

Limnoperna fortunei

The Ecology, Distribution and Control of a Swiftly Spreading Invasive Fouling Mussel

Springer

Editor
Demetrio Boltovskoy
Instituto de Ecología, Genética y Evolución
de Buenos Aires (IEGEBA)
Universidad de Buenos Aires-CONICET
Argentina

Invading Nature - Springer Series in Invasion Ecology
ISBN 978-3-319-13493-2   ISBN 978-3-319-13494-9 (eBook)
DOI 10.1007/978-3-319-13494-9

Library of Congress Control Number: 2015933521

Springer Cham Heidelberg New York Dordrecht London

Printed on acid-free paper

Springer is part of Springer Science+Business Media (www.springer.com)

Foreword

Alien invasive species are generally regarded as one of mankind's greatest ecological problems. Reports of species introductions and their myriad impacts have exploded since the early 1990s and include virtually all kingdoms and phyla. Included in this group is a small set of bivalve molluscs, including zebra (*Dreissena polymorpha*) and quagga (*D. rostriformis bugensis*) mussels. These two species have spread across much of Europe and North America, and were instrumental to efforts by managers in both the USA and Canada to develop better invasive species management strategies for preventing undesirable invasions and more effectively eradicate or limit dispersal of those that do succeed. British researchers recently identified the quagga mussel as the country's most high-risk potential invader, as spread of the species picks up across Europe.

A third member of the class Bivalvia and the topic of this volume—the golden mussel (*Limnoperna fortunei*)—has also been introduced abroad and is spreading quickly. Golden mussels are native to Southeast Asia, but have spread both in that region and throughout much of eastern and central South America. The latter invasions have occurred with lightning speed, with spread occurring over ~2000 km in ~20 years. Like the aforementioned dreissenid species, the golden mussel is an ecological engineer, radically altering many of the ecosystems it invades. Also like the dreissenid species, the golden mussel poses a significant biofouling problem to municipalities and industrial users of raw water. I became acutely aware of the pervasive nature of golden mussel biofouling while traveling in a remote section of the Pantanal and discovered mass colonization of native pictographs on rock walls lining the Paraguay River. The species also poses a very significant threat to man-made reservoirs and canals in China and elsewhere that were designed to move water from areas of abundance to those of greatest need.

Very clearly, a focused effort was required to assemble the most up-to-date and comprehensive information on this species. The book is edited by Demetrio Boltovskoy, a highly capable Argentinean researcher with more than 15 years' experience working with golden mussels. It covers virtually every aspect of golden mussels of interest to ecologists and managers. It is divided into sections, including those devoted to basic biology, environmental impact, behavior, comparisons to other biofouling molluscs, distribution and colonization, mitigation, and control.

The book's 40 authors are drawn from seven countries, with large contributions by researchers from Argentina, Brazil, and Japan. Individual chapters detail the anatomy, morphology, population genetics, population structure, reproductive cycles, interspecific interactions, and behavior of the species. Subsequent chapters focus on distribution and patterns of spread in South America, China, and Japan, and applied topics related to biofouling, control, and eradication.

This book will serve as the benchmark on the golden mussel as it continues its spread across Asia and South America, and more and more researchers and managers seek easily accessible and authoritative information on what to expect in their invaded system.

Windsor, Canada Hugh J. MacIsaac
Great Lakes Institute for Environmental Research,
University of Windsor,
Windsor, Ontario, Canada N9B 3P4

Preface

Ever since it started spreading to other countries, *Limnoperna fortunei* has become an issue of growing concern. In his pioneering work, Morton (1977) pointed out that "The fouling potential of *L. fortunei* cannot, as yet, be accurately forecast, but it is felt expedient to warn against artificial introduction of this mollusc elsewhere." This diagnosis proved correct shortly thereafter when *L. fortunei* was shown to have spread into Taiwan and Korea, and later into Japan and South America (see Part III in this volume). Introduction of the mussel in the latter two areas, which occurred at about the same time (around 1990), was clearly marked by a boom in the number of reports centered on this mussel (Fig. 1). Publications on *L. fortunei* soared from an average of <0.3 per year in 1950–1992, to over 20 per year in 1993–2012. By the end of 2013, there were ~ 500 works of various types devoted either solely or chiefly to the golden mussel. Journal articles made up around 60 % of this total, followed by conference reports, book chapters, theses, and miscellaneous works (web pages, internal reports, etc.) (Fig. 2).

Although such a volume of publications devoted to a single species may seem impressive, the number of works reporting new, peer-reviewed data in mainstream journals is much lower. Of the ca. 300 journal articles, only ~ 50 % have been published in widely available "white literature" (according to a recent survey by Barbosa 2014, for the period 1982–2012, the database of the Thomson Institute for Scientific Information—ISI contains 107 papers on *L. fortunei* published in 60 journals), whereas the other half corresponds to local or regional Asian or South American publications, largely in languages other than English (Fig. 3).

Significantly, much of this "grey literature" contains abundant useful information that is largely ignored by the international scientific community because of accessibility and language-related problems. Thus, one of the purposes of the present book is to critically summarize and disseminate a large amount of information that is not readily available. In order to ensure adequate coverage of widely dissimilar sources and languages, efforts were made to enlist authors from the different geographic areas occupied by *L. fortunei*.

This book centers specifically on *L. fortunei*, rather than on invasive aquatic bivalves in general. The latter have been profusely covered in several excellent

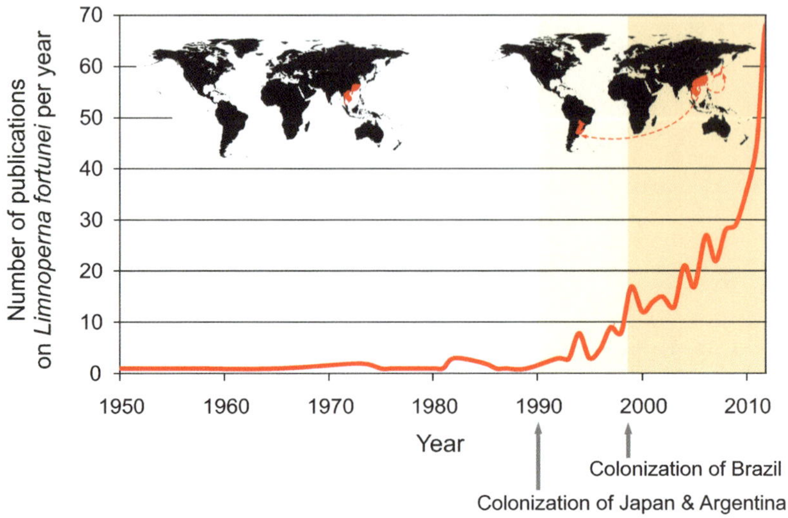

Fig. 1 Number of publications centered exclusively or chiefly on *Limnoperna fortunei*, published between 1950 and 2012

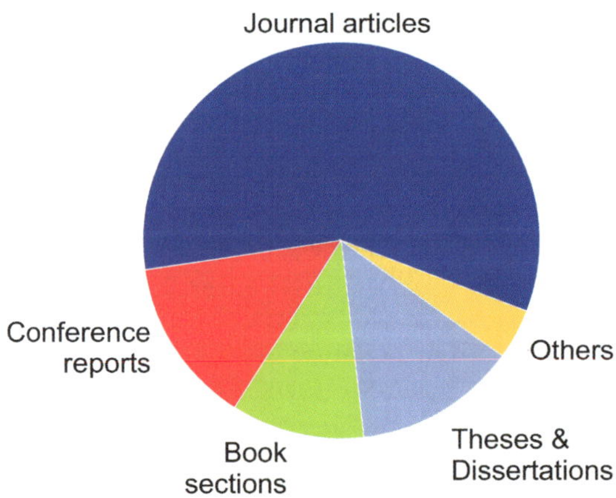

Fig. 2 Breakdown of publications shown in Fig. 1 according to source type

review works, so they are not discussed here, except in the context of direct comparisons with the golden mussel. Furthermore, the topics included herein are restricted to those for which data on *L. fortunei* are available, and these clearly do not cover all aspects of the biology and ecology of the mussel. Interestingly, however, even this limited volume of information contains disagreements. The reader will probably notice that different authors have different points of view and dissimilar outlooks on

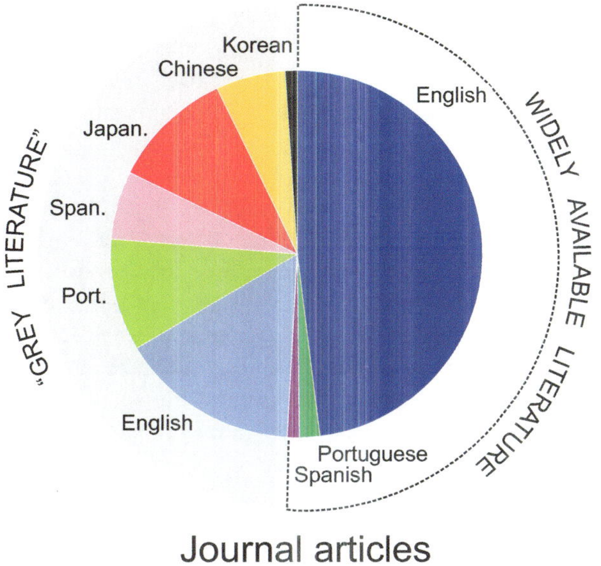

Fig. 3 Breakdown of journal articles published on *L. fortunei* up to 2013 according to source type and language. "Widely available literature" includes refereed journals indexed in most international databases (covered by SCOPUS). "Grey literature" includes refereed and un-refereed publications of limited distribution and not covered in most major international databases

the same issue. Aside from the fact that such discrepancies are unavoidable in most collective works, we deliberately avoided reconciling these disagreements in order to underscore the blanks and uncertainties in our current knowledge of this mussel.

Throughout the book, and particularly in the sections dealing with the impacts of *L. fortunei* on other organisms, care was taken to present a balanced overview of the effects that have effectively been investigated (rather than extrapolated from studies on other invasive mussels) in a dispassionate and nonjudgmental manner, avoiding repetition of general statements based on extrapolations from studies of other invasive species, in particular the zebra mussel. Much of the mainstream literature on alien species is focused on forcefully demonstrating that invasives are fundamentally different from indigenous species, and particularly that their effects are detrimental for the ecosystem (Davis et al. 2011). While this is clearly the case for many introduced species (Simberloff and signatories 2011; Simberloff et al. 2012; Paolucci et al. 2013), every introduction must be analyzed and weighed in its own right, rather than by extrapolation of knowledge from other invasives, or automatically labeled as harmful because the species involved is nonindigenous. Unfortunately, in the case of *L. fortunei* this has happened time and again. The following example illustrates this bias vividly. In a recent paper on the effects of climate variability on fisheries, Defeo et al. (2013) stated that "…NIS invasions have altered community structure and diversity in freshwater and estuarine ecosystems of Latin America, and negatively affected small scale fisheries. For example, the sustained increase of the Asiatic clams (*Corbicula fluminea* and *Limnoperna fortunei*) and the invasive whelk (*Rapana venosa*) in coastal/inshore ecosystems of South America generated drastic ecosystem effects that included the depletion of native species exploited by small scale fisheries, such as the blue mussel…." Aside from the fact that *Corbicula fluminea* and *L. fortunei* are typically freshwater organisms, practically absent in marine coastal/inshore ecosystems, the reference used by these authors to support their argument (Lercari and Bergamino 2011) does not mention *L. fortunei* at all, yet *L. fortunei* was included in the statement because, as is *C. fluminea* (whose negative influence on freshwater fisheries is debatable), it is an introduced species. Furthermore, the evidence available so far indicates that the presence of *L. fortunei* is probably beneficial, rather than harmful, for some exploited fish species (see "Trophic relationships of *Limnoperna fortunei* with larval fishes" and "Trophic relationships of *Limnoperna fortunei* with adult fishes" in this volume).

Numerous reports have shown that the same introduced species can have very dissimilar effects in different habitats, and even within the same habitat, with changing environmental conditions (Byers et al. 2002; D'Antonio and Hobbie 2005; Reise et al. 2006; Davis 2009; Ruokonen et al. 2014), let alone different (albeit functionally similar) species. Nevertheless, literature on *L. fortunei* is plagued with examples where results of studies showing the negative impacts of other species in vastly dissimilar environments (chiefly *D. polymorpha* in North America) are used as proof of a similar impact of *L. fortunei* in South America (Boltovskoy and Correa 2015). This approach hampers objective analyses and does little to advance our understanding of how this species interacts with its new environment.

Irrelevant, unoriginal, and unsubstantiated information is pervasive among all scientific fields, but "hot" topics seem to be particularly vulnerable. *L. fortunei* has become such a hot topic ever since it started interfering with the operation of power plants and its colonies became a salient feature of coastal freshwater areas. This has had a dual effect on published information on the golden mussel. On the one hand, it spurred a wealth of new data, particularly in the areas of population dynamics, geographic distribution, and ecology. On the other hand, however, it also stimulated the production of many reports with scarce—if any—original valuable information, centered on "crying wolf" and attributing to *L. fortunei*, without any supporting evidence, the environmental effects of *D. polymorpha* on European and North American waterbodies. Although some of these publications were careful to note that these effects were merely a possibility, they had a snowball effect whereby subsequent literature used them as proof of the putative impacts. This problem clearly reflects the fact that "… because positive results are more likely to be submitted and published, the invasion literature may be biased toward demonstrating that nonindigenous species have large ecological impacts" (Byers et al. 2002). Furthermore, the same bias seems to permeate grant submissions, conference reports, newspaper and magazine articles, consultant reports, thesis dissertations, book chapters, web pages, etc., where the importance of impacts is exaggerated in the hopes of getting funding or recognition. This phenomenon is likely also associated with the fact that the distinction between "pure" ecology and "applied" ecology, which prevailed in the 1950s and 1960s, has been blurring since the 1980s and ecologists have been increasingly concerned about justifying their research in a larger social context, particularly addressing the issue of how their work—funded by government money—benefits society at large (Davis 2006).

As noted by Davis (2006), during the last decade, invasion biology has developed along two distinct paths, vividly illustrated by the recent controversy of Davis et al. (2011) versus Simberloff and signatories (2011). The "Eltonian path" is characterized by the strongly conservation-oriented approach advocated by Elton (1958), while the "Asilomar path" is less focused on conservation/restoration issues. In their drive to prove their point, conservation-oriented researchers have often neglected an alternative approach whereby the introduced species, once established, is regarded as any native form, with both negative and positive effects on the colonized ecosystem. In addition to being inoperative on a practical level, the invasive vs. indigenous dichotomy can hamper objectivity and lead to nonsensical results (Valéry et al. 2013). Ignoring the fact that it is an alien form should most probably help us to assess its impacts more objectively (Thompson et al. 1995; Gurevitch and Padilla 2004; Davis et al. 2011). This should not imply a denial of the probability that nonnatives are more likely than natives to have strong impacts on local ecosystems (Simberloff et al. 2012; Paolucci et al. 2013), but the "invasive species approach" tends to lump all introduced species in the same bag thus focusing the assessment of their interactions on the traits that they share with other invasives (usually perceived as negative), rather than on those that characterize them best.

One of the most frequently marshaled arguments in association with the impact of introduced species is that their impact stems from the fact that they have no natural enemies in newly colonized areas (Van Driesche and Van Driesche 2000). While probably true for many nonnative species, this is probably not the case for *L. fortunei*, where >90% of the mussel's yearly production is lost to predation (Sylvester et al. 2007; Nakano et al. 2010). Admittedly, this assessment does not prove that *L. fortunei* is as vulnerable to predators in its invasive range as in its native range, yet it suggests that even some of the most deeply rooted tenets of invasion biology, such as the enemy release hypothesis (e.g., Colautti et al. 2004), cannot be applied indiscriminately to all invasives.

This discussion does not involve the pros and cons of deliberate species introductions, which mostly concern organisms directly used by man for food, shelter, medicine, ecosystem services, or aesthetic enjoyment (Ewel et al. 1999); it is hard to envision that *L. fortunei* will ever be the target of a deliberate introduction. Furthermore, basic precaution and the long list of introduced species that had devastating effects on native biota (Pimentel 2002; Simberloff 2003) clearly justify all possible efforts to keep biological invasions at bay or to eradicate them if feasible. However, once a nonnative species has been introduced and its eradication is out of the question (as is the case for *L. fortunei*), analyses of its interactions with the local biota should not be rooted in the notion that it is harmful.

As of 2014, in South America *L. fortunei*'s range seems still limited to the large Río de la Plata basin (Paraguay, Paraná, Uruguay rivers) and a few minor watersheds (Guaíba and Tramandaí, in Brazil, Mar Chiquita, in Argentina; see Chap. 19 in this volume). It has not been reported from the Amazon system, the next large South American watershed, but its colonization of this huge area is most probably inevitable. The proximity of the Amazon and the Río de la Plata basins makes the former very vulnerable to human transport of the golden mussel by watercraft, fishing, and fish culture gear, in particular fish cages, and any object that has been immersed in water and overgrown with mussels. Furthermore, the Amazon is navigable to ocean liners of virtually any tonnage, including ships with ballast water from infested ports along the Paraná-Uruguay-Río de la Plata waterways and the Guaíba basin, where compliance with international water ballast regulations is rather loosely enforced (Boltovskoy et al. 2011).

It is our intention that this book serves as a critical update and general reference for current knowledge on this important animal, and a guide to the many aspects that we still know almost nothing about. Indeed, after 30 years of relatively intensive research, the blanks are numerous and significant. Key aspects of the biology and ecology of *L. fortunei*, crucial for an appraisal of its impacts on the biota, have hardly been touched upon, such as its fertility, its metabolism and physiology, its natural enemies (other than fishes), its long-term effects on benthic communities, its direct and indirect impact on other filter-feeding organisms, etc. As detailed elsewhere in this book (see Chap. 7 in this volume), over two decades after *L. fortunei* started colonizing Japan and South America, only one attempt has been made at estimating its abundance across an entire water body. Density data available are almost invariably restricted to isolated, usually densely populated, spots, yielding little (if any) valuable information on the potential impact of these mussel beds. A significant void in our understanding of the impacts of the golden mussel is the absence of large-scale studies, both temporally and geographically. Almost all surveys on the relationships between *L. fortunei* and local organisms are laboratory or field studies of specific interactions that point at *possible* system-wide effects, but we do not know the extent to which these potential effects are actually changing preinvasion conditions. Difficulties in assessing the latter also stem from the paucity of background information, including long-term, comprehensive studies of the freshwater systems colonized by the mussel.

We hope that the critical update and summary of current knowledge contained in the following pages will help orient future research on *L. fortunei* as it spreads northwards.

Demetrio Boltovskoy

References

Barbosa FG (2014) The scientific literature on *Limnoperna fortunei* (Dunker 1857) from 1982 to 2012. Anais da Academia Brasileira de Ciências 86:1373–1384

Boltovskoy D, Correa N (2015) Ecosystem impacts of the invasive bivalve *Limnoperna fortunei* (golden mussel) in South America. Hydrobiologia 746:81–95

Boltovskoy D, Almada P, Correa N (2011) Biological invasions: assessment of threat from ballast-water discharge in Patagonian (Argentina) ports. Environ Sci Policy 14:578–583

Byers JE, Reichard S, Randall JM, Parker IM, Smith CS, Lonsdale WM, Atkinson IAE, Seastedt TR, Williamson M, Chornesky E, Hayes D (2002) Directing research to reduce the impacts of nonindigenous species. Conserv Biol 16:630–640

Colautti RI, Ricciardi A, Grigorovich IA, MacIsaac HJ (2004) Is invasion success explained by the enemy release hypothesis? Ecol Lett 7:721–733

D'Antonio CM, Hobbie S (2005) Plant species effects on ecosystem processes: insights from invasive species. In: Sax DF, Stachowicz JJ, Gaines SD (eds) Species invasions. Insights into ecology, evolution, and biogeography. Sinauer Associates, Sunderland, pp 65–84

Davis MA (2006) Invasion biology 1958–2005: the pursuit of science and conservation. In: Cadotte MW, McMahon SM, Fukami T (eds) Conceptual Ecology and Invasion Biology: reciprocal approaches to nature. Springer Verlag, Berlin, pp 35–64

Davis MA (2009) Invasion biology. Oxford University Press, New York, pp 1–244

Davis MA, Chew MK, Hobbs RJ, Lugo AE, Ewel JJ, Vermeij GJ, Brown JH, Rosenzweig ML, Gardener MR, Carroll SP, Thompson K, Pickett TA, Stromberg JC, Del Tredici P, Suding KN, Ehrenfeld JG, Grime JP, Mascaro J, Briggs JC (2011) Don't judge species on their origins. Nature 474:153–154

Defeo O, Castrejón M, Ortega L, Kuhn AM, Gutiérrez NL, Castilla JC (2013) Impacts of climate variability on Latin American small-scale fisheries. Ecol Soc 18:30

Driesche JV, Driesche RV (2000) Nature out of place. Biological invasions in the global age. Island Press, Washington, D.C., pp 1–380

Elton CS (1958) The ecology of invasions by animals and plants. Methuen, London, pp 1–196

Ewel JJ, O'Dowd DJ, Bergelson J, Daehler CC, D'Antonio CM, Gomez LD, Gordon DR, Hobbs RJ, Holt A, Hopper KR, Hughes CE, LaHart M, Leakey RRB, Lee WJ, Loope LL, Lorence DH, Louda SM, Lugo AE, McEvoy PB, Richardson DM, Vitousek PM (1999) Deliberate introductions of species: Research needs. Benefits can be reaped, but risks are high. BioScience 49:619–630

Gurevitch J, Padilla DK (2004) Are invasive species a major cause of extinctions? Trends Ecol Evol 19:470–474

Lercari D, Bergamino L (2011) Impacts of two invasive mollusks, *Rapana venosa* (Gastropoda) and *Corbicula fluminea* (Bivalvia), on the food web structure of the Río de la Plata estuary and nearshore oceanic ecosystem. Biol Invasions 13:2053–2061

Morton B (1977) The population dynamics of *Limnoperna fortunei* (Dunker 1857) (Bivalvia: Mytilacea) in Plover Cove Reservoir, Hong Kong. Malacologia 16:165–182

Nakano D, Kobayashi T, Sakaguchi I (2010) Predation and depth effects on abundance and size distribution of an invasive bivalve, the golden mussel *Limnoperna fortunei*, in a dam reservoir. Limnology 11:259–266

Paolucci EM, MacIsaac HJ, Ricciardi A (2013) Origin matters: Alien consumers inflict greater damage on prey populations than do native consumers. Divers Distrib 19:988–995

Pimentel D (ed) (2002) Biological invasions. Economic and environmental costs of alien plant, animal, and microbe species. CRC Press, Boca Raton, pp 1–369

Reise K, Olenin S, Thieltges DW (2006) Are aliens threatening aquatic coastal ecosystems? Helgol Mar Res 60:77–83

Ruokonen TJ, Karjalainen J, Hämäläinen H (2014) Effects of an invasive crayfish on the littoral macroinvertebrates of large boreal lakes are habitat specific. Freshw Biol 59:12–25

Simberloff D (2003) Confronting introduced species: a form of xenophobia? Biol Invasions 5:179–192

Simberloff D, signatories (2011) Non-natives: 141 scientists object. Nature 475:36

Simberloff D, Souza L, Nuñez MA, Barrios-Garcia MN, Bunn W (2012) The natives are restless, but not often and mostly when disturbed. Ecology 93:598–607

Sylvester F, Boltovskoy D, Cataldo D (2007) Fast response of freshwater consumers to a new trophic resource: Predation on the recently introduced Asian bivalve *Limnoperna fortunei* in the lower Parana River, South America. Austral Ecol 32:403–415

Thompson K, Hodgson JG, Rich TCG (1995) Native and alien invasive plants: more of the same? Ecography 18:390–402

Valéry L, Fritz H, Lefeuvre J-C (2013) Another call for the end of invasion biology. Oikos 122:1143–1146

Acknowledgments

I would like to express my sincere thanks to the authors of this book, whose diligence, support, patience, and cooperation were keys for covering a comprehensive selection of topics of interest for our current understanding of the mechanisms involved in the spread and impacts of *Limnoperna fortunei* around the world. I am particularly grateful to the reviewers who critically read the materials submitted for this volume, making valuable suggestions that greatly helped to improve the original manuscripts. Special thanks are due to Erik Thuesen, who style-edited the chapters included in this book. His assistance went far beyond his formal commitment to revise the texts prepared by nonnative English speaking authors for style, usage, grammar, and spelling, offering suggestions on contents, structure, sources of information, and illustrations. The Consejo Nacional de Investigaciones Científicas y Técnicas (CONICET), Argentina, provided financial support for editing this volume. I am particularly indebted to Dr. Mirtha Flawia, whose engagement was instrumental for the edition of this book. Technical assistance by Michiko Matsubara at various stages of this project is greatly appreciated.

The editor and the authors are grateful to the editors and publishing companies that allowed us to reproduce figures previously published in the following journals: Aquatic Sciences (Springer), Biodiversity (Taylor and Francis), Biological Invasions (Springer), Diversity and Distributions (Wiley), Environmental Science & Technology (American Chemical Society), Freshwater Biology (Wiley), Hydrobiologia (Springer), Japanese Journal of Benthology (Japanese Association of Benthology), Journal of Molluscan Studies (Oxford University Press), Journal of the American Water Works Association (The American Water Works Association), Limnology & Oceanography (Association for the Sciences of Limnology and Oceanography), Malacologia (The Field Museum), Molluscan Research (Magnolia Press), Qinghua Daxue Xuebao (Tsinghua University, China), Redes Editora (Brazil), Report of Yahagi River Institute, and Zoologia (Sociedade Brasileira de Zoologia).

The advice and editorial assistance provided by Dr. Daniel Simberloff, Editor of the Springer Series in Invasion Ecology—"Invading Nature," and Ms. Takeesha Moerland-Torpey and Ms. Judith Terpos, Senior Editorial Assistants at Springer Environmental Sciences, greatly helped to improve several aspects of the contents, organization, and layout of this book.

Demetrio Boltovskoy
Buenos Aires, Argentina

Contents

Part I Biology

Part II Ecology and Environmental Impact

Part III Distribution and Spread

Contributors

Juliana Américo Instituto de Biofísica Carlos Chagas Filho G2-050, Universidade Federal do Rio de Janeiro, 21044-020 Rio de Janeiro, Brazil

Takashi Baba Environmental Science graduate School, University of Shiga Prefecture, Hikone, Shiga 522-8533, Japan [present address: Japan Sea National Fisheries Research Institute,Fisheries Research Agency, 1-5939-22, Suido-cho, Chuou-ku, Niigata, 951-8121, Japan]

Demetrio Boltovskoy Instituto de Ecología, Genética y Evolución de Buenos Aires (IEGEBA), Facultad de Ciencias Exactas y Naturales, Universidad de Buenos Aires-CONICET, Ciudad Universitaria, 1428 Buenos Aires, and Museo Argentino de Ciencias Naturales 'Bernardino Rivadavia'-CONICET, Buenos Aires, Argentina

Rodrigo Brindeiro Instituto de Biofísica Carlos Chagas Filho G2-050, Universidade Federal do Rio de Janeiro, 21044-020 Rio de Janeiro, Brazil

Lyubov Burlakova The Research Foundation of The State University of New York, SUNY Buffalo State, Buffalo, NY 14222, USA

Mônica C. S. Campos Instituto Senai de Tecnologia em Meio Ambiente, Centro de Inovação e Tecnologia SENAI FIEMG - Campus CETEC, Av. José Cândido da Silveira 2000, Belo Horizonte, 31035536 MG, Brazil

Daniel Cataldo Instituto de Ecología, Genética y Evolución de Buenos Aires (IEGEBA), Facultad de Ciencias Exactas y Naturales, Universidad de Buenos Aires-CONICET, Ciudad Universitaria, 1428 Buenos Aires, and Museo Argentino de Ciencias Naturales 'Bernardino Rivadavia'-CONICET, Buenos Aires, Argentina

Renata Claudi RNT Consulting Inc., Picton, Ontario, Canada

Nancy Correa Servicio de Hidrografía Naval, Minsterio de Defensa, and Escuela de Ciencias del Mar, Instituto Universitario Naval, Argentina

Cristina Damborenea División Zoología Invertebrados, Museo de La Plata, Universidad Nacional de La Plata, del Bosque s/n, 1900 La Plata, and CONICET, Argentina

Gustavo Darrigran División Zoología Invertebrados, Museo de La Plata, Universidad Nacional de La Plata, Paseo del Bosque s/n, 1900 La Plata, and CONICET, Argentina

Francesco Dondero Department of Science and Technological Innovation, Università del Piemonte Orientale "Amodeo Avogrado", 15121 Alessandria, Italy

Stephen K. Hamilton Kellogg Biological Station and Dept. of Zoology, Michigan State University, 3700 E. Gull Lake Dr., Hickory Corners, MI 49060, USA

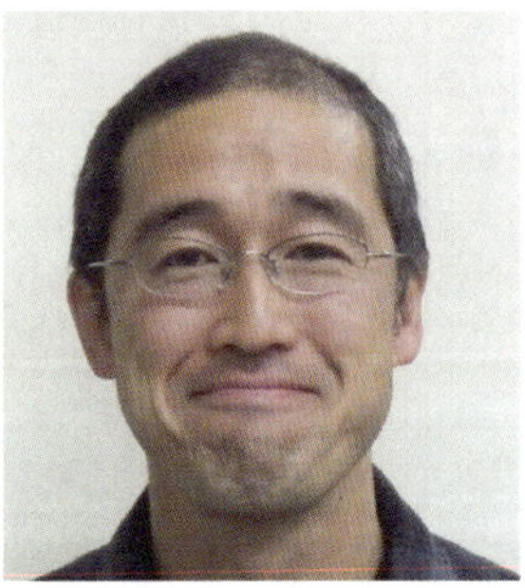

Kenji Ito National Institute for Agro-Environmental Sciences, Kannondai 3-1-3, Tsukuba, 7 305-8604, Japan

Keiji Iwasaki Institute for Natural Science, Nara University, 1500 Misasagicho, Nara, 631-8502, Japan

Susana José de Paggi Instituto Nacional de Limnología (INALI, Universidad Nacional del Litoral, Consejo Nacional de Investigaciones Científicas y Técnicas), Ciudad Universitaria, Paraje El Pozo S/N, Santa Fe 3000, Argentina

Alexander Y. Karatayev Great Lakes Center, SUNY Buffalo State, 1300 Elmwood Avenue, Buffalo, NY 14222, USA

Takuya Kobayashi Central Research Institute of Electric Power Industry (CRIEPI), 1646 Abiko, Chiba 270-1194, Japan

Hugh J. MacIsaac Great Lakes Institute for Environmental Research, University of Windsor, Windsor, Ontario, Canada N9B 3P4

María C. D. Mansur Universidade Federal do Rio Grande do Sul, Centro de Ecologia – CENECO, Av. Bento Gonçalves, 9500, setor 4, prédio 43411, CEP 91501-970, Porto Alegre, RS, Brazil

Luciana Montalto Instituto Nacional de Limnología (INALI, Universidad Nacional del Litoral-CONICET), and Facultad de Humanidades y Ciencias, Universidad Nacional del Litoral, Ciudad Universitaria, 3000 Santa Fe, Argentina

Brian Morton School of Biological Sciences, The University of Hong Kong, Hong Kong SAR, China

Daisuke Nakano Central Research Institute of Electric Power Industry (CRIEPI), 1646 Abiko, Chiba 270-1194, Japan

Takaomi Nomura Division of Applied Biology, Faculty of Textile Science and Technology, Shinshu University, Tokida 3-15-1, Ueda, Nagano Prefecture, Japan

Kousaku Ohkawa Division of Biological and Medical Fibers, Institute for Fiber Engineering (IFES), Interdisciplinary Cluster for Cutting Edge Research (ICCER), Shinshu University. Tokida 3-15-1, Ueda, Nagano Prefecture, Japan

Marcia D. Oliveira Embrapa Pantanal, Rua 21 de Setembro, 1880, Corumbá, MS, Brazil

Dianna K. Padilla Department of Ecology and Evolution, Stony Brook University, Stony Brook, NY 11794-5245, USA

Juan C. Paggi Instituto Nacional de Limnología (INALI, Universidad Nacional del Litoral, Consejo Nacional de Investigaciones Científicas y Técnicas), Ciudad Universitaria, Paraje El Pozo S/N, Santa Fe 3000, Argentina

Esteban M. Paolucci Instituto de Ecología, Genética y Evolución de Buenos Aires (IEGEBA), Facultad de Ciencias Exactas y Naturales, Universidad de Buenos Aires-CONICET, Ciudad Universitaria, 1428 Buenos Aires, and Museo Argentino de Ciencias Naturales 'Bernardino Rivadavia'-CONICET, Buenos Aires, Argentina

Pablo E. Penchaszadeh Museo Argentino de Ciencias Naturales 'Bernardino Rivadavia'-CONICET, Buenos Aires, Argentina

Pablo V. Perepelizin Museo Argentino de Ciencias Naturales „Bernardino Rivadavia"-CONICET, Av. Angel Gallardo 470, 1405 Buenos Aires, Argentina

Francisco Prosdocimi Instituto de Bioquímica Médica, Universidade Federal do Rio de Janeiro, 21044-020 Rio de Janeiro, Brazil

Mauro Rebelo Instituto de Biofísica Carlos Chagas Filho, Universidade Federal do Rio de Janeiro, 21044-020 Rio de Janeiro, Brazil

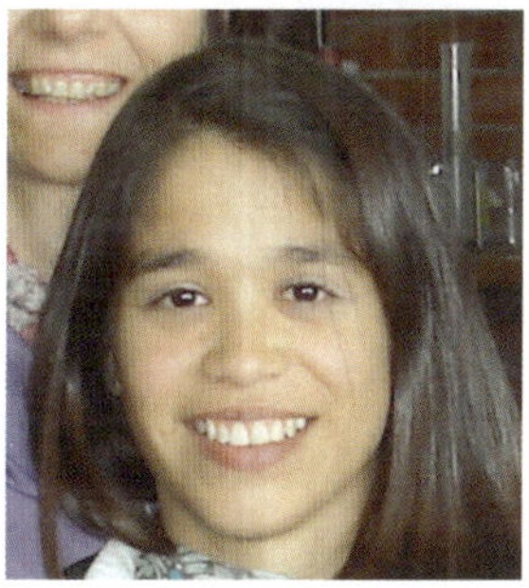

Florencia Rojas Molina Instituto Nacional de Limnología (INALI, Universidad Nacional del Litoral, Consejo Nacional de Investigaciones Científicas y Técnicas), Ciudad Universitaria, Paraje El Pozo S/N, Santa Fe 3000, Argentina

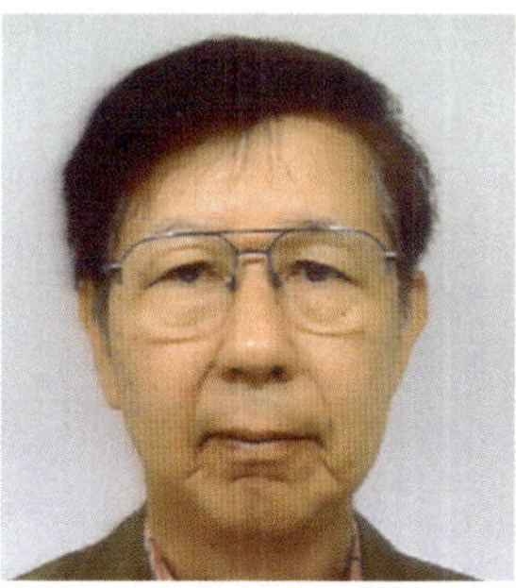

Isamu Sakaguchi Central Research Institute of Electric Power Industry (CRIEPI), 1646 Abiko, Abiko, Chiba 270-1194, Japan

Paula Sardiña Instituto de Investigaciones Biológicas y Biomédicas del Sur (INBIOSUR), Departamento de Biología, Bioquímica y Farmacia, Universidad Nacional del Sur-CONICET, San Juan 670, 8000 Bahía Blanca, Argentina

Francisco Sylvester Instituto de Ecología, Genética y Evolución de Buenos Aires (IEGEBA), Facultad de Ciencias Exactas y Naturales, Universidad de Buenos Aires-CONICET, Ciudad Universitaria, 1428 Buenos Aires, Argentina.
Current address: Instituto para el Estudio de la Biodiversidad de Invertebrados (IEBI), Facultad de Ciencias Naturales, Universidad Nacional de Salta, Av. Bolivia 5150, 4400 Salta, Argentina

Erik V. Thuesen Laboratory 1, The Evergreen State College, Olympia, Washington, 98505, USA

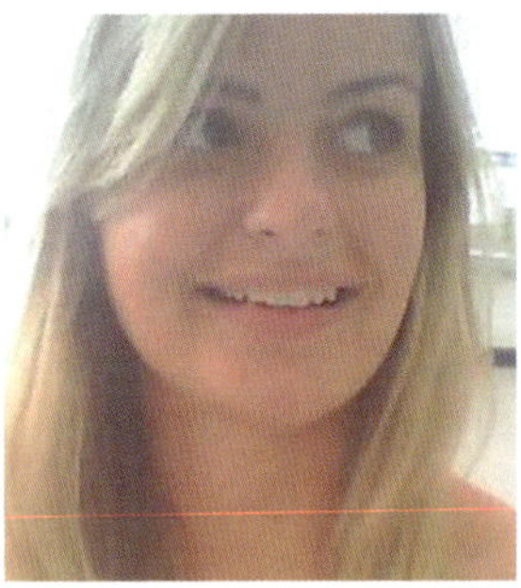

Marcela Uliano-Silva Instituto de Biofísica Carlos Chagas Filho G2-050, Universidade Federal do Rio de Janeiro, 21044-020 Rio de Janeiro, Brazil

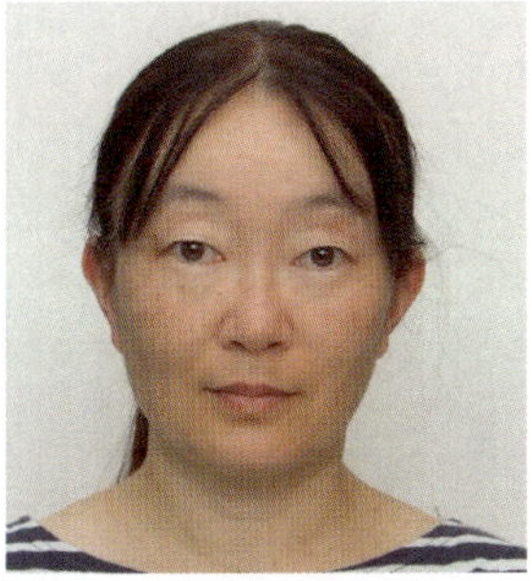

Misako Urabe Department of Ecosystem Studies, School of Environmental Science, University of Shiga Prefecture, Hikone, Shiga 522-8533, Japan

Mengzhen Xu State Key Laboratory of Hydroscience and Engineering, Tsinghua University, Beijing 100084, China

Part I
Biology

The Biology and Anatomy of *Limnoperna fortunei*, a Significant Freshwater Bioinvader: Blueprints for Success

Brian Morton

Abstract *Limnoperna fortunei* first came to scientific attention when it was introduced into the potable freshwater supply system of Hong Kong around 1968. The initial occurrence in Hong Kong was related to the commencement of supplies of water from the East River in China. Aspects of the biology and anatomy of *L. fortunei* were investigated in 1973 (Morton, Malacologia 12:265–281, 1973). Then, comparisons were made with other mytiloids and with the invasive *Dreissena polymorpha* (Dreissenidae), which *L. fortunei* resembles superficially. The suggestion of a relationship between the two taxa was discounted on anatomical grounds—both species being convergently heteromyarian, adapting them to similar lives in the freshwater habitats of Eastern and, subsequently, Western Europe and North America, and mainland Asia plus Japan and Taiwan and latterly South America, respectively. The present study re-examines aspects of the biology and anatomy of *L. fortunei* commensurate upon its recent range extensions, specifically with regard to its occupation of a variety of freshwater habitats in order to better understand the reasons for its opportunistic success. It is concluded that *L. fortunei* probably evolved from a brackish water either *Xenostrobus*-like or *Perna*-like ancestor—the question of *Limnoperna*'s ancestry still unresolved. Regardless, it was the evolution of the heteromyarian form in the, indisputedly mytiloidean, ancestor that opened up the hitherto bivalve-unoccupied hard-surface epibenthic environment for colonisation in both lentic and lotic freshwaters ecosystems. This was probably associated with the adoption of osmoregulation and efficient systems for the collection, selection and digestion of ample seston resources. The energy thereby obtained has been focussed into rapid fecundity at the expense of shell growth and the adoption, thereby, of a *r*-selected sexual strategy and life history trait.

Keywords *Limnoperna fortunei* · Golden mussel · Anatomy · Morphology · Organs · Phylogeny

---

B. Morton (✉)
School of Biological Sciences, The University of Hong Kong, Hong Kong SAR, China
e-mail: prof_bmorton@hotmail.co.uk

D. Boltovskoy (ed.), *Limnoperna fortunei*, Invading Nature - Springer Series in Invasion Ecology 10, DOI 10.1007/978-3-319-13494-9_1

Introduction

In recent years, the attentions of water supply engineers and biologists in Europe and North and South America have been drawn to the problems of fouling caused by introduced freshwater bivalve molluscs. Since the nineteenth century, *Dreissena polymorpha* (Pallas 1771) has expanded its range from an initial restricted area focussed on the Caspian and Black seas to one encompassing much of the European continent. Although the initial spread of this species may have started in the eighteenth century (Karatayev et al. 2007b), the subsequent dramatic range expansion is attributed to the construction of waterways for the transport of goods and raw materials during the Industrial Revolution. This is certainly the reason for the range expansion that *D. polymorpha* experienced in Great Britain in the nineteenth century and subsequently (Kerney and Morton 1970), and it is still being transported this way, along canals and rivers, in Ireland (Hayden and Caffrey 2013). In the twentieth century, however, *D. polymorpha* (McMahon 1982) and a second species, *Dreissena rostriformis bugensis* Andrusov, 1897 (Mills et al. 1996), were introduced into North America, probably both within the ballast water of ships arriving in the St. Lawrence Seaway around 1989 (see Chapter "Parallels and Contrasts Between *Limnoperna fortunei* and Species of *Dreissena*" in this volume). Subsequently, *D. rostriformis bugensis* has been back-introduced into Europe, but in the west, that is, the freshwater section of the Rhine–Meuse Estuary in The Netherlands in 2006 and, later, upstream into the River Rhine and its tributaries, the Main and Moselle in Germany, and the River Meuse in Belgium and France (Mathews et al. 2014; see Fig. 3 in Chapter "Parallels and Contrasts Between *Limnoperna fortunei* and Species of *Dreissena*"). Both dreissenid species possess a byssus and are thus adapted to the colonisation of solid surfaces, such as are found in the pipes and conduits of water supply systems, typically inaccessible to the endobenthic bivalve constituents of freshwater faunas, for example, representatives of the Unionoidea and Corbiculoidea.

That said, however, the Asian corbiculid *Corbicula fluminea* (Müller 1774) has been introduced into North American waterways and caused similar fouling problems as the two species of *Dreissena*. First reported upon in 1938 from the Columbia River in the state of Washington (Burch 1944), this species has since expanded its range to encompass all of the major North American river basins (McMahon 1982). The fouling problems caused by *C. fluminea* are somewhat different from those caused by *Dreissena* in that as an endobenthic species it became a particular nuisance when occurring in sands dredged for ultimate use in the manufacture of concrete. Although, it does clog pipes and power station condensers. As with the species of *Dreissena,* the rapid range expansion experienced by *C. fluminea* is attributable to man's development of artificial waterways for water supply purposes. *Corbicula fluminea* has been introduced into European waters being first reported upon from the Rhine in 1991 (Kinzelbach 1991). It has subsequently spread throughout much of Western Europe and invaded Great Britain being recorded first from the River Chet in the Norfolk Broads in 1998 (Howlett and Baker 1999). It has subsequently spread as far east as the Czech Republic (Beran 2000; 2006), colonising all the countries in-between. Like *Limnoperna fortunei*, *C. fluminea* has also been introduced into South America, being recorded (as *Corbicula manilensis*) from Brazilian rivers in 1978 by Veitenheimer-Mendes (1981) and from the Río de la Plata estuary, Uruguay (as *C. largillierti*), by Ituarte (1994), in 1982. Beasley et al. (2003) recorded *C. fluminea* as occurring first in the lower Amazon basin and the Pará and Tocantins rivers, Brazil, in 1997–1998.

Recently too, another corbiculid, that is, *Corbicula fluminalis* (Müller 1774) has been introduced into Venezuela (Martínez 1987; as *C. manilensis* (Philippi 1844), which is a junior synonym of *C. fluminalis*). This species also occurs in Asia, but is estuarine occupying the lower reaches of major rivers such as the Pearl River in southern China where it was first studied (Morton 1982). As an estuarine species, it is unlikely that *C. fluminalis* will impact water supply systems to the same extent as the wholly freshwater species of *Dreissena* and *C. fluminea*; although where such waters are tapped for industrial cooling water systems this species too may become a problem in its now, new, introduced range.

A similar problem has arisen with the Chinese pond mussel *Sinanodonta woodiana* (Lea 1834), which, in its native China, produces glochidia larvae that attach to fish hosts (Dudgeon and Morton 1984). Like *C. fluminea*, it has been introduced into the countries of Western (Austria, Belgium, Croatia, France, Germany, Italy, Portugal, Spain) and Central (Czech Republic, Hungary, Poland, Romania, Serbia, Slovakia, Ukraine) Europe, and has been identified from some locations in Indonesian and Caribbean islands. It is thought that the species has been introduced into such waters from China as glochidia attached to species of carp destined for aquaculture purposes in Europe and elsewhere (Douda et al. 2012).

The mytilid *L. fortunei* (Dunker 1857) occurs in the rivers of China and East Asia (see Chapter "Distribution and Spread of *Limnoperna fortunei* in China" in this volume) but aroused little interest until it too was introduced outside its native range into the potable freshwater supply system in Hong Kong around 1968 (Morton 1975) probably from the East River, itself a tributary of the Pearl where the species occurs naturally (Miller and McClure 1931). There is a dearth of early information on this animal, most records being found in old and, often, obscure journals. Early studies led to a plethora of names being erected for the species including *Dreissena siamensis* Morelet, 1866, *Limnoperna depressa* Brandt and Temcharoen, 1971, *Limnoperna lemeslei* Rochebrune, 1882, *Limnoperna supoti* Brandt, 1974, *Modiola cambodiensis* Clessin, 1889, *Modiola lacustris* Martens, 1875, *Mytilus martensi* Neumayer, 1898 and *Volsella fortunei* Dunker, 1857. Today, however, all these names of putative Asian species have been subsumed within *L. fortunei*, an account of the species' nomenclatural history being provided by Morton and Dinesen (2010).

Limnoperna fortunei is of interest, however, from a number of viewpoints. In the first instance it is, as a representative of the Mytiloidea, living in a unique habitat. Secondly, *L. fortunei* possesses a striking, albeit superficial, similarity to species of *Dreissena*. Finally, and most importantly, prior to 1968, *L. fortunei* was unstudied (other than taxonomically) and largely unknown outside China. The species does appear to be native to mainland China (Morton and Dinesen 2010). Holikoshi (1935), first recorded *L. fortunei* (as *Volsella (Limnoperna) lacustris* [v. Martens]) from Taiwan, and this record was subsequently accepted by Kuroda (1941) in his catalogue of the island's malacofauna. In the absence of reference specimens, however (Huang 2008), the species was not positively (re)discovered and recorded from Taiwan until >50 years later (Tan et al. 1987). *Limnoperna fortunei* was first found in Lake Biwa, Japan, in 1992 (Kimura 1994a; b) and is, today, known to be widespread in 10 of Japan's 47 prefectures (Tominaga et al. 2009; see Chapter "Colonization and Spread of *Limnoperna fortunei* in Japan" in this volume) and is now a pest of potable water supply systems (Magara et al. 2001; Goto 2002). Tominaga et al. (2009) also suggested, based on genetic studies, moreover, that introductions had occurred independently on at least two occasions but that several Japanese populations were established by dispersal of individuals from other non-native ones.

Around 1989, *L. fortunei* appeared in the Río de la Plata estuary, in Argentina (Pastorino et al. 1993; Darrigran and Pastorino 1995). By the early part of the twenty-first century, the species had spread to Rio Grande do Sul, Brazil (Santos et al. 2005) and then, in 2006, to Uruguay, Paraguay and Bolivia. It is a highly successful invasive species in South America and is expected to spread further (see Chapter "Colonization and Spread of *Limnoperna fortunei* in South America" in this volume).

The initial appearance of *L. fortunei* in the water supply complex of Hong Kong was considered by Morton (1975) to be a re-enactment of the earlier patterns of colonisation typical of the species of *Dreissena* and *C. fluminea* and, thus, that not only might this mytiloidean be detrimental to water supply systems throughout its initial range expansion into Hong Kong, Taiwan and, especially, Japan, but also South America. Moreover, it has the potential to be introduced elsewhere, notably North America and Europe (Oliveira et al. 2010), where, if successful, would add to the suite of already recognised problems engendered by alien freshwater bivalves.

This study of the biology and the anatomy of *L. fortunei* complements and adds to that of Morton (1973) and draws on other information published on the species subsequent to its range expansion outside the (natural) borders of China and proliferation in its alien range extensions. It seeks to explain this invasive success in terms of biology and anatomy although, ultimately, a greater understanding of the species will be needed as its spread continues inexorably.

The following glossary, intended for readers without a malacological background, offers brief definitions of the specialised terms used throughout this Chapter. Most of the features listed are illustrated in Fig. 5.

- Adductor muscle: one of normally two muscles that appose the shell valves against the opening force of the ligament; the insertion of these muscles in the shell interior is marked by a conspicuous adductor muscle scar.
- Branchial chambers: the supra-branchial chamber is the space above the ctenidia, among the folds of the two demibranchs (exhalant chamber); the infra-branchial chamber is the mantle cavity below the gills (inhalant chamber).

- Byssal groove: the groove along which the byssus passes post secretion and prior to attachment externally.
- Byssus: fibres produced by the byssal gland in the foot and used to anchor the animal to a substratum.
- Crystalline style: rod-shaped body in the stomach, produced by secretions within the style sac, with enzymatic properties that are released by rotation and trituration of ingested food particles against the stomach's gastric shield.
- Ctenidium: each of the gills, which serve for respiration, potential food collection and particle selection.
- Demibranch: one of the two folded lamellae on each side of the visceral mass, comprising inner and outer demibranchs which together form a ctenidium.
- Digestive diverticula: blind tubes opening laterally from the stomach wall and where intracellular digestion occurs.
- Dissoconch: the final juvenile and adult shell of a bivalve.
- Dorsal hood: pocket connected dorsally to the stomach.
- Druse: a clump or aggregation of mussels around either a stone or other hard object, held together by byssal threads.
- Exhalant siphon: siphon through which water and faeces from the mantle cavity are exhaled and expelled, respectively.
- Filibranchiate (eleutherorhabdic): a ctenidium characterised by loosely connected filaments held in place by interlocking cilia projecting from ciliated discs.
- Food-sorting caecum: a blind pouch connected to the stomach where food particles are selected for either consumption or rejection.
- Gastric shield: chitinous dorsal region of the stomach wall against which the crystalline style´s rotating action fragments food particles. Is probably also enzymatically active.
- Heteromyarian: with unequal sized adductor muscles. The posterior typically larger than the anterior.
- Hinge plate: the internally flattened area of the dorsal margin of the shells that bears the ligament.
- Hinge teeth: calcareous interlocking teeth and sockets that help keep the valves aligned during opening and closing (absent in *L. fortunei*).
- Homorhabdic: a ctenidial type characterised by a single type of filament.
- Inhalant aperture: the aperture for water intake into the mantle cavity. Typically separate from the pedal/byssal gape, but not in mytiloids.
- Labial palps: paired triangular particle-sorting lamellae situated on either side of the mouth.
- Ligament: elastic structure that connects the two shell valves at the hinge line. The ligament opens the valves when the adductor muscles relax, maintaining tonus.
- Mantle: outer layer of tissue lining the internal surface of the shell valves.
- Mantle margin: Typically three folded margin of the left and right mantle lobes which secrete the periostracum and the shell and may be selectively fused to form, for example, the exhalant siphon.

- Mesosomal lobes: lobes of the sac-like extension of the visceral mass and mantle containing the paired gonads.
- Nepioconch: the developmental stage of the shell after the planktonic prodissoconch stages and separated from the dissoconch, by a growth discontinuity.
- Opisthodetic: the ligament type restricted to the postero-dorsal margin of the shell beyond the umbones.
- Pallial line: scar on the interior of the shell marking the attachment of the pallial retractor muscles.
- Pallial sinus: an embayment in the posterior part of the pallial line that marks the attachment of the siphonal retractor muscles and defines the mantle cavity space into which the siphons can retract.
- Parivincular: dorsal ligament type.
- Pericardial gland: excretory organs that accumulate waste products and discharge them into the pericardial cavity, from where they are eliminated via the kidneys into the supra-branchial chamber.
- Pericardium: transparent sac or peritoneum, that contains the heart.
- Periostracum: outer organic layer of the shell and comprising three layers in the Mytiloidea.
- Prodissoconch: larval shell at the apex of the umbones, comprising two growth phases (prodissoconch I and II), separated by a growth line. The oldest region of the shell.
- Siphonal septum: sensory membrane at the apex of the inhalant aperture and beneath the exhalant siphon and restricting the size of the former.
- Statocyst: capsule-like sense organ conveying information about orientation.
- Style sac: cylindrical extension of the stomach that secretes the crystalline style.
- Typhlosole: an infolding along the inner wall of the intestine and stomach and whose function is to transport particles in an intestinal groove for either digestion, rejection or resorting.
- Umbo: rounded extremity of the bivalve shell that represents its oldest part, characterised by the prodissoconchs.
- Umbonal keel: prominent, angled or rounded, feature on the external shell surface that begins at the umbo and extends obliquely to the postero-ventral margin of each valve (=or rounded, feat

Biology

When living in relatively slow flowing waters, *L. fortunei* characteristically occurs in clumps, or nodules (sometimes referred to as druses), of individuals living bound together in thin monolayers (see Chapter "*Limnoperna fortunei* Colonies: Structure, Distribution and Dynamics" in this volume) by stout byssal threads.

The species prefers the deeper waters of freshwater bodies (Nakano et al. 2010; Brugnoli et al. 2011, and see below), but naturally occupies the shallower banks of colonised habitats because this is where water movements keep substrata free of settling silt. *Limnoperna fortunei* can also withstand periodic, but brief, natural bouts of emersion (Iwasaki 1997) but which may, however, be longer under experimental conditions (Montalto and Ezcurra de Drago 2003; see Chapter "Control of *Limnoperna fortunei* Fouling by Desiccation" in this volume). When settling, newly metamorphosed individuals are strongly thigmotaxic with a preference for the angled crevices between vertical walls. They are also attracted, as in many bivalves, by the presence of adult conspecifics (Sardiña et al. 2009) and the tendency to aggregate is reinforced by this behaviour. Juveniles of *L. fortunei* are also negatively phototaxic and positively geotaxic and Morton (1975) showed in reservoir experiments using three-dimensional settling panels that the greatest recruitment was (1) at depths of 7–10 m, (2) to the undersurfaces of the panels (negative phototaxis) and (3) the vertical crevices of upper surfaces (positive geotaxis; Fig. 4 in Morton 1975), as later demonstrated by Uryu et al. (1996) in the laboratory.

The strong negative phototaxis exhibited by *L. fortunei* under experimental conditions of strong light (Uryu et al. 1996) is interesting because no photosensory receptors have been identified for this species hitherto. This study will cast light on this problem (see below). The juvenile behaviour, however, helps explain how *L. fortunei* can survive in the pipelines and culverts of water supply systems. In fast-flowing waters, such as are found in some invaded rivers and potable water pipes, *L. fortunei* inhabits crevices and pits, although from these foci succeeding generations can spread out to cover more and more of any exposed surfaces. Its byssus makes it extremely difficult to dislodge in such situations. As with the two species of *Dreissena* and *C. fluminea*, therefore, the ecological effects of *L. fortunei* invasions are not only profound but ecosystem-wide (Boltovskoy et al. 2009; Cataldo et al. 2012; Boltovskoy and Correa 2015).

Anatomy

The Shell and Ligament

The shell of *L. fortunei* has four stages. Prodissoconch I and prodissoconch II are both free-swimming larval stages, the first reaching a length of between 115 and 117 µm (Cataldo et al. 2005). Prodissoconch II, the veliger, is larger (~320 µm), highly torsioned (~180°), with respect to subsequent stages and coloured rose-violet (Morton and Dinesen 2010). The nepioconch forms subsequently and may reach a shell length of ~1300 µm. This stage is sometimes referred to as the plantigrade stage because it possesses a long thin, pipe-like, crawling foot. The final shell stage, the dissoconch, is formed by the juvenile individual and becomes the permanent shell of the adult. This pattern of shell development is shared with some other mytiloidean lineages, including *Mytilus* and *Perna* (Ockelmann 1983). In addition to Fig. 2 in Cataldo et al. (2005), other illustrations of the *L. fortunei* larvae are provided by Choi and Kim (1985, their Figs. 4–8) and Pinheiro dos Santos et al.

(2005, their Figs. 2–10; see Chapter "Larval Development of *Limnoperna fortunei*" in this volume). The nepioconch grows progressively and typically rapidly into the juvenile. With growth, the distances moved by crawling juveniles decrease with increasing shell length (Uryu et al. 1996). Byssal attachment typically occurs when an habitat suitable for adult occupation has been chosen and the dissoconch quickly assumes the adult form.

Before it does this, however, Montalto and Rojas Molina (2014) have shown that the dissoconch, bysssally attached, juvenile shell of *L. fortunei* is characterised by an array of byssal setae. These have been considered to be of periostracal origin, that is, periostracal "hairs", but are in fact laid down by the foot (Ockelmann and Dinesen 2009). They were first identified for the shells of representatives of the mytiloid Dacrydiinae by Ockelmann (1983) and described in more detail by Ockelmann and Dinesen (2009) for species of *Adula* and *Mytilus* (also Mytiloidea). Most recently, they have been described for another mytiloid, *Modiolus modiolus* (Linnaeus 1758) by Dinesen and Morton (2014). They are the most noticeable features of the juvenile, nepioconch, shell of *M. modiolus* and, as in *L. fortunei,* predominantly occur postero-dorsally. These authors also showed that the byssal setae of *M. modiolus* are homologous to the distal regions of a byssal thread, that is, the "plaque foam" of Silverman and Roberto (2007). This distal region of a thread comprises a primary layer attached to a surface—the plaque foam—and a more proximal region of collagen that, yet further proximally, unites into a bundle (or stem) attached to the byssal retractor muscles (Silverman and Roberto 2007). The byssal setae thus terminate as a tapering thread of collagen, as seen in *L. fortunei* by Montalto and Rojas Molina (2014). It also seems possible that they are characteristic of the juvenile shells of most, if not all, mytiloids although this survey has never been undertaken. Similarly, the setae have often been considered to be a feature of the juvenile mytiloid shell (Ockelmann and Dinesen 2009; Dinesen and Morton 2014), although Montalto and Rojas Molina (2014) show that adult *L. fortunei* may possess them too. It may simply be that in many mytiloids, *M. modiolus* being a good example (Ockelmann and Dinesen 2009; Dinesen and Morton 2014), that when an adult size is reached, the foot can no longer reach the highest levels of the shell. With age too, the setae may be lost through abrasion.

Montalto and Rojas Molina (2014) review the possible functions of the *L. fortunei* juvenile's byssal setae and it seems likely that, as suggested for *M. modiolus* by Dinesen and Morton (2014), the principal function is defense although the threads themselves may camouflage such vulnerable shells but this would also come under the definition of predator avoidance. In a significant experiment, Wright and Francis (1984) scraped off the byssal setae from the shells of *M. modiolus* and showed that they became more susceptible to the drilling activities of the intertidal whelk *Nucella lapillus* (Linnaeus 1758). The setae thus either inhibit the drilling activity of the predator or prevent it gaining a good grip on the shell.

Subsequently, adult *L. fortunei* can grow to a shell length of about 45 mm, but the more usual size ranges from 20 to 30 mm, very few individuals proceeding past their second year of life (in Hong Kong) into a third (elsewhere) to allow this (Morton 1977). The general colour of the shell is golden brown, this being mostly attributable to the thick periostracum that covers the shell, which presumably has a mineralogy and structure typical of the Mytiloidea (Taylor et al. 1969) although this has never been studied. The umbones are close to the end of the valves and there are no hinge teeth. The shell is attached to the substratum by an array of byssal threads which can, in some circumstances, as in detached and subsequently experimentally manipulated *Mytilus galloprovincialis* Lamarck, 1819 (Morton 2011), be shed and regrown but also when, in nature, an alternative, presumably better, site for adult occupation is either searched for or chosen.

Morton (1973) showed that the overall dimensions of the shell of *L. fortunei* inhabiting Plover Cove Reservoir in Hong Kong were relatively regular as seen from the ratio of width:height:length, that is: 1:1.18 (±0.18):2.60 (±0.50). Similarly the ratios of shell width:shell length and shell height:shell length were calculated as being 0.38 (±0.06) and 0.45 (±0.06), respectively. One important feature of the shell of *L. fortunei*, however, is that it is exceptionally thin and brittle, adult individuals 35 mm in length having a mid-shell thickness of <200 μm, whereas *M. galloprovincialis* individuals of the same shell length are >1400 μm thick. This may of course be a consequence of either a lack of soluble calcium in the occupied freshwaters, as in Hong Kong, for example (see Table 1 in Morton 1975), as compared with seawater, or the low requirements for such salts in *L. fortunei* (see Fig. 4 in Karatayev et al. 2007b), or, again, both.

Figure 1 provides different views of a *L. fortunei* shell. The adult, dissoconch, shell of *L. fortunei* is roundly heteromyarian and distinctively inequilateral when seen from the right side (Fig. 1a) That is, the anterior margin is almost round whereas the posterior margin is more pointedly rounded. The umbones are terminal, almost at the apex of the rounded anterior end and the antero-dorsal margin is less steep than the posterior and almost straight. The shell is somewhat keeled, this form being created by the pattern of growth inflating the shell anteriorly and narrowing it posteriorly. The outer, impermeable, proteinaceous, periostracal layer of the shell is smooth and shiny and is thick where it curls inwards at the shell margin. Seen from the dorsal aspect (Fig. 1b), the shell is thus slightly inflated anteriorly and more narrowly pointed posteriorly. As a consequence, the dorsal valve margins are slightly sinusoidal, that is, the shell is somewhat inequivalve. The inconspicuous ligament of *L. fortunei* is internal and located along the outside of the dorsal ridge of the anterior shell septum. Ventrally (Fig. 1c), there is a narrow byssal notch. From the anterior aspect (Fig. 1d), the shell is diamond-shaped in outline and swollen ventrally with the ligament visible for the virtual height of the shell. From the posterior aspect (Fig. 1e), the shell is less wide, as also seen in Fig. 1b and c, and more pointed ventrally. The shell is widest (Fig. 1d and e) at the lowest one third of the dorso-ventral height of the shell.

Internally, the shell of *L. fortunei* (Fig. 2) is thin, transparent and nacreous. The internal shell layer is pearly white, only slightly lustrous, darker postero-dorsally and lighter antero-ventrally. This is caused by the nacre of the shells' interior being purple above and white below the keel. The ligament of *L. fortunei* is internal and relatively long. With elastic properties, the ligament acts in opposition to the counteractive forces of the paired adductor muscles, keeping the animal in a state

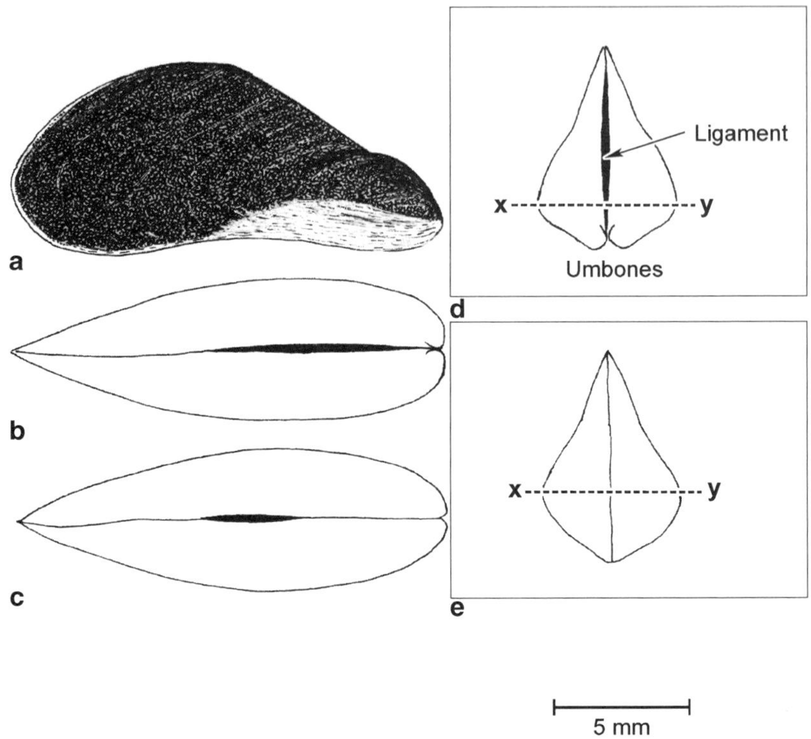

Fig. 1 Views of **a** the shell as seen from the right lateral aspect, **b** the dorsal, **c** the ventral, **d** the anterior and **e** the posterior aspects. *x----y* maximum shell width. (Modified after Morton and Dinesen 2010)

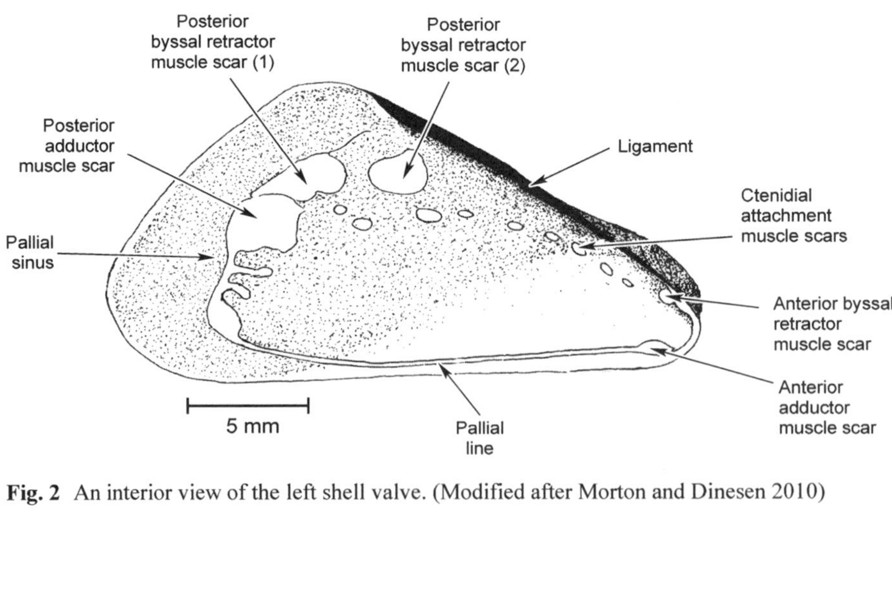

Fig. 2 An interior view of the left shell valve. (Modified after Morton and Dinesen 2010)

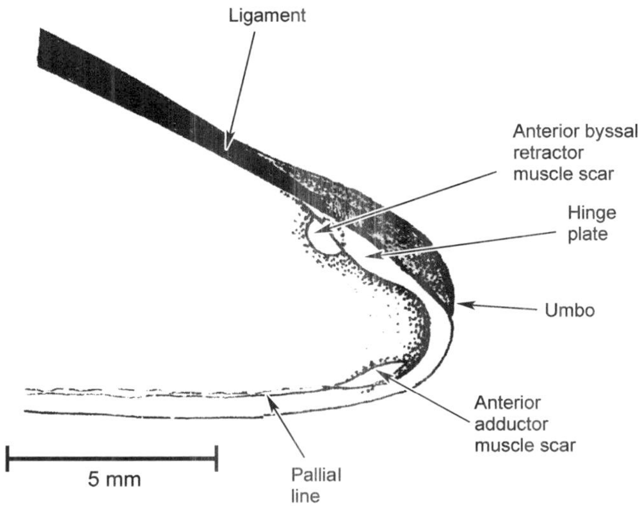

Fig. 3 An interior view of the anterior region of the left shell valve. (Modified after Morton and Dinesen 2010)

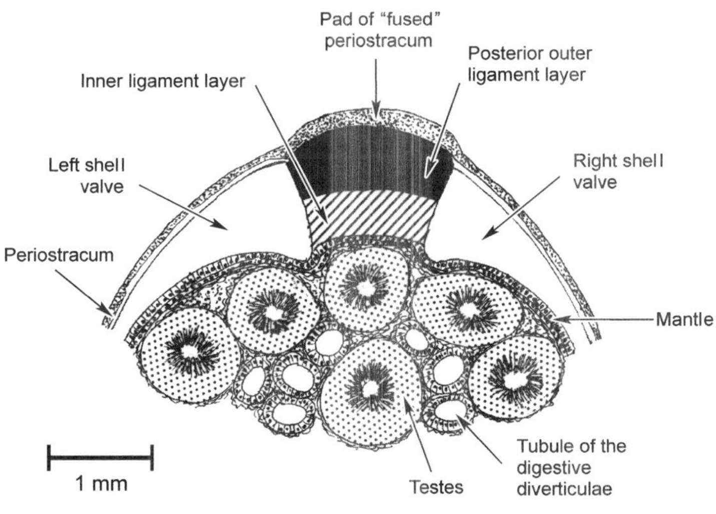

Fig. 4 A transverse section of the dorsal region of the shell showing the ligament and the underlying tissues

of tonus. With death, the muscles relax and the ligament thus keeps the valves separated. The muscle scars on the shell of *L. fortunei* comprise a large posterior adductor muscle and, anteriorly to it, two, separate, posterior byssal retractor muscles, labelled (1) and (2) in the figures. Internal to these is a posterior pedal retractor muscle scar but this is so small as to be virtually indistinguishable as a shell scar.

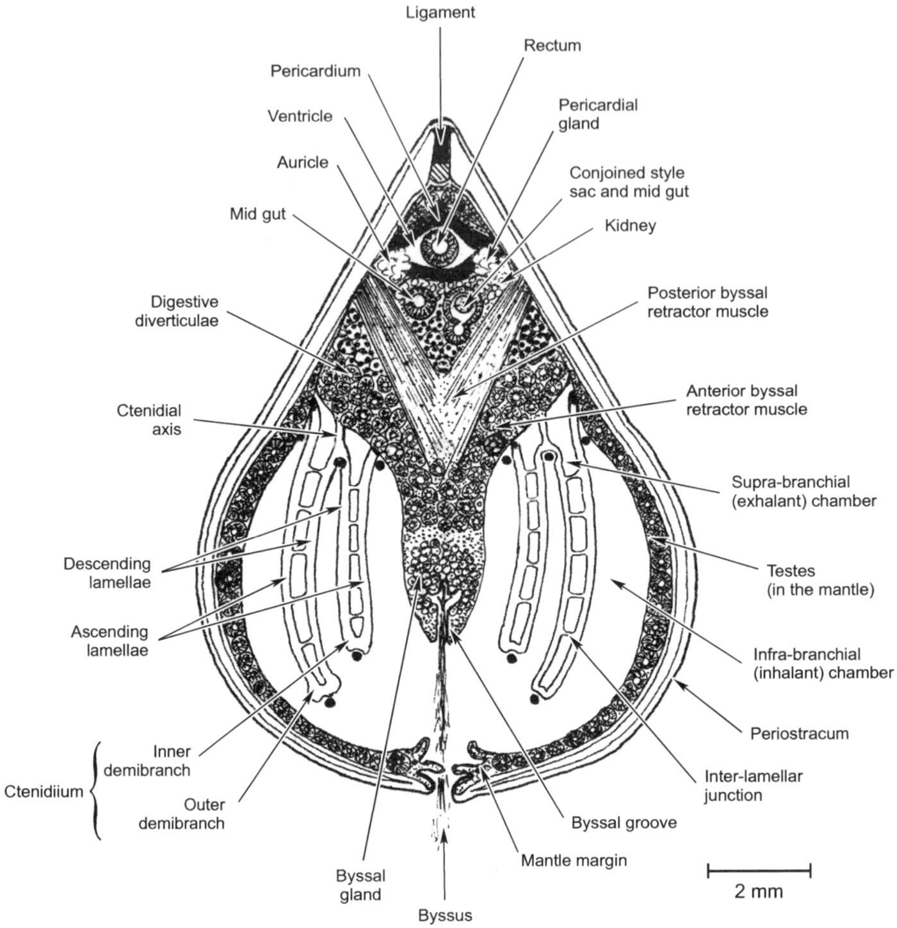

Fig. 5 A transverse section though an intact individual in the region of the heart identifying many of the more significant organs

The slight pallial sinus is thin and irregular although the pallial line is much thinner mid ventrally. There is a row of some nine ctenidial attachment scars of the ascending lamella of the outer demibranch that lead towards the scar of the anterior byssal retractor muscle, which is located just posterior to the umbo and half hidden by the hinge plate. There is a relatively large, kidney-shaped, anterior adductor muscle scar situated antero-ventrally.

A more detailed internal view of the anterior shell of *L. fortunei* shows no evidence of d-series hinge teeth and there are no antero-ventral folds or crenulations (Fig. 3). The shell is thus extremely simple, and this allows it to be separated from that of the only other freshwater mytiloid, *Sinomytilus harmandi* (Rochebrune 1882) which has an anterior septum upon which is located the scar of the anterior adductor muscle (Morton and Dinesen 2010). Each umbo lies slightly posterior to the anterior shell margin, which is smoothly rounded, that is, not crenulated as in some other mytiloids. A transverse section through the ligament and the dorsal region of the shell is illustrated in Fig. 4. The opisthodetic, paravincular ligament consists of two layers with staining reactions similar to those of *Mytilus edulis* (Linnaeus 1758; Trueman 1950; Beedham 1958), indeed of all representatives of the Mytiloidea (Yonge and Campbell 1968). The posterior outer ligament layer stains red and the inner ligament layer blue with both Mallory's triple stain and Masson's trichrome. Other mytiloids possess a ligament with a similar structure. In all these taxa, and in *L. fortunei*, the periostracum extends over the ligament, thereby adding another layer. This was termed fused periostracum (Owen et al. 1953) although, since periostracum must cover the entire shell, there is no fusion. The term is nevertheless retained herein simply to emphasise that it is present.

The shell of *L. fortunei* is illustrated in transverse section in Fig. 5. Its overall form accommodates the organs of the visceral mass dorsally. Ventrally, this leads to a single medial foot the base of which houses the byssal gland. This discharges its secretions into the byssal groove which forms, one-by-one, an array of threads that the foot plants onto the substratum thereby securing the animal, encased within its protective shell securely onto its chosen domicile. The shell is more swollen ventrally, providing (1) a flatter base, which achieves greater stability for it in lotic waters, and (2) allows accommodation for the ctenidia that are responsible for respiration and particle collection resulting from the feeding activities of this suspension feeding bivalve. The aragonitic shell of *L. fortunei* is encased within a shiny, proteinaceous, periostracum that is secreted by the outer surface of the outer fold of the mantle margin, which tracks the edges of both valves and, dorsally, secretes the ligament. The mantle, surrounding the body tissues in a protective cloak, also, as in many other mytiloids, contains elements of the gonads.

With this general plan of the body form in front of us, we can now proceed to a more detailed description of the anatomy of *L. fortunei.*

The Musculature and Byssus

The musculature of *L. fortunei* is seen from the right side in Fig. 6. The anterior adductor muscle is small and located on the anteroventral floor of the shell valves. In this respect *L. fortunei* is similar to other mytiloideans (White 1937; Wilson 1967). There is also a pair of thin anterior byssal retractor muscles arising from beneath the hinge plate and thus separate from the adductor muscle. In contrast to the anterior adductor, the posterior adductor muscle is large, circular and anteriorly adjoining byssal retractor muscles are divided into two major components. There is a small posterior pedal retractor muscle that has its origin anterior to the posterior byssal retractors. The pallial line, created by the pallial retractor musculature is generally thick especially posteriorly, but becomes thinner mid ventrally. The posterior byssal retractor is also illustrated connecting up with the byssus at the base of the foot within the visceral mass (Fig. 6). The foot has a distinct byssal groove.

Figure 7 is a left-right longitudinal section through the byssal apparatus at the base of the visceral mass of *L. fortunei*. Also seen in this section are the paired

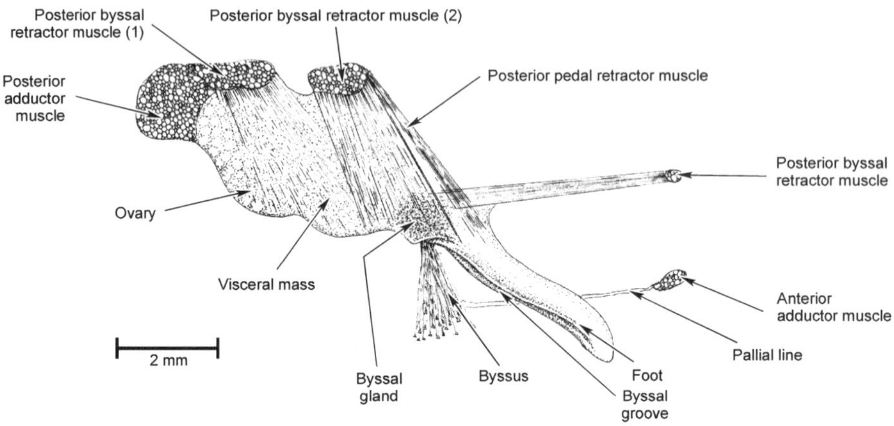

Fig. 6 The musculature and byssal apparatus as seen from the right side

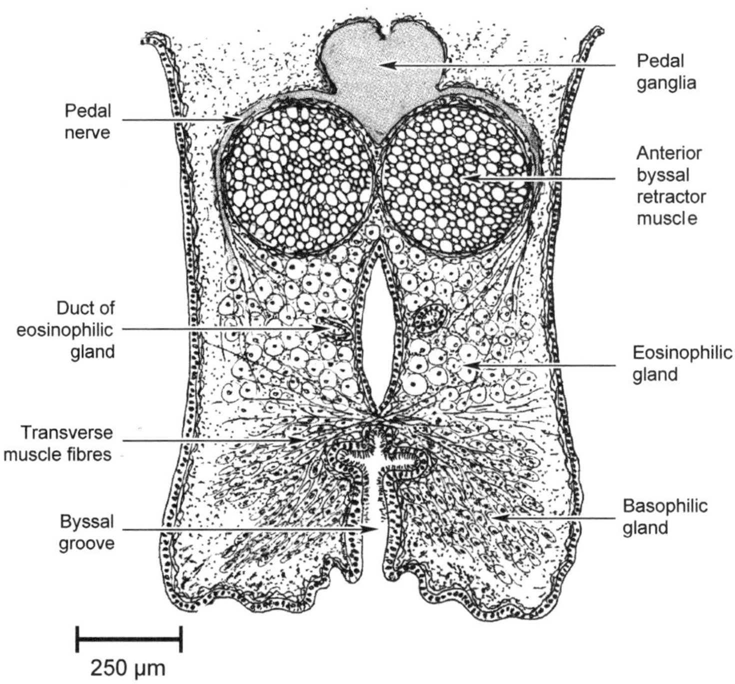

Fig. 7 A left-right longitudinal section through the byssal apparatus at the base of the visceral mass

anterior byssal retractor muscles and above them the pedal ganglia with lateral pedal nerves radiating into the foot. Unlike in many other bivalves, for example well-studied representatives of the Anomalodesmata (Morton 1985), no statocysts

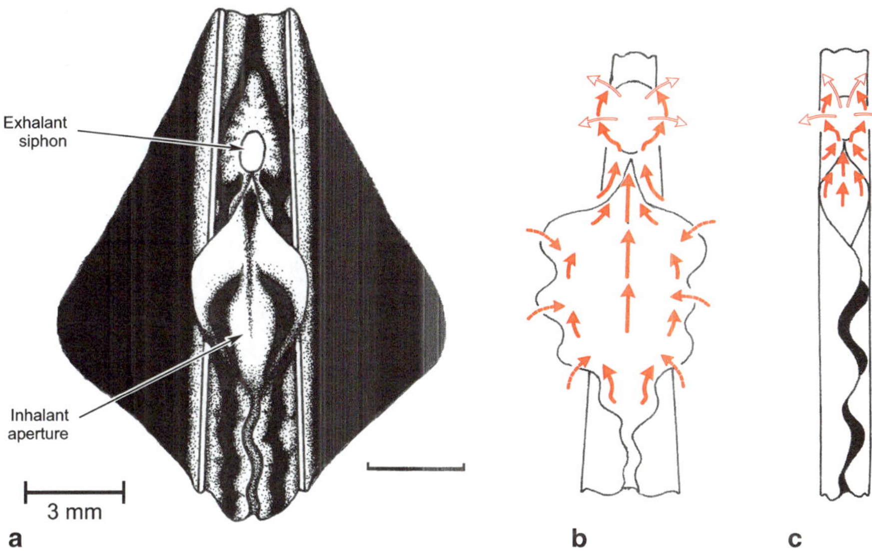

Fig. 8 **a** A posterior view of the animal showing the patterning on the mantle and exhalant siphon and inhalant aperture. The inhalant stream (*broken arrows*), exhalant stream (*open arrows*) and ciliary rejection currents (*solid arrows*) of **b** an actively filtering animal and **c** a disturbed animal are also shown. (Modified after Morton 1973; Malacologia, 12:270)

seem to occur dorsal to the pedal ganglia. As in other hitherto studied mytiloids, for example *Modiolus modiolus* (Linnaeus 1758; Dinesen and Morton 2014), the byssal groove is trefoil-shaped in section. The proteinaceous byssus is secreted by an eosinophilic byssal gland that comprises cells some 20 μm in length and arranged around the base of the byssal groove. This was termed the "white" or "collagen" gland in the foot of *M. edulis* by Brown (1952) and Pujol (1967). A second basophilic gland is arrayed sub-epithelially around the borders of the byssal groove and comprises elongate cells some 25 μm in length. These cells, the "purple" gland of brown (1952), probably secrete mucus, although Pujol (1967) thought it might be "enzymatic" and responsible for tanning the collagen produced by the "white" gland. Again, no specific studies of this structure have made for *L. fortunei*.

The Siphons

Figure 8a is a posterior view of *L. fortunei*, showing the pattern of pigmentation on the mantle and exhalant siphon and inhalant aperture. Externally, each mantle lobe is patterned with a brown stripe. These fuse dorsally to the exhalant siphon to form a single stripe. A similar brown stripe patterns the internal surfaces of the mantle lobes forming the inhalant aperture, and there is a dorsal median stripe on the inhalant aperture at the point of fusion of the mantle lobes forming the exhalant siphon (Fig. 8a). The exhalant siphon of *L. fortunei* is formed by fusion between the inner mantle folds only, this being type A (Yonge 1948; 1957). The inhalant aperture is not separated from the pedal/byssal aperture by fusion of the opposite mantle lobes but is separated functionally by their apposition. The inhalant stream (broken arrows), exhalant stream (open arrows) and ciliary rejection currents (solid arrows) of an actively filtering individual (Fig. 8b) and a disturbed one (Fig. 8c) are also shown. In both cases, as will be further described, the expulsion of pseudofaeces from the inhalant aperture is *via* its dorsal connection with the base of the exhalant siphon. Such a situation is typical of most representatives of the Mytiloidea (White 1937; Yonge 1955; Wilson 1967; Fankboner 1971), but atypical of more advanced siphonate bivalves. Neither the inhalant aperture nor the exhalant siphon bear either tentacles or papillae (Fig. 8a).

Figure 9a is a more detailed posterior view of junction between exhalant siphon and inhalant aperture of *L. fortunei* and highlights the siphonal septum. The septum is relatively small like that of *Modiolus modiolus* and unlike the large structure present in *M. galloprovincialis* (Dinesen and Morton 2014). The siphonal septum connects the ctenidia to the mantle at the point of fusion of the mantle lobes separating the exhalant siphon from the inhalant aperture. This septum effectively separates posteriorly the infra-branchial chamber from the supra-branchial. The septum is thought to act as a valve in other mytiloids (Fankboner 1971), controlling the volume of water entering the infra-branchial chamber. When filtering water, rates and, thus, efficiency of particle capture and retention are legislated by size and temperature (Sylvester et al. 2005) but also other factors including the relative amounts of seston and other potential food particles in the water column. When the animal is actively filtering and the siphons are extended, the septum is held near-vertically regulating the size of the inhalant stream. When the left and right folds of the inhalant aperture are withdrawn, however, the septum folds up left and right. It is, thus, a highly muscular structure in *L. fortunei* and, unlike other mytilids, bears five small papillae along its margin.

Figure 9b is a section through one of the papillae of the siphonal septum. The thick block of muscle fibres nearly fills the septum except marginally. The distal end of each papilla comprises a spherical head with an epithelium made up of what appear to be sensory cells each lined apically by microvilli. Internally, there is a nerve complex making contact with the sensory cells. The suggested function of the papillae will be discussed below.

The Mantle Margin

Mantle fusions occur dorsally above the exhalant siphon and between the exhalant siphon and inhalant aperture of *L. fortunei*. Mantle fusions, as with the siphons, are of the inner mantle folds only and thus of type A (Yonge 1957). Figure 10 is a transverse section through the left ventral mantle margin. The mantle margin, as in most bivalves, except for protobranchs and arcoids, comprises three folds: inner, middle and outer. The outer surface of the outer fold secretes the shell valve. The

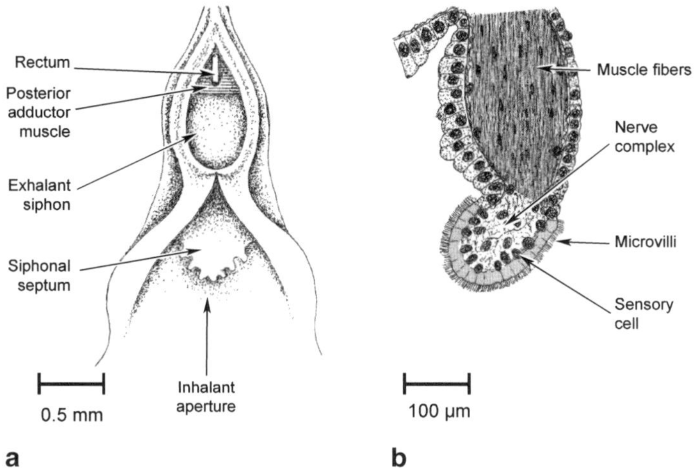

Fig. 9 **a** A more detailed posterior view of the siphonal septum between the exhalant siphon and inhalant aperture. **b** A section through one of the sensory papillae of the siphonal septum

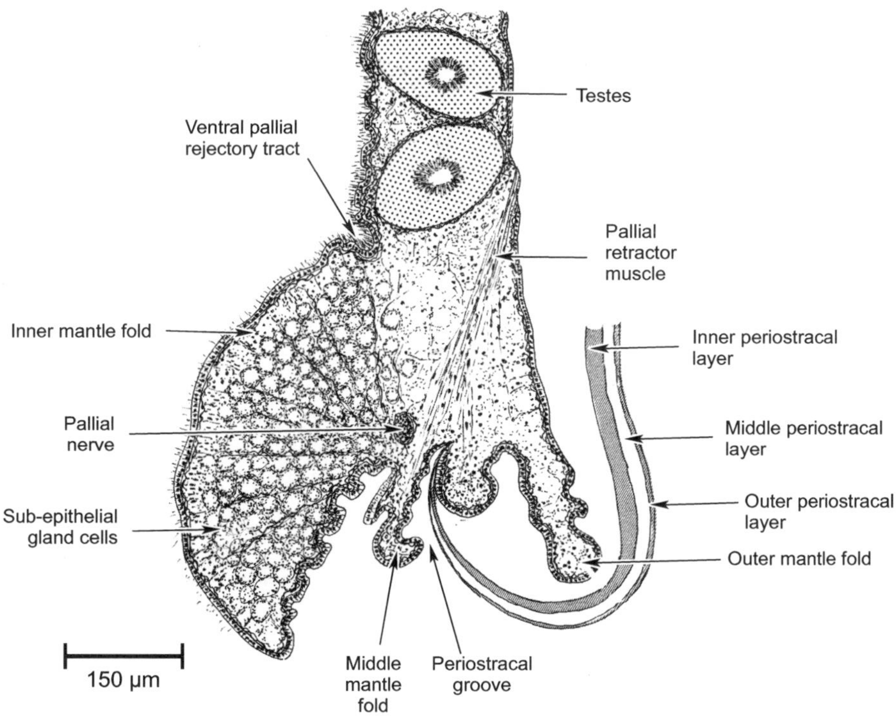

Fig. 10 A transverse section through the left ventral mantle margin

periostracum is secreted by the epithelia delimiting the periostracal groove, that is, the outer surface of the middle fold against the template of the inner surface of the outer fold. The periostracum is composed of three layers as in all mytiloideans studied hitherto and arises from the periostracal groove. The outer layer is thin (4 μm) and stains slightly grey with Heidenhain's haematoxylin but not at all with either Mallory's or Masson's stains. The middle layer is 20–25 μm thick at its greatest depth and stains light blue in Masson's trichrome. It is probably some kind of mucoid secretion. This layer does not, however, as does its counterpart in *M. edulis* (Fig. 1 in Beedham 1958), possess vacuoles. An inner laminated layer ultimately achieves a thickness of between 25 μm and 35 μm, and when first secreted stains red with both Mallory's and Masson's stains. Towards the margin of the shell valve, the outer laminations of this layer stain blue.

The inner fold of the mantle is ciliated densely, and at its junction with the general mantle surface there is a ventral pallial rejectory tract that transports unwanted particles posteriorly towards the inhalant aperture for eventual expulsion from the infra-branchial chamber. The swollen inner folds are highly glandular, the sub-epithelial cells are probably secreting mucus. The pallial retractor muscle, which attaches the mantle to the shell valves at the pallial line on each one, sends fibres mostly into the point of union of the middle and outer folds, that is, the base of the periostracal groove. There is a large pallial nerve. Distally, the mantle margin contains, as in some other mytiloids, much of the gonadal tissue of the animal, in this case the testes.

The Ciliary Currents of the Mantle

The ciliary currents of the left mantle lobe of *L. fortunei* are illustrated in Fig. 11. The ciliary currents of the mantle (including the siphons) and, as we shall see, the visceral mass and foot are all rejectory and serve to keep the mantle cavity free of either too large or unwanted particles. To achieve this, particles falling on to the mantle dorsally are passed anteriorly towards the mouth. Such particles, however, are all eventually transported ventrally, in a clockwise direction as seen from the right (counterclockwise when seen from the left) and, from the region around the mouth and labial palps are then passed posteriorly within the ventral pallial rejectory tract towards the inhalant aperture. Pseudofaeces are not concentrated at the base of the inhalant aperture to be expelled by the rapid adduction of the shell valves as, typically, in eulamellibranchs possessing a distinct siphon. Instead, the lobes of the inhalant aperture are highly mobile in *L. fortunei* and bear on their inner surfaces strong ciliary tracts, which pass the pseudofaeces dorsally towards the exhalant aperture as illustrated in Fig. 8b. When actively filtering, with the inhalant lobes fully expanded, water passes into the mantle cavity. The ciliary currents of the mantle take pseudofaeces towards the inhalant aperture against this incoming stream. Rapid closure of the shell valves forces water out of both apertures but, particularly, the exhalant siphon, thereby ejecting the pseudofaeces and the faeces, as we shall see. When the animal is disturbed, the shell valves only open partially

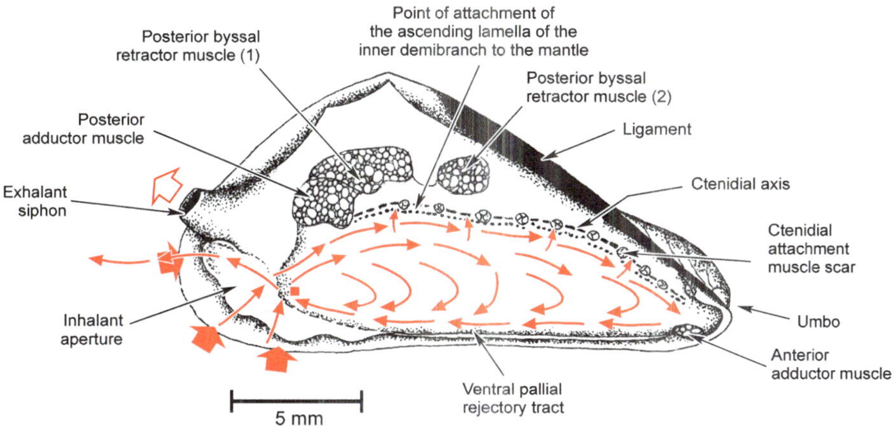

Fig. 11 The ciliary currents of the left mantle lobe

(Fig. 8c), but sufficiently to allow pseudofaeces to be similarly removed from the mantle cavity *via* a reduced inhalant aperture.

The Ctenidia and Labial Palps and Their Ciliary Currents

The organs of the mantle cavity of *L. fortunei* and their ciliary currents after removal of the right shell valve and mantle lobe are illustrated in Fig. 12. A comparable illustration of those of *M. edulis* were provided by Kellogg (1915, Fig. 18). The first detailed study of the bivalve ctenidia were undertaken by Ridewood (1903), but elaborated on in a series of classical studies by Daphne Atkins (died 1961) some of which are referred to herein. The ctenidia, which fulfill the dual roles of respiration and particle capture and transport, comprise two sub-equal demibranchs of which the outer is the longer dorso-ventrally. The upper, dorsal, margins of the ascending lamellae of the outer and inner demibranchs are attached to the mantle and the visceral mass, respectively, by ciliary fusions (Atkins 1937a). The ventral margin of the outer demibranch always lies tucked behind the incurving mantle margin with the associated, here thickened, periostracum. Like many other mytiloids also (Fankboner 1971), the outer demibranchs of *L. fortunei* are some five or six filaments shorter at their anterior ends than the inner ones. Fankboner (1971) states that "*a functional advantage for this anatomical reduction is unclear*". For *L. fortunei*, however, the advantage of this arrangement is clear in that it enables the ventral marginal food grooves of both demibranchs to be in contact with the sorting and, hence either acceptance or rejection functions of the inner surfaces of both inner and outer labial palps thereby greatly increasing the efficiency of particle selection by these structures. The ctenidial-labial palp junction of *L. fortunei* thus falls into Category I elucidated by Stasek (1963), and is typical of the Mytiloidea in general.

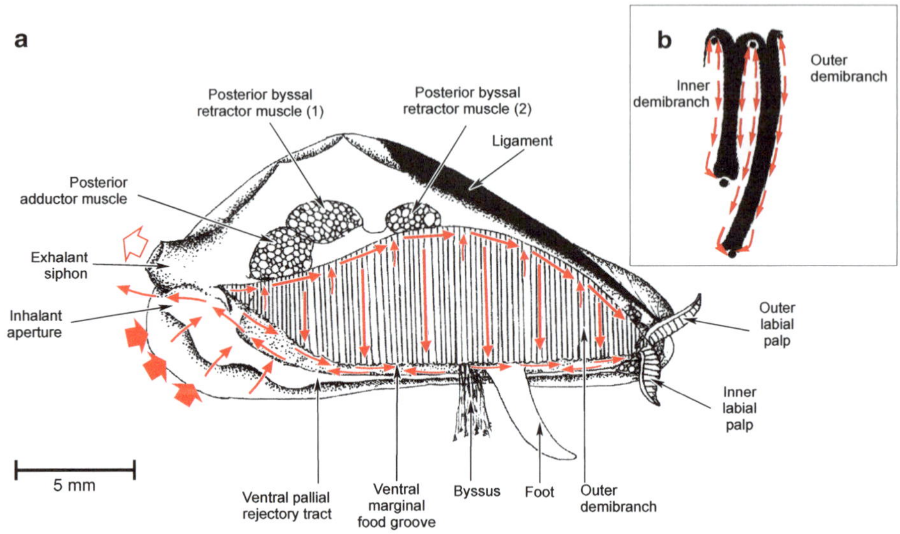

Fig. 12 **a** The organs of the mantle cavity and their ciliary currents after removal of the right shell valve and mantle lobe. **b** A transverse section through a single ctenidium showing the ciliary currents

The ctenidia of *L. fortunei* are flat, homorhabdic and filibranchiate (eleutherorhabdic). The ascending and descending lamellae of both demibranchs are cross-connected by inter-lamellar junctions, or unions (Fig. 13), Similarly, the dorso-ventrally aligned filaments which make up each lamella are connected laterally to each other by inter-filamentary junctions that maintain ctenidial cohesion. The junctions that cross-connect the individual filaments comprise ciliary discs as in other mytiloideans and as such are weak so that the filaments readily separate one from another when damaged. In lamellibranch bivalves, inter-filamentary union is achieved by tissue junctions that are much more robust.

The apices of the ctenidial filaments comprise a number of ciliary types that collectively fulfill the roles of filtration, particle entrapment and transportation. The currents through the ctenidia, that is, from the infra- to the supra-branchial chambers, are created by lateral cilia that are, thus, largely responsible for the forceful inhalant and exhalant streams into and out of the mantle cavity. The filtering apparatus itself is the responsibility of eulaterofrontal cirri, the fine structure of which have been illustrated and described for *M. edulis* by Owen (1974). Such a structure is typical of all studied mytiloideans, including *L. fortunei*, and all those bivalves that Atkins (1938) classified as the Macrociliobranchia, and which were refined histologically by Owen (1978). Other ciliary tracts on each filament head, principally the apical frontal cilia, are concerned with the transport either up or down of those particles flicked on to them by the eulaterofrontal cirri. It is the activities of these cilia that feed particles into the various ctenidial food grooves for onward transmission to and sorting by the labial palps.

The ciliation of the ctenidial surfaces of *L. fortunei* is of type B(I) (Atkins 1937b; Fig. 12b, 13a, b). Acceptance tracts are situated within the ventral marginal food grooves of both demibranchs, in the ctenidial axis and in the junctions of the as-

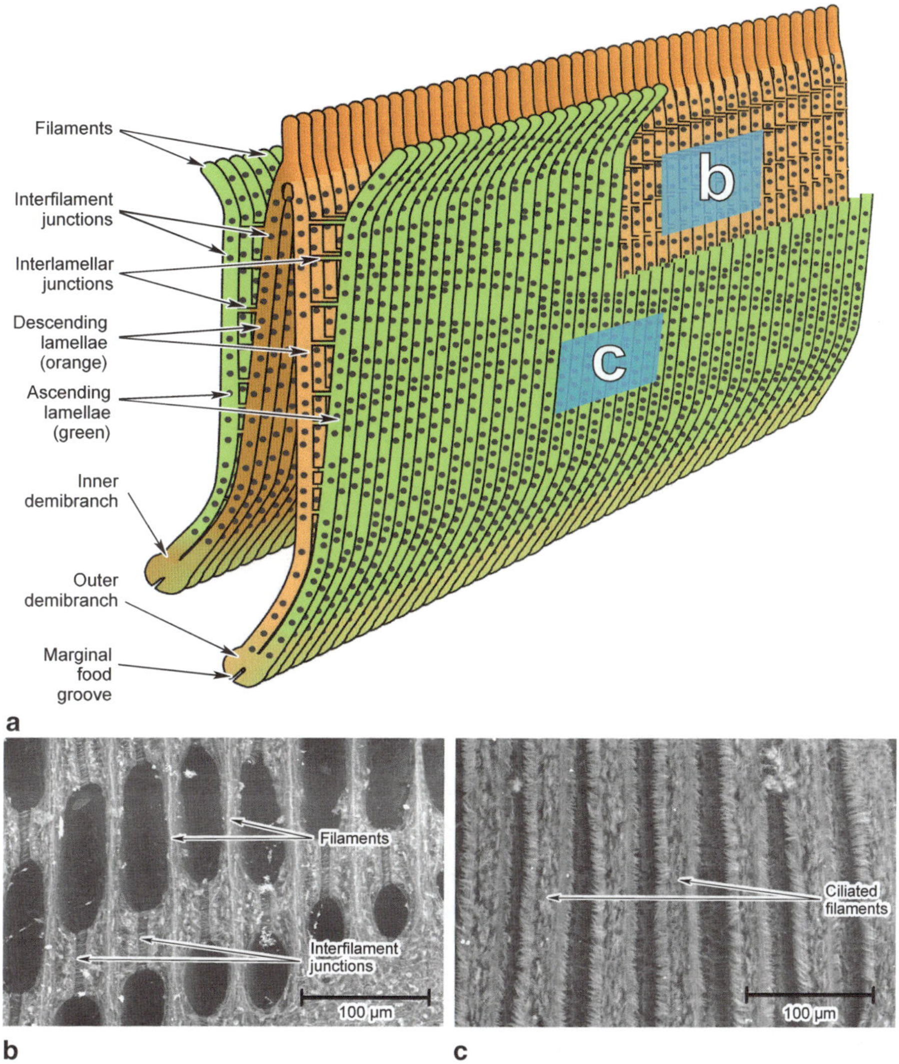

Fig. 13 **a** A schematic diagram of a ctenidium identifying salient features. **b** and **c** SEM microphotographs of the outer surface of the ascending lamella of the outer demibranch and the outward facing surface of the descending lamella of the outer demibranch, respectively. (Courtesy of Paolucci 2014)

cending lamella of the inner and outer demibranchs with the visceral mass and mantle, respectively. Only those particles arriving on the labial palps inside the ventral marginal food groove of the inner demibranchs, however, pass into the proximal oral groove and directly to the mouth. The ctenidial-labial palp junction of *L. fortunei*, as seen from the right side, is illustrated in Fig. 14. Particles arriving at the anterior end of the ctenidium *via* (1) the crests of the ventral marginal food grooves of both inner and outer demibranchs, (2) inside the ventral food groove of

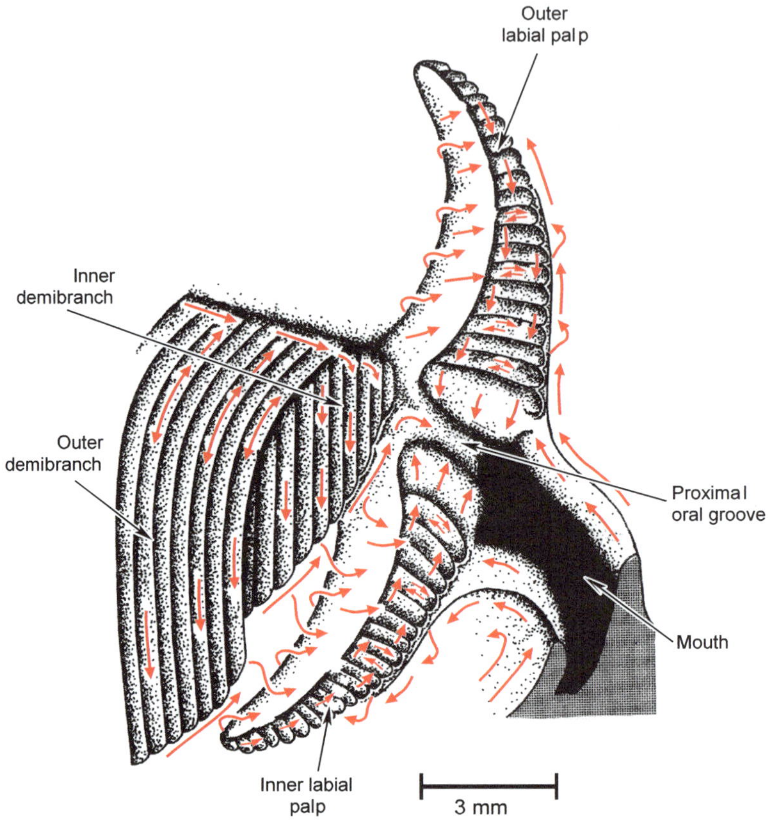

Fig. 14 The various ciliary currents of the labial palps and anterior end of the ctenidium of the right side. (Modified after Morton 1973; Malacologia, 12:271)

the outer demibranch and (3) in all three dorsal food grooves are subjected, before ingestion, to the ciliary selection currents of the labial palps. The abrupt termination of the outer demibranch, creating this unusually complicated sorting process, has not been observed in other bivalve taxa other than those of the Mytiloidea.

Particles are removed from the anterior ctenidial termini by the unridged portion of the labial palps, the ciliary currents of which subsequently pass the particles onto their ridged sorting region. This function is the attribute of the system of parallel ridges and grooves, which pass selected particles of a suitable nature and size over the crests of the ridges towards the proximal oral groove for ultimate ingestion. Too large and/or unwanted particles are passed laterally towards the opposite free edge of the palp for rejection. Recirculatory currents also exist. Details of the labial palp ciliation need not be elaborated upon since they are essentially the same as those described by Fankboner (1971) and are typical of mytiloids in general. The ciliary currents of the lips of the mouth are rejectory, passing unwanted material back to the labial palps for rejection along the prescribed course.

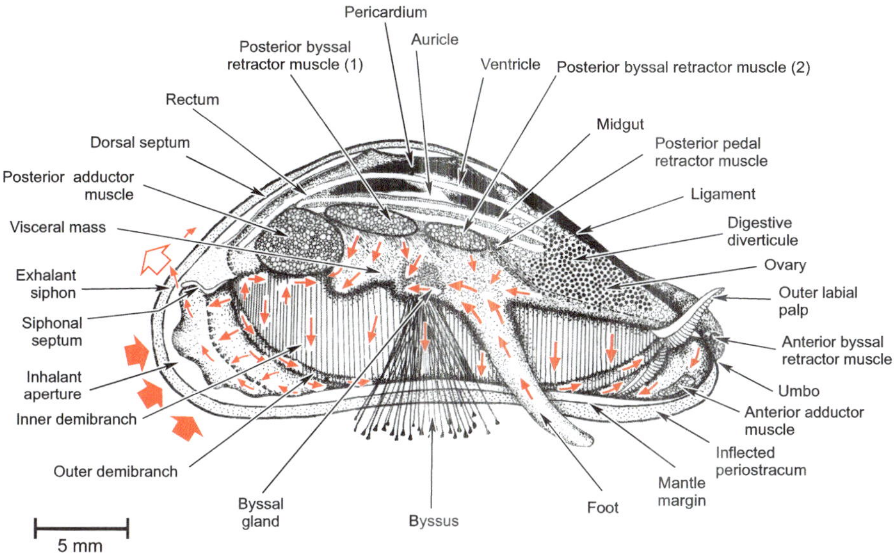

Fig. 15 The anatomy and ciliary currents of the foot and visceral mass after removal of the right ctenidium. The ciliary currents of the ascending lamella of the inner demibranch of the left ctenidium are also shown. (Modified after Morton 1973; Malacologia, 12:270)

The Ciliary Currents of the Visceral Mass and Foot

The ciliary currents of the surface of the visceral mass (Fig. 15) of *L. fortunei*, which near exclusively houses the gonads and byssal gland, pass particles downwards dorsally and then postero-ventrally to be concentrated at the postero-ventral tip of the visceral mass. From this point, they are presumably removed by the ventrally directed ciliary tracts of the ascending lamellae of the inner demibranchs. Being too large to enter the ventral marginal food grooves of these demibranchs, such material ultimately passes into the ventral, posteriorly directed, rejectory tracts of the mantle (Fig. 10; 11). The crawling and byssal thread-planting foot possesses ciliary currents that pass particles dorsally to join the posterior stream on the visceral mass and in this way these are also ultimately expelled from the mantle cavity. In Fig. 15, the ciliary currents of the ascending lamella of the inner demibranch of the left ctenidium, behind the visceral mass and foot, are also shown.

The Alimentary Canal

Figure 15 gives a general impression of the course of the intestine in *L. fortunei*, not within, but above and between the byssal retractor muscles and below the organs of the pericardium. The course of the intestine is illustrated more precisely in Fig. 16. Here, all other tissues except for the ventricle of the heart have been ignored and the

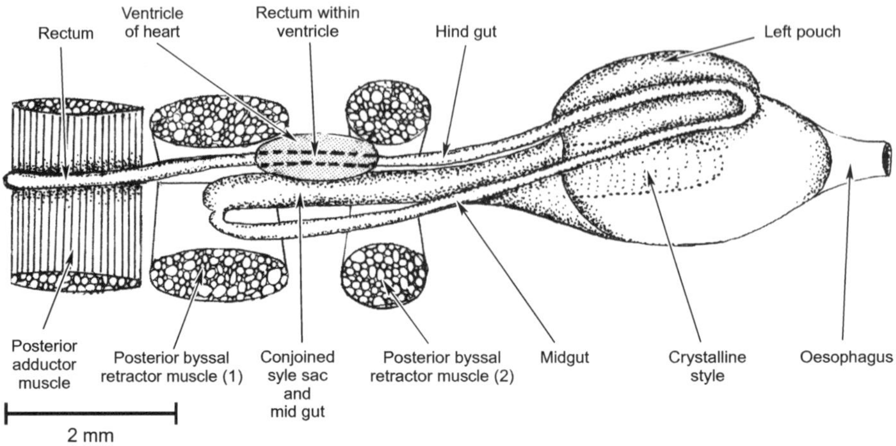

Fig. 16 A generalised dorsal view of the course of the intestine. All other tissues except for the ventricle of the heart have been ignored and the right byssal retractor muscle has been pulled to the right

right byssal retractor muscle blocks have been pulled to the right thereby exposing the intestine. The course of the intestine in *L. fortunei* is similar to that seen in other mytiloideans. The oesophagus passes upwards from the mouth, which lies between the anterior byssal retractor muscles and is closely applied to the dorsal surface of the anterior adductor muscle. The ciliated oesophagus opens into the stomach, which is located under the anterodorsal margin of the shell and is surrounded by the dark digestive diverticulae (Fig. 15). From the posterior end of the stomach arises the combined style sac and mid-gut, which passes backwards between both blocks of the posterior byssal retractor muscles. Just anterior to the posterior adductor muscle, the style sac terminates but the mid-gut, now separated from it, loops forwards alone to pass back between the posterior byssal retractors. The mid-gut loops again on the left side of the stomach (not the right as originally thought by Morton 1973), and turns posteriorly again to penetrate the ventricle of the heart, pass between the posterior pair of byssal retractor muscles, over the posterior adductor muscle, to terminate in an anus on the posterior face of this structure facing the exhalant siphon (Fig. 9). The histological structures of the style sac and intestine of *L. fortunei* are essentially the same as those described for *M. edulis* by Giusti (1971) and for all other mytiloideans hitherto studied.

The Stomach

The structure and ciliary currents of the interior of the stomach of *L. fortunei*, after opening by a horizontal incision in the right side, are illustrated in Fig. 17. The stomach is elongate and bears a close similarity to the stomachs of other mytiloideans (Graham 1949; Purchon 1957; Reid 1965; Dinamani 1967; Fankboner 1971)

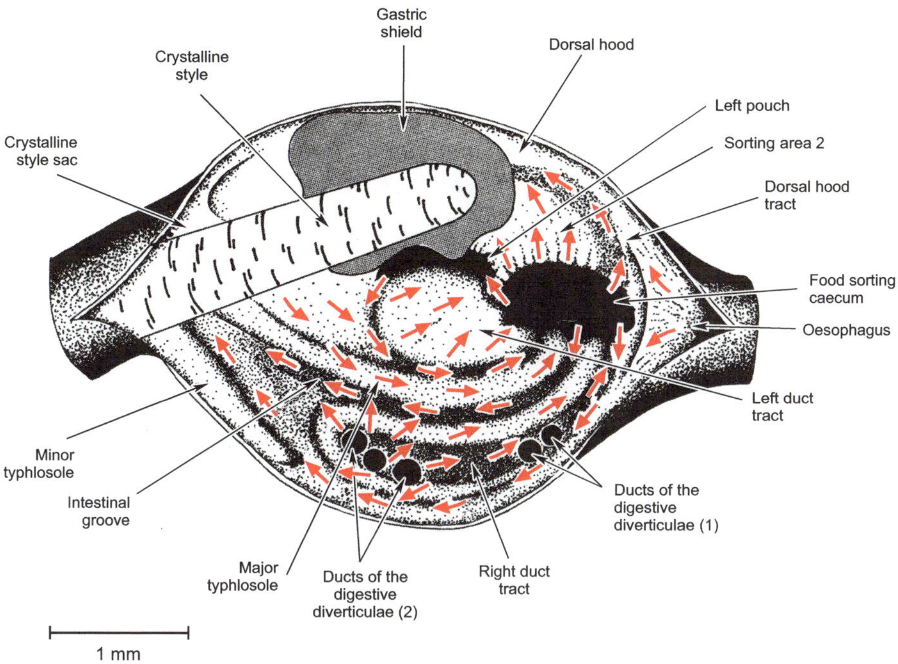

Fig. 17 The structure and ciliary currents of the interior of the stomach after opening by a horizontal incision in the right side. (Modified after Morton 1973; Malacologia, 12:274)

and thus belongs to type III and Section I of the stomach types elucidated by Purchon (1957) and Dinamani (1967), respectively. An attempt has been made in this description of the stomach of *L. fortunei* to combine the nomenclatorial systems of Purchon (1957), Reid (1965) and Dinamani (1967).

In *L. fortunei*, as in all bivalves, the floor of the stomach is dominated by the major typhlosole and associated intestinal groove, which arise in the style sac and pass forwards to penetrate the food-sorting caecum. The crystalline style is secreted in the style sac and, protruding into the stomach, rotates against the typically saddle-shaped gastric shield covering the left dorso-lateral wall of the stomach. The gastric shield sends a flare into the left pouch. The major typhlosole does not divide, as reported for *Adula falcata* Gould, 1851 by Fankboner (1971). The minor typhlosole also arises in the style sac and passes, for a short distance, along the right side of the stomach. On this side of the stomach too and associated with the right duct tract, are two groups of openings (1 and 2) into the ducts that lead to the digestive diverticula. Purchon (1957) considered the right duct tract to be a sorting area and termed it Sorting Area 3. A further sorting area (2) can be recognised dorsal to the entrance of the food-sorting caecum and separating this opening from the entrance to the left pouch. A sorting area of the left duct tract would appear to be the floor and walls of the left pouch.

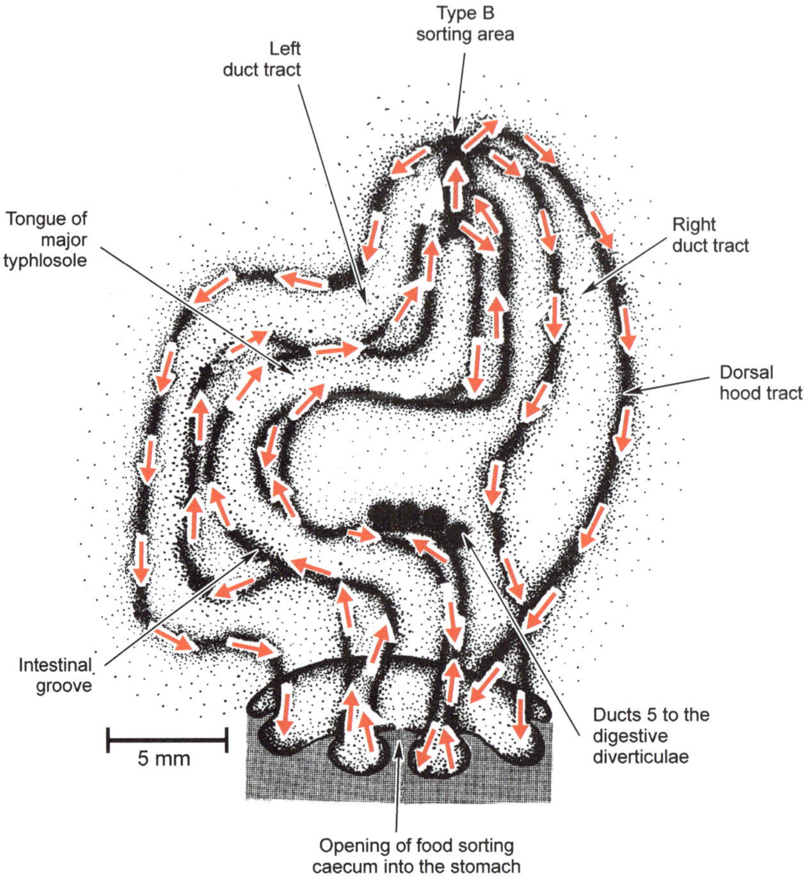

Fig. 18 The structure and ciliary currents of the food-sorting caecum of the stomach. (Modified after Morton 1973; Malacologia, 12:275)

The structure and ciliary currents on the left side of the stomach of *L. fortunei* constitutes the origin of what Reid (1965) has termed the left duct tract, which passes particles of food into the capacious food-sorting caecum (Fig. 18). The right duct tract also passes into the food-sorting caecum. This structure is a comparatively long finger-shaped pocket penetrated to its apex by a tongue of the major typhlosole. At the caecum's apex is a sorting area, which is of type B (Reid 1965), and found only in those bivalves that Purchon (1960; 1963) has grouped together as the Gastrotriteia and which is characteristic of the Mytiloidea (Reid 1965). The caecum also has a cluster of four ducts leading into the digestive diverticulae.

The structure and ciliary currents of the left pouch of the stomach of *L. fortunei* are illustrated in Fig. 19. This structure, which the gastric shield sends a flare into dorsally, has two further groupings of ducts leading to the digestive diverticula (3 and 4). Each of the sorting areas in the left pouch is a system of ridges and grooves, which Reid (1965) has identified as type A and which are found in all bivalves.

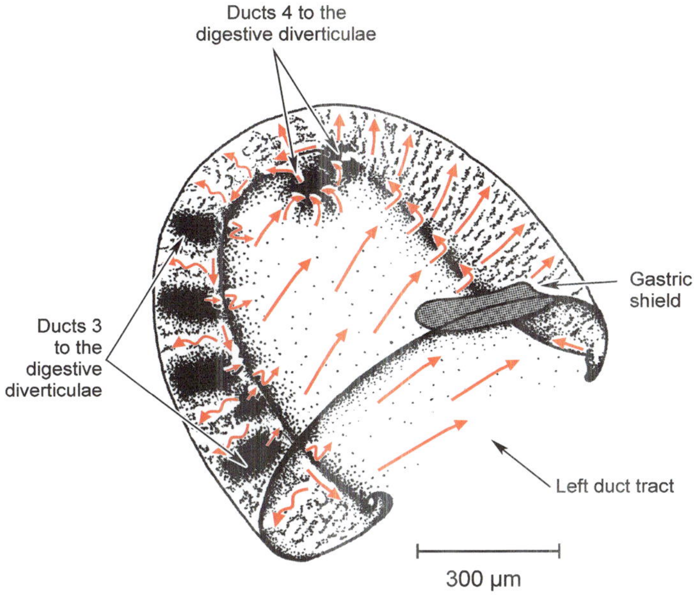

Fig. 19 The structure and ciliary currents of the left pouch of the stomach. (Modified after Morton 1973; Malacologia, 12:275)

The stomach of *L. fortunei* is a site of continuous food particle selection, although this species too, like *D. polymorpha*, may possess a diurnal rhythm of feeding and digestion (Morton 1969a), after primary sorting by the labial palps, and extra-cellular digestion by the slow dissolution of and release of enzymes from the slowly rotating crystalline style against the gastric shield which is probably also enzymatically productive (Halton and Owen 1968). Cilia on the crests of the major typhlosole and inner folds of the left and right duct tracts pass potential food material entering the stomach into the food-sorting caecum. Ciliary currents in the grooves of the inner folds of the left and right duct tracts and the incurrent fold of the intestinal groove also pass particles into the food-sorting caecum. At the apex of the caecum the B type sorting area (Reid 1965) sends acceptable particles of a suitable size into the outer folds of the left and right duct tracts which pass this material to the ducts of the digestive diverticula of the left pouch and right duct tract. Rejected particles pass out of the food-sorting caecum in the excurrent intestinal groove of the major typhlosole and pass into the mid-gut for ultimate defecation. Particles of intermediate size are probably recirculated by the dorsal hood tract passing them back to the dorsal hood and gastric shield. The minor typhlosole assists the major typhlosole in clearing the stomach of unwanted food into the mid-gut.

No appendix, as has been reported for other mytiloids by Reid (1965) and Fankboner (1971), could be identified in the stomach of *L. fortunei*. The basic structure of the ducts and the digestive tubules comprising the digestive diverticulae of *L. fortunei* bear a close similarity to those described by Owen (1955) for *M. edulis*.

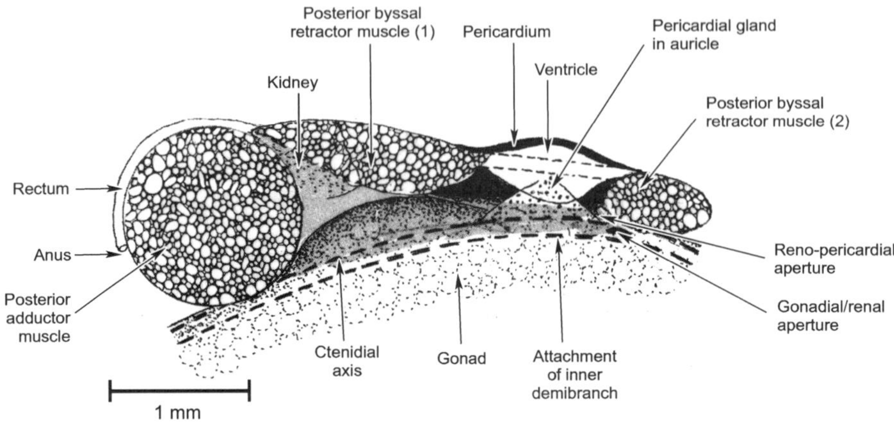

Fig. 20 The organs of the pericardium as seen from the right side. The course of the intestine, except for the rectum within the ventricle of the heart has been ignored

A review of the structures and processes involved in the feeding and digestion of the Bivalvia has been provided by Morton (1983).

The Organs of the Pericardium

The course of the rectum through the medial ventricle of the heart and between the two elements of the posterior byssal retractor and over the posterior adductor muscles has been described above. The organs of the pericardium of *L. fortunei*, as seen from the right side, are illustrated in Fig. 20. In this figure, the course of the intestine, except for the rectum within the ventricle of the heart, has been ignored. Posterior and ventral to the heart are the paired, pale-brown, kidneys. The reno-pericardial apertures of the kidneys are situated on the postero-ventral floor of the pericardium adjacent to the edge of the anterior-most blocks of the posterior byssal retractor muscles. The epithelia of the left and right auricles of the heart contain the similarly light brown pericardial gland.

The renal apertures, which are shared with the gonadal apertures, open onto the visceral mass to left and right beneath the posterior edge of the anterior-most blocks of the posterior byssal retractor muscles. The ascending lamella of the outer demibranch of the right ctenidium attaches to the mantle/shell at the junction between the visceral mass and mantle but is not illustrated here. The ctenidial axis and the point of union of the ascending lamella of the inner demibranch of the right ctenidium with the visceral mass are, however, illustrated in Fig. 20, and it can be seen that the renal/gonadal aperture opens into the supra-branchial chamber between these two attachment points. That is, excretory products and gametes are discharged into that component of the supra-branchial chamber situated between the ctenidial axis and the inner ctenidial demibranch.

Reproduction

Limnoperna fortunei is generally dioecious (Morton 1991), although there is one report of low-level hermaphroditism (Darrigran et al. 1998; see Chapter "Reproductive Output and Seasonality of *Limnoperna fortunei*" in this volume). The large paired gonads are located in mesosomal lobes in, primarily, postero-dorsal region of the visceral mass. As in *M. edulis* (White 1937), the gonadal tissues of *L. fortunei* also invade the mantle, everywhere, as illustrated in Figs. 4, 5 and 10.

Discussion

Limnoperna fortunei is a highly opportunistic species that has been introduced into many locations outside its mainland Chinese borders, notably, Taiwan, Japan and, most recently, South America. Morton and Dinesen (2010) also raised the intriguing possibility that *L. fortunei* might have been introduced into tropical Indochina (Cambodia, Laos, Thailand, Vietnam), hitherto considered part of its natural range, from China. Its present range in this region might thus reflect the pattern of past human migrations to and from China and the countries of South-east Asia.

Limnoperna fortunei first came to scientific, as opposed to conchological, attention in the late 1960s when it colonised the pipes, conduits and channels of part of the water supply system of Hong Kong (Morton 1975). *Limnoperna fortunei* is not alone in this respect. It is but the last of a string of freshwater (and to a lesser extent brackish water) invasions into non-native regions by bivalves most notably including *D. polymorpha* in Europe and latterly with *D. bugensis* into North America, and *C. fluminea* into North America and subsequently Europe and South America—all processes facilitated by transoceanic transport in the ballast water of container and other ships and the construction of interconnecting potable water supply systems (see Chapter "Distribution and Colonization of *Limnoperna fortunei:* Special Traits of an Odd Mussel" in this volume).

But, there is more to these successful invasions than human-mediated introductions. Just as with *D. polymorpha*, it would seem that *L. fortunei* is ideally adapted to a life in fast flowing (but not torrents) waters in the possession of a stout byssus and an heteromyarian, or mytiloid, shell form. Both can, however, also thrive in the relatively static waters of lakes and reservoirs (Morton 1969a) and in this habitat can cause problems of sedimentation resulting from their sheer numbers, pseudofaeces and faeces production and the accumulation of dead shells. It would seem that both of these species, despite their superficially specialised form, are liberal in their choice of habitat and are, hence, potentially detrimental at all stages of the water supply process and the industrial and potable resources arising from it.

The close similarity in choice of habitat and form existing between *L. fortunei* and *D. polymorpha* (especially, but also *D. bugensis*) was suggested to be related to some degree of phylogenetic affinity between the parent Mytiloidea and Dreissenoidea (Purchon and Brown 1969), respectively. Yonge and Campbell (1968) showed, however, that the similarities which exist between species of *Dreissena* and the similarly anteriorly septate mytiloid species of *Septifer* were due to convergence. Morton (1973, 1992) and Taylor et al. (1972) agreed with this view and further suggested that from both palaeontological and anatomical standpoints, the two taxa are wholly unrelated. It is envisaged, both for *Dreissena* and *Limnoperna*, however, that the neotenous retention of the larval byssus (Yonge 1962) in their respective ancestors resulted in the evolution of the heteromyarian form (Yonge and Campbell 1968) in both. Both too have subsequently exploited this condition, with the development of osmoregulatory powers, in the colonisation of freshwater systems.

Scarlato and Starobogatov (1979), placed the monospecific and monogeneric *Limnoperna* in its own sub-family—the Limnoperninae, a classification followed by Bieler et al. (2010). Although such a taxonomy seems highly dubious to this author, the search for an ancestor to *L. fortunei* must, nevertheless, be focussed on the Mytiloidea and Morton and Dinesen (2010) suggested that the genus is superficially similar to species of the brackish water *Xenostrobus*, a taxon formerly identified as *Modiolus* Lamarck, 1799. Such a superficial similarity prompted Beu (2006) to subsume *Xenostrobus* into junior synonomy with *Limnoperna*. Both have been considered to be representative of the Modiolinae, although the phylogeny of neither has been studied properly and there are anatomical differences between the two (see below), which are sufficiently significant to refute Beu's synonomy. Nevertheless, in some other respects, the two taxa are similar with regard to the possession of some comparable anatomical characters. In particular, both have simple thin modioliform shells lacking any anterior teeth and crenulations with sub-terminal umbones and no external sculpturing. The heteromyarian shells of both have an umbonal keel that is dark brown above and a paler yellow-brown below. The anterior byssal retractor muscle has its origin on the antero-dorsal roof of the shell in *L. fortunei* as in *Xenostrobus securis* and *X. pulex* (Wilson 1967). Similarly, the posterior byssal retractor muscle is divided into two blocks in both *L. fortunei* and *X. inconstans* Wilson, 1967 (but not in *X. securis* and *X. pulex* also first described by this author). Furthermore, and perhaps most significantly, both *X. securis* and *X. inconstans* live at the head of estuaries in Australia whereas other species, for example, *X. pulex* are marine (Wilson 1967). Ockelmann (1983) described two other species of *Xenostrobus*, that is *X. mangle* and *X. balani*, from estuarine habitats in Thailand.

It would thus seem, at least intuitively, possible that *Limnoperna* evolved from forms essentially similar to *Xenostrobus*. Thiele (1934) considered *Limnoperna* to have a modioline *sensu lato* ancestry. Conversely, the other freshwater, and also monospecific, Asian mytiloid, *S. harmandi*, was thought to be of mytiline sensu lato ancestry. If true, the freshwater environments of Indochina have been colonised twice at different times by two, monospecific, mytiloidean genera with different temporal origins. If true, the *Limnoperna* modioline/musculine lineage of the Mytiloidea dates back to the Devonian (~345–395 million years before present—mybp; Morton 1992). Conversely, the *Sinomytilus* mytiline line shares characters with several of the marine and estuarine species of the Mytilinae *sensu lato*, of which more modern taxa first appeared in the Permo-Triassic (~265–225 mybp; Morton 1992). *Limnoperna* and *Xenostrobus* are, thus, it is suggested, both representative of extremely ancient lineages. Hence, attractive as the scenario of a common ancestry for the above two genera might be (albeit both being very distantly related), there are anatomical differences between the two as noted above. For example, the heart of *L. fortunei* is located between the two blocks of the posterior byssal retractor muscles, whereas in all species of *Xenostrobus*, it is situated anterior to these muscles. In *L. fortunei* too, the mid gut loops over the left side of the stomach but, barely, over the right in all three species of *Xenostrobus* studied anatomically by Wilson (1967). Adding fuel to the ancestry debate, *L. fortunei* is, in other respects, reminiscent of the much larger and coastal species of *Perna*, notably with regard to the structure of the posterior byssal retractor muscle-pericardium complex and overall characteristics save for the absence of an anterior adductor muscle in *Perna viridis* (Linnaeus 1758; Morton 1987; Ockelmann 1995). The question of *Limnoperna*'s ancestry is, hence, unresolved although it undoubtedly resides in the modioline/musculine *sensu lato* lineage of the Mytiloidea. This subject is of ongoing personal research.

As if to highlight the above-hypothesised similarity between *Limnoperna* and *Xenostrobus*, however, Habe (1981) identified a subspecies of *L. fortunei*, that is, *L. fortunei kikuchii* (Fig. 4c in Morton 1997), first recorded from Japan sometime between 1974 and 1979 (Nishimura and Habe 1987) when it was introduced (supposedly) alongside consignments of *Corbicula* from China. It subsequently colonised the estuarine Shonai Inlet to Lake Hamana and excluded the similarly estuarine *Musculista senhausia* (Benson 1842; Abdel-Razek et al. 1993a, b). The ability of this sub-species to tolerate salinities of between 0 and 30‰ (Kimura et al. 1995), however, cast doubts about its stated identity and, subsequently Morton (1997) and Kimura and Sekiguchi (2009) confirmed it to be *X. securis* which has a near identical salinity tolerance of 1–31‰ (Wilson 1968). Conversely, *L. fortunei* tolerates only low salinities (Deaton et al. 1989) although, in the regularly fluctuating salinities of experimentally mimicked estuaries, it can survive periodic exposure to 14‰ (Sylvester et al. 2013), as must be the case occasionally in its natural habitat of the Pearl River in China for example (Miller and McClure 1931). *Xenostrobus securis* has now too been introduced widely into, for example, Korea (Shirafuji and Sato 2003) and is now present throughout much the Mediterranean being first recorded in 1991 from the Venice lagoon as *Xenostrobus* sp. (Sabelli and Speranza 1994; Mizzan 1999). Since its initial discovery, the species has spread throughout the Mediterranean and has most recently been recorded (as *Limnoperna securis*) from the Bay of Biscay marking its spread into the North Atlantic (Adarraga and Martínez 2011), probably in ballast water.

Hence, although no ancestor to *Limnoperna* can be identified with certainty in modern, let alone fossil, Mytiloidea, it seems reasonable that such a form should be sought among brackish-water relatives and their ancestors in turn. It has been established, for example, and it is perhaps significant, that species of *Dreissena* are closely related to the confamilial estuarine species of *Mytilopsis*, one of which, *M. sallei* (Récluz 1849) has, itself, become an invasive species albeit of brackish waters (Morton 1981). It would, thus, seem that *Dreissena* and *Limnoperna,* convergent as they are, represent the apices of two phyletic streams that have both adapted to life in freshwaters. Significantly, the hard surfaces found in freshwater systems in many parts of the world are not normally colonised by bivalves, most species being endobenthic, for example, representatives of the Unionoidea and Corbiculoidea. The hard surfaces niche was therefore a suitably vacant target for the ancestors of both *Dreissena* and *Limnoperna*. Significantly, within their own spheres of influence, both species would appear to be colonising this habitat as fast as it is artificially created for them.

But what features, specifically, not only adapt *L. fortunei* to life in freshwaters but have also facilitated its colonisation of waters external to its natural range? *Limnoperna fortunei* possesses ciliary tracts on the internal surfaces of the inhalant aperture, which carry pseudofaeces towards the exhalant siphon. The intermittent, posterior adductor muscle generated, rapid expulsion of water from the exhalant siphon blows these away together with the faeces. Living as it can do in fast-flowing waters, its siphons invariably facing the current, this process is a significant aspect of the morphology of *L. fortunei* since it enables the animals to feed and remove pseudofaeces at the same time but, more importantly, blows this waste material over the top of the animal and not straight out in front of it. This prevents the pseudofaeces and faeces from being taken back into the mantle cavity. In large colonies occupying more static waters with, for example, low Reynold's numbers, however, have illustrated how the mass of *L. fortunei* becomes covered with faeces and pseudofaeces (see Fig. 1 in Boltovskoy et al. 2009).

The outer demibranchs of the ctenidia of *L. fortunei* are unusually long dorso-ventrally. This adaptation gives a greater surface area for filtration and also places rejected particles travelling anteriorly on the crests of the ventral marginal food grooves of both demibranchs in much closer proximity to the rejection tracts of the mantle. The outer demibranch being longer dorso-ventrally but abruptly shorter antero-posteriorly relative to the inner demibranch also enables the labial palps to exert their selective influence upon all four of the gill lamellae. For an animal living, as *L. fortunei* can do, in a wide range of lentic and lotic waters experiencing large differences in suspended sediment loads, this ctenidial-labial palp relationship ensures that all particles reaching the anterior end of the ctenidia are potentially made available as food. Potentially because the animal has to balance its particle-collection and -sorting abilities for food against, in waters with high sediment loads, the opposite capability of removing inhaled but unwanted material surplus to requirements. And, thereby, avoiding the clogging of the mantle cavity and its organs. Paolucci et al. (2014) have shown that across 24 South American sites experiencing variations in suspended solid loads, *L. fortunei* showed phenotypic variation not just in shell morphology but also in relative ctenidial surface area. Also identified was a reduction in mean ctenidial ciliation with increasing suspended sediment loads suggesting a mechanism for balancing the aforementioned dietary and cleansing requirements (see Chapter "Colonization and Spread of *Limnoperna fortunei* in South America" in this volume). Although, more likely, but unmeasured, these functions are more particularly the role of the labial palps.

The alimentary system of *L. fortunei* is typical of the Mytiloidea in general, although the relatively large food-sorting caecum in this species, when compared with the short caeca of species described by Dinamani (1967), may indicate a greater selective facility in *L. fortunei* and, thus, ensure that all potential food material is utilised. This would be especially pertinent for such populations living in more oligotrophic waters. Conversely, in large silt-laden continental waters, the sorting and selection currents in the mantle cavity and stomach facilitate the removal of large amounts of unwanted material. The food-sorting caecum of *L. fortunei* also possesses ducts to the digestive diverticulae. These ducts, by increasing the total number of apertures to the digestive diverticulae, increase the capabilities of *L. fortunei* for channelling a greater number of particles into the digestive diverticulae and making its intracellular digestion much more efficient facilitating, in turn, the rapid growth characteristic of this species.

This second structural study of *L. fortunei* (that is, the present Chapter) has, however, identified another aspect of its anatomy hitherto un-noticed. The siphonal septum, unlike other mytiloids studied previously, possesses five sensory papillae. These are innervated, but not ciliary-based as one would expect if they were mechanoreceptors. Rather, the microvilli-fringed sensory cells are reminiscent of simple photoreceptors. In a preliminary experiment to determine if *L. fortunei* possesses a shadow reflex once acclimated to an electric light, a passing hand shadow did result in the siphons being partially retracted and valves shut slightly in some (<10 %) of individuals. In a second experiment under conditions of direct, bright, sunlight, of 38 individuals tested for a shadow reflex, 8 (21 %) closed their valves, 16 (42 %) retracted their siphons and 14 (37 %) gave no reaction (Romina Tokumon and Daniela Duchini, pers. comm.). Since the exposed general mantle edges of *L. fortunei* possess no papillae of any kind, it is possible that those on the siphonal septum are the source of these reactions. It is, moreover, unknown if *L. fortunei* possesses an endogenous diurnal rhythm of feeding and digestion as demonstrated for *D. polymorpha* by Morton (1969b). If so, and it would seem logical that is the case, then some kind of optical capability would be necessary (simple photosensory cells in the case of *D. polymorpha*) to modulate this rhythm. If the structures identified for *L. fortunei* are photosensory organs (in contrast to simple cells) this would represent an advance on, for example, even the more modern eulamellibranch *D. polymorpha*, a specialisation not hitherto identified in any other mytiloidean (Morton 2008) and an occurrence seen in only a few, more advanced, bivalve lineages.

Limnoperna fortunei is, thus, a relatively unspecialised mytilid in terms of its shell architecture, that is a lack of external ornamentation and internal teeth and crenulations, but those specialisations that do exist are concerned with greater efficiency in food collection and utilisation. There are quite obviously physiological specialisations, especially with regard to evolved osmoregulatory processes. In essence, however, *Limnoperna* is a typical representative of the Mytiloidea. *Dreissena polymorpha* is, similarly, anatomically unspecialised and, it has been suggested before (Morton 1982), that it is the retention of either primitive or, at least, simplified characters in a habitat where there has been a trend in other lamellibranchs towards greater and greater specialisation, for example, representatives of the Unionoidea, that makes the attribute of simplicity so successful. Such characters include, for example, a thin, plain, shell, a byssus, high fecundity and free-swimming larvae.

It is these anatomical features, combined with an *r*-expressed reproductive strategy (Morton 1991; see Chapter "Reproductive Output and Seasonality of *Limnoperna fortunei*" in this volume) and life history trait, which makes *L. fortunei* characteristic of opportunistic species. In such species, minimal energy is spent on shell thickness (particularly advantageous when competing for the colonisation of freshwaters poor in calcium, see above), the life cycle is short, just two, rarely three, years in the case of *L. fortunei*, but where all activity is focussed on food collection and digestion to maximise energy gain for reproduction. Such characteristics epitomise the success achieved by *L. fortunei* as a highly invasive species. The sexual strategy and life history trait of *L. fortunei* have previously been compared with those of *D. polymorpha* to explain the invasive successes of both species (Karatayev et al. 2007a; see Chapter "Parallels and Contrasts Between *Limnoperna fortunei* and Species of *Dreissena*" in this volume). Just as remarkable, however, is the fact that the two families have wholly separate phylogenetic origins and, thus, that such similarities are not just convergent but highlight the success of, ultimately, the byssally attached heteromyarian form in freshwaters in them both.

Acknowledgements I am grateful to the editors of Molluscan Research and Malacologia for permissions to reproduce herein redrawn and relabelled figures originally published in these journals (for details see figure legends).

References

Abdel-Razek FA, Chiba K, Kurokura H, Okamoto K, Hirano R (1993a) Distribution of *Limnoperna fortunei kikuchii* in Shonai Inlet, Lake Hamana. Suisan Zoshoku 41:89–95

Abdel-Razek FA, Chiba K, Kurokura H, Okamoto K, Hirano R (1993b) Life history of *Limnoperna fortunei kikuchii* in Shonai Inlet, Lake Hamana. Suisan Zoshoku 41:97–104

Adarraga I, Martínez J (2011) First record of the invasive brackish water mytilid *Limnoperna securis* (Lamarck, 1819) in the Bay of Biscay. Aquat Invasions 7:171–180

Atkins D (1937a) On the ciliary mechanisms and interrelationships of lamellibranchs. Part III. Types of lamellibranch gills and their food currents. Q J Microscop Sci 79:375–421

Atkins D (1937b) On the ciliary mechanisms and interrelationships of lamellibranchs. Part IV. Cuticular fusion, with special reference to the fourth aperture in certain lamellibranchs. Q J Microscop Sci 79:423–445

Atkins D (1938) On the ciliary mechanisms and interrelationships of lamellibranchs. Part VII. Laterofrontal cilia of the gill filaments and their phylogenetic value. Q J Microscop Sci 80:346–430

Beasley CR, Tagliaro CH, Figueiredo WB (2003) The occurrence of the asian clam *Corbicula fluminea* in the lower Amazon basin. Acta Amazonica 33:317–324

Beedham DE (1958) Observations on the non-calcareous component of the shell of the Lamellibranchia. Q J Microscop Sci 99:341–357

Beran L (2000) First record of *Corbicula fluminea* (Mollusca: Bivalvia) in the Czech Republic. Acta Soc Zool Bohem 64:1–2

Beran L (2006) Spreading expansion of *Corbicula fluminea* (Mollusca: Bivalvia) in the Czech Republic. Heldia 6:187–192
Beu AG (2006) Marine Mollusca of oxygen isotope stages of the last 2 million years in New Zealand. Part 2. Biostratigraphically useful and new Pliocene to recent bivalves. J R Soc New Zeal 36:151–338
Bieler R, Carter JG, Coan EV (2010) Part 2. Classification of bivalve families. Malacologia 52:113–184
Boltovskoy D, Correa N (2015) Ecosystem impacts of the invasive bivalve *Limnoperna fortunei* (golden mussel) in South America. Hydrobiologia 746:81–95
Boltovskoy D, Karatayev A, Burlakova L, Cataldo D, Karatayev V, Sylvester F, Mariñelarena A (2009) Significant ecosystem-wide effects of the swiftly spreading invasive freshwater bivalve *Limnoperna fortunei*. Hydrobiologia 636:271–284
Brown CH (1952) Some structural proteins of *Mytilus edulis*. Q J Microscop Sci 93:487–502
Brugnoli E, Dabezies MJ, Clemente JM, Muniz P (2011) *Limnoperna fortunei* (Dunker 1857) en el sistema de embalses del Rio Negro, Uruguay. Oecol Australis 15:576–592
Burch JQ (1944) Checklist of west American mollusks. Minutes of the Conchological Club of Southern California 38:18
Cataldo D, Boltovskoy D, Hermosa JL, Canzi C (2005) Temperature-dependent larval development rates of *Limnoperna fortunei* (Bivalvia: Mytilidae). J Mollus Stud 71:41–46
Cataldo D, Vinocur A, O'Farrell I, Paolucci E, Leites V, Boltovskoy D (2012) The introduced bivalve *Limnoperna fortunei* boosts *Microcystis* growth in Salto Grande Reservoir (Argentina): evidence from mesocosm experiments. Hydrobiologia 680:25–38
Choi SS, Kim JS (1985) Studies on the metamorphosis and the growth of larva in *Limnoperna fortunei*. Korean J Malacol 1:13–18 [In Korean]
Darrigran GA, Pastorino G (1995) The recent introduction of a freshwater Asiatic bivalve, *Limnoperna fortunei* (Mytilidae) into South America. The Veliger 38:171–175
Darrigran GA, Damborenea C, Penchaszadeh PE (1998) A case of hermaphroditism in the freshwater invading bivalve *Limnoperna fortunei* (Dunker, 1857) (Mytilidae) from Río de la Plata, Argentina. Iberus 16:99–104
Deaton LE, Derby JGS, Subhedar N, Greenberg MJ (1989) Osmoregulation and salinity tolerance in two species of bivalve mollusc: *Limnoperna fortunei* and *Mytilopsis leucophaeta*. J Exp Mar Biol Ecol 133:67–79
Dinamani P (1967) Variation in the stomach structure of the Bivalvia. Malacologia 5:225–268
Dinesen GE, Morton B (2014) Review of the functional morphology, biology and perturbation impacts on the boreal, habitat-forming horse mussel *Modiolus modiolus* (Bivalvia: Mytilidae: Modiolinae). Mar Biol Res 10:845–870
Douda K, Vrtílek M, Slavík O, Reichard M (2012) The role of host specificity in explaining the invasion success of the freshwater mussel *Anodonta woodiana* in Europe. Biol Invasions 14:127–137
Dudgeon D, Morton B (1984) Site selection and attachment duration of *Anodonta woodiana* (Bivalvia: Unionacea) glochidia on fish hosts. J Zool (Lond) 204:355–362
Fankboner PV (1971) The ciliary currents associated with feeding, digestion and sediment removal in *Adula* (*Botula*) *falcata* Gould 1851. Biol Bull 140:28–45
Giusti F (1971) The fine structure of the style sac and intestine in *Mytilus galloprovincialis* Lam. Proc Malacol Soc Lond 39:95–104
Goto Y (2002) Behavior of nuisance mussel, *Limnoperna fortunei*, in water supply facilities. Water Sci Technol 46:45–50
Graham A (1949) The molluscan stomach. T R Soc Edin 61:737–776
Habe T (1981) A catalogue of mollusca of Wakayama prefecture. The Province of Kii. I. Bivalvia, scaphopoda and cephalopoda. Pub Seto Mar Biol Lab 7:1–301
Halton DW, Owen G (1968) The fine structure and histochemistry of the gastric cuticle of the protobranchiate bivalve, *Nucula sulcata* Bronn. Proc Malacol Soc Lond 38:71–81

Hayden B, Caffrey JM (2013) First recording of the Asian clam (*Corbicula fluminea* (Müller, 1774)) from the River Shannon, with preliminary notes on population size and class distribution. Irish Nat J 32:29–31

Holikoshi Y (1935) Ecology of freshwater mollusca in Taiwan. T Nat Hist Soc Formosa 25:226–231 [In Japanese]

Howlett D, Baker R (1999) *Corbicula fluminea* (Müller); new to UK. J Conchol 36:83

Huang CW (2008) Population genetic structure of *Limnoperna fortunei* (Dunker, 1857) in Taiwan based on sequences of mitochondrial cytochrome C oxidase subunit I. MSc Thesis, National Changhua University of Education (Taiwan), pp 1–60

Ituarte CF (1994) *Corbicula* and *Neocorbicula* (Bivalvia: Corbiculidae) in the Paraná, Uruguay and Río de la Plata basins. The Nautilus 107:129–135

Iwasaki K (1997) Climbing behaviour and tolerance to aerial exposure of a freshwater mussel, *Limnoperna fortunei*. Venus (Jpn J Malacol) 56:15–25

Karatayev AY, Boltovskoy D, Padilla DK, Burlakova LE (2007a) The invasive bivalves *Dreissena polymorpha* and *Limnoperna fortunei* : parallels, contrasts, potential spread and invasion impacts. J Shellfish Res 26:205–213

Karatayev AY, Padilla DK, Minchin D, Boltovskoy D, Burlakova LE (2007b) Changes in global economies and trade: the potential spread of exotic freshwater bivalves. Biol Invasions 9:161–180

Kellogg JL (1915) Ciliary mechanisms of lamellibranches with descriptions of anatomy. J Morphol 26:625–701

Kerney MP, Morton B (1970) The distribution of *Dreissena polymorpha* (Pallas) in Britain. J Conchol 27:97–100

Kimura T (1994a) Morphological identification of *Limnoperna fortunei* (Dunker) and *Limnoperna fortunei kikuchii* Habe. Chiribotan (J Malacol Soc Jpn) 25:36–40 [In Japanese]

Kimura T (1994b) The earliest record of *Limnoperna fortunei* (Dunker) from Japan. Chiribotan (J Malacol Soc Jpn) 25:34–35 [In Japanese]

Kimura T, Sekiguchi H (2009) Spatial and temporal patterns of abundance of the exotic mytilid *Xenostrobus securis* and the native mytilid *Musculista senhousia* in the Lake Hamana, Japan. Biodivers Rec 2 e89:1–7

Kimura T, Kakuta I, Kurokura H (1995) Salinity tolerance and osmoregulation in freshwater and brackish water mytilids (Mytilidae, genus *Limnoperna*). Bull Soc Sea Water Sci, Jpn 49:148–152 [In Japanese]

Kinzelbach R (1991) Die Körbchen muscheln *Corbicula fluminalis, Corbicula fluminea* und *Corbicula fluviatilis* in Europe (Bivalvia: Corbiculidae). Mainzer Naturwissenschaftliches Archiv 29:215–228

Kuroda T (1941) A catalogue of molluscan shells from Taiwan (Formosa), with descriptions of new species. Memoirs of the faculty of science and agriculture. Taihoku Imperial University 22:65–216

Magara Y, Matsui Y, Goto Y, Yuasa A (2001) Invasion of the non-indigenous nuisance mussel, *Limnoperna fortunei*, into water supply facilities in Japan. J Water Supply: Res Technol—Aqua 50:113–124

Martínez ER (1987) *Corbicula manilensis* molusco introducido en Venezuela. Acta Científica Venezolana 38:384–385

Mathews J, van der Velde G, bij de Vaate A, Collas FPL, Koopman KR, Leuven RSEW (2014) Rapid range expansion of the invasive quagga mussel in relation to zebra mussel presence in The Netherlands and Western Europe. Biol Invasions 16:23–42

McMahon RF (1982) The ocurrence and spread of the introduced Asiatic freshwater clam, *Corbicula fluminea* (Muller) in North America: 1924–1982. The Nautilus 96:134–141

Miller RC, McClure FA (1931) The fresh-water clam industry of the Pearl River. Lingnan Sci J 10:307–322

Mills EL, Rosenberg G, Spidle AP, Ludyanskiy M, Pligin Y (1996) A review of the biology of the quagga mussel (*Dreissena bugensis*), a second species of freshwater dreissenid introduced to North America. Am Zool 36:271–286

Mizzan L (1999) Le specie alloctone del macrozoobenthos della Laguna di Venezia: il punto della situazione. Bollettino del Museo Civico di Storia Naturale di Venezia 49:145–177

Montalto L, Ezcurra de Drago I (2003) Tolerance to desiccation of an invasive mussel, *Limnoperna fortunei* (Dunker, 1857) (Bivalvia, Mytilidae), under experimental conditions. Hydrobiologia 498:161–167

Montalto L, Rojas Molina F (2014) Byssal hairs in the invasive Asian freshwater bivalve *Limnoperna fortunei* (Mytilidae) in the Paraná River system with comments on this species in South America. Molluscan Res 34:127–138

Morton B (1969a) Studies on the biology of *Dreissena polymorpha* Pall. (I). General anatomy and morphology. Proc Malacol Soc Lond 38:301–321

Morton B (1969b) Studies on the biology of *Dreissena polymorpha* Pall. (II). Correlation of the rhythms of adductor activity, feeding, digestion and excretion. Proc Malacol Soc Lond 38:401–414

Morton B (1973) Some aspects of the biology and functional morphology of the organs of feeding and digestion of *Limnoperna fortunei* (Dunker) (Bivalvia: Mytilacea). Malacologia 12:265–281

Morton B (1975) The colonization of Hong Kong's raw water supply system by *Limnoperna fortunei* (Dunker 1857) (Bivalvia: Mytilacea) from China. Malacol Rev 8:91–105

Morton B (1977) The population dynamics of *Limnoperna fortunei* (Dunker 1857) (Bivalvia: Mytilacea) in Plover Cove Reservoir, Hong Kong. Malacologia 16:165–182

Morton B (1981) The biology and functional morphology of *Mytilopsis sallei* (Recluz) (Bivalvia: Dreissenacea) fouling Visakhapatnam harbour, Andhra Pradesh, India. J Mollus Stud 47:25–42

Morton B (1982) Some aspects of the population structure and sexual strategy of *Corbicula* cf. *fluminalis* (Bivalvia: Corbiculacea) from the Pearl River, People's Republic of China. J Mollus Stud 48:1–23

Morton B (1983) Feeding and digestion in Bivalvia. In: Wilbur KM, Saleuddin ASM (eds) The Mollusca. Vol 5: Physiology, Part 2. Academic Press, New York, pp 65–147

Morton B (1985) Statocyst structure in the Anomalodesmata (Bivalvia). J Zool (Lond) 206:23–34

Morton B (1987) The functional morphology of the organs of the mantle cavity of *Perna viridis* (Bivalvia: Mytilidae). Am Malacol Bull 5:161–166

Morton B (1991) Do the Bivalvia demonstrate environment-specific sexual strategies? A Hong Kong model. J Zool (Lond) 223:131–142

Morton B (1992) The evolution and success of the heteromyarian form in the Mytiloida. In: Gosling EM (ed) The mussel *Mytilus* : Ecology, physiology, genetics and culture. Developments in aquaculture and fisheries. Elsevier, Amsterdam, pp 21–52

Morton B (1997) The aquatic nuisance species problem: a global perspective and review. In: D'Itri F (ed) Zebra mussels and aquatic nuisance species. Lewis Publishers, Boca Raton, pp 1–54

Morton B (2008) The evolution of eyes in the Bivalvia: new insights. Am Malacol Bull 26:35–45

Morton B (2011) Predator-prey-scavenging interactions between *Nucella lapillus* (Linnaeus, 1758), *Carcinus maenas* and *Eulalia viridis* all exploiting *Mytilus galloprovincialis* on a rocky shore recovering from tributyl-tin (TBT) pollution. J Nat Hist 45:2397–2417

Morton B, Dinesen G (2010) Colonization of Asian freshwaters by the Mytilidae (Bivalvia): a comparison of *Sinomytilus harmandi* from the Tonle-Sap River, Phnom Penh, Cambodia, with *Limnoperna fortunei*. Molluscan Res 30:57–72

Nakano D, Kobayashi T, Sakaguchi I (2010) Predation and depth effects on abundance and size distribution of an invasive bivalve, the golden mussel *Limnoperna fortunei*, in a dam reservoir. Limnology 11:259–266

Nishimura T, Habe T (1987) Chinese freshwater mussels mixed in the imported *Corbicula* clam. Chiribotan (J Malacol Soc Japan) 18:110–111

Ockelmann KW (1983) Descriptions of mytilid species and definition of the Dacrydiinae n. subfam. (Mytilacea-Bivalvia). Ophelia 22:81–123

Ockelmann KW (1995) Ontogenetic characters of mytilaceans. Phuket Mar Biol Centre Special Pub 15:85–88

Ockelmann KW, Dinesen GE (2009) Systematic relationship of the genus *Adula* and its descent from a *Mytilus* -like ancestor (Bivalvia, Mytilidae, Mytilinae). Steenstrupia 30:141–152

Oliveira MD, Hamilton SK, Jacobi CM (2010) Forecasting the expansion of the invasive golden mussel *Limnoperna fortunei* in Brazilian and North American rivers based on its occurrence in the Paraguay River and Pantanal wetland of Brazil. Aquatic Invasions 5:59–73

Owen G (1955) Observations on the stomach and digestive diverticula of the Lamellibranchia. Part I. The Anisomyaria and Eulamellibranchia. Q J Microscop Sci 96:517–537

Owen G (1974) Studies on the gill of *Mytilus edulis* : the eu-latero-frontal cirri. Proc R Society of Lond. Series B, 187:83–91

Owen G (1978) Classification and the bivalve gill. Philos T R Soc Lond, Series B, 284:377–386

Owen G, Trueman ER, Yonge CM (1953) The ligament in the Lamellibranchia. Nature 171:1–6

Paolucci EM, Sardiña P, Sylvester F, Perepelizin PV, Zhan A, Ghabooli S, Cristescu ME, Oliveira MD, MacIsaac HJ (2014) Morphological and genetic variability in an alien invasive mussel across an environmental gradient in South America. Limnol Oceanograph 59:400–412

Pastorino G, Darrigran GA, Martín SM, Lunaschi L (1993) *Limnoperna fortunei* (Dunker, 1857) (Mytilidae), nuevo bivalvo invasor en aguas del Río de la Plata. Neotropica 39:101–102

Pujol JP (1967) Formation of the byssus in the common mussel (*Mytilus edulis* L.). Nature 214:204–205

Purchon RD (1957) The stomach in the Filibranchia and the Pseudolamellibranchia. Proc Zool Soc Lond 129:27–60

Purchon RD (1960) The stomach in the Eulamellibranchia: stomach Types IV and V. Proc Zool Soc Lond 135:431–489

Purchon RD (1963) Phylogenetic classification of the Bivalvia with special reference to the Septibranchia. Proc Malacol Soc Lond 35:71–80

Purchon RD, Brown D (1969) Phylogenetic interrelationships among families of bivalve molluscs. Malacologia 9:163–171

Reid RGB (1965) The structure and function of the stomach in bivalved molluscs. J Zool (Lond) 147:156–184

Ridewood WG (1903) On the structure of the gills of the Lamellibranchia. Philos T R Soc, Series B, 211:147–284

Sabelli B, Speranza S (1994) Rinvenimento di *Xenostrobus* sp. (Bivalvia, Mytilidae) nella laguna di Venecia. Boll Malacol 29:311–318

Santos CP, Würdig Nl, Mansur MC (2005) Fases larvais do mexilhão dourado *Limnoperna fortunei* (Dunker) (Mollusca, Bivalvia, Mytilidae) na bacia do Guaíba, Rio Grande do Sul, Brasil. Revista Brasileira de Zoologia 22:702–708

Sardiña P, Cataldo D, Boltovskoy D (2009) Effects of conspecifics on settling juveniles of the invasive golden mussel, *Limnoperna fortunei*. Aquatic Sci 71:479–486

Scarlato OA, Starobogatov YI (1979) General evolutionary patterns and the system of the class Bivalvia. Trudy Zoologicheskogo Instituta, Akademiya Nauk SSSR, 80:5–38 [In Russian]

Shirafuji J, Sato S (2003) Benthic communities of tidal flats in Sacheon and Masan, Gyeongsang-namdo, Korea. Investigation report of tidal flat in cooperation with Japan and Korea. pp 20–25

Silverman HG, Roberto FF (2007) Understanding marine mussel adhesion. Mar Biotechnol 9:661–681

Stasek CR (1963) Synopsis and discussion of the association of ctenidia and labial palps in the bivalved Mollusca. The Veliger 6:91–97

Sylvester F, Dorado J, Boltovskoy D, Juárez A, Cataldo D (2005) Filtration rates of the invasive pest bivalve *Limnoperna fortunei* as a function of size and temperature. Hydrobiologia 534:71–80

Sylvester F, Cataldo D, Notaro C, Boltovskoy D (2013) Fluctuating salinity improves survival of the invasive freshwater golden mussel at high salinity: implications for the introduction of aquatic species through estuarine ports. Biol Invasions 15:1355–1366

Tan T-H, Pai J-Y, Hsha KC (1987) The recovery of fouling clam, *Limnoperna fortunei* from Taiwan. Bull Malacol, Republic of China, 13:97–100 [In Chinese]

Taylor JD, Kennedy WJ, Hall A (1969) The shell structure and mineralogy of the Bivalvia, introduction. Nuculacea-Trigonacea. Bull British Museum Nat Hist (Zool), Supplement 3:1–125

Taylor JD, Kennedy WJ, Hall A (1972) The shell structure and mineralogy of the Bivalvia. II. Lucinacea—Clavagellacea Conclusions. Bull British Museum Nat Hist (Zool) 22:255–294

Thiele J (1934) Handbuch der systematischen Weichtierkunde, Part III (Scaphopoda, Bivalvia, Cephalopoda). Familien Mytilidae. Verlag von Gustav Fisher, Jena, pp 797–801

Tominaga A, Goka K, Kimura T, Ito K (2009) Genetic structure of Japanese introduced populations of the golden mussel, *Limnoperna fortunei*, and the estimation of their range expansion process. Biodiversity 10:61–66

Trueman ER (1950) Observations on the ligament of *Mytilus edulis*. Q J Microscop Sci 91:225–235

Uryu K, Iwasaki K, Hinoue M (1996) Laboratory experiments on behaviour and movement of a freshwater mussel, *Limnoperna fortunei* (Dunker). J Mollus Stud 62:327–341

Veitenheimer-Mendes I (1981) *Corbicula manilensis*, (Philippi, 1844) molusco asiático, na bacia do Jacuí do Guaíba, Rio Grande do Sul, Brasil (Bivalvia, Corbiculidae). Iheringia (Série Zoologia) 60:63–74

White KM (1937) *Mytilus*. Liverpool Marine Biology Committee Memoirs 31:1–117

Wilson BR (1967) A new generic name for three recent and one fossil species of Mytilidae (Mollusca: Bivalvia) in Southern Australasia, with redescriptions of the species. Proc Malacol Soc Lond 37:279–295

Wilson BR (1968) Survival and reproduction of the mussel *Xenostrobus securis* (Lam.) (Mollusca-Bivalvia-Mytilidae) in a Western Australian estuary. I. Salinity tolerance. J Nat Hist 2:307–328

Wright MM, Francis L (1984) Predator deterrence by flexible shell extensions of the horse mussel *Modiolus modiolus*. The Veliger 27:140–142

Yonge CM (1948) Formation of siphons in lamellibranchs. Nature 161:198

Yonge CM (1955) Adaptation of rockboring in *Botula* and *Lithophaga* (Lamellibranchia, Mytilidae), with a discussion on the evolution of this habit. Q J Microscop Sci 96:383–410

Yonge CM (1957) Mantle fusion in the Lamellibranchia. Pubblicazione della Stazione Zoologica di Napoli 29:151–171

Yonge CM (1962) On the primitive significance of the byssus in the Bivalvia and its effects in evolution. J Mar Biol Assoc UK 42:113–125

Yonge CM, Campbell JI (1968) On the heteromyarian condition in the Bivalvia with special reference to *Dreissena polymorpha* and certain Mytilacea. T R Soc Edin 68:21–43

Larval Development of *Limnoperna fortunei*

Daniel H. Cataldo

Abstract Mature sperm cells of *Limnoperna fortunei* measure about 4 μm, and ova are typically spherical, 80–100 μm in diameter. Forty minutes after spawning, the first polar lobe appears, and the first division occurs 14 min later. Slightly over an hour after spawning, the second polar lobe appears and the second division yields a 4-cell stage. The third division occurs 90 min after spawning, and the fourth 115 min after spawning. Approximately 3.5 h after spawning (at 26 °C) the morula stage is reached. Six hours after spawning, the first trochophores appear (95–110 μm in length) at 28 °C. Subsequently, the prodissoconch I starts developing, initially as small rosette-shaped structures on the dorsal side of the trochophore. Straight-hinged veligers (115–160 μm) start appearing 24 h after spawning. These larvae start feeding externally and secrete the prodissoconch II. Umboned veligers (156–220 μm) are reached 287 (at 28 °C), 165 (25 °C) and 118 (20 °C) h after spawning. From there on, the larva reabsorbs its velum and develops a muscular, adhesive foot, yielding a plantigrade larva (250–405 μm), which shortly thereafter settles and attaches to the substrate. Development times are therefore strongly influenced by water temperature.

Keywords *Limnoperna fortunei* · Larvae · Veliger · Development · Morphology

Introduction

Freshwater bivalves have evolved different strategies to maximize survival and dispersion. While most marine species have free-living larvae, many freshwater species have a unique ectoparasitic larval stage, the glochidium, which attaches

---

D. H. Cataldo (✉)
Facultad de Ciencias Exactas y Naturales, Instituto de Ecología, Genética y Evolución de Buenos Aires (IEGEBA), Universidad de Buenos Aires-CONICET, Ciudad Universitaria, 1428 Buenos Aires, Argentina
e-mail: daniel@ege.fcen.uba.ar

Museo Argentino de Ciencias Naturales 'Bernardino Rivadavia'-CONICET, Buenos Aires, Argentina

D. Boltovskoy (ed.), *Limnoperna fortunei,* Invading Nature - Springer Series in Invasion Ecology 10, DOI 10.1007/978-3-319-13494-9_2

to the gills of fish. During early stages of development, these larvae are incubated in marsupia in the adults. Once these stages have developed into the glochidial stage, they are released into the water where they need to quickly reach a host fish in order to develop into a juvenile. While this strategy necessitates finding the right fish to complete development, it insures that larvae are not washed out to sea. Other freshwater species incubate their offspring to avoid being lost to unfavourable downstream environments, and larval development is completed in specialized gill pouches in the adult. Once the miniature adult is large enough, it falls directly to the bottom (e.g. Corbiculidae). In this brooding strategy, the risk of expatriation is decreased, but dispersal opportunities are sacrificed.

In contrast, *Limnoperna fortunei* develops through a series of planktonic larval stages. This has the benefit of effective and rapid downstream dispersal and the ability to conquer new environments, but it risks expatriation of larvae to areas unsuitable for survival, including estuaries and the sea. Evolution suggests that for the vast majority of freshwater organisms, in the balance between enhanced dispersion and risk of expatriation into hostile environments, drawbacks from expatriation are more important than advantages of dispersion, and therefore free-swimming larvae of marine ancestors have been mostly lost. Nevertheless, *L. fortunei* seems to have benefited from having retained a free-swimming larval stage (see Chapter "Distribution and Colonization of *Limnoperna fortunei*: Special Traits of an Odd Mussel" in this volume).

Development of *L. fortunei* is typical of planktotrophic larvae, and characterizes many mytilids and dreissenids (including the zebra mussel *Dreissena polymorpha*). The initial trochophore stage is followed by a veliger and, subsequently, by D-shaped larval stages (at which time prodissoconch 1 and 2 are secreted), followed by the newly settled juvenile (the dissoconch stage) (Ockelmann 1995). Organisms, such as *L. fortunei* that utilize a planktotrophic larval development strategy typically start as small eggs and require a long larval development period. These larvae feed on plankton and their development is prolonged compared to species with lecithotrophic larvae. This favours long distance dispersion of planktotrophic species both in their natural environment and through the ballast water of ships on the high seas.

Methodological Approaches

Description of mollusc larval stages can be carried out (1) on the basis of field-collected materials, (2) on the basis of larvae obtained in the laboratory inducing spawning of ripe adults and following fertilization and development under controlled conditions or (3) a combination of the above, tracking the development of field-collected larvae in the laboratory.

The first descriptions of the larval stages of the golden mussel were undertaken on larval specimens collected with plankton nets and whose development was subsequently followed in the laboratory (Choi and Kim 1985; Choi and Shin 1985). Santos et al. (2005) analyzed larval stages collected bimonthly in Guaíba Lake in

Rio Grande do Sul, Brazil. Ezcurra de Drago et al. (2006) also described larval development stages from field-collected samples undertaking studies in 1997–2000 at various sites along the Middle Paraná River (31°38'S, 60°40'W). While field-collected samples allow for the adequate description of the morphology of each developmental stage, they are not appropriate for determining the time required for each stage to transition to the next.

Cataldo et al. (2005) analyzed the embryonic development of *L. fortunei* by inducing adult specimens to spawn in the laboratory and identifying the various larval stages and tracking the time needed to reach each developmental stage. In their study, Cataldo et al. (2005) used three experimental temperatures, 20, 25 and 28 °C, which generally span the temperatures during the reproductive period of the mussels in South America.

The majority of experimental studies on bivalves have used serotonin as the agent to induce the production of gametes (Matsutani and Nomura 1982; Braley 1985; Ram and Nichols 1993; Vanderploeg et al. 1996); however, serotonin does not produce satisfactory results in *L. fortunei*. Cataldo et al. (2005) used an antifouling molluscicide commercially known as Spectrus CT1300 (n-alkyl dimethylbenzyl ammonium chloride, a quaternary ammonium) (Cataldo et al. 2003; see Chapter "Chemical Strategies for Control of the Golden Mussel (*Limnoperna fortunei*) in Industrial Facilities"in this volume), which proved to be very effective at concentrations of 0.75 ppm (active ingredient), yielding ripe gametes in over 90 % of the experimental beakers 30–180 min after exposure of the adults to the chemical. In the remaining 10 % of mussels, it took up to 8 h for gametes to appear (Cataldo et al. 2005).

Analysis of developmental times for the larval stages of this species, in particular the time it takes for larvae to reach settlement stage, has both theoretical and applied interest. Learning more about the basic biology of *L. fortunei* allows for a better understanding of the mechanisms governing the dispersion of this species and also allows for comparisons with other molluscs in order to highlight common behavioural patterns based on shared constraints. For the purposes of controlling this pest in industrial installations, it is potentially useful to know the origin of the populations that are actually seeding the individuals fouling pipes and intakes, and knowledge of the duration of each developmental stage at different temperatures has important implications for developing treatment methods. Knowledge of these larvae is also necessary for early detection and monitoring of colonization by *L. fortunei*.

Developmental Stages

Larval development of *L. fortunei* can be divided into two main stages. The first stage comprises nonshelled development from fertilization until the formation of the trochophore larva. The second stage is characterized by shelled forms from veliger to plantigrade larva, at which point the animal is capable of binding to the substrate.

Gametes and Nonshelled Developmental Stages

Mature sperm cells measure about 4 μm (excluding the tail) and are highly mobile (Fig. 1b). Ova, of typically spherical form, are 80–100 μm in diameter (Choi and Shin 1985; Cataldo et al. 2005) (Fig. 1a). These cells are still diploid upon release and chromatic reduction takes place in the medium by the production of polar bodies (Fig. 1c).

Segmentation is similar to that observed in many invertebrates and in most molluscs, and complete spiral cleavage yields a characteristic trochophore larva. Segmentation starts about 40 min after fertilization, producing the first polar lobe, and very shortly thereafter, the first cellular division takes place (Fig. 1d,e and f). After a few minutes, the first polar lobe is resorbed by one of the daughter cells (Fig. 1f). At 26 °C, this first division, yielding two uneven blastomeres, occurs 54 min after fertilization. Approximately 11 min later (slightly over 1 h after spawning), the second polar lobe appears (Fig. 1g) and the second division starts, yielding a 4-cell stage (Fig. 1h). The third division takes place along the equatorial plane, separating four micromeres from four macromeres; at 26 °C, this 8-celled stage occurs 90 min after fertilization (Fig. 1i). Twenty-five minutes later (115 min after fertilization), the fourth division yields the 16-celled stage (Fig. 1j). Approximately 3.5 h after fertilization (at 26 °C) the morula stage is reached (Fig. 1k).

The morula is ciliated and has limited, poorly coordinated movement, only occasionally leaving the bottom and venturing into the water column. Six hours and twenty minute after fertilization, the first trochophores (95–110 μm in length) appear (Fig. 1L. These active larvae have a well-developed apical tuft of cilia that allows for well-coordinated swimming. Most of these larvae wander freely in the water column and rarely rest on the bottom. Shells appear as small rosette-shaped structures on the dorsal side of the trochophore (Fig. 1m). The first shell, prodissoconch I, is secreted by the shell gland, and begins to split into two, slowly coating the soft tissues of the larva. The nutrition of the larva is supplied exclusively from yolk reserves until the trochophore stage. The transition period between the trochophore larva and the next stage—the veliger larva—is also referred to by some authors as a pre-veliger larva with incomplete shells (Fig. 1n; Ezcurra de Drago et al. 2006).

Shelled Developmental Stages

The first shelled stage is termed the veliger larva, D larva, or straight-hinged larva (since the dorsal margin of the shell is straight). This stage starts when the valves completely cover the body of the animal. This larva has a fully developed velum located anteriorly and is provided with a strip of cilia and a central group of flagella (Ezcurra de Drago et al. 2006; Fig. 1o and p). Once the velum has completely developed, the larva begins to feed on plankton. From this phase onwards, the shell is secreted by the mantle (rather than by the larval shell gland), and the prodissoconch II is formed. This transition is marked by the appearance of growth lines, which define

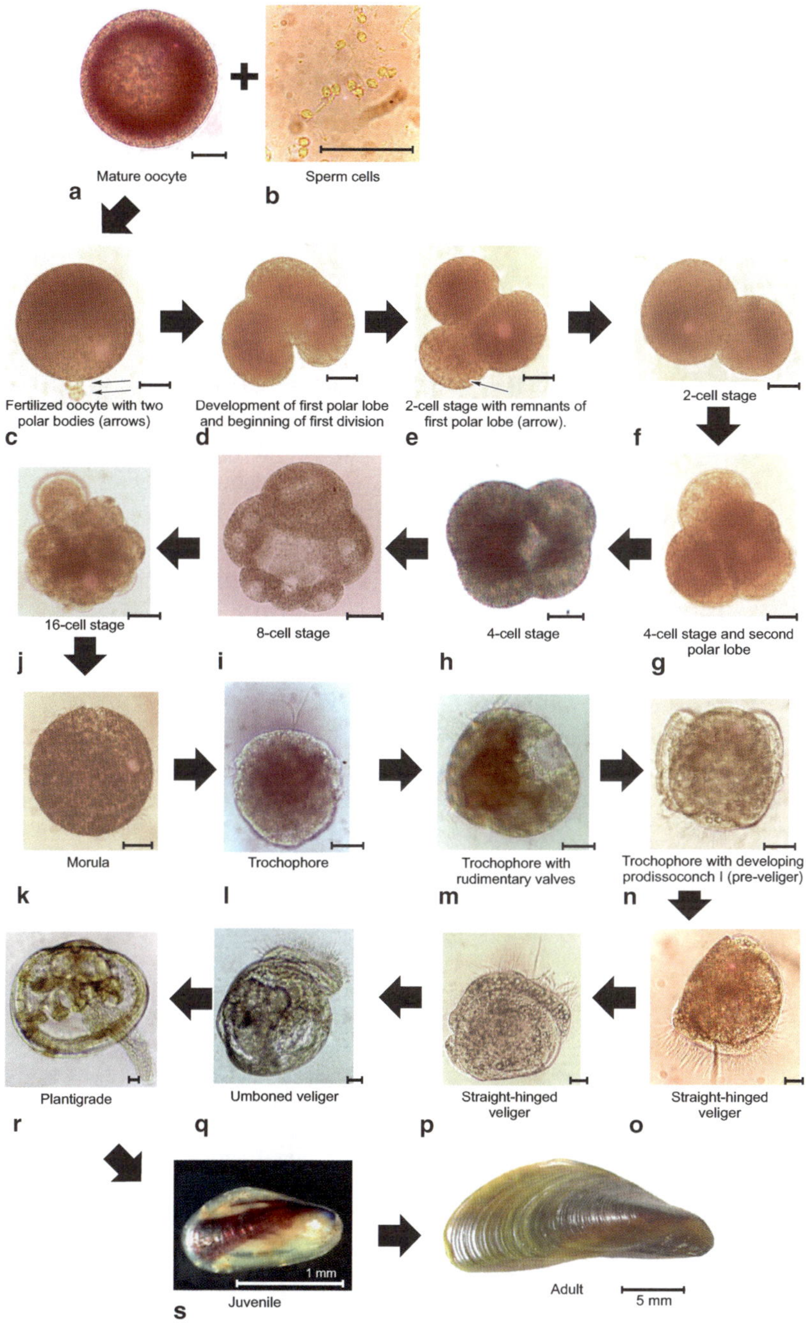

Fig. 1 Larval stages of *L. fortunei* (scale bars are 25 μm, unless otherwise noted). (Modified from Cataldo et al. 2005)

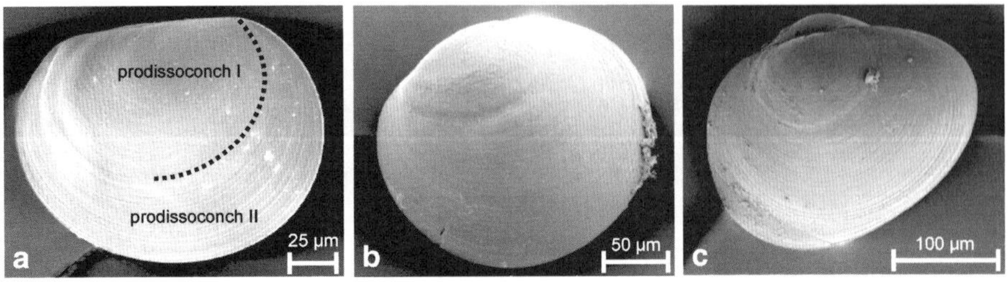

Fig. 2 SEM photographs of the shell of a straight-hinged larva (**a**), umboned larva (**b**) and plantigrade larva (**c**). (Adapted from Cataldo et al. 2005)

the start of the prodissoconch II (Fig. 2a). Straight-hinged larvae are 115–180 µm in length (Choi and Kim 1985; Cataldo et al. 2005; Santos et al. 2005).

The dorsal margin of the shell, which is initially straight, gradually bulges and the umbo appears as a progressively more conspicuous bump, giving origin to umboned veligers. These larvae are 190–230 mm in length (Choi and Kim 1985; Cataldo et al. 2005; Santos et al. 2005; Fig. 1q and 2b). During this stage, the prodissoconch II completes its growth, whereas the prodissoconch I remains restricted to the dorsal margin and forms part of each umbo (Fig. 2b).

As larvae develop, swimming activity becomes progressively slower and the animals tend to spend more time on the bottom of the vessel. Shortly before settling, each larva reabsorbs its velum and develops a muscular, adhesive foot, giving rise to a plantigrade larva (around 250–405 µm in length; Choi and Kim 1985; Cataldo et al. 2005; Santos et al. 2005; Fig. 1rr and 2c). At this stage, the only means of locomotion is the foot and the valves begin to elongate. The organism completes its internal development and the siphons and gills are easily observed. This stage is associated with exploratory behaviour of the substrate and concludes with the attachment of the byssal threads.

Effects of Temperature on the Larval Development of *L. fortunei*

The effect of temperature on the rate of larval development was analyzed by Cataldo et al. (2005) through induced spawning, fertilization of gametes and incubation of eggs under controlled laboratory conditions at three temperatures (20, 25 and 28 °C). As expected, the fastest developmental rates were observed at the highest temperature (28 °C), and decreasing at 25 °C and further at 20 °C (Fig. 3, lower panel). Differences in developmental times for all stages surveyed were almost twice as high between 20 and 25 °C as they were between 25 and 28 °C (Fig. 3). Size, on the other hand, varied little with temperature (Cataldo et al. 2005).

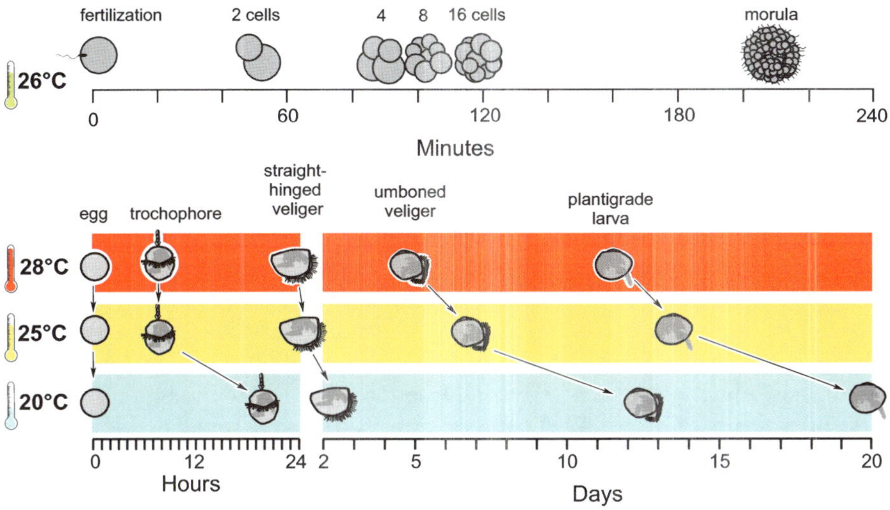

Fig. 3 Developmental times of *L. fortunei* at 26 °C from fertilization to morula (*upper panel*), and at three different temperatures to plantigrade larva (*lower panel*) (figures are not to scale). (Based on data from Cataldo et al. 2005)

A 20°C (which is 3–5 °C above the lower thermal limit for reproduction of the species in the lower Paraná River delta; see Chapter "Reproductive Output and Seasonality of *Limnoperna fortunei*" in this volume; Cataldo and Boltovskoy 2000), the trochophore larval stage is reached ~20 h after fertilization. The stage of straight-hinged veliger appears after 45 h. The umboned veliger larva develops after 11 days and is actively swimming. Swimming behaviour slowly becomes less vigorous, and larvae tend to descend towards the bottom transforming from a swimming to a crawling phase. Beginning at day 20, the foot develops, and the plantigrade larval stage is reached (Fig. 3).

At 25 °C, the growth rate is significantly higher than that observed at 20 °C. Trochophore larvae develop within 6–7 h after fertilization, while the 'D' form (D-shaped or straight-hinged) appears 26 h after fertilization. From the 4th day, the hinge bends leading to the umboned stage, and the foot is developed beginning on the 13th day, about 7 days sooner than at 20 °C (Fig. 3).

A 28 °C (which are typically the highest summer water temperatures in the lower delta of the Paraná River and the Río de la Plata estuary), development is about twice as fast as at 20 °C, and approximately 20 % faster than at 25 °C. At 6 h, the trochophore larva develops, and the straight-hinged veliger stage appears just after the 1st day. The umbo is evident beginning on the 4th day, which is 2 days earlier than at 25 °C, and a full week before those incubated at 20 °C. On day 11, organisms have a functional foot, and they are exploring the substrate by crawling on the bottom and walls of the chamber.

The information reported by Cataldo et al. (2005) for 28 °C overlaps the developmental times in Choi and Kim (1985) for the straight-hinged stage (23 h) (see Fig. 3). However, the times reported by Choi and Kim (1985) to reach the umboned larval stage (10 days) and the pediveliger stage (18 days) are considerably longer than those reported by Cataldo et al. (2005) (~5 and 11 days, respectively). The study by Choi and Kim (1985) used larvae that were collected from the plankton, and hence the elapsed times following fertilization were not exactly known. Similarly, the water temperatures of their laboratory experiments were not reported. The laboratory study of Cataldo et al. (2005) clearly showed that the velum is fully developed by the time the dorsal side of the animal shows a straight hinge between the two valves, however in Choi and Kim's (1985) study, early D-shaped larvae were still devoid of a velum; they only observed a velum in middle D-shaped larvae. This discrepancy may be due to artefacts induced through plankton sampling, whereby large proportions of net-sampled larvae are stressed or dead, and they show little or no mobility since their vela are retracted and inconspicuous.

Comparison of the developmental rates of *L. fortunei* with those reported for the zebra mussel, *D. polymorpha*, indicates that both are roughly similar. The zebra mussel reaches the D-shaped stage in about 30–70 h (Sprung 1987; Nichols 1993; Stoeckel et al. 1996), whereas the golden mussel reaches this stage between 24 and 50 h (Cataldo et al. (2005). Vanderploeg et al. (1996) reported that settlement of *D. polymorpha* occurs at 15–22 days, and the plantigrade larval stage of *L. fortunei* appears on days 11–20 (Cataldo et al. 2005). The developmental rates reported for some marine mytilids are within the ranges found for *L. fortunei* (e.g. *Perna viridis*, Tan 1975; Siddall 1980; *Mytilus platensis*, Penchaszadeh 1980; *Modiolus modiolus*, Schweinitzd and Lutz 1976), whereas they are significantly slower for other marine mytilids. For example, it can take *Mytilus edulis* up to 35 days to reach the veliger stage, and it may take over 6 months for it to complete metamorphosis (Bayne 1976). These comparisons are not well refined, since the modulating effects of temperature have not always been accounted for in these studies, but most marine bivalves with free-swimming larvae seem to have slower development rates. Over 80 % of the 37 marine bivalves surveyed by Thorson (1961) have development times longer than those of *L. fortunei* (at 28 °C), and only 5 % have faster rates. Accelerated development may be an adaptation to colonize freshwater environments. Indeed, whereas marine benthic organisms may gain significant advantage from extended larval periods (e.g. Scheltema 1986), many freshwater animals can incur the danger of being flushed out to the ocean unless settling occurs more-or-less rapidly. Estuarine larvae use vertical migration to overcome seaward transport (e.g. Cronin 1982), but this is not feasible in the turbulent conditions of streams and rivers. Furthermore, the near-bottom high-saline waters involved in estuarine upstream water transport (Guerrero et al. 1997; Acha et al. 2008), are unfit for the survival of *L. fortunei* (Sylvester et al. 2013). Thus, it is conceivable that the relatively short development times of *L. fortunei* have evolved to overcome the expatriation hazards associated with planktonic larvae.

Mortality Rates

In their laboratory study of larval development, Cataldo et al. (2005) noticed very high mortality rates for *L. fortunei*, around 80–90 %. However, mortality was not even throughout development, and it was highest during the transition from the straight-hinged to the umboned veliger stage. High mortality rates during the transition between these two stages of development have also been reported for other bivalves (Wada 1968).

By tracking cohorts of zebra mussel during their downstream drift in the Illinois River (USA) and through laboratory rearing experiments, Schneider et al. (2003) concluded that the mortality of the planktonic larvae of *D. polymorpha* was significantly higher during the transition from straight-hinged to umboned veligers, than during any stage before or after this period. Larval mortality can respond to many different stressors, including scarce or inadequate food supply, pollution, predation, advective sinking in the water column, etc. (Morgan 1995). However, most of these stressors should affect all larval stages similarly. This suggests that peak mortalities are associated with the major changes in larval morphology and anatomical reorganizations, including the formation of a digestive tract and accompanying peaks in metabolic activity, that take place during the transition from the straight-hinged to umboned stage (Sprung and Widdows 1986; Fujimura et al. 1995; Schneider et al. 2003). These shifts may cause physiological or alimentary stress and result in higher mortality (Schneider et al. 2003).

Concluding Remarks

Industrial plants and water treatment facilities become infested with *L. fortunei* when free-swimming larvae pass through protecting grids and filters of the intake pipes. Afterwards, larvae settle out onto these same grids and filters as well as pipes, heat exchangers and other surfaces, and cause severe fouling problems (see Chapter "Impacts of *Limnoperna fortunei* on Man-Made Structures and Control Strategies: General Overview" in this volume). Since residence times of water are low (usually < 1 h) in cooling systems, only very late plantigrade larvae are retained within plant components, whereas earlier stages are released with the effluent water. The development times reported by Cataldo et al. (2005) indicate that the dense populations usually present on the outer and inner hard surfaces of the plants have no impact with regards to further infestation of the plant's inner components. Fouling populations could seed internal surfaces only if water flow regimes within or around plant intakes are highly irregular and abundant 'dead-water sites' with very high water residence times are present. On the other hand, it is possible to estimate the location of seeding populations by using these development times and taking into account water temperature and flow characteristics. For example, since both the Paraná and the Uruguay rivers have mean flow speeds of about 0.3 m/s, the seeding populations for the biofouling-affected plants located in Buenos Aires along the Río de la Plata estuary are ~250 km upstream in the summer, when water temperatures are above 25 °C, and ~500 km upstream in the autumn and spring, when water temperatures are below 20 °C. Thus, fine-tuning our knowledge of temperature-dependent development times can contribute to the assessment of the fouling mechanisms and help to develop sound control measures.

References

Acha ME, Mianzan H, Guerrero R, Carreto J, Giberto D, Montoya N, Carignan M (2008) An overview of physical and ecological processes in the Rio de la Plata Estuary. Continental Shelf Res 28:1579–1588

Bayne BL (1976) The biology of mussel larvae. In: Bayne BL (ed) Marine mussels: their ecology and physiology. Cambridge University Press, New York, pp 81–120

Braley RD (1985) Serotonin-induced spawning in giant clams (Bivalvia: Tridacnidae). Aquaculture 47:321–325

Cataldo D, Boltovskoy D (2000) Yearly reproductive activity of *Limnoperna fortunei* (Bivalvia) as inferred from the occurrence of its larvae in the plankton of the lower Paraná River and the Río de la Plata estuary (Argentina). Aquat Ecol 34:307–317

Cataldo D, Boltovskoy D, Pose M (2003) Toxicity of chlorine and three non-oxidizing molluscicides to the invasive pest mussel *Limnoperna fortunei*. J Am Water Works Assoc 95:66–78

Cataldo D, Boltovskoy D, Hermosa JL, Canzi C (2005) Temperature-dependent larval development rates of *Limnoperna fortunei* (Bivalvia: Mytilidae). J Molluscan Stud 71:41–46

Cronin TW (1982) The estuarine retention of larvae of the crab *Rhithropanopeus harrisi*. Estuarine, Coastal Shelf Sci 15:207–220

Choi SS, Kim JS (1985) Studies on the metamorphosis and the growth of larva in *Limnoperna fortunei*. Korean J Malacol 1:13–18 [In Korean]

Choi SS, Shin CN (1985) Study on the early development and larvae of *Limnoperna fortunei*. Korean J Malacol 1:5–12 [In Korean]

Ezcurra de Drago I Montalto L Oliveros OB (2006) Desarrollo y ecología larval de *Limnoperna fortunei*. In: Darrigran G, Damborenea C (eds) Bio-invasión del mejillón dorado en el continente americano. Editorial de la Universidad de La Plata, La Plata, pp 85–93

Fujimura T, Wada K, Iwaki T (1995) Development of digestive system of the pearl oyster larvae, *Pinctada fucata*. Venus (Jpn J Malacol) 54:203–223

Guerrero R, Acha EM, Framiñan MB, Lasta CA (1997) Physical oceanography of the Río de la Plata Estuary, Argentina. Continental Shelf Res 17:727–742

Matsutani T, Nomura T (1982) Induction of spawning by serotonin in the scallop *Patinopecten yessoensis* (Jay). Marine Biol Lett 3:353–358

Morgan SG (1995) Life and death in the plankton: larval mortality and adaptation. In: McEdward L (ed) Ecology of marine invertebrate larvae. CRC Press, Boca Raton, pp 279–322

Nichols SJ (1993) Maintenance of the zebra mussel (*Dreissena polymorpha*) under laboratory conditions. In: Nalepa TF, Schloesser D (eds) Zebra mussels: biology, impacts, and control. Lewis Publishers, Boca Raton, pp 733–747

Ockelmann KW (1995) Ontogenetic characters of mytilaceans. Phuket Marine Biol Centre Special Publ 15:85–88

Penchaszadeh PE (1980) Ecología larvaria y reclutamiento del mejillón del Atlántico Suroccidental. *Mytilus platensis* d'Orbigny. Cahiers de Biologie Marine 21:169–179

Ram JL, Nichols SJ (1993) Chemical regulation of spawning in the zebra mussel (*Dreissena polymorpha*). In: Nalepa TF, Schloesser D (eds) Zebra mussels: biology, impacts, and control. Lewis Publishers, Boca Raton, pp 307–314

Santos CP, Würdig Nl, Mansur MC (2005) Fases larvais do mexilhão dourado *Limnoperna fortunei* (Dunker) (Mollusca, Bivalvia, Mytilidae) na bacia do Guaíba, Rio Grande do Sul, Brasil. Revista Brasileira de Zoologia 22:702–708

Scheltema RS (1986) Epipelagic meroplankton of tropical seas: its role for the biogeography of sublittoral invertebrate species. In: Pierrot-Bults AC, Spoel S, Zahuranec BJ, Johnson RK (eds) Pelagic Biogeography. UNESCO Press, Paris, pp 242–249

Schneider DW, Stoeckel JA, Rehmann CR, Douglas Blodgett KD, Sparks RE, Padilla DK (2003) A developmental bottleneck in dispersing larvae: implications for spatial population dynamics. Ecol Lett 6:352–360

Schweinitzd EH, Lutz RA (1976) Larval development of the northern Horse mussel, *Modiolus modiolus* (L.) including a comparison with the larvae of *Mytilus edulis* L. as an aid in planktonic identification. Biol Bull 150:348–360

Siddall SE (1980) A clarification of the genus *Perna* (Mytilidae). Bull Marine Sci 30:858–870.

Sprung M (1987) Ecological requirements of developing *Dreissena polymorpha* eggs. Archiv Hydrobiologie 79(Suppl.):69–86

Sprung M, Widdows J (1986) Rate of heat dissipation by gametes and larval stages of *Mytilus edulis*. Marine Biol 91:41–45

Stoeckel JA, Camlin L, Blodgett KD, Sparks R (1996) Growth rates of zebra mussel veligers in the Illinois river: implications for larval dispersal and settlement patterns. In: Sixth International Zebra Mussel and other Aquatic Nuisance Species Conference, Dearborn (USA)

Sylvester F, Cataldo D, Notaro C, Boltovskoy D (2013) Fluctuating salinity improves survival of the invasive freshwater golden mussel at high salinity: implications for the introduction of aquatic species through estuarine ports. Biol Invasions 15:1355–1366

Tan WH (1975) Egg and larval development in the green mussel, *Mytilus viridis* Linnaeus. Veliger 18:151–155

Thorson G (1961) Length of pelagic larval life in marine bottom invertebrates as related to larval transport by ocean currents. In: Sears M (ed) Oceanography. American Association for the Advancement of Science, Washington DC, pp 455–474

Vanderploeg HA, Liebig JR, Gluck AA (1996) Evaluation of different phytoplankton for supporting development of zebra mussel larvae (*Dreissena polymorpha*): the importance of size and polyunsaturated fatty acid content. J Great Lakes Res 22:36–45

Wada SK (1968) Mollusca. In: Kume M, Dan K (eds) Invertebrate embryology, NOLIT. Publishing House, Belgrade, pp 485–525

Parasites of *Limnoperna fortunei*

Takashi Baba and Misako Urabe

Abstract General information about bucephalid trematodes (Digenea) and the biology of two bucephalid species, *Parabucephalopsis parasiluri* and *Prosorhynchoides ozakii*, that are parasitic in *Limnoperna fortunei*, are reviewed in this chapter. Both of these species have been introduced into Japan with *L. fortunei*. The results of fish sampling in the field, and field and laboratory experiments showed that 27 fish species (5 families) and 13 fish species (5 families) are the second intermediate hosts of *Pa. parasiluri* and *Pr. ozakii*, respectively. The annual prevalence of *Pa. parasiluri* in *L. fortunei* (shell length ≥ 15 mm) in the Uji and Yodo Rivers in Japan fluctuated from 2 to 18 % during 2001 and 2008. Heavy infections of *Pa. parasiluri* metacercariae cause hemorrhages in the fins, skin, and eyes of some freshwater fishes in Japan, indicating negative impacts on these species. On the other hand, larval bucephalids castrate *L. fortunei* and may depress their population growth.

Keywords *Limnoperna fortunei* · Golden mussel · Parasites · Trematodes · Fish · Digenea · Bucephalidae

T. Baba (✉)
Environmental Science Graduate School, University of Shiga Prefecture, 2500, Hassaka-cho, Hikone, Shiga 522-8533, Japan
e-mail: dragonfly-5910@hotmail.co.jp

Japan Sea National Fisheries Research Institute, Fisheries Research Agency, 1-5939-22, Suido-cho, Chuou-ku, Niigata 951-8121, Japan

M. Urabe
Department of Ecosystem Studies, School of Environmental Science, University of Shiga Prefecture, Hikone, Shiga 522-8533, Japan
e-mail: urabe@ses.usp.ac.jp

D. Boltovskoy (ed.), *Limnoperna fortunei*, Invading Nature - Springer Series in Invasion Ecology 10, DOI 10.1007/978-3-319-13494-9_3

Introduction

One of the economic and ecological damages caused by the golden mussel *Limnoperna fortunei* stems from the fish disease caused by parasites of the mussel. Lake Biwa in Japan was invaded by *L. fortunei* in 1992, and the mussel colonized the outlet of the lake, the Uji River, around 1994 (Nakai 1995). Outbreaks of fish disease caused by the trematode *Parabucephalopsis parasiluri* (parabucephalopsiosis) have occurred several times since 1999 in the Uji River (Urabe et al. 2001; Ogawa et al. 2004; Urabe et al. 2009; Baba and Urabe 2011b). In this chapter, we review general information about the trematodes of *L. fortunei* and other related mussels based on the published studies, and summarize the biology of two bucephalid species that were introduced in Japan and their impacts on native fish species.

Trematode Parasites of *Limnoperna fortunei*

Morphological Characteristics and Life Cycle of Bucephalid Trematodes (Digenea)

All trematodes recorded from *L. fortunei* and other related mussels belong to the family Bucephalidae (Digenea, Trematoda, and Platyhelminthes). The trematodes in this family are characterized by having a mouth located in the middle of the body. Bucephalid trematodes need three host species to complete their life cycle: bivalves, fishes, and piscivorous fishes. It is thought that eggs are ingested by their first intermediate hosts that are bivalves belonging to the orders Mytiloida, Veneroida, Ostreoida, and Unionoida (Yamaguti 1975; Francisco et al. 2012). Miracidia hatch in the alimentary canals of bivalves, develop into mother sporocysts, and infiltrate all host tissues. Mother sporocysts of bucephalids are branched tubules with many swellings (Fig. 1). Many cercariae with two long tails are produced asexually in the sporocysts (Fig. 2). Cercariae swim out from bivalves and penetrate the skin of fishes. They invade into the muscle and nervous tissues of fishes and metamorphose into metacercariae. Typically, the effect of trematode metacercariae on second intermediate hosts is slight because metacercariae do not exploit them. However, bucephalid metacercariae are often harmful to fishes and cause disease with heavy infections (Matthews 1973; Baturo 1978). The definitive hosts of bucephalids are piscivorous fishes and one species of amphibian (Wang and Wang 1998). Definitive hosts prey on infected fish and ingested metacercariae develop into adults in their intestines.

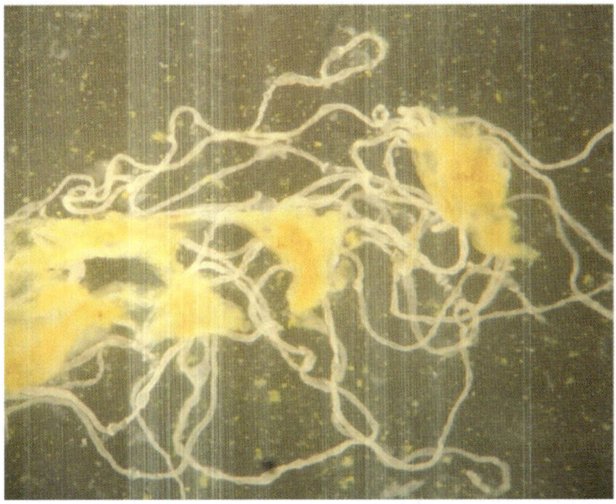

Fig. 1 Mother sporocysts of *Parabucephalopsis parasiluri*

Fig. 2 Cercaria of *Prosorhynchoides ozakii*

Bucephalid Trematodes Reported from Freshwater Areas in China and Southeast Asia

Thirty bucephalid species have been reported from freshwater and brackish water fishes in China (Heilongjiang, Jiangsu, Hubei, Sichuan, Zhejiang, Fujian, Jiangxi, Hunan, Guizhou, Guangdong, Guangxi, and Yunnan Provinces). Most of the collection localities are in the southern part of China, which is the native region of *L. fortunei* (Feng and Wang 1995; Wang and Wang 1998) (see Chapter "Distribution and spread of *Limnoperna fortunei* in China" in this volume). Little is known about the first intermediate hosts of these bucephalids. The first intermediate hosts are known for only two species, *Parabucephalopsis prosthorchis* and *Dollfustrema foochowensis*. Both of them use *Limnoperna fortunei* (=*L. lacustris*) as the first intermediate host (Tang and Tang 1976; Nakai 1995).

There are some additional records of bucephalids from freshwater fishes in Southeast Asia where *L. fortunei* is native, which suggests that the mussel may also be their first intermediate host. *Prosorhynchoides siamensis*, *Pr. chiangkhongensis*, *Pr. philippinorum*, *Prosorhynchoides* sp. 1, *Prosorhynchoides* sp. 2, *Dollfustrema* sp., *D. bagarii*, *Rhipidocotyle* sp., and *Bucephaloides philippinorum* have been reported from some freshwater fishes in the north of Thailand (Khamboonruang et al. 2006; Purivirojkul 2009). *Bucephalopsis karvei* was collected from *Xenentodon cancila* (garfish, a species of needlefish) caught in the Bhumipol Dam reservoir in the west of Thailand (Sey and Wongsawad 2004) and from the same species caught in Laos (Scholz 1991). Bucephalid cercariae were also obtained from *Limnoperna* spp. in the Lampao River in the east of Thailand (Krailas et al. 2008).

Parabucephalopsis parasiluri *and* Prosorhynchoides ozakii *Introduced into Japan*

In Japan, an introduced bucephalid, *Pa. parasiluri*, which uses *L. fortunei* as the first intermediate host (Fig. 3; Urabe et al. 2007), parasitizes the intestines (mainly the rectum) of its definitive host, the Lake Biwa catfish *Silurus biwaensis* (Siluridae). Their morphological characteristics were described in detail by Urabe et al. (2007) (Fig. 4). About 19,000 individuals were recorded from one catfish (78.5 cm in standard length) caught in the Uji River (Urabe et al. 2007). In the indigenous area, *Pa. parasiluri* has been recorded in Sichuan, Guizhou, and Fujian Provinces in China. The definitive hosts in China are *Silurus asotus*, *S. meridionalis* (Siluridae), *Liobagrus marginatus* (Amblycipitidae), and *Sinilabeo rendahli* (Cyprinidae) (Table 1; Urabe et al. 2007). The first intermediate host is unknown in China.

Twelve fish species were recorded as natural second intermediate hosts in Japan, and another 15 species were confirmed to be susceptible under experimental conditions (Table 2). The morphological characteristics of metacercariae were described in detail by Ogawa et al. (2004). Metacercariae can parasitize the entire body of a fish, but they show particularly high infestation densities at the base of caudal fin and axial muscles (Urabe 2004). In some cases, the number of metacercariae per fish reaches several thousands. About 5000 metacercariae were reported from one individual of *Squalidus chankaensis tsuchigae* (Cyprinidae), and about 10,000 metacercariae were reported from one individual of *Opsariichthys platypus* (Cyprinidae) (Fig. 5; Urabe et al. 2001).

Another bucephalid species, *Prosorhynchoides ozakii*, which also uses *L. fortunei* as the first intermediate host (Baba et al. 2012), parasitizes the intestines (mainly the midsection) of *Silurus asotus* and *S. biwaensis* in Japan (Fig. 4; Urabe et al. 2007; Baba and Urabe 2011b). The morphological characteristics of adults and metacercariae have been described in detail (Ogawa et al. 2004; Urabe et al. 2007). Thirteen fish species were recorded as natural second intermediate hosts in Japan

Fig. 3 *Limnoperna fortunei* infested with many cercariae of *Parabucephalopsis parasiluri*. The mussel shells were cracked for parasite examination. (From Baba and Urabe 2011a)

Fig. 4 Stained whole mounts of adults of **a** *Parabucephalopsis parasiluri* and **b** *Prosorhynchoides ozakii*. (From Baba and Urabe 2011a)

(Table 2). The metacercariae are found at high densities in muscles. *Pr. ozakii* is distributed indigenously in China (Sichuan, Guizhou, and Fujian Provinces), Korea, and Vietnam (Ozaki 1928; Moravec and Sey 1989; Wang and Wang 1998; Thuy and Buchmann 2008). *Pelteobagrus vachellii* (Bagridae), *Saurogobius dobryi* (Gobiidae), and *Pangasianodon hypophthalmus* (Pangasiidae) are reported as definitive hosts in these countries (Table 1; Ozaki 1928; Moravec and Sey 1989; Wang and

Table 1 List of the definitive hosts of *Parabucephalopsis parasiluri* and *Prosorhynchoides ozakii* in Japan and other countries

Family	Fish species	*Parabucephalopsis parasiluri*	*Prosorhynchoides ozakii*	Unidentified immature bucephalid worm	References
Cyprinidae	*Sinilabeo rendahli*	1			Wang and Wang (1998)
Amblycipitidae	*Liobagrus marginatus*	1			Wang and Wang (1998)
Siluridae	*Silurus asotus*	1	2	3	Ozaki (1928), Wang (1985), Wang and Wang (1998), Urabe et al. (2001), Baba and Urabe (2011b)
	S. biwaensis	2	2		Urabe et al. (2007)
	S. meridionalis	1			Wang and Wang (1998),
Bagridae	*Pelteobagrus vachellii*		1		Moravec and Sey (1989)
Pangasiidae	*Pangasianodon hypophthalmus*		1		Thuy and Buchmann (2008)
Centrarchidae	*Micropterus salmoides*			3	Urabe et al. (2001)
Gobiidae	*Saurogobius dobryi*		1		Moravec and Sey (1989)

1 hosts in the native region, *2* hosts in the introduced area (Japan), *3* incidental hosts from which only immature worms have been recorded

Wang 1998; Thuy and Buchmann 2008). The first and second intermediate hosts are unknown in its indigenous regions.

Silurus biwaensis, which is the definitive host of both *Pa. parasiluri* and *Pr. ozakii*, is endemic to the Lake Biwa water system. In general, colonizing non-native areas is difficult for many trematode species because they have complicated life cycles (Torchin et al. 2003). However, its morphological characteristics suggest that *S. biwaensis* is closely related to *S. meridionalis*, which is one of the definitive hosts of *Pa. parasiluri* in its native area (Kobayakawa 1994). This may be why at least *Pa. parasiluri* can use *S. biwaensis*, which is not present in China, as a definitive host. Because *Pa. parasiluri* does not use other fish species as definitive hosts in Japan (it is uncertain why it does not parasitize *S. asotus* in Japan, in contrast to China), it will not be able to colonize water systems other than the Lake Biwa water system. In contrast, *Pr. ozakii* can use both *S. biwaensis* and *S. asotus* as definitive hosts. Because *S. asotus* is widely distributed in Japan and East Asia, *Pr. ozakii* has the potential to invade and colonize other water systems.

Table 2 List of the second intermediate hosts of *Parabucephalopsis parasiluri* and *Prosorhynchoides ozakii*

Family	Fish species	*Pa. parasiluri*	*Pr. ozakii*	References
Cyprinidae	*Candidia temminckii*	2		Baba and Urabe (2011b)
	Candidia sieboldii	2		Baba and Urabe (2011b)
	Opsariichthys platypus	1	1	Urabe et al. (2001)
	Opsariichthys uncirostris	1	1	Urabe et al. (2007)
	Hemigrammocypris neglectus	2		Baba and Urabe (2011b)
	Rhynchocypris logowskii steindachneri	2		Baba and Urabe (2011b)
	Rhynchocypris oxycephalus jouyi	2		Baba and Urabe (2011b)
	Ischikauia steenackeri	2		Urabe et al. (2008)
	Gnathopogon caerulescens	2		Baba and Urabe (2011b)
	Gnathopogon elongatus	2		Urabe et al. (2001)
	Pseudorasbora parva	1	1	Urabe et al. (2001)
	Pseudorasbora pumila pumila	2		Baba and Urabe (2011b)
	Pseudogobio esocinus	1	1	Urabe et al. (2007)
	Squalidus chankaensis tsuchigae	1	1	Urabe et al. (2001)
	Hemibarbus labeo	1	1	Urabe et al. (2007)
	Cyprinus carpio	2		Baba and Urabe (2011b)
	Carassius sp.	1	1	Urabe et al. (2007)
	Carassius buergeri grandoculis	2		Baba and Urabe (2011b)
	Carassius cuvieri	2		Urabe et al. (2001)
	Tanakia limbata	2		Urabe et al. (2001)
	Acheilognathus rhombeus	1	1	Baba and Urabe (2011b)
	Acheilognathus tabira tabira	1	1	Baba and Urabe (2011b)
Siluridae	*Silurus asotus*		1	Baba and Urabe (2011b)
Bagridae	*Pelteobagrus nudiceps*		1	Baba and Urabe (2011b)
Adrianichthyidae	*Oryzias* sp.	2		Baba and Urabe (2011b)
Mugilidae	*Mugil cephalus*	1		Urabe et al. (2007)
Centrarchidae	*Lepomis macrochirus*	1	1	Urabe et al. (2007)
Gobiidae	*Rhinogobius kurodai*	2		Baba and Urabe (2011b)
	Tridentiger brevispinis	1	1	Urabe et al. (2007)

1 Natural hosts, *2* Experimental hosts

Fig. 5 *Opsariichthys platypus* (Cyprinidae) heavily infected by *Parabucephalopsis parasiluri* metacercariae, showing hemorrhaging in the eyes and caudal fins. (From Baba and Urabe 2011a)

Fish Disease Caused by a Heavy Infection of Parabucephalopsis parasiluri *Metacercariae*

Heavy infections of *Pa. parasiluri* metacercariae cause hemorrhages in the fins, skin, and eyes of fishes (Ogawa et al. 2004; Urabe et al. 2008). Heavily infected fish swim lethargically (Ogawa et al. 2004). In Japan, these symptoms emerged in *Opsariichthys platypus, O. uncirostris,* and *Squalidus chankaensis tsuchigae* under natural conditions (Ogawa et al. 2004; Urabe et al. 2008). Field experiments demonstrated that three cyprinids, namely *Ischikauia steenackeri*, *Cyprinus carpio*, and *Carassius buergeri grandoculis,* also hemorrhaged due to parabucephalopsiosis (Urabe et al. 2008; Baba and Urabe 2011b). A laboratory experiment showed that some fish are killed by parabucephalopsiosis. For example, one individual of *Oryzias* sp. (Adrianichthyidae) (SL 24.6 mm) died within 24 h when it was exposed to 3010 *Pa. parasiluri* cercariae in a plastic tank with 0.3 L of water (Baba and Urabe 2011b). Two hundred and fourteen metacercariae were found in the caudal fin of the dead fish. Thus, *Pa. parasiluri* obviously has negative impacts on fish populations. The effects of *Pr. ozakii* on fishes have not been investigated, but it is likely that they have a similar harmful influence.

Infection Dynamics of Bucephalids in L. fortunei *in the Uji River*

Most of the bucephalid larvae that infect *L. fortunei* in the Uji River are surmised to be *Pa. parasiluri* (Urabe et al. 2009). The parasite's life cycle is thought to be annual (Urabe et al. 2009). Sporocysts of *Pa. parasiluri* are found in *L. fortunei* in all months of the year. Cercariae start to develop in sporocysts in October (water temperature ca. 20 °C). The ratio of mussels harboring cercariae to all infected mussels then reaches 81.8 % in October. The number of cercariae is about 4560 to 14,360 per host (Urabe et al. 2001). Parabucephalopsiosis often breaks out in January, indicating that the cercariae of *Pa. parasiluri* are shed from *L. fortunei* in winter (water temperature ca. 7 °C; Urabe et al. 2009).

Between 2001 and 2008, the annual prevalence of *Pa. parasiluri* in *L. fortunei* (shell length≥15 mm) in the Uji and Yodo rivers fluctuated from 2 to 18% (Baba and Urabe 2011a). The prevalence of *Pa. parasiluri* in *L. fortunei* in autumn is negatively correlated with the average discharge from the Amagase Dam (located upstream) during the preceding January (Urabe et al. 2009). This suggests that a large amount of discharge from the dam in January sweeps out free cercariae in the water, and subsequently reduces the abundance of metacercariae in second intermediate hosts, reduces the abundance of adults in definitive hosts, and in turn reduces the prevalence of *Pa. parasiluri* in *L. fortunei* in the following autumn.

Tanaka et al. (2004) investigated the effects of bucephalids (*Pa. parasiluri* and *Pr. ozakii*) on the growth and reproduction of *L. fortunei*. Average growth rate measured as shell length over 44 days was not different between infected and uninfected mussels. However, the gonads of uninfected mussels were filled with spermatocytes or oocytes, whereas those of infected mussels were filled with sporocysts and lacked any gametes. Therefore, larval bucephalids do not affect the growth of their hosts but castrate them, and may depress the population growth of *L. fortunei*.

Concluding Remarks

Fish diseases caused by bucephalids introduced with *L. fortunei* have negative effects on native fish populations. Although there are no statistical catch data of *Opsariichthys platypus* or *Squalidus chankaensis tsuchigae,* because such small cyprinid fish are not commercially important, local fishermen feel that populations of *O. platypus* declined in the Uji River after the outbreak of parabucephalopsiosis in 2000 (Uji River fishermen's cooperative, personal communication). However, this does not exclude the possibility that some factors other than parabucephalopsiosis caused the decline of the fish population.

When the mussel was introduced in South America in larval form (most probably with ballast water: Darrigran and Pastorino 1995), the seeding population was likely free of parasites, even when discharged as juveniles, having completed their development underway, in the ballast water tanks. On the other hand, some bucephalid species, including *Pa. parasiluri*, *Pr. ozakii* and others, may invade South America via living fish exported from East Asian freshwater areas. As the life cycles of most bucephalids in the native area of *L. fortunei* are still unknown, caution is required regarding all freshwater fish species. Therefore, a strict quarantine system is needed when importing living fish for aquaculture or for aquaria from the native and introduced areas of bucephalids. For example, fish for import could be restricted to those cultivated in natural/artificial water systems without *L. fortunei*.

Acknowledgements We thank Dr. K. Ogawa, Meguro Parasitological Museum, for revising this manuscript.

References

Baba T, Urabe M (2011a) Examination methods for the bucephalids which use *Limnoperna fortunei* as the first intermediate host. Rep Yahagi River Inst 15:97–101 [In Japanese]

Baba T, Urabe M (2011b) Expanding distribution of pathogenic fish trematodes *Parabucephalopsis parasiluri* and *Prosorhynchoides ozakii* in the Lake Biwa water system and their new host record as second intermediate and definitive hosts. Bull Kansai Organ Nat Conserv 33:117–124 [In Japanese]

Baba T, Nakamura D, Hosoi M, Urabe M (2012) Molecular identification of larval bucephalids, *Prosorhynchoides ozakii* and *Parabucephalopsis parasiluri*, infecting the golden mussel, *Limnoperna fortunei*, by PCR-RFLP. J Parasitol 98:669–673

Baturo B (1978) Larval bucephalosis in artificially heated lakes of the Konin region, Poland. Acta Parasitol Pol 30:307–321

Darrigran GA, Pastorino G (1995) The recent introduction of a freshwater Asiatic bivalve, *Limnoperna fortunei* (Mytilidae) into South America. The Veliger 38:171–175

Feng W, Wang W (1995) Descriptions on three new digenetic species of fishes. Acta Hydrobiol Sinica 19:304–310 [In Chinese]

Francisco CJ, Hermida MA, Santos MJ (2012) *Prosorhynchus crucibulum* (Digenea: Bucephalidae) miracidium morphology and its passive transmission pattern. Parasite 19:277–280

Khamboonruang P, Karnasuta P, Purivirojkul W, Sirikanchana P (2006) Parasites of PlaKae *Bagarius bagarius* (Hamilton-Buchanan) from Mekong River at Amphoe Chiang Khong Chiang Rai Province. In: 44th Kasetsart University Annual Conference, Kasetsart (Thailand)

Kobayakawa M (1994) Siluridae. In: Research Group for Natural History of Lake Biwa (ed) The natural history of Lake Biwa. Yasaka Shobo, Tokyo, pp 221–234 [In Japanese]

Krailas D, Chotesaengsri S, Pattaradussadee N, Notesiri N, Dechruksa W (2008) Bucephalid (Gasterostome) cercariae obtained from freshwater clams in Thailand. J Trop Med Parasito 31:70–76

Matthews RA (1973) The life-cycle of *Bucephalus haimeanus* Lacaze-Duthiers, 1854 from *Cardium edule* L. Parasitology 67:341–350

Moravec F, Sey O (1989) Some trematodes of freshwater fishes from North Vietnam with a list of recorded endohelminths by fish hosts. Folia Parasitol 36:243–262

Nakai K (1995) Intrusion of the freshwater mytilid mussel, *Limnoperna fortunei* (Dunker, 1857), into Japan. Bull Kansai Organ Nat Conserv 17:49–56 [In Japanese]

Ogawa K, Nakatsugawa T, Yasuzaki M (2004) Heavy metacercarial infections of cyprinid fishes in Uji River. Fish Sci 70:132–140

Ozaki Y (1928) Some gasterostomatus trematodes of Japan. Jpn J Zool 2:35–60 [In Japanese]

Purivirojkul W (2009) Fish parasite diversity in the Mekong River in the north of Thailand. KKU Sci J 37(Supplement):62–70

Scholz T (1991) Observations on trematode fauna of freshwater fish in Laos. Helminthologia 28:125–130

Sey OW, Wongsawad P (2004) Trematodes from fish in the Bhumipol dam, Tak province, Thailand. Southeast Asian J Trop Med Public Health 35(Supplement):284–287

Tanaka H, Deno T, Kondo T (2004) Effect of gasterostome trematode infection on growth and reproduction in *Limnoperna fortunei* (Bivalvia: Mytilidae). Mem Osaka Kyoiku Univ Ser III, Nat Sci Appl Sci 53:1–7

Tang C, Tang Z (1976) Studies on nine species of gasterostomes from Fujian, with observations on the life cycles of two species. Acta Zool Sinica 22:263–274 [In Chinese]

Thuy DT, Buchmann K (2008) Intestinal trematodes *Prosorhynchoides ozakii* (Bucephalidae; Bucephalinae) in pond-cultured catfish *Pangasianodon hypophthalmus* in the Mekong Delta (Vietnam). Bull Eur Assoc Fish Pathol 28:186–193

Torchin ME, Lafferty KD, Dobson AP, McKenzie VJ, Kuris AM (2003) Introduced species and their missing parasites. Nature 421:628–630

Urabe M (2004) Parabucephalopsis. In: The Ecological Society of Japan (ed) Handbook of alien species in Japan. Chijin Shokan, Tokyo, p 216 [In Japanese]

Urabe M, Ogawa K, Nakatsugawa T, Imanishi Y, Kondo T, Okunishi T, Kaji Y, Tanaka H (2001) Newly recorded gasterostome trematode (Digenea: Bucephalidae) in the Uji River: the life history, distribution and damage to fishes. Bull Kansai Organ Nat Conserv 23:13–21 [In Japanese]

Urabe M, Ogawa K, Nakatsugawa T, Nakai K, Tanaka M, Wang G (2007) Morphological description of two bucephalid trematodes collected from freshwater fishes in the Uji River, Kyoto, Japan. Parasitol Int 56:269–272

Urabe M, Tanaka M, Nakamura D (2008) The pathogenic fish trematode *Parabucephalops isparasiluri*, spreading to the Seta River and Lake Biwa. Bull Kansai Organ Nat Conserv 30:45–48 [In Japanese]

Urabe M, Nakai K, Nakamura D, Tanaka M, Nakatsugawa T, Ogawa K (2009) Seasonal dynamics and yearly change in the abundance of metacercariae of *Parabucephalopsis parasiluri* (Trematoda: Bucephalidae) in the second intermediate host in the Uji-Yodo River, central Japan. Fish Sci 75:63–70

Wang P (1985) Notes on some species of gasterostome trematodes of fishes mainly from Fujian Province. J Fujian Norm Univ Nat Sci 4:73–83 [In Chinese]

Wang G, Wang W (1998) Taxonomy and keys to the bucephalid species in China, with descriptions of three new species. Acta Hydrobiol Sinica 22:100–110 [In Chinese]

Yamaguti S (1975) A synoptical review of life histories of digenetic trematodes of vertebrates with special reference to the morphology of their larval forms. Keigaku Publishing Co., Tokyo, pp 1–590

The Genetics of the Golden Mussel (*Limnoperna fortunei*): Are Genes Related to Invasiveness?

Marcela Uliano-Silva, Juliana A. Américo, Rodrigo Brindeiro, Francesco Dondero, Francisco Prosdocimi and Mauro F. Rebelo

Abstract Successful invasive species often share some ecological traits, such as rapid growth and rapid sexual maturation, both of which are characteristic of *Limnoperna fortunei*. In addition, phenotypic plasticity, i.e., the capability to express different phenotypes as a response to diverse environmental challenges, may play a fundamental role in the geographic expansion of many invasive species, including the golden mussel. Little is known about the genetics of *L. fortunei*, but the first transcriptome for *L. fortunei* has recently been sequenced and gene–environment relationships that are likely associated with the successful invasions of this species have begun to be elucidated. Over 24,000 transcripts have been functionally annotated, and results suggest the expansion of the gene families' heat shock protein 70 and cytochrome P450. This may indicate that *L. fortunei* has a broad genetic repertoire that confers it an advantage to deal with stressors presented in new locations. Several other key genes such as byssus proteins, immune system-related genes, and antioxidant enzymes have been characterized and are now available for gene

M. Uliano-Silva (✉) · J. A. Américo · M. F. Rebelo
Instituto de Biofísica Carlos Chagas Filho G2-050, Universidade Federal do Rio de Janeiro, Rio de Janeiro 21941-902, Brazil
e-mail: marcela.uliano@biof.ufrj.br

J. A. Américo
e-mail: juliana@biof.ufrj.br

M. F. Rebelo
e-mail: mrebelo@biof.ufrj.br

R. Brindeiro
Instituto de Biologia, Universidade Federal do Rio de Janeiro, Rio de Janeiro 21044-020, Brazil
e-mail: robrinde@biologia.ufrj.br

F. Dondero
Department of Science and Technological Innovation, Università del Piemonte Orientale "Amodeo Avogrado", 15121 Alessandria, Italy
e-mail: francesco.dondero@mfn.unipmn.it
F. Prosdocimi
Instituto de Bioquímica Médica, Universidade Federal do Rio de Janeiro, Rio de Janeiro 21044-020, Brazil
e-mail: fprosdocimi@bioqmed.ufrj.br

D. Boltovskoy (ed.), *Limnoperna fortunei*, Invading Nature - Springer Series in Invasion Ecology 10, DOI 10.1007/978-3-319-13494-9_4

expression studies. A genome project is being carried out in Brazil to characterize the entire set of genes and gene families of *L. fortunei*. This will expand our current knowledge of the genetic characteristics of the invasion helping to forecast new invasion events and develop biotechnology-based strategies to control the infestation.

Keywords *Limnoperna fortunei* · Golden mussel · Genetics · Phenotypic plasticity · Transcriptome · Adaptation

Phenotypic Plasticity

Around 5 t of ballast water are transported annually around the globe discharging more than 10,000 alien species in new environments every day. From this total, only a small fraction ends up establishing and colonizing a new environment (Carlton and Geller 1993; David and Gollasch 2008). What makes some alien species become invasive and others not?

Some ecological traits are well-known to be related to success in colonizing a new environment, such as short life spans, early sexual maturation, and high reproductive rates (Lockwood et al. 2007; Davis 2009), all of which are characteristic of *Limnoperna fortunei* (Karatayev et al. 2007; Uliano-Silva et al. 2013). Nonetheless, such traits may not suffice for successful colonization of a new region if the alien species is faced with a number of new stressors to which it is not adapted. The genetic substrate of invasive populations, upon which natural selection operates, could be of primary importance for the success, or lack thereof, of a biological invasion (Cox 2002). Among the components of this genetic substrate, genetic plasticity is essential insofar as it is responsible for the environmental tolerance of the invading species while it struggles to occupy a new and challenging environment (Franks and Munshi-South 2014). Phenotypic plasticity is the ability of a species to express multiple alternative phenotypes from a single genotype under a variety of different environment conditions (Stearns et al. 1991). Evolutionary studies have shown that phenotypic plasticity confers some advantages, such as the maintenance of genetic diversity, higher adaptation rates, and reduction of the bottleneck effect when species are facing the challenges of a transitional environment (Stearns et al. 1991).

Shortly after invading South America (Pastorino et al. 1993) and Japan (Kimura 1994), *L. fortunei* became a major nuisance for many industrial facilities, as well as a conspicuous component of local fauna, which fostered studies on this mussel (see Preface in this volume). In recent years, evidence has accumulated on the ecological and economic harm caused by this species (Magara et al. 2001; Uliano-Silva et al. 2013; Boltovskoy and Correa 2015), granting it the status of an “ecosystem engineer” (Darrigran and Damborenea 2011). Successful invasions are thought to be largely associated with the ability of *L. fortunei* to colonize waterbodies with a wide range of environmental conditions (Karatayev et al. 2007; Oliveira et al. 2011; Uliano-Silva et al. 2013). This may suggest that the golden mussel’s success has benefited from the ability to express alternative phenotypes to cope with the many environmental challenges of the Asian and South American lotic and lentic bodies of water colonized (Uliano-Silva et al. 2013). However, the genetics of *L. fortunei* was not investigated until late 2013, when its first transcriptome was sequenced, assembled, and annotated.

First Transcriptome Survey for *L. fortunei*

Uliano-Silva et al. (2014) performed the first transcriptome survey of *L. fortunei.* A transcriptome is the total set of genetic transcripts in a given organism, which can vary with intracellular and external environmental conditions. The transcriptome reflects the genes that are being actively expressed at a given time. In this study (Uliano-Silva et al. 2014), only mRNA was sequenced, covering the genes expressed in the tissues of the gills, adductor muscle, digestive gland, foot, and mantle. Using next-generation sequencing technology (Roche 454 GS Junior), 84,063 partial gene sequences were sequenced and assembled, where 1351 were complete full-length genes, and more than 24,000 expressed transcripts were functionally annotated. This extensive material has provided information that can be used to investigate the gene–environment relationships of *L. fortunei* during the colonization process (Uliano-Silva et al. 2014).

Invading and colonizing new environments present major challenges and is stressful for organisms (Lee 2002). A range of gene families is known to become highly expressed at times of cellular stress helping to avoid homeostatic imbalance (Evans and Hofmann 2012). Examples of such gene families are the antioxidant enzymes, which neutralize or repair damage caused by free radicals in the cell (Sies 1997), the molecular chaperones that prevent protein denaturation under heat and several other cellular stressful settings (Clark and Peck 2009), and the family of cytochrome P450, which are phase 1 biotransformation enzymes that transform harmful xenobiotics to a more polar molecule facilitating its excretion from the cell (Teunissen et al. 1992).

Molecular Chaperones, HSP70

One of the first gene families closely investigated in *L. fortunei* comprises the HSP70 chaperones. These genes have recently been shown to be important in the Pacific oyster *Crassostrea gigas* (Zhang et al. 2012). The sequenced genome of this bivalve, also an exotic species in many areas of the world, including southern South America, where it is cultured for human consumption (Melo et al. 2010), has 88 copies of the HSP70 gene. This is significantly more than previously found for the same gene in other species (e.g., 39 in sea urchins, 17 in humans; Zhang et al. 2012). Furthermore, the expression of HSP70 genes were induced at least 15-fold in oysters exposed to several stressors, including heat and metal pollution. Exposure to heat and air increased up to 2000-fold the expression of five *C. gigas* HSP70s. This gene expansion and their notable induction were attributed to the remarkable ability of this intertidal species to cope with severe stress when exposed to air during low tides (Zhang et al. 2012).

The importance of this gene family to cope with stressful situations in the oyster *C. gigas,* suggests that it may also be important for the invasive success of *L. fortunei*. The transcriptome of the golden mussel has at least 55 different isoforms of HSP70, which is markedly higher than in humans and sea urchins. However, the transcriptome characterization of gene families is not a fail-safe approach, and it may underestimate the number of different isoforms of HSP70 present in the genome of the golden mussel.

Preliminary analysis of *L. fortunei* HSP70s supports the hypothesis that this gene family is indeed related to successful invasions. Analysis of mollusc HSP70 phylogeny showed that two *L. fortunei* HSP70s isoforms are evolutionarily related to the expansion observed in *C. gigas* (Uliano-Silva et al. 2014). These results also showed that all the other *L. fortunei* HSP70s are phylogenetically related to several other HSP70 isoforms that are expressed under a variety of circumstances in other bivalves.

The fact that *L. fortunei* has an extensive repertoire of HSP70s indicates that the challenge of invading new environments can be facilitated by the modulation of these anticellular stress genes (Uliano-Silva et al. 2014).

Cytochrome P450

Another group of genes that are worth investigating in relation to the ability to withstand stressful environmental challenges is the cytochrome P450 (CYP) gene family. These monooxygenase enzymes are chiefly involved in catalyzing the biotransformation of xenobiotics and hydrophobic compounds into more polar forms facilitating their excretion from the cell and are present in several other biochemical pathways (Teunissen et al. 1992). Each isoform is known to catalyze specific reactions. For example, the isoform CYP3A participates in the biosynthesis of cholesterol and steroid hormones, and is responsible for metabolizing about 80% of all man-made drugs (Li et al. 2008). CYP2A is involved in the biotransformation of polychlorinated biphenyls (PCBs) (Fernandez-Salguero and Gonzalez 1995), once widely employed in various industrial processes and now banned in most countries. Several studies have reported the role of cytochromes in the maintenance of homeostasis in bivalves exposed to chemical compounds (Mello et al. 2012), and domestic sewage (Bainy et al. 2000).

After having annotated the genes through transcriptome sequencing, work on *L. fortunei* has centered on investigating whether the cytochrome P450 profile would allow it to invade, settle, and adapt to specific habitats. Transcripts representing the 24 CYP isoforms of *L. fortunei* were combined with the CYP gene sequences for all

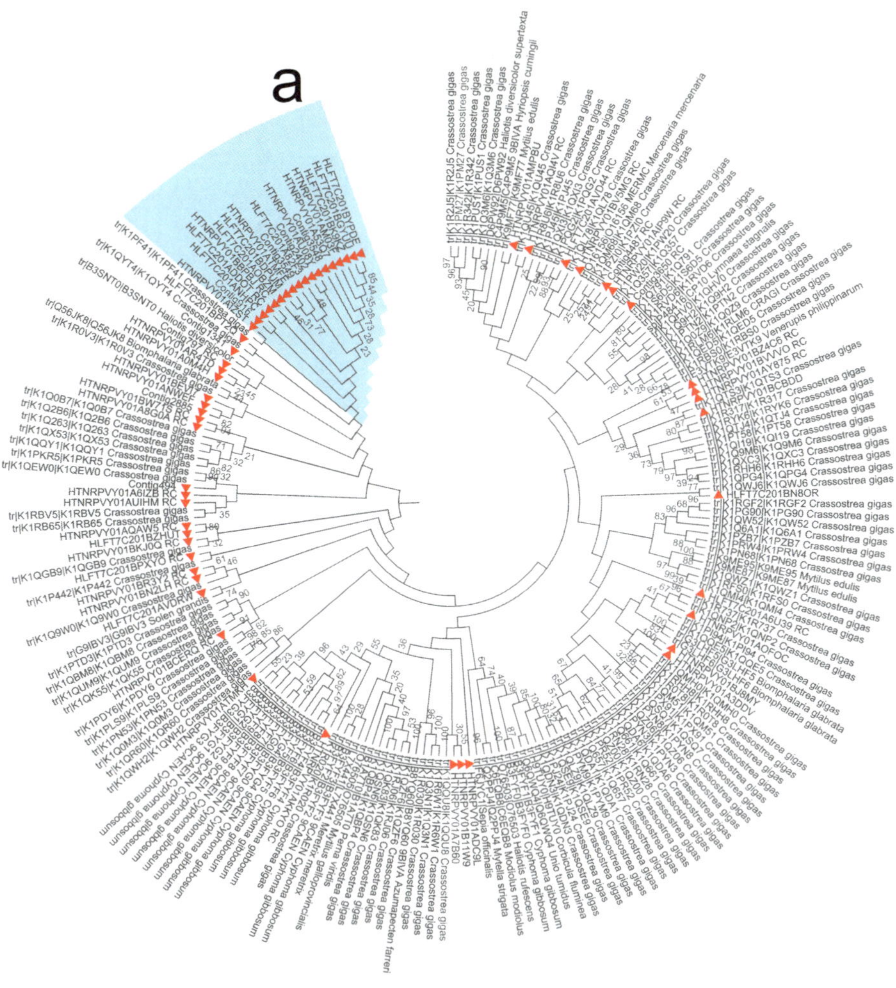

Fig. 1 Consensus phylogenetic tree of Mollusca cytochrome P450 family members. All 66 cytochrome P450 sequences described in the transcriptome of *L. fortunei* were combined with all 143 cytochrome P450 gene sequences available for the 19 molluscs in the UNIPROT database for the alignment and construction of the phylogenetic tree. The tree was built using the maximum likelihood method and bootstrapping (100 pseudoreplicates, values less than 30 % are not shown). Sequences of *L. fortunei* are marked with *red triangles*. Detail (**a**) highlights the possible expansion of the CYP3A genes found in *L. fortunei*

molluscs available online to construct a phylogenetic tree (Fig. 1). The figure shows that there are *L. fortunei* CYPs phylogenetically related to several *C. gigas* CYP isoforms throughout the tree, thus confirming the possible similarity of the CYP profile in the genomes of both species. However, a possible source of bias in this analysis is that the CYP sequences of *C. gigas* are overrepresented in this molluscan phylogeny, because the genome of *C. gigas* is one of only two bivalve genomes to have been sequenced. Interestingly, *L. fortunei* has a set of CYP3As that do not relate phylogenetically to any CYPs of other bivalves, but they are strongly associated with each other (Fig. 1, detail a). This suggests an expansion in the CYP3A gene family in *L. fortunei* that has not yet been described for any other bivalve. If genome sequencing confirms this expansion, the next step will be to elucidate whether or not this CYP profile, more robust than for other species, allows *L. fortunei* to cope with the stress of challenging environments, such as the acid waters of the Amazon system (Oliveira et al. 2010).

Other Genes Potentially Important for Controlling the Spread and Impacts of *L. fortunei*

In addition to gene families related to resilience under cellular stress, there are other important genes in the context of invasion dynamics of the golden mussel. A salient trait of *L. fortunei* is its ability to attach to hard objects using its byssal threads (see Chapter "Distribution and Colonization of *Limnoperna fortunei*: Special Traits of an Odd Mussel" in this volume). The byssal plaque has strong crosslinks between neighboring proteins, with a high content of DOPA, a modified amino acid, and metal atoms, ensuring firm adhesion even in the presence of water (Ohkawa et al. 2001; Lee et al. 2011; see Chapter "Control of *Limnoperna fortunei* Fouling: Antifouling Materials and Coatings" in this volume). Two byssus proteins, Mepf1 and Mepf2, were found in the transcriptome of *L. fortunei*; their expression patterns are a promising topic of research that may help in the search of methods to mitigate fouling in industrial facilities (Uliano-Silva et al. 2014).

The immune system is another key factor allowing bivalves to survive environmental challenges. While viral and fungal infectious diseases represent a serious threat to shrimp farming worldwide, cultures of *C. gigas* seem immune to these problems. The oyster thrives in aquaculture facilities in southern Brazilian where waters are highly contaminated with viruses (human adenovirus, noroviruses, hepatitis A, and JC Polyomavirus) and fecal coliform bacteria (Souza et al. 2012). The transcriptome of *L. fortunei* shows the presence of at least eight genes involved in the signaling pathway of toll-like receptors (Uliano-Silva et al. 2014) indicating that *L. fortunei* possesses precursors of an adaptive immune system as shown for other invertebrates (Hibino et al. 2006; Miller et al. 2007; Philipp et al. 2012). This would confer an advantage when invading environments with a variety of contaminants.

Limnoperna fortunei is more resistant to chemical control than other fouling mussels, such as *Dreissena polymorpha* (Cataldo et al. 2003; see Chapters "Parallels and Contrasts Between *Limnoperna fortunei* and Species of *Dreissena*" and "Chemical Strategies for Control of the Golden Mussel (*Limnoperna fortunei*) in Industrial Facilities" in this volume). It thrives under a wide range of conditions (Karatayev et al. 2010); but one of the few catastrophic events that seems to strongly depress its numbers, occasionally wiping out entire populations, is the "dequada," an extensive anoxic event that occurs after periodic floods in the Pantanal wetland of Brazil (Oliveira et al. 2010). These anoxic episodes are responsible for massive kills, but some local organisms have special adaptations to cope with them. The fish *Piaractus mesopotamicus*, for example, has an enhanced basal activity of the peroxide-depredate enzyme, GPx, that helps it support oxidative stress (Cunha Bastos et al. 2007). Records of massive kills of *L. fortunei* in association with dequadas in the Pantanal wetland may indicate that the mussel does not have a robust antioxidant system. This assumption is supported by the apparent lack of expansions of antioxidant gene families in the transcriptome of *L. fortunei* (Uliano-Silva et al. 2014), as also noted for the genome of *C. gigas* (Zhang et al. 2012). The profile of the antioxidant genotype of *L. fortunei* needs further investigation, but if the above assumptions prove correct, the antioxidant system may be an important target for the development of control tools against *L. fortunei* biofouling (see Chapter "Control of *Limnoperna fortunei* Fouling by Oxygen Deprivation" in this volume).

Concluding Remarks

Preliminary results on the transcriptome of *L. fortunei* indicate that the mussel's invasive success may be intimately linked to its phenotypic plasticity (Uliano-Silva et al. 2014). However, the ultimate goal of this approach is acquiring adequate knowledge of *L. fortunei* genetics through the sequencing of its entire genome. The assessment of coding regions and the number of genes and gene families of a species, as revealed by transcriptome sequencing, may be incomplete because it only characterizes genes expressed in the organism at the time of collection of the corresponding tissue samples. The only reliable way to characterize genes and gene families, other transcripts (e.g., interference RNA, micro RNA), as well as various noncoding regions, is through the sequencing and assemblage of the species' genome. The genome can also characterize the "taxonomically restricted genes" (TGRs), which represent 10–20 % of the genes of a species and are not obscured by similarities with phylogenetically close organisms.

Phenotypic plasticity is ultimately dependent upon the genotype of a species. The range of gene families and their expression is what will confer the organism the capability to adapt to stressful environments. The goal of work currently underway is to relate this genetic information with the spread of *L. fortunei* outside of its native range. Assembling the genome of *L. fortunei* will expand our current knowledge of the genetic characteristics of biological invasions, help to forecast new invasion events, and aid in the development of biotechnology-based strategies to control infestations. Genome sequencing of the oyster *C. gigas* has shown the potential of this approach to clarify the relationship between the genotype and the life-habits of a species. Zhang et al. (2012) concluded that the genome of *C. gigas* is about 637 Mb in size. Our preliminary results show that the genome of *L. fortunei* is slightly larger, ~800 Mb (unpublished data). The next step is to determine how much of this represents novel genes, and how much of it is noncoding DNA.

Acknowledgements Thanks are due to the students who assisted us in the transcriptome sequencing runs, Bianca Duarte and Gabriel Gonçalves. This work was financed by the Conselho Nacional de Pesquisa (CNPq, Brazil). We would also like to thank Igor da Costa for programming the PerlScripts used in the transcriptome work. Finally, we thank all the people who made the sequencing runs of the Golden Mussel Genome Project possible through the crowd-funding campaign.

References

Bainy AC, Almeida EA, Muller IC, Ventura EC, Medeiros ID (2000) Biochemical responses in farmed mussel *Perna perna* transplanted to contaminated sites on Santa Catarina Island, SC, Brazil. Mar Environ Res 50:411–416

Boltovskoy D, Correa N (2015) Ecosystem impacts of the invasive bivalve *Limnoperna fortunei* (golden mussel) in South America. Hydrobiologia 746:81-95

Carlton JT, Geller JB (1993) Ecological roulette: the global transport of nonindigenous marine organisms. Science 261:78–82

Cataldo D, Boltovskoy D, Pose M (2003) Toxicity of chlorine and three non-oxidizing molluscicides to the invasive pest mussel *Limnoperna fortunei*. J Am Water Works Assoc 95:66–78

Clark MS, Peck LS (2009) HSP70 heat shock proteins and environmental stress in Antarctic marine organisms: a mini-review. Mar Genomics 2:11–18

Cox GW (2002) Alien species and evolution. The evolutionary ecology of exotic plants, animals, microbes and interacting native species. Island Press, Washington DC, pp 1–393

Cunha Bastos VL, Salles JB, Valente RH, León IR, Perales J, Dantas RF, Albano RM, Bastos FF, Cunha Bastos J (2007) Cytosolic glutathione peroxidase from liver of pacu (*Piaractus mesopotamicus*), a hypoxia-tolerant fish of the Pantanal. Biochimie 89:1332–1342

Darrigran GA, Damborenea C (2011) Ecosystem engineering impact of *Limnoperna fortunei* in South America. Zool Sci 28:1–7

David M, Gollasch S (2008) EU shipping in the dawn of managing the ballast water issue. Mar Pollut Bull 56:1966–1972

Davis MA (2009) Invasion biology. Oxford University Press, New York, pp 1–244

Evans TG, Hofmann GE (2012) Defining the limits of physiological plasticity: how gene expression can assess and predict the consequences of ocean change. Philos Trans R Soc Lond Ser B Biol Sci 367:1733–1745

Fernandez-Salguero P, Gonzalez FJ (1995) The CYP2A gene subfamily: species differences, regulation, catalytic activities and role in chemical carcinogenesis. Pharmacogenetics 5:S123–S1288

Franks SJ, Munshi-South J (2014) Go forth, evolve and prosper: the genetic basis of adaptive evolution in an invasive species. Mol Ecol 23:2137–2140

Hibino T, Loza-Coll M, Messier C, Majeske AJ, Cohen AH, Terwilliger DP, Buckley KM, Brockton V, Nair SV, Berney K, Fugmann SD, Anderson MK, Pancer Z, Cameron RA, Smith LC, Rast JP (2006) The immune gene repertoire encoded in the purple sea urchin genome. Dev Biol 300:349–365

Karatayev AY, Boltovskoy D, Padilla DK, Burlakova LE (2007) The invasive bivalves *Dreissena polymorpha* and *Limnoperna fortunei*: parallels, contrasts, potential spread and invasion impacts. J Shellfish Res 26:205–213

Karatayev AY, Burlakova LE, Karatayev VA, Boltovskoy D (2010) *Limnoperna fortunei* versus *Dreissena polymorpha*: population densities and benthic community impacts of two invasive freshwater bivalves. J Shellfish Res 29:975–984

Kimura T (1994) The earliest record of *Limnoperna fortunei* (Dunker) from Japan. Chiribotan (J Malacol Soc Jpn) 25:34–35 [In Japanese]

Lee CE (2002) Evolutionary genetics of invasive species. Trends Ecol Evol 17:386–391

Lee BP, Messersmith PB, Israelachvili JN, Waite JH (2011) Mussel-inspired adhesives and coatings. Annu Rev Mater Res 41:99–132

Li D, Yang XL, Zhang SJ, Lin M, Yu WJ, Hu K (2008) Effects of mammalian CYP3A inducers on CYP3A-related enzyme activities in grass carp (*Ctenopharyngodon idellus*): possible implications for the establishment of a fish CYP3A induction model. Comp Biochem Physiol Part C Toxicol Pharmacol 147:17–29

Lockwood J, Hoopes M, Marchetti M (2007) Invasion ecology. Blackwell, Malden, pp 1–304

Magara Y, Matsui Y, Goto Y, Yuasa A (2001) Invasion of the non-indigenous nuisance mussel, *Limnoperna fortunei*, into water supply facilities in Japan. J Water Supply Res Technol Aqua 50:113–124

Mello DF, Oliveira ES, Vieira RC, Simoes E, Trevisan R, Dafre AL, Barracco MA (2012) Cellular and transcriptional responses of *Crassostrea gigas* hemocytes exposed in vitro to brevetoxin (PbTx-2). Mar Drugs 10:583–597

Melo CMR, Silva FC, Gomes CHAM, Solé-Cava AM, Lazoski C (2010) *Crassostrea gigas* in natural oyster banks in southern Brazil. Biol Invasions 12:441–449

Miller DJ, Hemmrich G, Ball EE, Hayward DC, Khalturin K, Funayama N, Agata K, Bosch TC (2007) The innate immune repertoire in cnidaria–ancestral complexity and stochastic gene loss. Genome Biol 8:R59

Ohkawa K, Ichimiya K, Nishida A, Yamamoto H (2001) Synthesis and surface chemical properties of adhesive protein of the Asian freshwater mussel, *Limnoperna fortunei*. Macromol Biosci 1:376–386

Oliveira MD, Hamilton SK, Jacobi CM (2010) Forecasting the expansion of the invasive golden mussel *Limnoperna fortunei* in Brazilian and North American rivers based on its occurrence in the Paraguay River and Pantanal wetland of Brazil. Aquat Invasions 5:59–73

Oliveira MD, Calheiros DF, Jacobi CM, Hamilton SK (2011) Abiotic factors controlling the establishment and abundance of the invasive golden mussel *Limnoperna fortunei*. Biol Invasions 13:717–729

Pastorino G, Darrigran GA, Martín SM, Lunaschi L (1993) *Limnoperna fortunei* (Dunker, 1857) (Mytilidae), nuevo bivalvo invasor en aguas del Río de la Plata. Neotropica 39:101–102

Philipp EE, Kraemer L, Melzner F, Poustka AJ, Thieme S, Findeisen U, Schreiber S, Rosenstiel P (2012) Massively parallel RNA sequencing identifies a complex immune gene repertoire in the lophotrochozoan *Mytilus edulis*. PLoS One 7:e33091

Sies H (1997) Oxidative stress: oxidants and antioxidants. Exp Physiol 82:291–295

Souza DS, Ramos AP, Nunes FF, Moresco V, Taniguchi S, Leal DA, Sasaki ST, Bicego MC, Montone RC, Durigan M, Teixeira AL, Pilotto MR, Delfino N, Franco RM, Melo CM, Bainy AC, Barardi CR (2012) Evaluation of tropical water sources and mollusks in southern Brazil using microbiological, biochemical, and chemical parameters. Ecotoxicol Environ Saf 76:153–161

Stearns S, de Jong G, Newman B (1991) The effects of phenotypic plasticity on genetic correlations. Trends Ecol Evol 6:122–126

Teunissen Y, Geraerts WP, van Heerikhuizen H, Planta RJ, Joosse J (1992) Molecular cloning of a cDNA encoding a member of a novel cytochrome P450 family in the mollusc *Lymnaea stagnalis*. J Biochem 112:249–252

Uliano-Silva M, Fernandes FFCF, de Holanda IBB, Rebelo MF (2013) Invasive species as a threat to biodiversity: the golden mussel *Limnoperna fortunei* approaching the Amazon river basin. In: Allodi S (ed) Exploring themes on aquatic toxicology. Research Signpost, Kerala, pp 1–14

Uliano-Silva M, Américo JA, Brindeiro R, Dondero F, Prosdocimi F, Rebelo MF (2014) Gene discovery through transcriptome sequencing for the invasive mussel *Limnoperna fortunei*. PLoS One 9:e102973

Zhang G, Fang X, Guo X, Li L, Luo R, Xu F, Yang P, Zhang L, Wang X, Qi H, Xiong Z, Que H, Xie Y, Holland PW, Paps J, Zhu Y, Wu F, Chen Y, Wang J, Peng C, Meng J, Yang L, Liu J, Wen B, Zhang N, Huang Z, Zhu Q, Feng Y, Mount A, Hedgecock D, Xu Z, Liu Y, Domazet-Loso T, Du Y, Sun X, Zhang S, Liu B, Cheng P, Jiang X, Li J, Fan D, Wang W, Fu W, Wang T, Wang B, Zhang J, Peng Z, Li Y, Li N, Chen M, He Y, Tan F, Song X, Zheng Q, Huang R, Yang H, Du X, Chen L, Yang M, Gaffney PM, Wang S, Luo L, She Z, Ming Y, Huang W, Huang B, Zhang Y, Qu T, Ni P, Miao G, Wang Q, Steinberg CE, Wang H, Qian L, Liu X, Yin Y (2012) The oyster genome reveals stress adaptation and complexity of shell formation. Nature 490:49–54

Reproductive Output and Seasonality of *Limnoperna fortunei*

Demetrio Boltovskoy, Brian Morton, Nancy Correa, Daniel Cataldo, Cristina Damborenea, Pablo E. Penchaszadeh and Francisco Sylvester

Abstract Young *Limnoperna fortunei* mature sexually from 5–6 to ~15 mm. The species is generally dioecious, with approximately equal numbers of males and females and very small (<0.6%) proportions of hermaphrodites. The gametogenic cycle has been described for both Asian and South American populations, recognizing between four and five reproductive phases. Gonadal cycles based on histological sections yielded somewhat dissimilar results for different areas. In Hong Kong, two yearly peaks in reproductive output were detected. In South America, mature sperm and ova have been recorded year round and several irregularly spaced spawning events have been observed, as well as more or less continuous breeding punctuated by peaks in spring and at the end of the summer. Reproductive studies based on changes in the abundance of larvae in the water column have been carried out in South America and in Japan. In tropical and subtropical South America, larval output is more or less continuous for 6–10 months of the year, often with a major peak in spring–early summer, and a smaller one in the late summer–autumn. In Japan, at considerably lower water temperatures, larval production is limited to 1–2 months centered around summer. Apparent disagreements between results based on histological data and on larval counts stem from the fact that while the latter integrate the reproductive output of extensive mussel beds dispersed over large areas, histological evidence pinpoints with high precision the ripening and spawning of isolated mussel clusters. Aside from water temperature, several other factors (pH, salinity, dissolved oxygen, suspended solids, chlorophyll a, flood–drought cycles) have been proposed as reproductive triggers, but actual associations have not been demonstrated. Peak larval densities can exceed 20,000 ind./m^3, but, normally, values range around 6000 ind./m^3, showing major fluctuations within short periods, as well as changes as a function of time elapsed post colonization, and availability of substrata suitable for adult occupation. Microcystin-producing cyanobacterial blooms can kill *L. fortunei* larvae.

D. Boltovskoy (✉) · D. Cataldo · F. Sylvester
Facultad de Ciencias Exactas y Naturales, Instituto de Ecología, Genética y Evolución de Buenos Aires (IEGEBA),Universidad de Buenos Aires-CONICET, Ciudad Universitaria, 1428 Buenos Aires, Argentina
e-mail: demetrio@ege.fcen.uba.ar

D. Boltovskoy · D. Cataldo
Museo Argentino de Ciencias Naturales 'Bernardino Rivadavia'-CONICET, Buenos Aires, Argentina

D. Boltovskoy (ed.), *Limnoperna fortunei,* Invading Nature - Springer Series in Invasion Ecology 10, DOI 10.1007/978-3-319-13494-9_5

Keywords *Limnoperna fortunei* · Golden mussel · Reproduction · Seasonal cycles · Cohorts

Introduction

The reproductive cycle of *Limnoperna fortunei* is probably one of the most intensively studied aspects of the species' biology. This interest is partly due to the fact that reproduction is perceived as one of the main biological traits of organisms in general and, again partly, because being an important fouling nuisance in industrial installations, strategies for its mitigation and control depend largely on the time frame over which such infrastructures are vulnerable to infestation. This occurs through the settlement of its planktonic larvae and therefore coincides with periods when the animal is reproductively active.

Seasonality in the reproduction of *L. fortunei* has been investigated using different approaches: (1) assessment of gonadal maturation, often with the aid of histological thin sections, (2) studies of temporal changes in the density of the species larvae in the plankton, (3) analyses of recruitments onto either natural or artificial substrata, and (4) various combinations of the above (Table 1).

B. Morton
School of Biological Sciences, The University of Hong Kong,
Hong Kong SAR, China
e-mail: prof_bmorton@hotmail.co.uk

N. Correa
Servicio de Hidrografía Naval (Ministerio de Defensa),
and Escuela de Ciencias del Mar (Instituto Universitario Naval),
Av. Montes de Oca 2124, 1271 Buenos Aires, Argentina
e-mail: ncorrea@hidro.gov.ar

D. Cataldo
e-mail: daniel@ege.fcen.uba.ar

C. Damborenea
División Zoología Invertebrados, Museo de La Plata,
Universidad Nacional de La Plata and CONICET, Paseo del Bosque s/n,
1900 La Plata, Argentina
e-mail: cdambor@fcnym.unlp.edu.ar

P. E. Penchaszadeh
Museo Argentino de Ciencias Naturales 'Bernardino Rivadavia'-CONICET,
Av. Angel Gallardo 470, 1405 Buenos Aires, Argentina
e-mail: pablop@retina.ar

F. Sylvester
Facultad de Ciencias Naturales,Instituto para el Estudio de la Biodiversidad de Invertebrados (IEBI), Universidad Nacional de Salta, Av. Bolivia 5150, 4400 Salta, Argentina
e-mail: fsylvester@natura.unsa.edu.ar

Table 1 Summary of surveys on reproductive seasonality in *Limnoperna fortunei*

Area	Waterbody	Latitude, Longitude	Months covered (period)	No. of samples or observations	Type of data	Reprod. peaks (months)	Source
Plover Cove (Hong Kong SAR, China)	Reservoir	22.47°N, 114.23°E	29 (Oct 1971 to Feb 1974)	28	R	Jan–Apr, Jun–Jul, Sep	Morton (1977)
Plover Cove (Hong Kong SAR, China)	Reservoir	22.47°N, 114.23°E	29 (Oct 1971 to Feb 1974)	28	G	Jan, Feb, Jun, Jul	Morton (1982)
Uji River (Japan)	River	34.93°N, 135.92°E	Sep 1994 to Oct 1995	12	G R	Jun–Sep	Iwasaki and Uryu (1998)
Lake Ohshio (Japan)	Reservoir	36.22°N, 138.88°E	13 (Aug 2007 to Nov 2008)	49	L	Jul–Sep	Nakano et al. (2010a)
Lake Takenuma (Japan)	Reservoir	36.23°N, 139.02°E	18 (Jun 2007 to Nov 2008)	49	L	Aug–Sep	Nakano et al. (2010a)
Yahagi River (Japan)	River	35.32°N, 137.14°E	22 (May 2007 to Feb 2010)	34	L	Aug	Hamada (2011)
Rio de la Plata (Argentina)	Estuary	34.87°S, 57.81°W	29 (Jul 1992 to Nov 1994)	20	G R	Apr–Jul, Oct	Darrigran et al. (1999)
Paraná de las Palmas (Argentina)	River	33.96°S, 59.21°W	17 (Sep 1997 to Jan 1999)	68	L R	Nov–Dec	Boltovskoy and Cataldo (1999), Cataldo and Boltovskoy (2000)
Rio de la Plata (Argentina)	Estuary	34.56°S, 58.41°W	15 (Oct 1997 to Dec 1998)	63	L	Oct–Feb	Cataldo and Boltovskoy (2000)
Rio de la Plata (Argentina)	Estuary	34.87°S, 57.81°W	13 (Mar 1994 to Mar 1995)	9	R	Mar, May, Jun, Nov	Maroñas et al. (2003)
Rio de la Plata (Argentina)	Estuary	34.87°S, 57.81°W	20 (Aug 1998 to Mar 2000)	16	G R	Jan, Jul, Aug, Nov	Darrigran et al. (2003)
Riacho Santa Fé (Argentina)	River	31.68°S, 60.72°W	24 (Jan 1998 to Dec 1999)	24	L	Nov–Apr	Cepero (2003)

Table 1 (continued)

Area	Waterbody	Latitude, Longitude	Months covered (period)	No. of samples or observations	Type of data	Reprod. peaks (months)	Source
Itaipú (Brazil-Paraguay)	Reservoir	25.41°S, 54.59°W	36 (Mar 2002 to Dec 2004)	34	L	Sep–Jan	Canzi et al. (2005)
Riacho Santa Fé (Argentina)	River	31.63°S, 60.67°W	13 (Dec 1999 to Dec 2000)	11	L	Sep–Mar	Ezcurra de Drago et al. (2006)
Carapachay River (Argentina)	River	33.40°S, 58.60°W	18 (Dec 2002 to Jun 2004)	12	L R	Oct–Apr	Sylvester (2006)
Yacyretá, Paraná River (Argentina–Paraguay)	Reservoir	27.48°S, 56.74°W	26 (Apr 1999 to May 2001)	27	L R	Nov–Feb	Darrigran et al. (2007)
Paraná River (Brazil)	River	34.57°S, 54.59°W	13 (Jan 2005 to Feb 2006)	12	L	Oct–May	Pestana et al. (2008)
Lago Guaiba (Brazil)	Lake	30.21°S, 51.20°W	16 (Sep 2002 to Dec 2003)	24	L R	Oct–Dec	Santos et al. (2008)
Embalse de Río Tercero (Argentina)	Reservoir	32.23°S, 64.44°W	13 (Mar 2005 to Mar 2006)	50	L	Dec–Apr	Boltovskoy et al. (2009b)
Itaipú (Brazil–Paraguay)	Reservoir	25.41°S, 54.59°W	13 (Mar 2002 to Apr 2003)	262	L	Nov–Mar	Boltovskoy et al. (2009b)
Colastiné River (Argentina)	River	31.66°S, 60.60°W	12 (Sep 2004 to Sep 2005)	34	L	Oct–Apr	Rojas Molina (2010)
Riacho Santa Fe (Argentina)	River	31.64°S, 60.69°W	12 (Sep 2004 to Sep 2005)	33	L	Oct–Apr	Rojas Molina (2010)
Río Santiago (Argentina)	River	34.85°S, 57.89°W	25 (Mar 2007 to Mar 2009)	14	L R	a	Bonel (2011), Bonel et al. (2013)
Coronda River (Argentina)	River	31.69°S, 60.74°W	22 (Apr 2007 to Feb 2009)	15	L R	a	Bonel (2011), Bonel et al. (2013)
Itaipú (Brazil–Paraguay)	Reservoir	25.41°S, 54.59°W	87 (Mar 2002 to Sep 2009)	87	L R	Oct–Mar	Mata (2011)

Table 1 (continued)

Area	Waterbody	Latitude, Longitude	Months covered (period)	No. of samples or observations	Type of data	Reprod. peaks (months)	Source
Paraguay River (Brazil)	River	19.60°S, 57.43°W	47 (Jan 2004 to Nov 2007)	47	L R	Oct–Feb	Oliveira et al. (2011)
Miranda River (Brazil)	River	19.57°S, 57.25°W	47 (Jan 2004 to Nov 2007)	47	L R	Aug–Sep	Oliveira et al. (2011)
Paraguay River (Brazil)	River	18.98°S, 57.70°W	47 (Jan 2004 to Nov 2007)	47	L R	Nov–Jan	Oliveira et al. (2011)
Salto Grande (Argentina–Uruguay)[b]	Reservoir	31.27°S, 57.94°W	104 (Jun 2004 to Jan 2013)	367	L	Oct–Apr	Boltovskoy et al. (2013)

Type of data, *G* gonadal maturation and/or thin sections, *L* abundance of larvae in the water column, *R* analyses of recruitment

[a] Too many gaps in coverage

[b] 2010 data only (other years have atypical reproduction trends due to extensive cyanobacterial blooms, see text)

Examination of the literature indicates that larval densities are generally well coupled with recruitment of settled juveniles, but gonadal maturation studies can yield somewhat dissimilar results. This outcome, which may at first seem puzzling, is understandable when one takes into account that the two parameters reflect different aspects of the same reproductive process (see below).

Reproductive Strategy: Evidence from Gonadal Cycles

The reproductive cycle of *L. fortunei* in Asia was described by Morton (1982), augmenting an earlier study (Morton 1977) of recruitment and population dynamics of the species in Plover Cove reservoir in Hong Kong at latitude 22°N. The reservoir was constructed by the damming of a coastal inlet, pumping out the seawater and allowing it to fill from the natural catchment and newly arriving water from China. The species was first recorded from Plover Cove reservoir in 1969, although it had probably arrived in Hong Kong's potable water supply around 1965–1966 (Morton 1977).

Analysis of histological sections of the gonads of *L. fortunei* from Plover Cove reservoir was carried out from October 1971 to February 1974 inclusive (Morton 1982). As with all mytiloids, because the intestine (other than the stomach) does not occur in the visceral mass, this plus the mantle tissues beneath the shell dorsal to their ventral margins (Morton 1973) are occupied wholly by the paired, left and right, maturing gonads (see Chapter "The Biology and Anatomy of *Limnoperna fortunei*, a Significant Freshwater Bio-Invader: Blueprints for Success", and Figs. 4, 15 therein). *L. fortunei* is dioecious; in Hong Kong, not one hermaphrodite was identified from the ~300 individuals encompassing a range of shell sizes in the 29-month histological study by Morton (1982). Later studies based on South American populations, however, identified low (<0.6%) proportions of simultaneous hermaphroditism (e.g., Darrigran et al. 1998, in the Río de la Plata estuary; Uliana and Callil 2006, in the Paraguay River).

When ripe, male and female gametes are discharged via left and right gonadal apertures situated at the surface of the visceral mass in the supra-branchial chamber of the outer demibranches of the ctenidia. They are then released via the exhalant siphon for external fertilization.

In Hong Kong, *L. fortunei* grows fast, reaching a shell length between 15 and 16 mm in about the first 5 months post recruitment and, by 10 months, individuals are certain to become sexually mature. In South America, sexual maturity was observed to vary seasonally beginning at 5–6 mm in winter–spring, and at 7–10 mm in autumn (Darrigran et al. 1999). In Hong Kong, during the first year of life, individuals are predominantly (>65%) females (Morton 1982), but in South America, a slight predominance of males has been observed (45% males to 40% females, the remainder being represented by sexually undifferentiated individuals; Darrigran et al. 1999).

The gametogenic cycle of *L. fortunei* in Hong Kong was interpreted from histological sections (Morton 1982) and categorized into five phases as follows (Fig. 1):

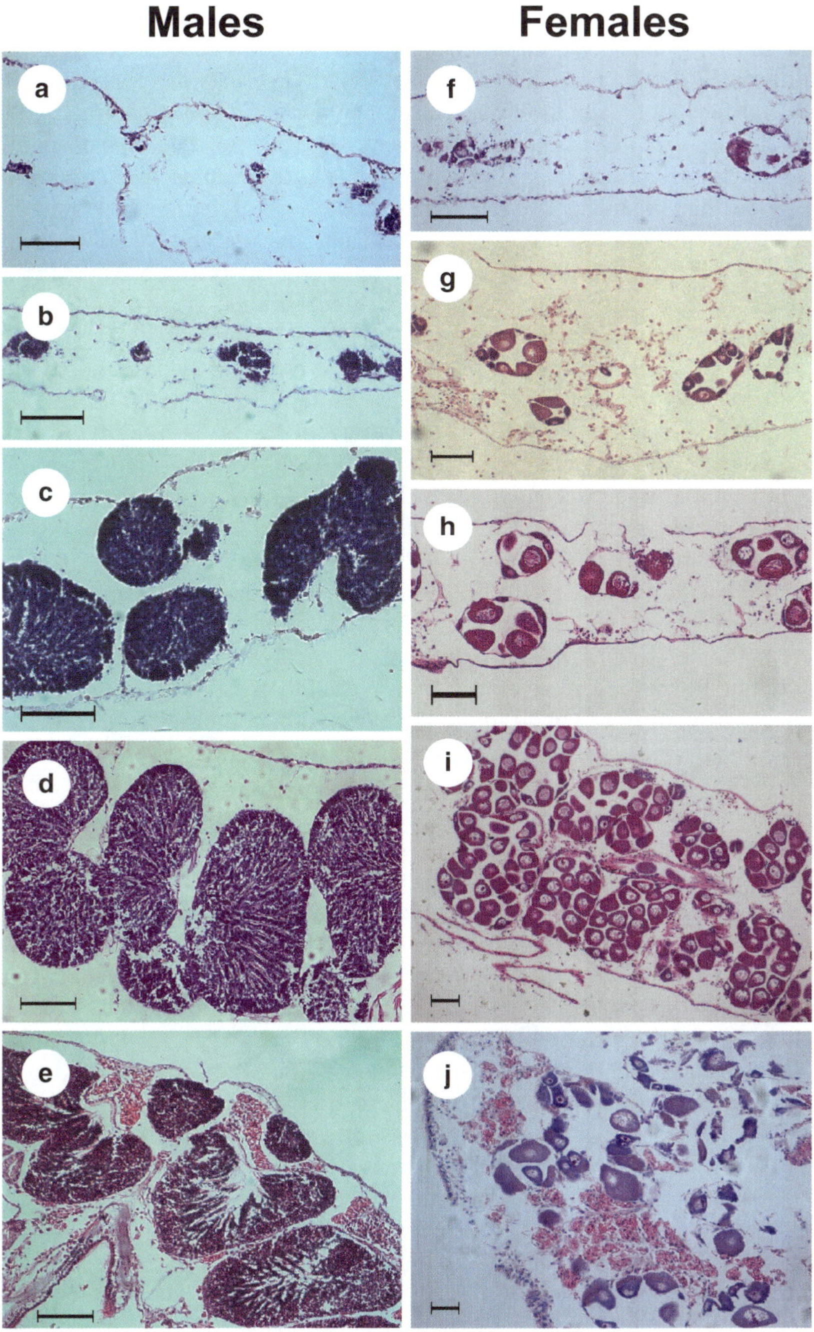

Fig. 1 Photomicrographs of the five stages of gametogenesis in males and females of *Limnoperna fortunei*. **a**, **f** Primordia; **b**, **g** developing; **c**, **h** maturing; **d**, **i** mature; **e**, **j** partially spent with resorption of remaining undischarged gametes and simultaneously new growing ova attached to the epithelial wall of the ovarian follicle **(j)**. Scale bars: 100 μm

(1) inactive, (2) developing, (3) maturing, (4) mature, and (5) spent. In a similar study based on Brazilian populations, four gametogenic stages for both sexes were defined (Callil et al. 2012). In Hong Kong, during January–February and July–August, both male and female gonads were inactive and comprised small, primordial, seminiferous tubules and ovarian follicles, respectively (Fig. 1). The tubules comprised a thin germinal epithelium producing round, lightly staining, primary spermatocytes ~5 μm in diameter. The follicles are comprised of germinal epithelium from which small oocytes, each with a clear, vesicular nucleus and a distinct nucleolus, were developing. In March and September, the seminiferous tubules had enlarged and comprised primary and more darkly staining secondary spermatocytes 2.5 μm in diameter with a few spermatids between 1.5 and 2.0 μm in diameter occurring in the follicle lumina (Fig. 1). At this time too, ovarian follicles still possessed narrow lumina, although the epithelial walls of these were producing broadly stalked oogonia with a maximum diameter of 10 μm (Fig. 1). By April–May and October, gametogenesis was progressing and the seminiferous tubules and ovarian follicles had increased in size substantially (Fig. 1). In the former, there were considerably more spermatids being produced and a few spermatozoa. In the latter, the oogonia had attained an average diameter of 30 μm but were still attached to the epithelial wall by a narrower, more pinched stalk. The testes and ovaries were considered to be mature from May to June and November to December. At these times, the seminiferous tubules comprised many primary and secondary spermatocytes and spermatids, but the central lumina of each tubule was full of slipper-shaped spermatozoa arranged in radial chords and forming lamellae with their heads projecting towards the surrounding and the still active germinal epithelium (Fig. 1). Coincidentally, the follicles were packed with, now detached and rounded, ova some 60 μm in diameter and each with a darkly staining nucleus10 μm in diameter (Fig. 1). These mature ova possessed only a little yolk and can thus be classified as oligolecithal. By December–January and July, gonads were spent (Fig. 1). They comprised empty tubules and follicles undergoing a generalized pattern of size reduction and resorption of the remaining gametogenic primordia and un-discharged gametes. Knowing, however, that another cycle of gametogenesis will soon begin in each adult surviving into a new life-cycle phase or maturing juvenile, it is clear that the spent phase lasts but for a short period.

The gonadal cycle exhibited by *L. fortunei* in Hong Kong comprised two phases of reproductive activity each year. During January–February and July–August, the gonads were in a state of regression. These months coincided with the coldest (16–17 °C, winter) and hottest (27–28 °C, summer) water temperatures in the reservoir. As temperatures subsequently warmed and cooled, respectively, gametogenesis proceeded, ultimately resulting in gonadal maturity from May to June and November to December. These two phases were followed by gamete release.

In contrast to this pattern, Choi and Shin (1985) estimated that, in South Korea, at a latitude of ~38°N, *L. fortunei* showed a unimodal pattern of reproduction over the course of a year with sperm and ova being released in a single event between July and August for a period of only about 15–20 days. This is in contrast to the situation in subtropical Hong Kong where, as identified by Morton (1977, 1982), two cycles of gametogenic activity were recorded each year, although recruitment, albeit with summer and winter peaks, occurred for 10 months of the year. Also, South American subtropical and tropical populations have been observed to recruit for up to >10 months of the year (Fig. 2; Boltovskoy et al. 2009b). In tropical Asia, it is possible that *L. fortunei* reproduces year round as well, although this has never been studied.

Working on populations from the Río de la Plata estuary shortly after the species was first discovered in the area, Darrigran et al. (1999) recorded mature sperm year round and five major spawning events between July 1992 and November 1994: in September to October 1992, December 1992 to January 1993, May to July 1993, April to June 1994, and October to November 1994. Subsequent results of Darrigran et al. (2003) and Damborenea and Penchaszadeh (2006) indicated that, in the Río de la Plata estuary, *L. fortunei* produces sperm and oocytes continuously throughout the year, with spawning peaks in spring (September–November) and at the end of summer (February–March).

In Brazil, diverse situations have been described, from uninterrupted breeding throughout the entire year to punctuated breeding with interruptions and more or less extended resting periods (Table 1).

Reproductive Strategy: Evidence from Temporal Series of Larval Densities

Studies of temporal changes in the densities of larvae in the water column spanning at least 1 year have been carried out repeatedly, both in Japan and in South America (Table 1). A few studies extended over more than 1 year, and at least two surveys monitored larval abundances over periods of 7 (Mata 2011) and >9 years (Boltovskoy et al. 2013; Fig. 2).

In South America, the months of highest larval output vary from site to site and from year to year (Fig. 2), but the most usual pattern involves a spring–early summer peak around November–December, and a second, less pronounced, peak in February–March (Fig. 3), seasonally similar to the situation in Hong Kong. Months barren of larvae are either few or absent. Around 80% of the temporal series performed (Table 1) show that there are at least ten larvae per cubic meter of water for 11–12 months of the year. The remaining 20% of the studies identified larvae in 7–10 months of the year.

In Japan, larval production is also associated with the summer, but the reproductive period is considerably shorter. In the reservoirs and rivers investigated, larvae were absent from the water column for 7–8 months of the year, and peak reproduction was clearly concentrated within 1–2 months (Magara et al. 2001; Nakano et al. 2010a; Hamada 2011; Figs. 2 and 3). In the Uji River (Japan), *L. fortunei* produced recruits during 4 months of the year (June–September; Iwasaki and Uryu 1998).

The highest recorded larval densities of *L. fortunei* are around 100,000 ind./m^3 (Darrigran et al. 2007; Nakano et al. 2010a), but such values are exceptional. During months of peak reproduction (November through March in the southern

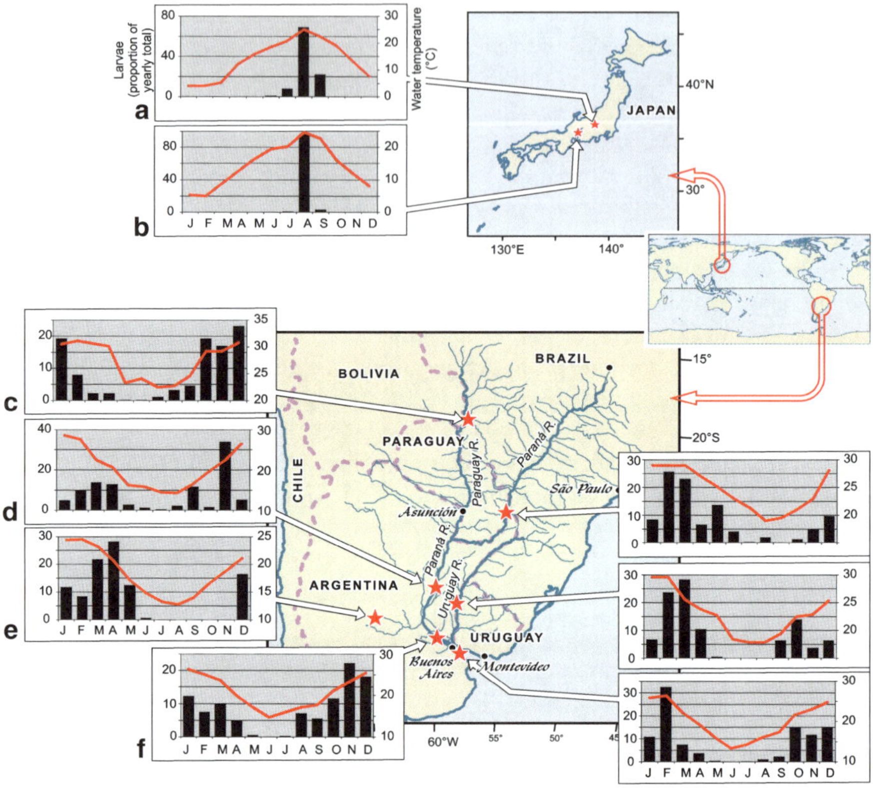

Fig. 2 Reproductive cycles of *Limnoperna fortunei* in different waterbodies, as shown by the seasonal occurrence of its larvae in the plankton. *Bars* denote proportions of the yearly total recorded each calendar month (January to December, left *y*-scale; some values are interpolated; overlapping periods for consecutive years are averaged). *Red line* indicates mean surface water temperature (right *y*-scale). **a** Yahagi River (35.32°N, 137.14°E), 34 samples collected between May 2007 and February 2010 (based on data from Hamada 2011). **b** Lake Ohshio (36.22°N, 138.88°E), 49 samples collected between August 2007 and November 2008 (based on data from Nakano et al. 2010a, data courtesy of Daisuke Nakano). **c** Paraguay River (19.60°S, 57.43°W), 47 samples collected between January 2004 and November 2007 (based on data from Oliveira et al. 2011). **d** Riacho Santa Fe (tributary of the Middle Paraná River) (31.64°S, 60.68°W), 33 samples collected between September 2004 and September 2005 (based on data from Rojas Molina 2010). **e** Embalse de Río Tercero (32.23°S, 64.44°W), 40 samples collected between March 2005 and March 2006 (based on data from Boltovskoy et al. 2009b). **f** Lower Paraná delta (33.96°S, 59.21°W), 68 samples collected between September 1997 and January 1999 (based on data from Boltovskoy and Cataldo 1999). **g** Itaipú reservoir (Upper Paraná River) (25.41°S, 54.59°W), 12 samples collected between January 2005 and December 2005 (based on data from Mata 2011). **h** Salto Grande reservoir (Uruguay River) (31.27°S, 57.94°W), 37 samples collected between January 2010 and December 2010 (based on data from Boltovskoy et al. 2013). **i** Río de la Plata estuary (34.56°S, 58.41°W), 63 samples collected between October 1997 and December 1998 (based on data from Cataldo and Boltovskoy 2000)

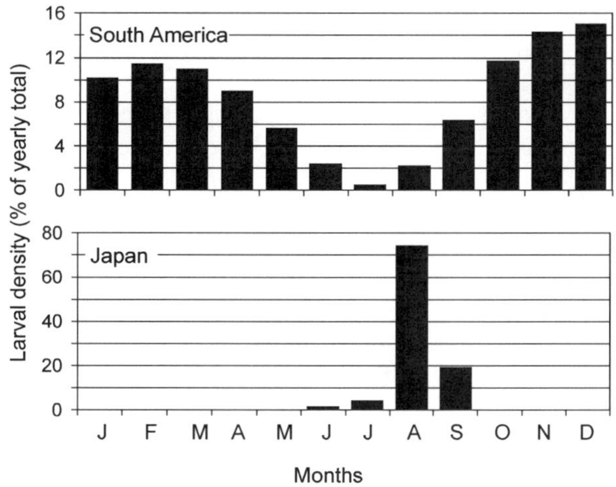

Fig. 3 Seasonal occurrence of *Limnoperna fortunei* larvae in the water column in South America (mean values based on 25 surveys) and in Japan (mean values based on three surveys)

hemisphere), larval densities normally range between 6000 and 7000 ind./m^3, whereas in June–September, they decline to <1000 ind./m^3. Data are more scarce for Japan, but summer peak larval densities are probably higher than those recorded in South America (e.g., around 20,000–35,000 ind./m^3; Magara et al. 2001; Nakano et al. 2010a), but with no larvae for most of the remaining months in the year.

The larval densities described above are considerably lower than the maxima recorded for *Dreissena polymorpha,* which can exceed 500,000 ind./m^3 (Garton and Haag 1993; Smit et al. 1993), suggesting that the zebra mussel has a higher instantaneous fecundity. This does not, however, necessarily imply a higher overall fecundity, because the reproductive period of *D. polymorpha* is often considerably shorter (Karatayev et al. 2007; Boltovskoy et al. 2009b; see Chapter "Parallels and Contrasts Between *Limnoperna fortunei* and Species of *Dreissena*" in this volume). Adult population densities, on the other hand, seem either similar or slightly higher for *L. fortunei* (Karatayev et al. 2010).

Differences Between Results: Gonadal Maturation Versus Larval Cycles

Different methods (larval abundance trends vs. gonadal maturation) have produced different results concerning the timing of the periods of highest reproductive output: Larval cycles point to an almost continuous spring-to-autumn reproduction with two more or less conspicuous peaks around December and February–March (in South America), or a single peak in the middle of summer (Japan, Korea; Fig. 3). Gonadal studies, in turn, identify two or more discrete pulses, sometimes even in the middle of winter (Morton 1982; Darrigran et al. 1999, 2003). These differences are most probably due to the fact that the two types of observations reveal different aspects of the same process. That is, larval cycles integrate the reproductive output of extensive mussel beds dispersed over large areas, whereas gonadal data pinpoint with high precision the ripening and spawning of more or less isolated individuals or spatially restricted clusters.

Larvae of *L. fortunei* collected at any one location are produced by adults dispersed over large areas upstream from the collection site. The time required for a fertilized egg to reach the pediveliger stage, when it is ready to settle, is ca. 20 days at 20 °C (Cataldo et al. 2005; see Chapter "Larval Development of *Limnoperna fortunei*" in this volume) and probably longer at lower temperatures (*L. fortunei* reproduces at water temperatures as low as 15–17 °C; Morton 1977; Cataldo and Boltovskoy 2000; Brugnoli et al. 2011). At current velocities of ~3–5 km/h, not uncommon in many rivers, larvae caught at any one site could thus have originated from as far as 2000 km upstream. A plankton sample therefore typically integrates the reproductive output of hundreds to millions of *L. fortunei* beds located upstream from the sampling site.

Gonadal maturation, on the other hand, is based on the analysis of a few individuals collected over a small area, usually the same throughout the entire cycle analyzed. As suggested for other bivalves (Garton and Haag 1993), while restricted beds of *L. fortunei* may spawn synchronously, when larger areas are considered, the lack of reproductive synchronicity between populations results in a continuous breeding season and discrete spatfalls are not recognizable in the plankton (Boltovskoy et al. 2009b). As a consequence, both punctuated and continuous reproduction modes have been reported by the same authors when using histological data and larval counts in plankton samples, respectively (Darrigran et al. 1999, 2002).

While inducing *L. fortunei* to spawn using artificial stimuli in laboratory conditions is fairly easy (Cataldo et al. 2005), detecting the timing, frequency, and intensity of spawning events of restricted beds in the field poses serious difficulties and has not been achieved hitherto. With the aid of an experimental device that mimicked natural conditions, Cataldo and Boltovskoy (unpublished) sampled the prefiltered water, bathing a cluster of ca. 2000 adult individuals in order to detect spawning. Although the experiment was carried out at a time of peak reproduction, after a week of uninterrupted filtration, no ova were observed. Admittedly, the observational period was limited, but the result seems to support the notion that reproduction of discrete clusters is not continuous, as suggested by the presence of larvae in the plankton, but punctuated, as deduced from gonadal maturation studies. Given the clear advantages of epidemic over asynchronous spawning (particularly for a species whose distribution encompasses many thousands of kilometers and, due to its requirement for hard substrata, is extremely patchy), it is hard to envision how *L. fortunei* manages to produce such impressive numbers of offspring if it lacks mechanisms for synchronizing the liberation of its gametes. Although epidemic spawning is characteristic of marine invertebrates, and is less common in freshwater organisms (Olive 2002), *L. fortunei* is a member of a typically marine bivalve superfamily (Mytiloidea) and, as such, shares many traits with most of its marine relatives, notably a rather extended free-swimming larval stage.

It is concluded that although both of these methods to identify spawning (gonadal thin sections and larval counts) yield useful data, the information conveyed by them is dissimilar. Studies based on the assessment of gonadal maturation focus on the reproductive behavior of individuals and, by extension, of the species. They convey precise information on the number and timing of reproductive events in any one breeding season and are, probably, the most promising paths for studies of fecundity. Larval counts, on the other hand, are more meaningful in terms of the overall ecology of *L. fortunei* and of the biota that interacts with it. For fouling control studies, the information conveyed by larval counts is more useful as it better allows the pinpointing of periods when industrial installations are particularly vulnerable to fouling by this species.

Triggers of Reproduction

"What triggers spawning?" is a key question associated with surveys of the reproductive behavior of *L. fortunei.* The variable that has been recurrently observed to covary with reproduction is water temperature (Boltovskoy et al. 2009b; Fig. 2). Several authors have suggested that 15–18 °C is the threshold value below which production of larvae is near zero (Morton 1977; Choi and Shin 1985; Cataldo and Boltovskoy 2000; Nakano et al. 2010a; Brugnoli et al. 2011), but this threshold is obviously not applicable to tropical waterbodies where water temperatures never fall below these values. Even in these locations, however, there usually is a well-defined period of reproductive relaxation, which invariably occurs during the coolest months (Canzi et al. 2005; Boltovskoy et al. 2009b; Mata 2011; Oliveira et al. 2011; Fig. 2c and g). For Salto Grande reservoir (Argentina–Uruguay), where surface water temperatures vary from 14 to 32 °C, densities of *L. fortunei* larvae have been estimated weekly between June 2004 and 2014. Lowest larval densities occurred between ~1 June and 15 August, at temperatures around 14–24 °C (mean: 18 °C). However, of the 87 plankton samples collected between 1 June and 15 August (2004 through 2013), only ten contained no larvae at all (i.e., <0.5 larvae/m^3). Boltovskoy et al. (2009b) suggested that, regardless of ambient temperature, *L. fortunei* needs a resting period in order to resume gametogenesis, as has been reported for other invertebrate species (McMahon and Bogan 2001). Interestingly, this behavior differs from that of *D. polymorpha* in artificially heated reservoirs and laboratory conditions at temperatures permanently >12 °C, where the species has been reported to produce larvae year round (Stanczykowska 1977; Lewandowski 1982b; Nichols 1993). Nevertheless, in the temperate and warm waterbodies of southern North America, spawning does not begin until water temperatures rise to >20 °C (Nichols 1996). In southern England, larval settlement of *D. polymorpha* begins at a temperature of ~15 °C (Morton 1969).

While there is little doubt that temperature plays a key role in the reproductive cyclicity of *L. fortunei,* there most probably are other factors that also influence both the timing and the intensity of spawning events. Their association with reproduction, however, is less clear and mostly based on circumstantial evidence (unless they range outside of the normal levels of tolerance of the species, in which case they affect all vital traits, including reproduction; Nakano et al. 2010a). In his studies of Plover Cove reservoir, Morton (1982) remarked that pH, salinity, and dissolved oxygen may be associated with gametogenesis and spawning. His results suggested that spawning takes place at times of high temperatures and low dissolved oxygen levels in summer and at low temperatures with high dissolved oxygen levels in winter. Oliveira et al. (2011) argued that, in addition to temperature, dissolved oxygen, and calcium levels, pH, water velocity, suspended solids, and chlorophyll *a* concentrations may be instrumental in modulating larval numbers in the upper Paraguay River. Changing conditions associated with flood–drought cycles characteristic of many large South American rivers have also been suggested as important factors involved in the reproduction of *L. fortunei,* particularly in areas where temperatures are high year around (Callil et al. 2012).

The availability of food, which was found to be important in determining the frequency of reproduction in *D. polymorpha* (Gist et al. 1997), may also play a role in the extension of the reproductive period of *L. fortunei*. Boltovskoy et al. (2009b) noted that, in practically all the South American waterbodies surveyed, the winter phase of reproductive relaxation spanned 1.8–3 months, with the exception of the reservoir Embalse de Río Tercero, where larvae were virtually absent from the waterbody for almost half the year (Fig. 2e). Water temperatures in this reservoir were not significantly different from those of other sites, suggesting that this factor is unlikely to account for such a reproductive difference. On the other hand, suspended particulate organic matter in Embalse de Río Tercero (around 0.5–1 mg/L) was much less abundant than in the other waterbodies (around 8 mg/L), which may point to food supply as a significant factor that affects reproductive output.

Vertical Distribution of Larvae

A few studies have investigated the vertical distribution of *L. fortunei* larvae in the water column. In rivers, turbulent flow normally precludes the development of vertical stratification and physical, chemical, and biological properties, including *L. fortunei* larvae and other plankton, are evenly distributed from top to bottom (Fig. 4). In lentic environments, on the other hand, contrasting situations have been described. In Palmar reservoir (Uruguay), Brugnoli et al. (2011) did not identify significant differences in the concentrations of *L. fortunei* veligers at different depths. In contrast, in Lake Ohshio (Japan), veligers were more abundant near the bottom (ca. 18 m) and at mid-depths than in the uppermost 2-m layer (Nakano et al. 2010a). Differential colonization rates at different depths of the same waterbody could also point to differences in preferred larval depth (Morton 1977), but interpretation of these patterns is complicated by possible dissimilarities in the survival of recruits, including the effects of predation (Nakano et al. 2010b). In any case, it would be reasonable to expect that at least older larvae, that is, those nearing the settling stage, are more abundant closer to their final destination—the epibenthos—than elsewhere in the water column (as observed in laboratory conditions, see Chapter "Larval Development of *Limnoperna fortunei*" in this volume).

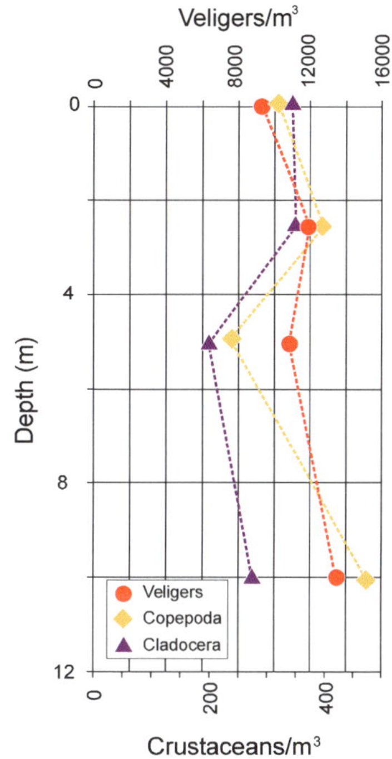

Fig. 4 Densities of *Limnoperna fortunei* larvae, copepods, and cladocerans at different depths in the Paraná de la Palmas River (Lower Paraná River delta; 34.29°S, 58.56°W) recorded on 3 March 2003 (unpublished authors' data)

Factors Affecting Larval Densities

In addition to seasonality, several other factors can modulate larval densities in the water column. Some of these have been investigated, including distance from suitable substrata, time after colonization, the effects of cyanobacterial blooms, and sample-to-sample variability.

Distance from Suitable Substrata

Although adult *L. fortunei* can use a variety of substrata for attachment (see Chapter "*Limnoperna fortunei* Colonies: Structure, Distribution and Dynamics" in this volume), hard, immobile surfaces, like stone and wood, are normally favored. Such substrata, especially immobile boulders, are often scarce in natural lotic and lentic systems. In South America, in particular, the main waterbodies colonized by *L. fortunei* are large, lowland, riverine systems, whose streambeds are characterized by fine, unconsolidated sediments, unfit for colonization by *L. fortunei*. Over large river stretches, therefore, the only hard substrata available are man-made structures associated with populated areas, such as piers, groynes, pilings, breakwaters, and revetments. Larval densities may thus be higher in the vicinity of these structures than farther away. Figure 5 shows the results of a transect in the Río de la Plata estuary, off Buenos Aires,

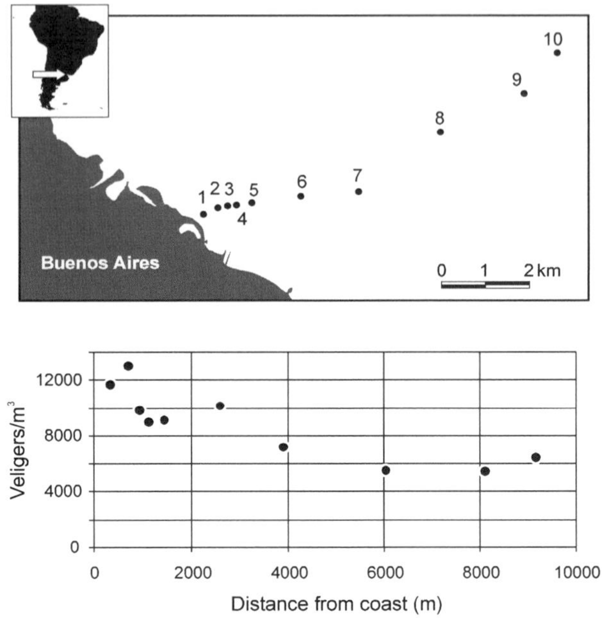

Fig. 5 Larval densities recorded along a transect perpendicular to a coast modified by a concrete coastal revetment densely populated by adult mussels off Buenos Aires, on 11 December 2002. (unpublished authors' data)

roughly perpendicular to a coastal sector modified entirely by a concrete revetment. Larval numbers fell from around 12,000 ind./m^3 next to the man-made concrete wall (densely populated by adult individuals) to 6000 ind./m^3 at 6–10 km from this substratum (the bottom in this area is silt-mud). While this difference is well within the range of the sample-to-sample variability (Fig. 6), the decreasing trend is consistent, suggesting that distance from the coast plays a significant role in influencing population numbers. This effect is probably of little importance when analyzing seasonal trends in larval abundance at a fixed site, but site-to-site comparisons may be affected significantly.

Time After Colonization

Larval concentrations also depend strongly on the time elapsed after initial colonization of the waterbody. Data from Itaipú reservoir, which was colonized by *L. fortunei* around 2000 (Zanella and Marenda 2002), indicate that mean annual larval densities were only 106 ind./m^3 in 2002, 512 ind./m^3 in 2003, and 2000–2500 ind./m^3 in 2004–2005, suggesting that the spread of beds upstream from the sampling site was responsible for this increase (Canzi et al. 2005). In 2006, larval densities fell to ca. 1400 ind./m^3 and varied thereafter at around 1000–1500 ind./m^3 until 2008 (Mata 2011; Fig. 7). This decline and subsequent plateau could point to stabilization of the adult population densities in the reservoir, although the time elapsed since

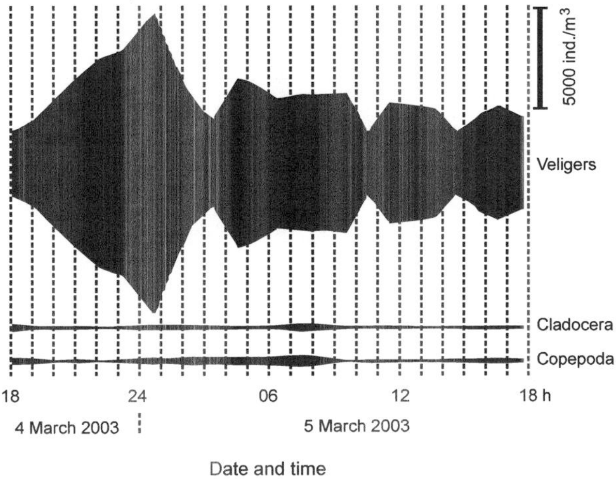

Fig. 6 Densities of *Limnoperna fortunei* larvae, copepods, and cladocerans recorded throughout a 24-h period in Honda River (Lower Paraná River delta; 34.32°S, 58.53°W) on 4–5 March 2003 (unpublished authors' data)

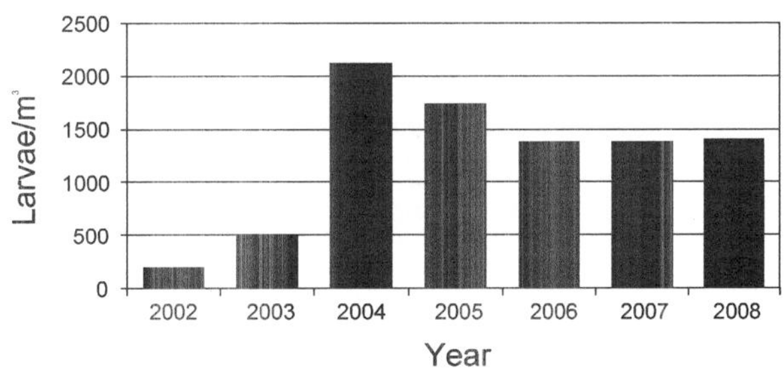

Fig. 7 Mean densities of *Limnoperna fortunei* larvae in Itaipú reservoir between 2002 and 2008 (based on data from Canzi et al. 2005; Pestana et al. 2008; Boltovskoy et al. 2009b; Mata 2011)

initial colonization appears short for this process. Populations of invasive bivalves, including *D. polymorpha,* have been suggested to follow a so-called boom–bust cycle (Stanczykowska 1977), stabilizing at around 7–12 years post introduction (Burlakova et al. 2006; Boltovskoy et al. 2009a, 2009b). Nevertheless, several other population modes have also been described, some of them involving large cyclical or irregular fluctuations (Strayer and Malcom 2006).

Although the scarce evidence available suggests that larval densities increase in response to adult population growth, no direct proof of a significant correlation between the numbers of adults and those of their planktonic larvae has yet been provided. This relationship may not be as trivial as it seems. Several surveys of *D. polymorpha* have failed to establish a clear relationship between adult population size and larval density (Stanczykowska 1977; Garton and Haag 1993; Nalepa et al. 1995), with some authors suggesting that larval numbers in the plankton may be governed by other factors, such as the age structure of the adult population (Lewandowski 1982a) and water levels (Smirnova et al. 1993).

The Effects of Cyanobacterial Blooms

Another factor that has been shown to impact reproductive periods and larval numbers significantly is the development of cyanobacterial blooms. On the basis of 9 years of observational data in Salto Grande reservoir (on the Uruguay River, Argentina–Uruguay), Boltovskoy et al. (2013) showed that reproduction by *L. fortunei* was interrupted during dry summers (January–April), coinciding with periods of peak *Microcystis* spp. growth and low water discharge levels— which favor the buildup of algal biomass. Conversely, wet summers with high discharge rates were characterized by low densities of *Microcystis* spp. and high numbers of *L. fortunei* larvae in the water column (Fig. 8). Laboratory experiments showed that microcystin-LR (produced by toxic strains of several Cyanobacteria) is highly toxic to *L. fortunei* larvae, eliminating 58–100 % of individuals in 48 h at 10–20 µg/L (Boltovskoy et al. 2013). Cyanobacterial blooms in this reservoir are particularly strong in late summer-autumn, often suppressing the December/January to April reproductive pulse completely (Fig. 9). While associations between microcystin and larval mortality are beyond doubt, the toxin may also affect other aspects of *L. fortunei* reproduction, such as gamete production and survival, fertilization, and hatching. Suppression of reproduction due to high microcystin concentrations is most likely in subtropical and tropical lakes and reservoirs (Soares et al. 2013), especially when

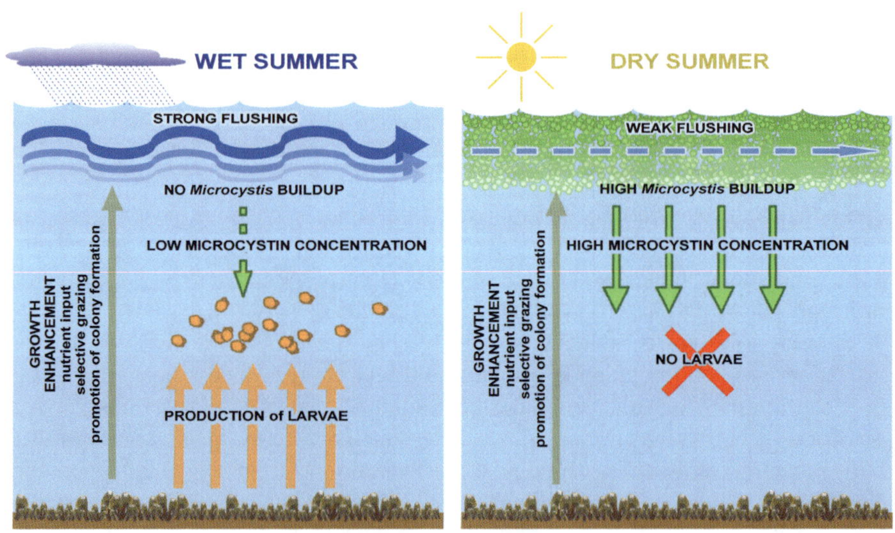

Fig. 8 Schematic diagram of cause–effect relationships between *Microcystis* spp. blooms and *Limnoperna fortunei* reproduction in Salto Grande reservoir (Argentina–Uruguay). During rainy summers, high discharge rates preclude *Microcystis* spp. buildup and microcystin levels are low in the water, allowing for normal production of larvae. During dry summers, weak flushing and strong thermal vertical stratification favor *Microcystis* spp. growth and high microcystin concentrations in the water column, which kill *L. fortunei* larvae precluding reproduction. (Modified after Boltovskoy et al. 2013, with permission from John Wiley & Sons)

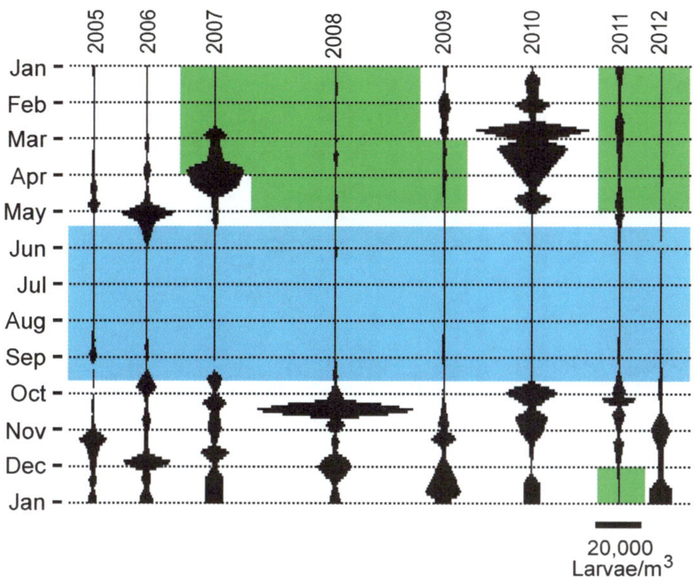

Fig. 9 Densities of *L. fortunei* larvae in Salto Grande reservoir between 2005 and 2012. *Green shading* denotes periods of extensive *Microcystis* spp. blooms (no data for *Microcystis* are available in 2005 and 2006); *light blue* shading denotes water temperatures below 21 °C (mean for 2005–2012; based on data from Boltovskoy et al. 2013)

their drainage areas include agricultural land yielding high nutrient loads. Rivers, on the other hand, rarely host strong blooms, and would thus be less likely to show atypical larval abundance cycles.

In addition to the above, several other factors have been reported to affect the reproductive output of *L. fortunei*. Trematode parasites can inhibit gamete production (Tanaka et al. 2004; see Chapter "Parasites of *Limnoperna fortunei*" in this volume). Stress-inducing factors, like pollution and low dissolved oxygen levels, can impact many of the species' vital activities and processes, including reproduction (e.g., Morton 1977; Nakano et al. 2010a; Bonel 2011; Oliveira et al. 2011; Bonel et al. 2013).

Sample-to-Sample Variability

Agreement between surveys indicates that the overall trends in the larval abundance profiles of *L. fortunei* do reflect actual density variations throughout the year but, as with most planktonic abundance estimates, variability is high. The profiles shown in Fig. 6 illustrate abundance estimates for veligers, copepods, and cladocerans obtained at hourly intervals on 4 and 5 March 2003 in the Lower Paraná River delta. Numbers of *L. fortunei* larvae varied between ca. 2000 and almost 16,000 ind./m^3, or slightly less than one order of magnitude. This variability does not seem to reflect the result of reproductive pulses from different *L. fortunei* beds distributed upstream from the sampling site, because copepods and cladocerans showed a similar range of variation. All these samples were collected using a submersible centrifugal pump (rather than towed flow-metered net), thus ensuring precise estimates of the volume of water filtered and precluding any possibility of net clogging. Furthermore, all counts were conducted twice in order to minimize methodological bias. These results underscore the need for intensive, densely packed (in space and/or in time) sampling programs when investigating reproductive output trends.

Concluding Remarks

Comparison of annual recruitment cycles from several climatic zones leaves little doubt that temperature has an overwhelming importance on the reproductive process. Remarkably, summer maxima differ moderately between the areas where the reproductive cycles of *L. fortunei* have been investigated, but winter minima are quite dissimilar. Lowest values are those where winter minima are around 0-4 °C (Choi and Shin 1985; Nakano et al. 2010a; Table 2). Here, >90 % of the reproductive output is concentrated within 1–2 summer months (July and August); there are no larvae in the water column for the rest of the year. In records from the Uji River (Japan), at somewhat higher water temperatures (ca. 13–26 °C), *L. fortunei* produced recruits during 4 months of the year (June–September) (Iwasaki and Uryu 1998). In the southernmost South American localities investigated (Lower Paraná River delta, Río de la Plata estuary), where lowest water temperatures are ~ 10 °C, larvae are produced for around 5–7 months of the year, but there often are two more or less well-defined peaks—a major one shortly after the winter trough, in November or December, and a second, usually smaller one, around March. Further north in the Río de la Plata watershed (Upper Paraná River, Paraguay River), winter temperature minima are higher (13–14 °C) and the corresponding reproductive relaxation period is shorter. Finally, in subtropical Asia (Hong Kong), where water temperatures vary seasonally between 15 and 32 °C, gonadal activity is present throughout most of the year and, in addition to the summer peak, there also is a peak in February (northern hemisphere winter; Morton 1977). These differences in reproductive activity are summarized in Fig. 10 and show that reproduction is confined to a single summer peak in the northern temperate zone (Korea, Japan). Most interestingly, however, the data show that in the northern (Hong Kong) and southern (South American) subtropics, there are roughly two reproductive peaks, but both occurring in their respective (and opposite) early spring and late summer seasons.

Contrasts between Korea and Japan, on the one hand, and South America, on the other, are very sharp (Fig. 2), but trends anticipated from the corresponding thermal regimes are not always as clear-cut. For example, water temperatures of the Yahagi River (Japan, 10–26 °C) are similar to those of the Middle Paraná River (Argentina, 10–29 °C), yet larval production is restricted to 1 month in the former, while

Table 2 Temporal concentration of the reproductive activity of *Limnoperna fortunei* in various Asian and South American waterbodies characterized by different water temperatures, as indicated by the number of months needed to account for ≥90 % of the annual output of larvae when monthly values are sorted in descending order (based on the "production half-time" concept of Berger and Wefer 1990)

Site	Latitude	Lowest water temp. (°C)	Highest water temp. (°C)	Month(s) of the year when % of annual recruitment is > 20 % of the annual total	≥90 % of the yearly larval output occurs in (months)	References
Plover Cove reservoir (Hong Kong)[a]	22°N	15	32	Feb, Jul	9	Morton (1977)
Yahagi River (Japan)	35°N	10	26	Aug	1	Hamada (2011)
Lakes Ohshio and Takenuma (Japan)	36°N	4	27.5	Aug	2	Nakano et al. (2010a)
Paldang reservoir (South Korea)	37°N	2	29	Aug	2	Choi and Shin (1985)
Paraguay River (Brazil)	19°S	22.5	31.2	Jan, Nov, Dec	6–9	Oliveira et al. (2011)
Miranda River (Brazil)	19°S	22.5	31.2	Aug, Sep	6	Oliveira et al. (2011)
Upper Paraná River (Brazil–Paraguay)	25°S	18	29	Feb, Mar, Apr, May, Oct, Nov, Dec	4–8	Canzi et al. (2005), Pestana et al. (2008), Boltovskoy et al. (2009b), Mata (2011)
Upper Paraná River, Yacyretá reservoir (Argentina–Paraguay)	27°S	15	32	Feb, Dec	7	Darrigran et al. (2007)
Middle Paraná River (Argentina)	30°S	10	29	Mar, Oct, Nov, Dec	6–8	Cepero (2003), Ezcurra de Drago et al. (2006), Rojas Molina (2010), Bonel (2011)
Guaíba Lake (Brazil)	30°S	12	27.5	Oct	8	Santos et al. (2008)
Uruguay River, Salto Grande reservoir (Argentina–Uruguay)[b]	31°S	18	30	Feb, Mar	7	Boltovskoy et al. (2013)

Table 2 (continued)

Site	Latitude	Lowest water temp. (°C)	Highest water temp. (°C)	Month(s) of the year when % of annual recruitment is > 20 % of the annual total	≥90 % of the yearly larval output occurs in (months)	References
Embalse de Río Tercero reservoir (Argentina)	32°S	12.8	24.5	Mar, Apr	5	Boltovskoy et al. (2009b)
Río Negro River, Palmar reservoir (Uruguay)	33°S	12.5	27	Apr	8	Brugnoli et al. (2011)
Lower Paraná River delta (Argentina)	34°S	13.3	26.4	Nov, Dec	8	Boltovskoy and Cataldo (1999), Sylvester (2006)
Río de la Plata estuary (Argentina)	34°S	13.3	27.6	Feb	6–8	Cataldo and Boltovskoy (2000), Bonel (2011)

[a] Estimated recruitment values
[b] 2010 data only (other years have atypical reproduction trends due to extensive cyanobacterial blooms, see text)

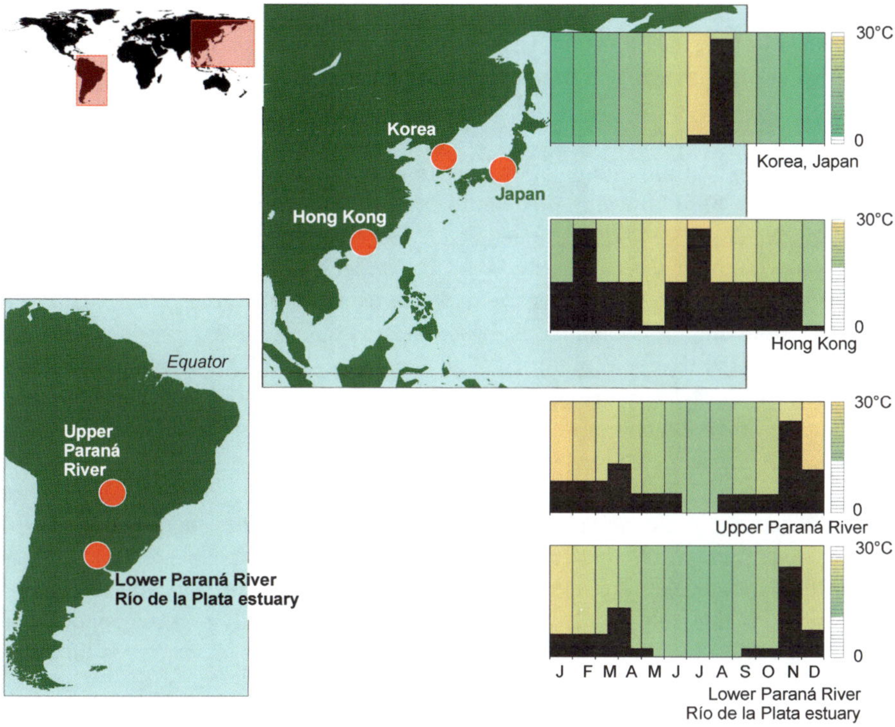

Fig. 10 Proposed relationship between latitude, water temperature, and reproductive activity of *Limnoperna fortunei*. Histograms indicate relative strength of recruitment throughout the year. *Bars* next to histogram denote water temperature (see text for explanation)

it extends more than 6–8 months in the latter (Table 2). At similar thermal regimes (around 13–26 °C), larvae are present during only 4 months in the Uji River (Japan; Iwasaki and Uryu 1998), and up to 8 months in the Río de la Plata estuary (Table 2). Departures from this temperature-dependent reproduction mode are also evident when comparing several South American sites. For example, in the reservoir Embalse de Río Tercero (Argentina), the reproductively active period is at least 3–4 months shorter than expected from its water temperature span (Table 2). Boltovskoy et al. (2009b) suggested that a lower food supply is possibly responsible for the shorter reproductive period at this waterbody. In the Miranda River (tributary of the Paraguay River), where water temperatures are always above 20 °C, larvae are absent for half of the year (Table 5.1.2), which may be a response to extended anoxic events and/or low calcium concentrations (Oliveira et al. 2011). In Salto Grande reservoir (Uruguay River, Argentina–Uruguay), extensive blooms of toxic cyanobacteria can suppress reproduction almost entirely, regardless of water temperature (Boltovskoy et al. 2013; Figs. 8 and 9).

These results suggest that, in addition to temperature, many factors can play key roles in shaping the reproductive cycle of *L. fortunei*. Ultimately, however, the reproductive cycle, in coordination with aspects of its overall physiological tolerances, determines the success of the species not only in its native environment but also in its introduced ranges. In particular, the short life span of *L. fortunei,* rapid sexual maturity, and fast growth designate it as an *r*-selected species in contrast to the more long-lived, slow-growing, and similarly freshwater representatives of the Unionidae, which follow *K*-selected life history strategies (Morton 1991).

The heteromyarian, triangular, basally flattened, body form associated with byssal attachment has allowed *L. fortunei* to exploit the epibenthic niche of both lentic and lotic waters—previously unoccupied by representatives of the Bivalvia. These freshwater habitats were similarly invaded by species of Dreissenidae, notably *D. polymorpha,* and thus provide us a remarkable example of convergent evolution. In both mytiloid and dreissenid cases, however, it is their similar reproductive strategies of fast growth, rapid maturity, high fecundity, broadcast spawning, external fertilization, brief larval lives, and recruitment to established colonies (probably mediated by parental pheromones: Sardiña et al. 2009) that are undoubtedly responsible for the success of them both.

At present, the natural and introduced ranges of both *L. fortunei* and *D. polymorpha* do not overlap and although we can but hope that they never do, they most likely inevitably will. In that scenario, it will be fascinating to research the outcome (see Chapter "Parallels and Contrasts Between *Limnoperna fortunei* and species of *Dreissena*" in this volume). In the modern world, increasingly, it is the opportunists who are coming to dominate our global ecosystems, and nowhere else is this more obvious than in freshwaters.

Acknowledgments This work was partially financed by grants from the University of Buenos Aires, Argentina (UBA X-020 and 20020100100035) and from the Argentine Agencia Nacional de Promoción Científica y Tecnológica, Argentina (PICT 2007 1968) to DB.

References

Berger WH, Wefer G (1990) Export production: seasonality and intermittency, and paleoceanographic implications. Palaeogeogr Paleoclimatol Paleoecol 89:245–254

Boltovskoy D, Cataldo D (1999) Population dynamics of *Limnoperna fortunei*, an invasive fouling mollusc, in the lower Paraná River (Argentina). Biofouling 14:255–263

Boltovskoy D, Karatayev A, Burlakova L, Cataldo D, Karatayev V, Sylvester F, Mariñelarena A (2009a) Significant ecosystem-wide effects of the swiftly spreading invasive freshwater bivalve *Limnoperna fortunei*. Hydrobiologia 636:271–284

Boltovskoy D, Sylvester F, Otaegui A, Leytes V, Cataldo D (2009b) Environmental modulation of the reproductive activity of the invasive mussel *Limnoperna fortunei* in South America. Austral Ecol 34:719–730

Boltovskoy D, Correa N, Bordet F, Leites V, Cataldo D (2013) Toxic *Microcystis* (cyanobacteria) inhibit recruitment of the bloom-enhancing invasive bivalve *Limnoperna fortunei*. Freshw Biol 58:1968–1981

Bonel N (2011) Demografía experimental de un bivalvo invasor, *Limnoperna fortunei* (Dunker, 1857), en la cuenca del Plata. PhD Thesis, Universidad Nacional de La Plata (Argentina), pp 1–180

Bonel N, Solari LC, Lorda J (2013) Differences in density, shell allometry and growth between two populations of *Limnoperna fortunei* (Mytilidae) from the Río de la Plata basin, Argentina. Malacologia 56:43–58

Brugnoli E, Dabezies MJ, Clemente JM, Muniz P (2011) *Limnoperna fortunei* (Dunker 1857) en el sistema de embalses del Rio Negro, Uruguay. Oecologia Australis 15:576–592

Burlakova LE, Karatayev AY, Padilla DK (2006) Changes in the distribution and abundance of *Dreissena polymorpha* within lakes through time. Hydrobiologia 571:133–146

Callil CT, Gomes ALT, Soares ACPV (2012) A gametogênese em *Limnoperna fortunei* (Dunker, 1857). In: Mansur MCD, Santos CP, Pereira D, Padula Paz IC, Leite Zurita ML, Raya Rodriguez MT, Vilar Nehrke M, Aydos Bergonci PE (eds) Moluscos límnicos invasores no Brasil. Biologia, prevenção, controle. Redes Editora, Porto Alegre, pp 111–118

Canzi C, Bortoluzzi L, Rodriguez Fernandez D (2005) Ocorrência e situação atual do mexilhão dourado (*Limnoperna fortunei*) no reservatório da central hidrelétrica de Itaipu. In: I Simpósio Brasileiro sobre Espécies Exóticas Invasoras, Brasilia (Brazil)

Cataldo D, Boltovskoy D (2000) Yearly reproductive activity of *Limnoperna fortunei* (Bivalvia) as inferred from the occurrence of its larvae in the plankton of the lower Paraná River and the Río de la Plata estuary (Argentina). Aquat Ecol 34:307–317

Cataldo D, Boltovskoy D, Hermosa JL, Canzi C (2005) Temperature-dependent larval development rates of *Limnoperna fortunei* (Bivalvia: Mytilidae). J Molluscan Stud 71:41–46

Cepero E (2003) Control de *Limnoperna fortunei*. Infestación de plantas potabilizadoras de agua. In: 13 Congreso Argentino de Saneamiento y Medio Ambiente, Buenos Aires (Argentina)

Choi SS, Shin CN (1985) Study on the early development and larvae of *Limnoperna fortunei*. Korean J Malacol 1:5–12 [In Korean]

Damborenea MC, Penchaszadeh P (2006) Biología reproductiva de *Limnoperna fortunei*. In: Darrigran G, Damborenea MC (eds) Bio-invasion del mejillón dorado en el continente americano. Editorial de la Universidad de La Plata, La Plata, pp 71–84

Darrigran GA, Damborenea C, Penchaszadeh PE (1998) A case of hermaphroditism in the freshwater invading bivalve *Limnoperna fortunei* (Dunker, 1857) (Mytilidae) from Río de la Plata, Argentina. Iberus 16:99–104

Darrigran GA, Penchaszadeh PE, Damborenea C (1999) The reproductive cycle of *Limnoperna fortunei* (Dunker, 1857) (Mytilidae) from a neotropical temperate locality. J Shellfish Res 18:361–365

Darrigran GA, Penchaszadeh PE, Damborenea C, Greco N (2002) Abundance and distribution of golden mussel (*Limnoperna fortunei*) larvae in a hydroelectric power plant in South America. In: 11th International Conference on Aquatic Invasives Species, Alexandria (USA)

Darrigran GA, Damborenea C, Penchaszadeh PE, Taraborelli C (2003) Reproductive stabilization of *Limnoperna fortunei* (Bivalvia, Mytilidae) after ten years of invasion in the Americas. J Shellfish Res 22:141–146

Darrigran GA, Damborenea C, Greco N (2007) An evaluation pattern for antimacrofouling procedures: *Limnoperna fortunei* larvae study in a hydroelectric power plant in South America. Ambio 36:575–579

Ezcurra de Drago I, Montalto L, Oliveros OB (2006) Desarrollo y ecología larval de *Limnoperna fortunei*. In: Darrigran G, Damborenea C (eds) Bio-invasión del mejillón dorado en el continente americano. Editorial de la Universidad de La Plata, La Plata, pp 85–93

Garton DW, Haag WR (1993) Seasonal reproductive cycles and settlement patterns of *Dreissena polymorpha* in western Lake Erie. In: Nalepa TF, Schloesser DW (eds) Zebra mussels: biology, impacts, and control. Lewis Publishers, Boca Raton, pp 111–128

Gist DH, Miller MC, Brence WA (1997) Annual reproductive cycle of the zebra mussel in the Ohio River: a comparison with Lake Erie. Archiv Hydrobiologie–Beiheft Ergebnisse der Limnologie 138:365–379

Hamada M (2011) Relationship between water temperature and period of occurrence and attachment in planktonic larvae of golden mussel *Limnoperna fortunei*, in the Yahagi River. Rep of Yahagi River Inst 15:45–54 [In Japanese]

Iwasaki K, Uryu Y (1998) Life cycle of a freshwater Mytilid mussel, *Limnoperna fortunei*, in Uji River, Kyoto. Venus (Japanese J Malacol) 57:105–113

Karatayev AY, Boltovskoy D, Padilla DK, Burlakova LE (2007) The invasive bivalves *Dreissena polymorpha* and *Limnoperna fortunei* : parallels, contrasts, potential spread and invasion impacts. J Shellfish Res 26:205–213

Karatayev AY, Burlakova LE, Karatayev VA, Boltovskoy D (2010) *Limnoperna fortunei* versus *Dreissena polymorpha* : population densities and benthic community impacts of two invasive freshwater bivalves. J Shellfish Res 29:975–984

Lewandowski K (1982a) On the variable numbers of the mollusc *Dreissena polymorpha* (Pall.). Wiad Ekologiczne 28:141–154

Lewandowski K (1982b) The role of early developmental stages in the dynamics of *Dreissena polymorpha* (Pall.) (Bivalvia) populations in lakes. II. Settling of larvae and the dynamics of number of settled individuals. Ekologia Polska 30:223–286

Magara Y, Matsui Y, Goto Y, Yuasa A (2001) Invasion of the non-indigenous nuisance mussel, *Limnoperna fortunei*, into water supply facilities in Japan. J Water Supply: Res Technol—Aqua 50:113–124

Maroñas ME, Darrigran GA, Sendra ED, Breckon G (2003) Shell growth of the golden mussel *Limnoperna fortunei* (Dunker 1857) (Mytilidae) in the Rio de la Plata, Argentina. Hydrobiologia 495:41–45

Mata FAR (2011) Abundância e distribuição temporal de *Limnoperna fortunei* Dunker, 1857 (Mollusca, Bivalvia) e os impactos da incrustação em usinas geradoras de energia elétrica. MSc Thesis, Universidade Federal de Ouro Preto (Brazil), pp 1–91

McMahon RF, Bogan AE (2001) Mollusca: Bivalvia. In: Thorn JH, Covich AP (eds) Ecology and classification of North American freshwater invertebrates. Academic Press, San Diego, pp 331–429

Morton B (1969) Studies on the biology of *Dreissena polymorpha* Pall. (III). Population dynamics. Proc Malacol Soc Lond 38:471–182

Morton B (1973) Some aspects of the biology and functional morphology of the organs of feeding and digestion of *Limnoperna fortunei* (Dunker) (Bivalvia: Mytilacea). Malacologia 12:265–281

Morton B (1977) The population dynamics of *Limnoperna fortunei* (Dunker 1857) (Bivalvia: Mytilacea) in Plover Cove Reservoir, Hong Kong. Malacologia 16:165–182

Morton B (1982) The reproductive cycle in *Limnoperna fortunei* (Dunker, 1857) (Bivalvia: Mytilidae) fouling Hong Kong's raw water sypply system. Oceanologia Limnologia Sinica 13:312–324

Morton B (1991) Do the Bivalvia demonstrate environment-specific sexual strategies? A Hong Kong model. J Zool Lond 223:131–142

Nakano D, Kobayashi T, Sakaguchi I (2010a) Differences in larval dynamics of golden mussel *Limnoperna fortunei* between dam reservoirs with and without an aeration system. Landsc Ecol Eng 6:53–60

Nakano D, Kobayashi T, Sakaguchi I (2010b) Predation and depth effects on abundance and size distribution of an invasive bivalve, the golden mussel *Limnoperna fortunei*, in a dam reservoir. Limnology 11:259–266

Nalepa TF, Wojcik JA, Fanslow DL, Lang GA (1995) Initial colonization of the zebra mussel (*Dreissena polymorpha*) in Saginaw Bay, Lake Huron: population recruitment, density, and size structure. J Great Lakes Res 21:417–434

Nichols SJ (1993) Spawning of zebra mussels (*Dreissena polymorpha*) and rearing of veliger under laboratory conditions. In: Nalepa TF, Schloesser DW (eds) Zebra mussels: biology, impacts, and control. Lewis Publishers, Boca Raton, pp 315–329

Nichols SJ (1996) Variations in the reproductive cycle of *Dreissena polymorpha* in Europe, Russia, and North America. Am Zool 36:311–325

Olive PJW (2002) Reproduction and life cycles in invertebrates. In: Tickle C (ed) Encyclopaedia of life sciences. Wiley, New York, pp 1–9

Oliveira MD, Calheiros DF, Jacobi CM, Hamilton SK (2011) Abiotic factors controlling the establishment and abundance of the invasive golden mussel *Limnoperna fortunei*. Biol Invasions 13:717–729
Pestana D, Pie MR, Ostrensky A, Boeger WA, Andreoli C, Franceschi F, Lagos P (2008) Seasonal variation in larval density of *Limnoperna fortunei* (Bivalvia, Mytilidae) in the Iguaçu and Paraná rivers, in the region of Foz do Iguaçu, Paraná, Southern Brazil. Braz Arch Biol Technol 51:607–612
Rojas Molina F (2010) Efectos del molusco invasor *Limnoperna fortunei* (Dunker) sobre el zooplancton del paraná medio. PhD Thesis, Universidad de Buenos Aires (Argentina), pp 1–205
Santos CP, Mansur MCD, Wurdig NI (2008) Variaçoes no comprimento dos individuos de uma populacão do mexilhão dourado, *Limnoperna fortunei* (Mollusca: Bivalvia: Mytilidae), ao longo do ano, na Praia do Veludo, Lago Guaiba, Rio Grande do Sul, Brasil. Braz J Biol 25:389–396
Sardiña P, Cataldo D, Boltovskoy D (2009) Effects of conspecifics on settling juveniles of the invasive golden mussel, *Limnoperna fortunei*. Aquat Sci 71:479–486
Smirnova NF, Biochino GI, Vinogradov GA (1993) Some aspects of the zebra mussel (*Dreissena polymorpha*) in the former European USSR with morphological comparisons to Lake Erie. In: Nalepa TF, Schloesser D (eds) Zebra mussels: biology, impacts, and control. Lewis Publishers, Boca Raton, pp 217–226
Smit H, bij de Vaate A, Reeders HH, van Nes EH, Noordhuis R (1993) Colonization, ecology, and positive aspects of zebra mussels (*Dreissena polymorpha*) in The Netherlands. In: Nalepa TF, Schloesser D (eds) Zebra mussels: biology, impacts, and control. Lewis Publishers, Boca Raton, pp 55–77
Soares MCS, Huszar VLM, Miranda MN, Mello MM, Roland F, Lürling M (2013) Cyanobacterial dominance in Brazil: distribution and environmental preferences. Hydrobiologia 717:1–12
Stanczykowska A (1977) Ecology of *Dreissena polymorpha* (Pall.) (Bivalvia) in lakes. Polskie Archiwum Hydrobiologii 24:461–530
Strayer DL, Malcom HM (2006) Long-term demography of a zebra mussel (*Dreissena polymorpha*) population. Freshw Biol 51:117–130
Sylvester F (2006) Biología alimentaria y ecología del molusco invasor *Limnoperna fortunei* (Mytilidae) en el Paraná inferior y Río de la Plata. PhD Thesis, Universidad de Buenos Aires (Argentina), pp 1–160
Tanaka H, Deno T, Kondo T (2004) Effect of gasterostome trematode infection on growth and reproduction in *Limnoperna fortunei* (Bivalvia: Mytilidae). Memoirs of Osaka Kyoiku University, Series III. Nat Sci Appl Sci 53:1–7
Uliana ER, Callil CT (2006) Avaliação do ciclo sexual do mexilhão dourado, *Limnoperna fortunei* (Dunker, 1857), no Rio Paraguai, Porto Esperança—MS. BS Thesis, Universidade Federal de Mato Grosso (Brazil), pp 1–33
Zanella O, Marenda LD (2002) Ocorrência de *Limnoperna fortunei* na Central Hidrelétrica de Itaipu. In: V Congresso Latinoamericano de Malacologia, São Paulo (Brazil)

Population Dynamics and Growth of *Limnoperna fortunei*

Daisuke Nakano, Takuya Kobayashi and Isamu Sakaguchi

Abstract During its first year of life, *Limnoperna fortunei* grows from ~10 to >30 mm in shell length, primarily depending on water temperature. Two-year-old individuals attain 20–30 mm, and 30 mm is usually the largest size, although specimens up to >50 mm in size have been reported. The life span is 2–3 years. Water temperature, including the season of each cohort, is the most important factor that determines the growth rate, but other constraints can play important roles too, including calcium concentrations, pollution, food availability and intraspecific competition.

Keywords Estimate methods · Mortality · Growth rate · Cohorts · Size · Biomass

Introduction

Population dynamics and individual growth of *Limnoperna fortunei* are of major importance both in the context of fundamental science (ecology), and for practical applications (biofouling control). The growth rate of *L. fortunei* depends on environmental conditions, and mussel size affects its impacts on the ecosystem through processes such as nutrient cycling, respiration, consumption of organic matter and excretion (see Part 2 in this volume). Growth of mussels on screens, piping and channels of water intake facilities and the size attained by adult individuals are of major importance for gauging and mitigating the damage produced by its fouling (see Chapter "Impacts of *Limnoperna fortunei* on Man-made Structures and Control Strategies: General Overview" in this volume). Understanding the mechanisms that modulate individual growth and population dynamics of *L. fortunei* allows for the definition of suitable methods for implementing antifouling strategies, including the

D. Nakano (✉) · T. Kobayashi · I. Sakaguchi
Central Research Institute of Electric Power Industry (CRIEPI),
1646 Abiko, Chiba 270-1194, Japan
e-mail: d-nakano@criepi.denken.or.jp

T. Kobayashi
e-mail: t-koba@criepi.denken.or.jp

D. Boltovskoy (ed.), *Limnoperna fortunei*, Invading Nature - Springer Series in Invasion Ecology 10, DOI 10.1007/978-3-319-13494-9_6

time period and frequency of treatment. Several surveys have estimated *L. fortunei* growth by using size distribution data over a given time offset, usually a year. These studies have produced useful results, but limitations associated with low-sampling frequencies have often resulted in gaps and imprecise information that obscure interpretation of the cohorts involved and, hence, conclusions on individual growth. Furthermore, although cohort analysis is a powerful technique for the interpretation of population data, it requires that reproduction events be more or less discrete in time (Edmondson and Winberg 1971; Sparre and Venema 1998). Continuous reproduction modes, as is the case for *L. fortunei* in many tropical and subtropical areas (see Chapter "Reproductive Output and Seasonality of *Limnoperna fortunei*" in this volume), lack the pulses whose identification in size-frequency analyses permits the definition of growth parameters. Sessile species with extended reproduction periods offer an alternative possibility: the deployment of experimental substrata (Boltovskoy and Cataldo 1999; Nakano et al. 2011). This technique allows eliminating all size classes older than the one starting at the time of deployment of the substrata, thus unequivocally pinpointing a zero age class whose monitoring through time allows assessment of the growth of the species. On the other hand, identifying the effects of environmental factors on growth is best achieved through laboratory studies under controlled conditions, an approach which has rarely been used with *L. fortunei*.

This chapter reviews our present understanding of the population dynamics, growth and mortality of *L. fortunei* in Asia and South America, highlighting similarities and differences between these invaded areas.

Methods Used for Population Dynamics Studies and Growth Estimates

For field estimates of recruitment and growth of *L. fortunei*, the most frequently used method is cohort analysis. This technique consists of the identification of individuals born at approximately the same time (a cohort), and tracking their increase in size through time. Recruitment events are indicated by peaks in the relative abundance of recently settled mussels, usually < 1 mm in shell length, and by peaks in the abundance of veligers in the water column. This method is particularly suitable for populations that have a restricted reproductive period, thus allowing precise separation of generations. In contrast, when reproduction is more or less continuous during extended periods (months), interpretation of successive cohorts is complicated. Furthermore, although at birth (or settlement on a substrate, in the case of byssate mussels) the size of all members of the same cohort is practically identical (Fig. 1, first histogram of upper panel), individual differences in growth increase with time hindering unequivocal identification of the cohort and of its mean size (Fig. 1, last histogram of upper panel). Increasing sampling frequency can alleviate some of these problems, but pinpointing individual cohorts may still pose major difficulties. This is especially complicated when working on samples obtained from natural

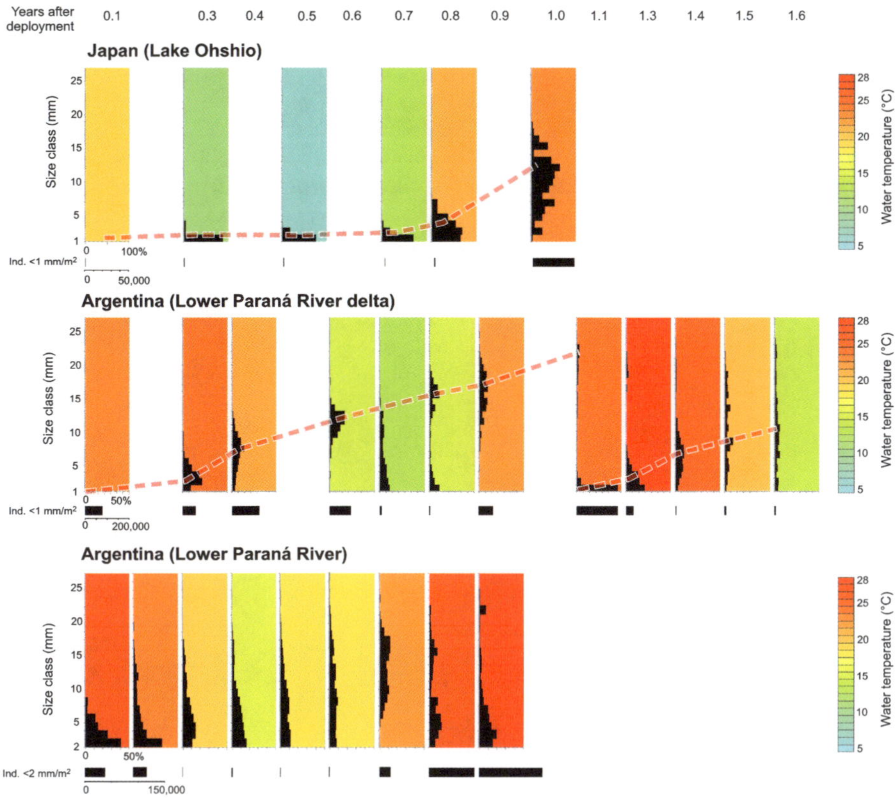

Fig. 1 Changes in the size structure of *L. fortunei* colonizing artificial substrata in Lake Ohshio, Japan (based on data from Nakano et al. 2011), in the delta of the Lower Paraná River, Argentina (based on data from Sylvester et al. 2007) and in the Lower Paraná River (based on data from Boltovskoy and Cataldo 1999). *Dashed red line* suggests evolution of cohorts. Deployment dates were in Japan: 31 August 2008, in Argentina: 6 November 2002 (delta) and 20 January 1998 (Paraná River)

substrata whose time of exposure to colonization is not known (Iwasaki and Uryu 1998; Maroñas et al. 2003).

A widely employed alternative that allows circumventing some of these problems is the use of artificial substrata deployed ad hoc in a water body. This approach allows determining unequivocally a zero age class because all subsequent samples (normally several substrata are deployed, retrieving them one by one at preset intervals) contain no mussels older than the lapse between deployment and retrieval. This facilitates identification of at least one—the first—cohort recruited on the substrata. Nevertheless, when reproduction is continuous, in subsequent samplings this first cohort may be increasingly difficult to separate from subsequent settlement. A variety of artificial substrata have been used by researchers (Fig. 2), all of which offer the mussels hard, colonizable surfaces with different orientation (vertical, horizontal, upper and lower surfaces). Because predation of settled individuals

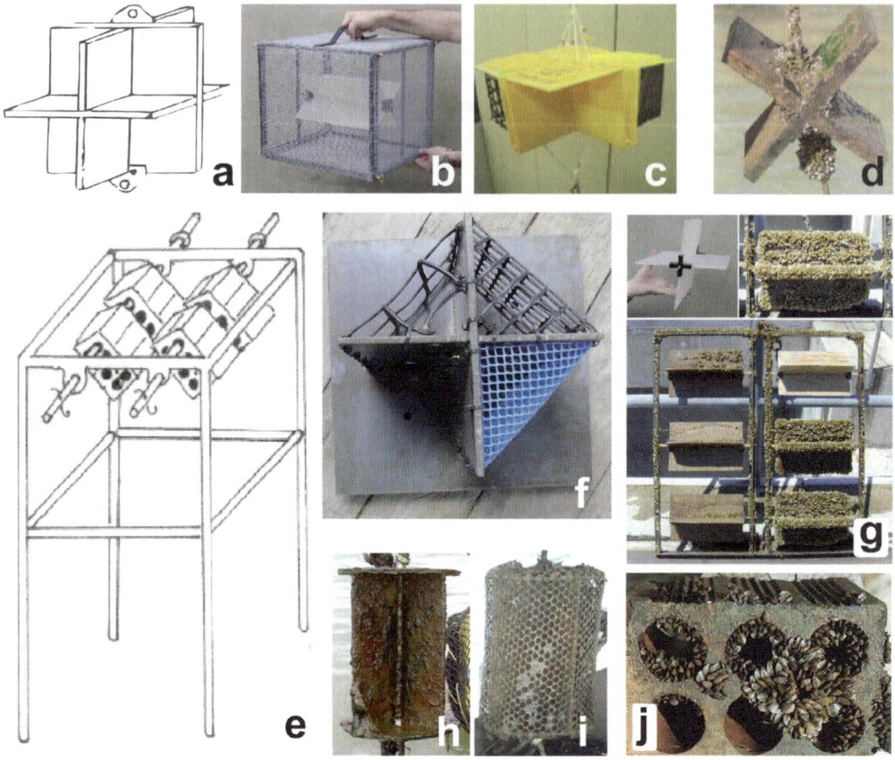

Fig. 2 Various types of artificial substrata and experimental enclosures used for surveys on growth and population dynamics of *L. fortunei*. **a** Morton (1977), **b** courtesy of G. Darrigran, **c** Nakano et al. (2010), **d** Mansur et al. (2009), **e** Santos et al. (2008), **f** Sylvester (2006), **g** Courtesy of G. Darrigran, **h, i** Boltovskoy and Cataldo (2003), **j** Mansur et al. (2003). Notice that some of the designs include surfaces protected from predators by plastic nettings (**b, c, f, i**)

can eliminate over 90 % of the mussels (Sylvester et al. 2007; Nakano et al. 2010), several studies have used unprotected sectors alongside sectors protected by mesh nettings of different sizes in order to exclude predators (Fig. 2c, f, e, h, i).

Protected enclosures (Darrigran et al. 2011; Nakano, unpublished data) and laboratory experiments (Pestana 2006) have also been used for estimating growth of postsettlement individuals (Fig. 2b and 3).

While studies based on artificial substrata have yielded more solid results than those based on natural ones, they still have important limitations. In areas where *L. fortunei* has a single, short, reproductive period per year, as in Korea and Japan (see Chapter "Reproductive Output and Seasonality of *Limnoperna fortunei*" in this volume), identification of cohorts is usually relatively simple. On the other hand, in most of South America, where reproduction can span over 10 months of the year, the first settlers on the substrata can be difficult to separate from subsequent

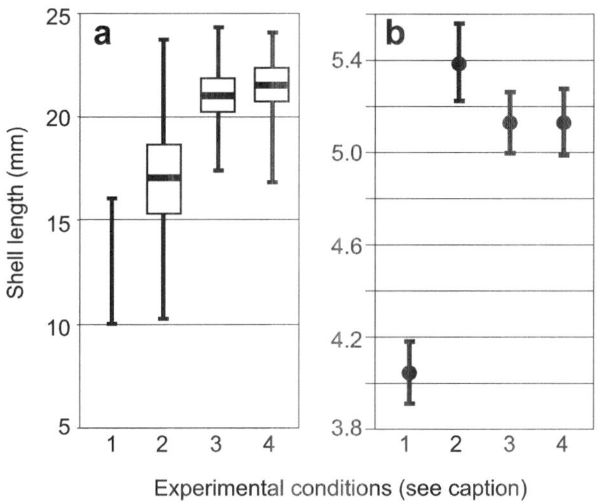

Fig. 3 Contrasting results of the effects of conspecifics on growth of *L. fortunei*. **a** Growth of caged mussels 10–16 mm in length after 104 days in Lake Ohshio, Japan, for a total of 104 days, at water temperatures ranging between 13.6 and 27.7 °C. (*1*) Initial size range of experimental specimens, (*2*) final size of specimens stocked at high densities (140 small mussels per cage), (*3*) final size of specimens stocked at low densities (10 small mussels per cage), (*4*) final size of small specimens (10 ind. per cage) stocked jointly with large specimens (25–32 mm, 10 ind. per cage). Graph shows range of values, the first and third quartiles (*box*) and the median (*band inside the box*). (From Nakano, unpublished.) **b** Size attained by *L. fortunei* specimens 3 months after recruitment on tiles barren of conspecifics (*1*), and on tiles with low (*2*), medium (*3*) and high (*4*) densities of adult conspecifics (800, 4000 and 12,000 mussels/m^2, respectively), in the delta of the lower Paraná River (Argentina), at water temperatures around 25–26 °C. (Based on data from Sardiña et al. 2009)

recruits. Another major problem is associated with the protective mesh precluding mussel predation. Meshes with large openings allow access to small-sized predators (presumably fish), whereas those with small openings are better at protecting the mussels from predation but eventually clog with the mussels' faeces and pseudofaeces, and the resulting siltation results in high mussel mortalities (Sylvester et al. 2007).

Population Dynamics and Growth Rates

In general terms, most field studies in subtropical water bodies agree in that *L. fortunei* grows to a size of ~20 mm in its first year of life, attaining ~30 mm during the second year, this being the typical size of large, adult individuals. However, specimens > 50 mm in length have occasionally been recorded (Karatayev et al. 2010). The life span is estimated at 2–3 years. Growth is fastest during warmer months (which is when reproduction is most active; see Chapter "Reproductive Output and

Seasonality of *Limnoperna fortunei*" in this volume), and slows down significantly during the winter. Differences between sites, however, are large, and depend on several factors.

Figure 4 illustrates growth curves for several Asian and South American water bodies, suggesting that growth rates are primarily dependent on temperature and the length of time with high water temperatures. In the Upper Paraná River, where water temperatures are above 20 °C year around, *L. fortunei* can reach over 35 mm in its first year of life (Fig. 4d). Farther south, in the Lower Paraná River, where water temperatures are around 10–28 °C, mussels grow to ~20 mm in 1 year (Fig. 4e, f, h, i). A similar growth rate was also described for Hong Kong (Fig. 4c), where water temperatures vary seasonally from 15 to 31 °C. In Japan, at temperatures ranging from 4 to 26 °C, 1-year-old individuals can attain only 10 mm in length (Fig. 4a; see Table 1).

Slower growth rates at lower temperatures were also found by all field surveys that incorporated seasonality in their estimates, or whose raw data were herein reprocessed taking this variability into account (i.e., the von Bertalanffy seasonal growth formula, VBSGF; Fig. 4a, b, c, f). Slower growth rates have also been documented in laboratory experiments (Fig. 5).

Environmental Drivers of Growth Rates

Growth rates reported in field experiments can vary significantly as a function of the time window employed. Subsequent cohorts may show dissimilar growth rates depending on the time of the year (and, therefore, the water temperature) during the first months after settlement (Fig. 4c, d, g, h, i). Highest growth rates are those of recruits settling in the spring and early summer, whereas slowest rates are characteristic of mussels born in the autumn and winter. Even in tropical and subtropical areas, where temperatures are high year around, seasonal differences in growth are marked. On the basis of data from Bela Vista Reservoir (Upper Paraná River), Belz et al. (2010) concluded that the growth coefficient (k) of the summer (December) cohort (cohort 4 in Fig. 4d) that settled at a temperature around 26 °C, was ca. 50 % higher than that of the winter (July) cohort (cohort 12 in Fig. 4d) that settled at ca. 20 °C (k=3.2 and 2.2, respectively). Some surveys have reported the opposite situation, i.e., higher k values for winter than for spring or summer cohorts (e.g., Spaccesi 2013; Fig. 4h, i; Table 1), but the confounding effects of using populations from natural substrata may have obscured identification of the cohorts.

Like most animals, golden mussels grow slower as they age. During their first year of life, they usually reach ~20 mm, but add only ~10 mm during their second year (Fig. 4). The results of Pestana (2006) clearly illustrate the effects of temperature and age on growth rates, showing that small mussels (2.5–8.0 mm) grow proportionally almost 100 times more than large ones (>24 mm) (Fig. 5).

On the basis of data from Lake Ohshio (Japan), Nakano et al. (2011) investigated the effects of water temperature, dissolved oxygen, chlorophyll a concentrations,

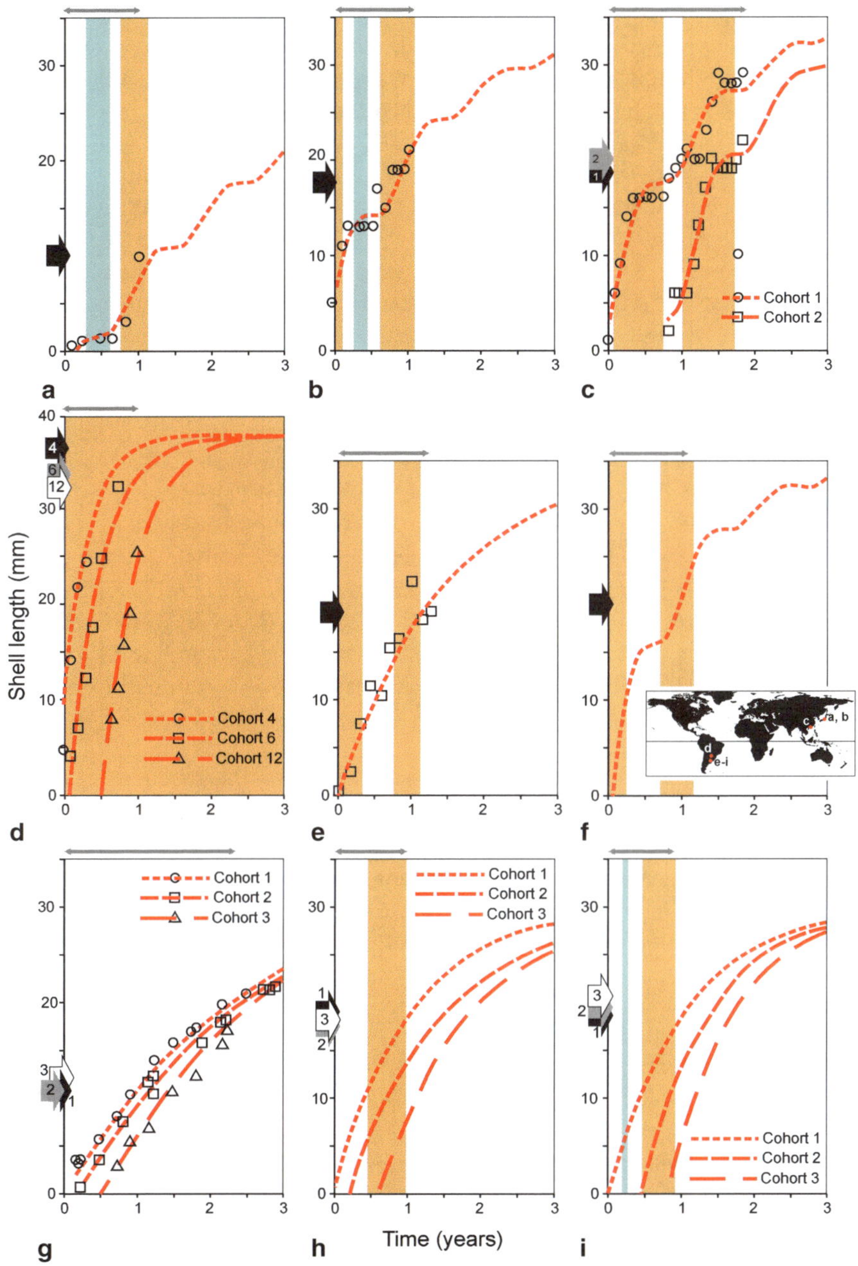

Fig. 4 Growth curves of *Limnoperna fortunei* in various water bodies of Asia and South America. **a** Lake Ohshio, Japan (Nakano et al. 2011), **b** Uji River, Japan (Iwasaki and Uryu 1998), **c** Plover Cove Reservoir, Hong Kong (Morton 1977), **d** Bela Vista Reservoir, Brazil (Belz et al. 2010), **e** Carapachay River (delta of the Lower Paraná River, Argentina; Sylvester et al. 2007), **f** Lower Paraná River, Argentina (Boltovskoy and Cataldo 1999), **g** Bagliardi Beach (1992–1994; middle Río de la Plata estuary, Argentina; Maroñas et al. 2003), **h** Bagliardi Beach (2001–2002; middle

conductivity and turbidity on the growth of *L. fortunei*. They concluded that in this lake only water temperature was positively correlated with the growth of the golden mussel. However, while temperature has an overwhelming influence on the growth of *L. fortunei*, it is not the only variable that can affect growth.

The golden mussel can tolerate very low calcium concentrations (see Chapter "Parallels and Contrasts Between *Limnoperna fortunei* and Species of *Dreissena*" in this volume), but the growth is hindered in Ca-deficient water bodies. Dos Santos et al. (2007) deployed cages with field-collected individuals 7–10 mm in length in three water bodies associated with the Paraguay River (Brazil) with different concentrations of Ca. After 150 days, mussels from the site with highest calcium concentrations (18 mg/L) grew significantly more (0.09 mm/d) than those at the other two sites (4–6 mg Ca/L, 0.02–0.05 mm/d). These results align with those of Hincks and Mackie (1997), showing that Ca concentrations are significantly associated with survival and growth of zebra mussels.

The golden mussel survives in highly polluted waters, but, again, pollution impacts its growth. Bonel et al. (2013) compared population traits of *L. fortunei* in two South American rivers differing in pollution levels (as indicated by dissolved oxygen, pH and conductivity), concluding that growth rates were significantly lower at the more polluted site (Río Santiago River), than in the more pristine one (Coronda River).

Food availability has also been shown to affect growth of byssate freshwater mussels (Schneider 1992; Dorgelo 1993; McMahon 1996), including *L. fortunei*. Bergonci et al. (2012) deployed artificial substrata at two locations in the Jacuí River-Guaíba Lake system (southern Brazil) 4 km apart. After 11 months, substrata were recovered and all mussels were measured and weighed. Organisms from the site with higher concentrations of organic matter had significantly more biomass and were larger than those from the other location, suggesting that differences in food availability were responsible for the dissimilar growth rates observed. The lower particulate organic matter concentrations found in the reservoir Embalse de Río Tercero (central Argentina) compared to those of the Paraná River system were suggested to result in shorter reproductive periods of *L. fortunei* (Boltovskoy et al. 2009).

Effects of Intraspecific Competition on Growth Rates

Life in densely packed clusters of conspecifics, as that of gregarious, sessile mussels, has many advantages but also drawbacks (Bertness and Grosholz 1985; Okamura 1986; Côté and Jelnikar 1999). Among the drawbacks, cannibalism of ready-to-settle larvae and competition for food and space are probably the most important. In order to investigate the effects of neighbouring conspecifics on growth

Río de la Plata estuary, Argentina; Spaccesi 2013), **i** Punta Indio (middle Río de la Plata estuary, Argentina; Spaccesi 2013). *Arrows at top* of each graph denote observational period. *Arrows along y*-axis indicate shell length at 1 year of age for each of the cohorts considered. *Light-blue shading* denotes water temperatures < 10 °C, orange for temperatures > 20 °C. *Inset map* shows the geographic range of growth studies

Table 1 Results of studies of the population dynamics and growth of *Limnoperna fortunei* in various water bodies of Asia and South America. (see Fig. 4 for the corresponding growth curves)

Site	Lake Ohshio (Japan)	Uji River (Japan)	Plover Cove reservoir (Hong Kong)	Bela Vista reservoir (Brazil)	Carapachay River (Argentina)	Lima (Argentina)	Bagliardi beach (Argentina)	Bagliardi beach (Argentina)	Punta Indio (Argentina)
River system	Tone River	Yodo River	Dong River	Paraná River	Paraná River	Paraná River	Río de la Plata estuary	Río de la Plata estuary	Río de la Plata estuary
Latitude	36.22°N	34.93°N	22.28°N	25.40°S	33.38°S	33.95°S	34.87°S	34.92°S	35.27°S
Longitude	138.87°E	135.92°E	144.25°E	54.58°W	58.58°W	59.20°W	57.80°W	57.78°W	57.22°W
k[a]	0.3	0.7	0.6–0.7	2.0-3.2	0.7	1.0	0.3–0.4	0.8-1.0	1.0-1.1
Sampling period	2008–2009	1994–1995	1972–1974	2004–2005	2002–2004	1998	1993–1994	2001–2002	2001–2002
Temp. range (°C)	4.2–23.6	7.0–26.0	15.0–29.0	20.0–32.0[c]	13.0–28.0	12.9–26.5	14.0–24.0	10.7–27.9	8.5–28.6
Months > 20 °C	4	6	8	12	8	6	ND	6	6
Months < 10 °C	4	2	0	0	0	0	ND	0	1
Substrate	Artificial	Natural	Artificial	Natural	Artificial	Artificial	Natural	Natural	Natural
Method[b]	VBSGF	VBSGF	VBSGF	VBGF	VBGF	VBSGF	VBGF	VBGF	VBGF
Reference	Nakano et al. (2011)	Iwasaki and Uryu (1998)	Morton (1977)	Belz et al. (2010)	Sylvester et al. (2007)	Boltovskoy and Cataldo (1999)	Maroñas et al. (2003)	Spaccesi (2013)	Spaccesi (2013)

[a] *k*, the growth coefficient of the von Bertalanffy growth formula, which reflects how fast the maximum size is reached;
[b] VBGF von Bertalanffy growth formula (growth rate differences between seasons disregarded); VBSGF von Bertalanffy seasonal growth formula (growth rate differences between seasons considered);
[c] based on data from the nearby Itaipu Reservoir

Fig. 5 Growth after 179 days as a function of initial size in mussels held in captivity at 18 and 25 °C. Values are based on 45 individuals for each size class, of which 53–89 % survived through the experimental period. (Based on data from Pestana 2006)

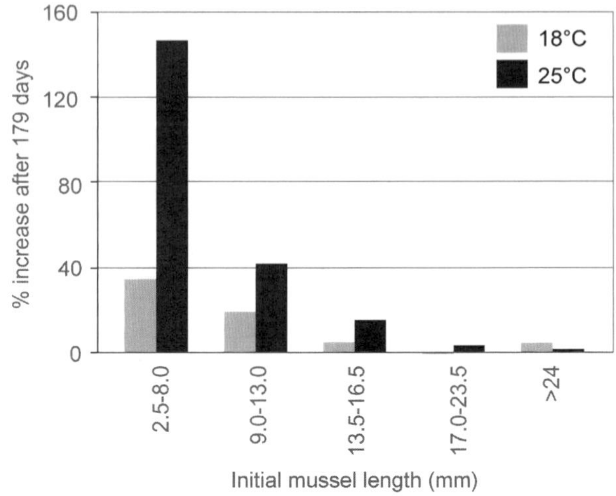

of *L. fortunei*, Nakano (unpublished data) conducted a series of experiments in Lake Ohshio (Japan). Specimens of *L. fortunei* 10–16 mm in length were stocked in cages 10 × 10 × 12 cm in size made of plastic netting with a mesh size of 5 mm. Experimental conditions included low density of small mussels (10 specimens per cage), high density of small mussels (140 mussels per cage), and small (10) + large (25–32 mm, 10) individuals together. Cages were kept in situ at 5 m depth for 104 days. Mortality at the end of the experiment was very low (2.7 %). Comparison of mean shell lengths of 10 small labelled individuals showed that growth in densely populated cages (4.4 ± 1.1 mm) was lower than growth in cages with fewer mussels (7.7 ± 0.8 and 8.1 ± 0.5 mm, in cages with small and small + large mussels, respectively; Fig. 3a). Statistically, end values of low- versus high-stocking densities differed significantly ($P < 0.0001$, t-test), suggesting that intraspecific competition, probably for food and space, may affect growth. However, the impact of adverse conditions associated with high densities seemed to differ widely between individuals, as the range of final lengths in high-density cages was much higher than that in the low density ones (Fig. 3a).

Information on the zebra mussel indicates that there are significant differences between food availability and the quality of the interstitial water at different depths in the colony (Burks et al. 2002; Tuchman et al. 2004). Although in contrast to *D. polymorpha*, *L. fortunei* does not build multilayered mussel beds (see Chapter "*Limnoperna fortunei* Colonies: Structure, Distribution and Dynamics" and Fig. 8 therein), it is conceivable that even in single-layered colonies denser aggregates involve more growth-limiting conditions, especially in stagnant waters (e.g., from higher concentrations of waste products, more oxygen consumption, and lower availability of food).

Despite the fact that high densities may affect mussel growth, the benefits of gregarious behaviour must largely outweigh their drawbacks. Recruitment of golden mussels is significantly higher in areas occupied by older conspecifics than in areas barren of mussels (Sardiña et al. 2009). Furthermore, growth of recruits was also found to be enhanced by the presence of conspecific adults (Sardiña et al. 2009; Fig. 3b). This disagreement with the data illustrated in Fig. 3a may indicate that relationships between growth and density vary with mussel age, and/or that there are threshold density levels above which the interaction becomes negative.

Mortality

Data on larval mortality of the golden mussel indicate that around 80–90 % of the larvae die before reaching the settling stage, with highest mortalities occurring during the transition from the straight-hinged to the umboned veliger stage (Cataldo et al. 2005; see Chapter "Larval Development of *Limnoperna fortunei*" in this volume). Highest mortalities of settled individuals occur during the earliest stages of growth; comparison of size-frequency distributions indicates that > 93 % of the mussels < 1 mm in size die before reaching 2 mm (Sylvester et al. 2007). These values are roughly comparable to those reported for zebra mussels (up to 99 %; Lewandowski 1982; Sprung 1989). For animals, > 1 mm mortality drops sharply, with ca. 80 % of the mussels 2 mm in length surviving to 20–23 mm (Sylvester et al. 2007). Thus, approximately 2 % of the animals that reach the settling stage survive until first reproduction (at about 7 mm, cf. Darrigran et al. 1999), and only 0.5 % survive the first year of life. Mortality rates increase sharply during the winter, with smaller mussels being affected the most (Sylvester et al. 2007).

Size–biomass Relationships

Growth increases both the shell length and the biomass of *L. fortunei*. Relationships between shell length, total wet mass (including the shell), wet tissue weight and dry tissue weight, were estimated by Sylvester (2006) on the basis of organisms collected in the delta of the Lower Paraná River (Fig. 6). Interestingly, at a shell length of about 10–15 mm, there seems to be a break in the size–weight relationship, whereby data below and above these length values are better correlated when analyzed separately than in bulk (Fig. 6a, b, c). These expressions are useful for general purposes, but relationships between size and biomass can vary significantly between water bodies and times of the year, depending strongly on feeding conditions. Pestana (2006), for example, reported that, after ~2 months, the condition index (i.e., the ratio of tissue weight to shell length) of starved *L. fortunei* was about six times lower than that of fed specimens.

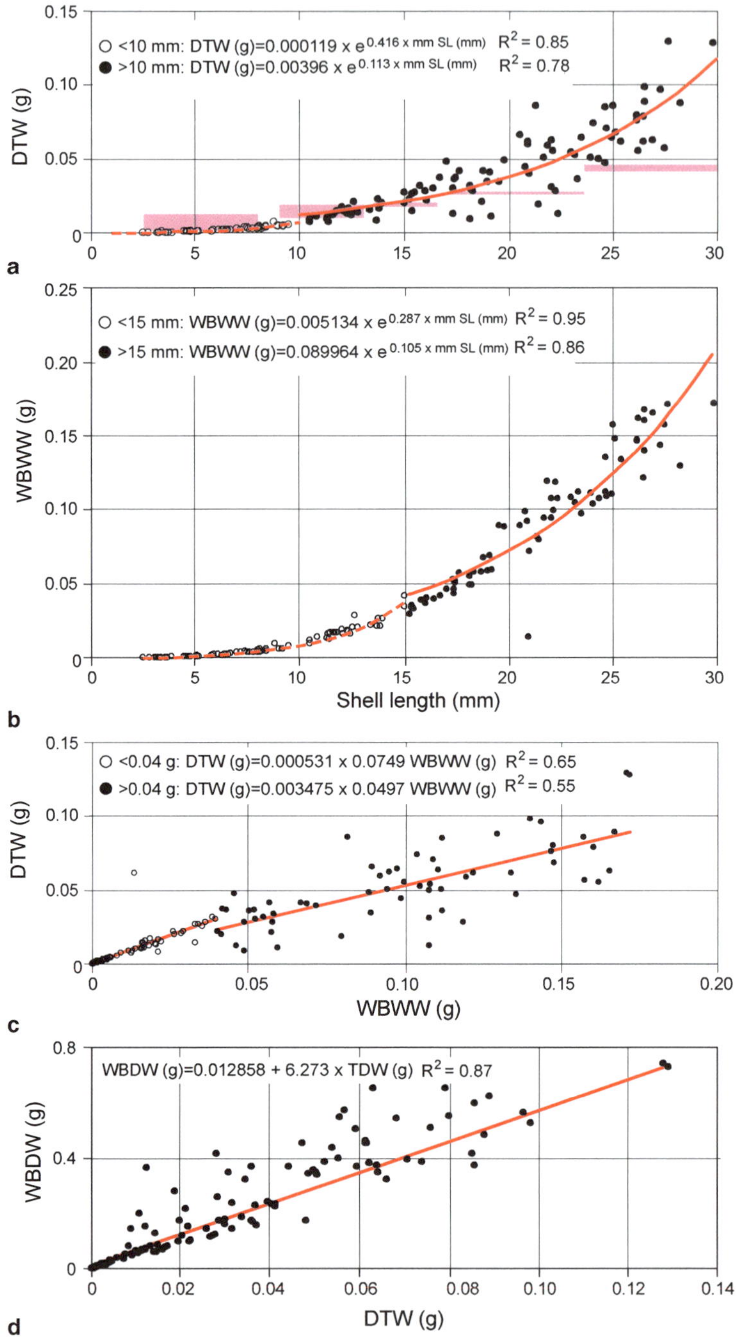

Fig. 6 Relationships between shell length (*SL*), dry tissue weight (*DTW*), whole body dry weight (*WBDW*) and whole body wet weight (*WBWW*) for small and large *L. fortunei* specimens (from Sylvester 2006). *Pink shadings* in panel **a** denote SL vs. DTW ranges reported by Pestana (2006)

References

Belz CE, Darrigran GA, Bonel N, Mader Netto OS (2010) Density, recruitment, and shell growth of *Limnoperna fortunei* (Mytilidae), an invasive mussel in tropical South America. J Freshw Ecol 25:227–233

Bergonci PEA, Santos CP, Vilar Nehrke M, Pereira D, Mansur MCD (2012) *Limnoperna fortunei* na bacia hidrográfica do baixo Rio Jacuí e Lago Guaíba, Rio Grande do Sul. In: Mansur MCD, Santos CP, Pereira D, Padula Paz IC, Leite Zurita ML, Raya Rodriguez MT, Vilar Nehrke M, Aydos Bergonci PE (eds) Moluscos límnicos invasores no Brasil. Biologia, prevenção, controle. Redes Editora, Porto Alegre, pp 193–200

Bertness MD, Grosholz E (1985) Population dynamics of the ribbed mussel, *Geukensia demissa* : the costs and benefits of an aggregated distribution. Oecologia 67:192–204

Boltovskoy D, Cataldo D (1999) Population dynamics of *Limnoperna fortunei*, an invasive fouling mollusc, in the lower Paraná River (Argentina). Biofouling 14:255–263

Boltovskoy D, Cataldo D (2003) Estudios de evaluación, monitoreo y análisis de medidas de control de incrustaciones de moluscos bivalvos (*Limnoperna fortunei*) en las instalaciones de la central hidroeléctrica Itaipu. Infome final de trabajos. Central Hidroelectrica Itaipú, CHI, pp 1–281

Boltovskoy D, Sylvester F, Otaegui A, Leytes V, Cataldo D (2009) Environmental modulation of the reproductive activity of the invasive mussel *Limnoperna fortunei* in South America. Austral Ecol 34:719–730

Bonel N, Solari LC, Lorda J (2013) Differences in density, shell allometry and growth between two populations of *Limnoperna fortunei* (Mytilidae) from the Río de la Plata basin, Argentina. Malacologia 56:43–58

Burks RL, Tuchman NC, Call CA (2002) Colonial aggregates: effects of spatial position on zebra mussel responses to vertical gradients in interstitial water quality. J N Am Benthol Soc 21:64–75

Cataldo D, Boltovskoy D, Hermosa JL, Canzi C (2005) Temperature-dependent larval development rates of *Limnoperna fortunei* (Bivalvia: Mytilidae). J Molluscan Stud 71:41–46

Côté IM, Jelnikar E (1999) Predator-induced clumping behaviour in mussels (*Mytilus edulis* Linnaeus). J Exp Mar Biol Ecol 235:201–211

Darrigran GA, Penchaszadeh PE, Damborenea C (1999) The reproductive cycle of *Limnoperna fortunei* (Dunker, 1857) (Mytilidae) from a neotropical temperate locality. J Shellfish Res 361–365

Darrigran GA, Bonel N, Colautti D, Cazzaniga NJ (2011) An alternative method to assess individual growth of the golden mussel (*Limnoperna fortunei*) in the wild. J Freshw Ecol 26:1–9

Dorgelo J (1993) Growth and population structure of the zebra mussel *(Dreissena polymorpha*) in Dutch Lakes differing in trophic state. In: Nalepa TF, Schloesser D (eds) Zebra mussels: biology, impacts, and control. Lewis Publishers, Boca Raton, pp 79–94

Dos Santos CL, Xavier IMG, Oliveira MD, Jacobi CM, Eilers V (2007) Influência da concentração de cálcio na taxa de crescimento do mexilhão dourado (*Limnoperna fortunei*) em ambientes do Pantanal Sul-Matogrossense. In: 11 Congresso Brasileiro de Limnologia, Rio de Janeiro (Brazil)

Edmondson WT, Winberg GG (1971) A manual on methods for the assessment of secondary production fresh waters. Blackwell, Oxford, pp 1–358

Hincks SS, Mackie GL (1997) Effects of pH, calcium, alkalinity, hardness, and chlorophyll on the survival, growth, and reproductive success of zebra mussel (*Dreissena polymorpha*) in Ontario lakes. Can J Fish Aquat Sci 54:2049–2057

Iwasaki K, Uryu Y (1998) Life cycle of a freshwater Mytilid mussel, *Limnoperna fortunei*, in Uji River, Kyoto. Venus. Jpn J Malacol 57:105–113

Karatayev AY, Burlakova LE, Karatayev VA, Boltovskoy D (2010) *Limnoperna fortunei* versus *Dreissena polymorpha* : Population densities and benthic community impacts of two invasive freshwater bivalves. J Shellfish Res 29:975–984

Lewandowski K (1982) On the variable numbers of the mollusc *Dreissena polymorpha* (Pall.). Wiadomosci Ekologiczne 28:141–154

Mansur MCD, Santos CP, Darrigran G, Heydrich I, Callil CT, Cardoso FR (2003) Primeiros dados quali-quantitativos do mexilhão-dourado, *Limnoperna fortunei* (Dunker), no Delta do Jacuí, no Lago Guaíba e na Laguna dos Patos, Rio Grande do Sul, Brasil e alguns aspectos de sua invasão no novo ambiente. Revista Brasileira de Zoologia 20:75–84

Mansur MCD, Pereira D, Santos CP, Bergonci PEA, Thormann BM, Takeda AM (2009) Colonização de substrato artificial pelo mexilhão dourado, *Limnoperna fortunei* (Dunker, 1857) (Bivalvia, Mytiloida, Mytilidae), no Delta do Rio Jacuí (RS, Brasil). Biotemas 22:75–80

Maroñas ME, Darrigran GA, Sendra ED, Breckon G (2003) Shell growth of the golden mussel *Limnoperna fortunei* (Dunker 1857) (Mytilidae) in the Rio de la Plata, Argentina. Hydrobiologia 495:41–45

McMahon RF (1996) The physiological ecology of the zebra mussel, *Dreissena polymorpha*, in North America and Europe. Am Zool 36:339–363

Morton B (1977) The population dynamics of *Limnoperna fortunei* (Dunker 1857) (Bivalvia: Mytilacea) in Plover Cove Reservoir, Hong Kong. Malacologia 16:165–182

Nakano D, Kobayashi T, Sakaguchi I (2010) Predation and depth effects on abundance and size distribution of an invasive bivalve, the golden mussel *Limnoperna fortunei*, in a dam reservoir. Limnology 11:259–266

Nakano D, Kobayashi T, Endo N, Sakaguchi I (2011) Growth rate and settlement of *Limnoperna fortunei* in a temperate reservoir. J Molluscan Stud 77:142–148

Okamura B (1986) Group living and the effects of spatial position in aggregations of *Mytilus edulis*. Oecologia 69:341–347

Pestana D (2006) Aspetos bioecológicos do mexilhao dourado *Limnoperna fortunei* (Bivalvia, Mytilidae) (Dunker, 1857). PhD Thesis, Universidade Federal de Curitiba (Brazil), pp 1–123

Santos CP, Mansur MCD, Wurdig Nl (2008) Variaçoes no comprimento dos individuos de uma populacão do mexilhão dourado, *Limnoperna fortunei* (Mollusca: Bivalvia: Mytilidae), ao longo do ano, na Praia do Veludo, Lago Guaiba, Rio Grande do Sul, Brasil. Braz J Biol 25:389–396

Sardiña P, Cataldo D, Boltovskoy D (2009) Effects of conspecifics on settling juveniles of the invasive golden mussel, *Limnoperna fortunei*. Aquat Sci 71:479–486

Schneider DW (1992) A bioenergetics model of zebra mussel, *Dreissena polymorpha*, growth in the Great Lakes. Can J Fish Aquat Sci 49:1406–1416

Spaccesi FG (2013) Abundance, recruitment, and shell growth of the exotic mussel *Limnoperna fortunei* in the Río de la Plata (Argentina). Zool Stud 52:1–11

Sparre P, Venema SC (1998) Introduction to tropical fish stock assessment. Part 1: Manual. Food and Agriculture Organization of the United Nations, Rome, pp 1–407

Sprung M (1989) Field and laboratory observations of *Dreissena polymorpha* larvae: abundance, growth, mortality and food demands. Archiv für Hydrobiologie 115:537–561

Sylvester F (2006) Biología alimentaria y ecología del molusco invasor *Limnoperna fortunei* (Mytilidae) en el Paraná inferior y Río de la Plata. PhD Thesis, Universidad de Buenos Aires (Argentina), pp 1–160

Sylvester F, Boltovskoy D, Cataldo D (2007) Fast response of freshwater consumers to a new trophic resource: predation on the recently introduced Asian bivalve *Limnoperna fortunei* in the lower Parana River, South America. Austral Ecol 32:403–415

Tuchman NC, Burks RL, Call CA, Smarrelli J (2004) Flow rate and vertical position influence ingestion rates of colonial zebra mussels (*Dreissena polymorpha*). Freshw Biol 49:191–198

Limnoperna fortunei Colonies: Structure, Distribution and Dynamics

Nancy Correa, Paula Sardiña, Pablo V. Perepelizin and Demetrio Boltovskoy

Abstract Settlement of new recruits of *Limnoperna fortunei* occurs preferentially on areas already colonized by conspecifics, and on surfaces with well-developed periphytic biofilms. Hard substrata (immobile rocks, wood) are preferred by the mussel, but colonization can also take place on muddy areas stabilized by roots or fibrous debris, on floating and submerged plants, and on mussel shells, crustaceans, etc. Colonization starts in crevices, angles and other sites inaccessible to large predators, but it often extends over open areas as well. Mussel beds rarely exceed 7–10 cm in thickness, with most adults being at least partially attached to the substrate. Juveniles often settle on larger shells. Densities of over 200,000 ind./m^2 have been reported occasionally, but such high numbers are invariably dominated by specimens <2 mm in length. Densities of adult mussels (>5-7 mm) are usually below 10,000 ind./m^2. The only site where densities were estimated over an entire

N. Correa (✉)
Servicio de Hidrografía Naval (Ministerio de Defensa), and Escuela de Ciencias del Mar (Instituto Universitario Naval), Av. Montes de Oca 2124, 1271 Buenos Aires, Argentina
e-mail: ncorrea@hidro.gov.ar

P. Sardiña
Departamento de Biología, Bioquímica y Farmacia, Instituto de Investigaciones Biológicas y Biomédicas del Sur (INBIOSUR), Universidad Nacional del Sur-CONICET, San Juan 670, 8000 Bahía Blanca, Argentina
e-mail: psardina@gmail.com

P. V. Perepelizin
Museo Argentino de Ciencias Naturales 'Bernardino Rivadavia'-CONICET, Av. Angel Gallardo 470, 1405 Buenos Aires, Argentina
e-mail: perepelizin@macn.gov.ar

D. Boltovskoy
Facultad de Ciencias Exactas y Naturales, Instituto de Ecología, Genética y Evolución de Buenos Aires (IEGEBA), Universidad de Buenos Aires-CONICET, Ciudad Universitaria, 1428 Buenos Aires, Argentina
e-mail: demetrio@ege.fcen.uba.ar

D. Boltovskoy
Museo Argentino de Ciencias Naturales 'Bernardino Rivadavia'-CONICET, Buenos Aires, Argentina

D. Boltovskoy (ed.), *Limnoperna fortunei*, Invading Nature - Springer Series in Invasion Ecology 10, DOI 10.1007/978-3-319-13494-9_7

water body, the reservoir Embalse de Río Tercero, yielded an average of 959 ind./m^2. Mussel colonies are usually most abundant and dense along the coastal fringe, where rock outcrops are common. Deeper areas are covered with clay and silt, and are therefore unfit for mussel colonization. Data at hand are still insufficient for describing multiannual trends in mussel abundance in South America; however, ancillary evidence suggests that, after having peaked 7–10 years after introduction, densities have been waning. Size structure of individuals in mussel colonies depends strongly on the time of the year. During periods of peak recruitment (spring to late summer) juveniles < Size structure of individual >2 mm in length can represent >90% of the population, whereas during the winter they normally account for 10-15%.

Keywords *Limnoperna fortunei* · Golden mussel · Colonies · Recruits · Substrate · Density · Size structure · Biomass · Multiannual cycles

Settlement of Recruits

The ecological and economic impacts of *Limnoperna fortunei* are due in part to life history traits typical of its marine ancestors, but very unusual among freshwater animals. Unlike most freshwater bivalves, *L. fortunei* possesses a series of free-swimming larval stages, the last one of which, the pediveliger, can either swim using its velum or crawl using its foot (Cataldo and Boltovskoy 2000; Cataldo et al. 2005; see Chapter "Larval Development of *Limnoperna fortunei*" in this volume). Upon receiving the proper cues, the pediveliger will settle onto an appropriate surface and secrete byssal threads; once anchored, it will complete its metamorphosis to become a postveliger or plantigrade mussel.

Settlement is an active process in which the pediveliger selects the site and substrate on which to settle (Rodriguez et al. 1993). Pediveligers of *L. fortunei* have been shown to recruit preferentially to sites providing some kind of protection. In his pioneering work on *L. fortunei*, Morton (1977) reported that in Plover Cove Reservoir, Hong Kong, *L. fortunei* larvae always prefer to settle into crevices or joints, as compared to open surfaces, and the same was observed by Boltovskoy and Cataldo (1999) and by Sylvester et al. (2007) for larvae settling onto experimental frames in the Lower Paraná River. This behaviour has been extensively addressed in studies of marine sessile invertebrates and is thought to be effective in escaping predation and dislodgement by physical disturbances such as waves and currents (Walters and Wethey 1996).

Aggregates of conspecifics and surfaces covered with a dense biofilm are also preferred by *L. fortunei* for settlement over clean substrata (Morton 1977, Sardiña et al. 2009; Balazote Oliver 2011). Many studies attest to the fact that larvae of marine mussels and other sessile invertebrates are influenced by specific chemical cues to settle and metamorphose within conspecific colonies (Burke 1986; Had-

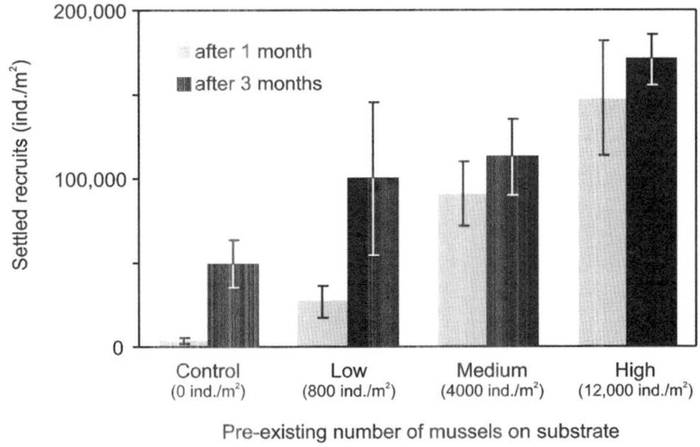

Fig. 1 *Limnoperna fortunei* recruits settled on artificial tiles after 1 and 3 months of deployment in the lower Paraná River delta (*bars* denote means of three replicates, error bars are SE). Tiles were deployed on 28 December 2007. (Based on data from Sardiña et al. 2009)

field and Paul 2001; Tamburri et al. 2008). While settlement cues have not yet been identified in *L. fortunei*, there is considerable evidence from field experiments (see below) for the existence of site-specific chemical cues that promote larval settlement in this species.

Using artificial tiles in a 3-month field experiment, Sardiña et al. (2009) found significantly higher numbers of recruits on tiles with conspecific adults than on tiles without conspecifics, and a positive relationship between the number of recruits and the number of adults on the tiles was also observed (Fig. 1). However, a density-dependent response was also detected when the population appeared to reach the carrying capacity (ca. 112,000–170,000 ind./m^2). As the experiment progressed, the rate of larval settlement on tiles highly covered by mussels diminished (Fig. 1). It was suggested that settlement cues provided by attached individuals (adults and newly established settlers) induced larvae to settle preferentially on substrata with conspecifics, as reported for *Dreissena polymorpha* by Chase and Bailey (1996). Sardiña et al. (2009) hypothesized that these chemical cues have threshold concentrations above which settlement is hindered signaling that the site is no longer advantageous for establishment, for example, because of intraspecific competition for limited food resources when population density is too high. Such a mechanism was also suggested for *D. polymorpha* (Hebert et al. 1991; Wood 2013), as well as for many marine invertebrates (Browne and Zimmer 2001).

The presence of a biofilm was also found to enhance settlement of *L. fortunei* postveligers. In a field experiment, artificial tiles on which a biofilm had developed after exposure underwater in laboratory conditions for different periods of time (0, 1, 2, and 4 weeks) were immersed in the lower Paraná River for 2 weeks to test the response of *L. fortunei* larvae to the presence and age of the biofilm. Larvae were found to recruit more actively on tiles initially covered with heavy biofilm (2 and 4 weeks old) than on tiles with weak biofilm (1 week old) or no biofilm at the time of deployment (Fig. 2; Balazote Oliver 2011). This behaviour mimics settlement of *D. polymorpha* larvae under similar conditions (Wainman et al. 1996; Kavouras and Maki 2003).

Fig. 2 Number of *Limnoperna fortunei* recruits settled on artificial substrata after 2 weeks in November 2009 in the lower delta of the Paraná River. Substrata were previously exposed to periphytic colonization for periods of 0–4 weeks. (Based on data from Balazote Oliver 2011)

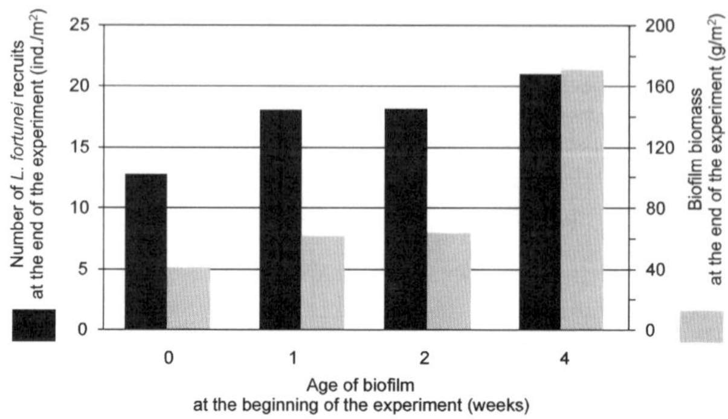

The above experiments, confirmed by subsequent studies (e.g. Nakano et al. 2010; Nakano et al. 2011) elsewhere, indicate that recruitment patterns are governed by two potentially synergistic mechanisms: (1) Conspecifics and biofilms promote larval settlement through specific settlement cues. These chemical cues may not only be released to the medium, but may also act on contact of a larva with a sessile conspecific or biofilm. This conclusion is supported by the results of Morton (1977) and Uryu et al. (1996), who reported strong thigmotaxis in *L. fortunei* larvae, stressing the importance of stimulus of contact for larval settlement. (2) Conspecifics and biofilms provide protection to the newly settled larvae, and thus survival is enhanced compared to barren areas.

These two mechanisms are intimately linked, since enhanced survival rates and other fitness payoffs (e.g. fertilization success in the case of gregarious settlement) would result in the evolutionary acquisition of mechanisms that attract larvae toward a surface covered by conspecifics or biofilms (Sardiña et al. 2009).

Types of Substrata Colonized

Highest mussel densities occur on hard, immobile substrata. In many areas, such substrata are associated with man-made structures, including piers, spur dikes, groynes, pilings, breakwaters, revetments, rock armors, gabions, quay walls, boat hulls, etc., for which reason mussel densities in the vicinity of populated sites are often higher than elsewhere (especially in areas dominated by soft, unconsolidated substrata), and are therefore a poor indicator of overall population numbers.

Colonization by *L. fortunei*, however, is not restricted to hard substrata. In the Paraná River delta mussel clusters occur on soft, silty bottom stabilized with reed or rush roots and fibrous plant debris (Fig. 3a; Boltovskoy et al. 2006). Along the coasts of Salto Grande Reservoir (Argentina/Uruguay), *L. fortunei* thrives on soft silty-sandy areas covered by a thin hardened crust (Fig. 3d). Plants may constitute important sites for attachment, including reed and rush roots (Fig. 3b; Mansur et al. 2003),

Fig. 3 *Limnoperna fortunei* on different substrata. **a** On muddy bottom, attached to loose fibers of plant debris and roots (lower delta of the Paraná River); **b** attached to reed (*Scirpus californicus*) stems and roots (lower delta of the Paraná River); **c** forming a druse around a large *L. fortunei* specimen (Embalse de Río Tercero reservoir); **d** on hardened crust overlaying soft sediments (Salto Grande Reservoir); **e** attached to roots of water hyacinth (*Eichhornia crassipes*) (Middle Paraná River); **f** Attached to a larger bivalve (Unionidae) (Embalse de Río Tercero reservoir); **g** and **h** entirely covering a GRP (fiberglass reinforced plastic) boat hull (Embalse Río Tercero reservoir); **i** and **j** on tree remains recovered from the bottom of Salto Grande Reservoir. (**a** and **d** from Boltovskoy et al. 2006; **e** from Rojas Molina 2010)

roots, rhizomes and stolons of the water hyacinth (*Eichhornia crassipes*, *E. azurea*; Callil et al. 2006; Marçal and Callil 2008, 2012; Rojas Molina 2010; Rojas Molina et al. 2010; Ohtaka et al. 2011; Fig. 3e), waterweed (*Egeria densa*) stems and leaves (Alvarenga et al. 2005), bahiagrass (*Paspalum* sp.; Darrigran and Ezcurra de Drago 2000), hydrilla or water thyme (*Hydrilla verticillata*; Michelan et al. 2014), etc. Although densities on these plants are usually comparatively low (Table 1), they are most probably very important for the dispersion of the species, especially in areas where hard surfaces are scarce, as in most of the habitats associated with the Paraná and Paraguay rivers.

Mussels and other freshwater invertebrates provided with hard shells or exoskeletons are also used by *L. fortunei* for settling. Shu and Wu (2005) reported that 35% of the bivalves (*Arconaia lanceolata, Larnprotula leai, Larnprotula caveata* and *Larnprotula rochechouarti*) of Poyang Lake (China) are "infected" by the mussel, with an average of 6.6 mussels per clam. Both in Asia and in South America, *L. fortunei* has been observed on several bivalves (*Anodontites trapesialis, A. trapezeus, A. tenebricosus, Diplodon koseritzi, Corbicula fluminea, Leila blainvilliana*), gastropods (*Pomacea canaliculata*), crustaceans (*Aegla platensis, Trichodactylus borellianus*), and even freshwater sponges (*Trochospongilla* sp.) (Darrigran and Ezcurra de Drago 2000; Darrigran 2002; Mansur et al. 2003; Ezcurra de Drago et al. 2004; Lopes et al. 2009; Karatayev et al. 2010; Ohtaka et al. 2010; Rojas Molina and Williner 2013). These associations between *L. fortunei* and live substrata may sometimes represent a significant negative impact for the organisms "infected" (as has been suggested for *D. polymorpha*, e.g., Schloesser et al. 1996; see Chapter "Relationships of *Limnoperna fortunei* with Benthic Animals" in this volume). For the mussel, they can also be of significance; although population densities recorded are low, the availability of isolated hard objects for attachment in areas otherwise barren of adequate settling surfaces may represent important seeding spots or stepping stones for further dispersion.

As opposed to *Dreissena* species, where empty shells have been observed to represent an important source of substrate (Strayer et al. 1996; Burlakova et al. 2006; Strayer and Malcom 2006), dead conspecifics and other bivalve and gastropod remains are seldom significant for *L. fortunei* in South America. This difference with *D. polymorpha* is likely due to the fact that calcium concentrations in South American inland waters are normally much lower than those in Europe and North America (Boltovskoy et al. 2006; Karatayev et al. 2007), and therefore dead mollusc shells dissolve before they are colonized. In some water bodies, dissolution of the shells is so fast that it often precedes decomposition of the soft parts, as indicated by the occurrence of dead, softened shells embedded in abundant organic remains (Boltovskoy et al. 2009b).

Table 1 Maximum densities of *L. fortunei* reported by different surveys

Max. dens. reported (ind./m^2)	Site	Substrate	Reference [comments]
85	São Gonçalo channel (Brazil)	Mud	Lopes and Vieira (2012) [probably on hard-surfaced objects lying on the bottom]
100	Pond in Buenos Aires city (Argentina)	Waterweed (*Egeria densa*) stems and leaves	Alvarenga et al. (2005)
108	Upper Paraguay River (Brazil)	Roots of anchored water hyacinth (*Eichhornia azurea*)	Callil et al. (2006)
2500	Middle Paraná River (Argentina)	Artificial (plastic)	Cepero (2003)
3616	Upper Paraguay River (Brazil)	Roots of water hyacinth (*Eichhornia crassipes*)	Marçal and Callil (2012) [mean on E. crassipes for 15 lakes: 1327 ind./m^2]
22,400	Itaipú Reservoir (Brazil/Paraguay)	Artificial (plastic)	Belz (2006)
31,900	Río Santiago (Río de la Plata estuary, Argentina)	Artificial (plastic netting)	Bonel (2011)
50,000	Salto Grande Reservoir (Argentina/Uruguay)	Rock	Boltovskoy et al. (2006)
65,700	Coronda River (Argentina)	Artificial (plastic netting)	Bonel (2011)
82,151	Río de la Plata estuary (Argentina)	Mudrock	Darrigran and Pastorino (1993)
93,000	Itaipú Reservoir (Brazil/Paraguay)	Artificial (wood)	Takeda and Fujita (2012)
119,000	Itaipú Reservoir (Brazil/Paraguay)	Artificial (dam structures)	Mata (2011)
120,000	Jacuí River delta (Brazil)	Artificial (steel)	Kapusta and Fagondes de Freitas (2012)
124,000	Upper Paraguay River (Brazil)	Rock outcrops	Oliveira and Calheiros (2012)
143,500	Lagoa dos Patos (Brazil)	Rush (*Scirpus californicus*) rhizomes	Mansur et al. (2003)
150,000	Bagliardi Beach, Río de la Plata estuary (Argentina)	Mudrock	Darrigran and Ezcurra de Drago (2000)

Table 1 (continued)

Max. dens. reported (ind./m^2)	Site	Substrate	Reference [comments]
170,000	Río de la Plata estuary (Argentina)	Mudrock	Darrigran et al. (2003)
203,000	Lower Paraná River delta (Argentina)	Artificial (PVC)	Sylvester (2006), Sylvester et al. (2007) [92 % below 1 mm]
701,000	Bagliardi Beach, Río de la Plata estuary (Argentina)	Artificial (stone revetments)	Spaccesi and Rodrigues Capitulo (2012) [annual mean based on 12 samples: 227,000]

Location and Structure of Mussel Colonies

Almost invariably, colonization of a new substrate starts in the crevices, holes, corners, angles and other less accessible areas. This tendency to form clusters is not restricted to the larvae (Morton 1977; see above), but is also displayed by dislodged adults, which normally crawl around for some time before reattaching, reattachment being more frequent in the angles than elsewhere (Uryu et al. 1996; see Chapter "Behavior and Taxis of Young and Adult *Limnoperna fortunei*" in this volume). In many cases, colonization is restricted to these protected sites and, regardless of the age of the mussel bed, does not extend beyond them (Fig. 4). In others, however, colonization starts there but eventually covers the entire surface available (Fig. 3g, h; 4 and 5). Aggregation in mussels and other sessile organisms, both freshwater and marine, has been described for numerous species. Experimental studies indicate that this gregarious behaviour confers better protection against water movements and predators, increases the amount of surface available for attachment, and improves the chances of successful fertilization (Côté and Jelnikar 1999; Cheung et al. 2004; Kobak et al. 2010).

Colonization of unprotected, widely open areas is sometimes clearly associated with the abundance and diversity of predators. In most areas of the Paraná River, including its lower delta and the Río de la Plata estuary, dense *L. fortunei* beds develop on many hard substrata, but boat hulls are never colonized (regardless of the presence, age and type of their antifouling coatings), with the exception of restricted crevices and angles around propellers, submerged components of rudder mechanisms and scupper pipe fittings. On the other hand, in the reservoir Embalse de Río Tercero, boat hulls, especially those whose antifouling coating has not been maintained for some time, are completely overgrown (Fig. 3g and h). A major difference between these two habitats is that while the Paraná River hosts anywhere between 200 and >500 fish species (Bonetto 1998; López et al. 2008), in the reservoir only 13 species have been recorded (Freyre et al. 1983). Furthermore, most of the ca. 50 fishes known to consume adult mussels (see Chapter "Trophic Relationships of *Limnoperna fortunei* with Adult Fishes" in this volume and review in Boltovskoy and Correa 2015) are absent from Embalse de Río Tercero. In the Lower Paraná delta, predators (chiefly fishes) eliminate up to 95 % of the mussel

Fig. 4 Different degrees of development of mussel beds on pilings along the coast of the Luján River (lower delta of the Paraná River). (From Boltovskoy et al. 2006)

Fig. 5 Mussel beds on rock outcrops in the reservoir Embalse de Río Tercero in 2005 and 8 years later, in 2013. (Courtesy of Miguel Hechem)

biomass (Sylvester et al. 2007), which suggests that the lower predation pressure in Embalse de Río Tercero allows mussel beds to develop in unprotected areas (but see below).

Nevertheless, predation is most probably not the only deterrent to mussel colonization. In the Luján River (Lower Paraná delta), concrete pilings located a few hundred meters apart host very different population densities (Fig. 4; Boltovskoy et al. 2006). Widely dissimilar degrees of mussel coverage and extremely high patchiness unassociated with more or less obvious causes, like differences in predation pressure or substrate availability, seem to be common throughout the range of the species. In an attempt to pinpoint the factors responsible for the uneven distribution of mussel beds on mudrock substrata along the coast of the Río de la Plata estuary (Fig. 6), Boltovskoy et al. (unpublished data) monitored changes in nine fixed areas 70 × 70 cm in size for 36 months. Aside from a general trend toward decreasing mussel densities with increasing air exposure (Fig. 7), none of the variables considered (small-scale topographic differences, insolation, substrate tilt, wave exposure, etc.) were associated with mussel coverage. Thus, there probably are complex biotic and abiotic interactions, as well as intrinsic factors, whose significance still eludes our comprehension.

Fig. 6 *L. fortunei* beds on mudrock along the coast of the upper Río de la Plata estuary

Several authors investigated *L. fortunei*'s preferences for attachment as a function of the orientation of the substrate, but the results are still inconclusive. Morton (1977) suggested that a combination of depth-dependent geotactic and phototactic responses may be responsible for dissimilar settling rates on vertical and horizontal upward- and downward-facing surfaces of experimental substrata. Uryu (1996) performed a series of laboratory experiments which also suggest that geotaxis and phototaxis affect settlement. These experiments yielded interesting data on the behaviour of the mussels, but they may not necessarily constitute an adequate proxy of actual settling rates in nature. For example, horizontal upward-facing surfaces often host much lower mussel densities than vertical and horizontal downward-facing

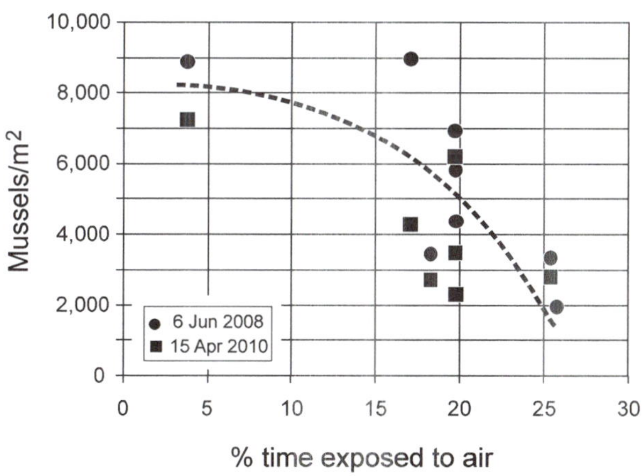

Fig. 7 Relationship between height of substrate (coastal mudrock along the Río de la Plata estuary) and *Limnoperna fortunei* densities on two different sampling dates. Substrate height is given as the mean proportion of overall time when the corresponding site is exposed to air (i.e. above the waterline). Exposure to air is based on a long-term series of historical water-level readings (rather than on calculated tide values, which are very strongly influenced by wind direction, intensity and duration). *Dotted line* indicates suggested trend

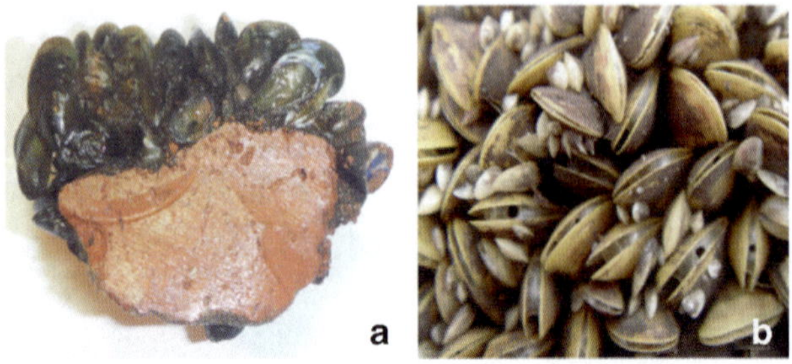

Fig. 8 Cross-section of a rock covered by mussels (**a**) and top view of a mussel bed showing small juveniles attached to the shells of larger individuals (**b**)

ones because the former retain more sediments (Xu et al. 2013b). Also in pipes, when the lower half accumulates clay and silt, colonization is restricted to the sides and roof of the ducts.

The height of mussel beds can occasionally exceed 10 cm (Xu et al. 2013b), but normally it is around 5–7 cm. Unlike the zebra mussel, whose colonies can attain a vertical thickness of up to 20–30 individuals (Burks et al. 2002), in golden mussel colonies most adults tend to be at least partly attached to the substrate, and although the irregular distribution of the shells does not allow defining the number of mussel layers involved precisely, large (> 7–10 mm) specimens supported only by underlying shells are comparatively few (Fig. 8a). On the other hand, juveniles up to 3–4 mm in length are often attached only to the sides of larger conspecifics (Fig. 8b).

As shown by studies of the zebra mussel, this spatial organization may partly reflect the fact that veligers tend to settle onto the surface of colonies, which results in a vertical stratification with larger individuals at the bottom and younger ones at the top (Burks et al. 2002). However, there probably is also active migration of the smaller specimens from deep and intermediate positions toward the top of the mussel bed. Burks et al. (2002) found significant differences between the quality of the interstitial water at the bottom, middle and top of colonies of *D. polymorpha*, whereby oxygen was the lowest and NO_2–N (but not NO_3–N) were the highest at the base. Presumably in response to this gradient, as well as the relative scarcity of food at the base of the colony (Tuchman et al. 2004), mussels were observed to migrate upwards, with significantly higher relocation rates for the smaller (< 13 mm) individuals. Concomitantly, mortality rates were higher in the bottom layers of the colony than at the surface. This suggests that as *L. fortunei* individuals grow in size, they are gradually displaced down toward the bottom of the mussel bed. By this time their capacity to detach, migrate and reattach has diminished, while their tolerance to adverse conditions has increased allowing them to survive in this harsher environment. Although environmental conditions at the bottom of the mussel bed are less favourable, direct attachment to the substrate should be more favourable than attachment to the shells of other mussels, since there is a risk of dislodgement when the substrate mussels die.

Differences in colony structure between *L. fortunei* (single-layered) and *D. polymorpha* (multi-layered) may also be due to the very fast postmortem dissolution of dead golden mussel shells in the Ca-poor waterbodies colonized (see Chapter "Parallels and Contrasts Between *Limnoperna fortunei* and Species of *Dreissena*" in this volume). Dead individuals are but a very small proportion of the population (usually less than 5 %), which indicates that dislodgment, dissolution and destruction of dead mussels is fast.

Densities

Reported *L. fortunei* densities on natural and artificial substrata are extremely variable (Table 1). Differences are primarily associated with type of substrate, but also with other factors, including time after initial colonization, season, depth, water quality, etc. Curiously, the highest value reported in the literature is from Bagliardi Beach, in the Río de la Plata estuary, where *L. fortunei* was first detected in South America (Pastorino et al. 1993). Here, on granite revetments, in February 2002, Spaccesi and Rodrigues Capítulo (2012) recorded over 700,000 ind./m^2 (Table 1). The second highest estimate (203,000 ind./m^2) is that of Sylvester (2006) on PVC artificial substrata in the lower delta of the Paraná River.

Assuming that the footprint of a 5–25 mm shell is ca. 10–70 mm^2, and that no space is left between shells, 1 m^2 of substrate can theoretically accommodate around 15,000–100,000 animals. Thus, extremely high densities are obviously largely due to the overwhelming dominance of tiny recent recruits below 1–2 mm in length. Densities above 30,000–50,000 ind./m^2 are almost invariably strongly dominated by very small mussels (Fig. 9), and seasonal changes in mussel densities are chiefly a reflection of recruitment processes (Fig. 10). For example, the second highest density record (203,000 ind./m^2) corresponds to a sample recovered in December 2003 from an artificial substrate where 97 % of the specimens were below 2 mm in length (Sylvester et al. 2007).

Despite the fact that mussel densities have been estimated numerous times in different areas and on different substrata (Table 1), their usefulness as an indicator

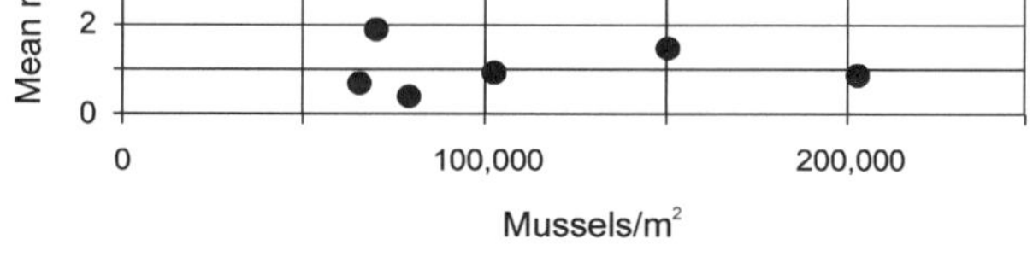

Fig. 9 Relationship between mean mussel size and overall mussel density in 11 samples from artificial substrata deployed in the Lower Paraná River delta between December 2002 and June 2004. (Based on data from Sylvester 2006)

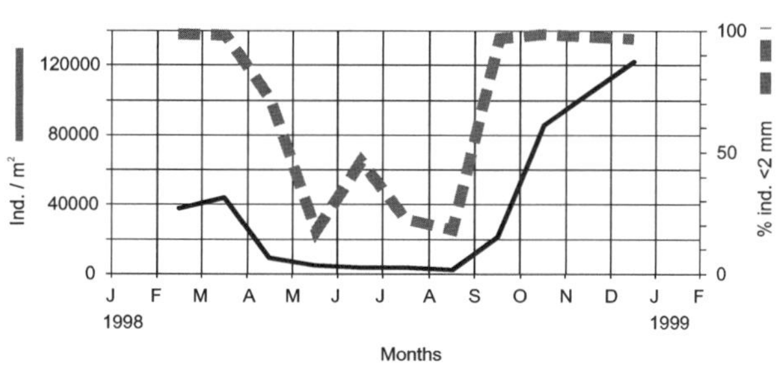

Fig. 10 Changes in mussel total densities and in the proportions of individuals <2 mm in length on artificial substrata deployed in the Lower Paraná River between 20 Jan 1998 and 17 Dec 1998 (Based on data from Boltovskoy and Cataldo 1999)

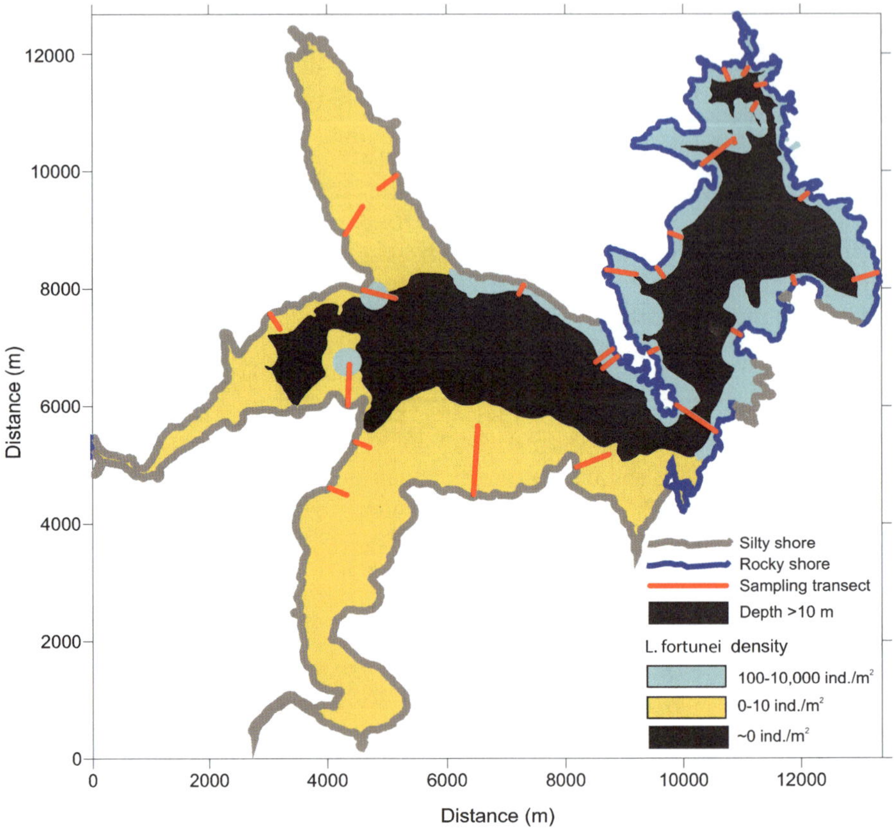

Fig. 11 Distribution of *Limnoperna fortunei* beds in Embalse Río Tercero reservoir. (Based on data from Boltovskoy et al. 2009a)

of the ecological importance of the bivalve is very limited. Indeed, practically all these figures refer to abundances over very restricted areas, usually less than 1 m^2 in size, and the sites in question are not selected at random, but because they are densely covered by mussels. Assessment of average densities over large areas is complicated by the fact that beds of *L. fortunei* have an extremely patchy distribution (Fig. 6). This seems to be associated not only with the uneven distribution of available substrate, but also with some other less obvious traits (see above).

The only work that attempted to produce density estimates over a large area (an entire water body) is that of Boltovskoy et al. (2009a), in a 47 km^2 reservoir (Embalse de Río Tercero, Argentina). Densities were assessed on the basis of diver-collected samples along 25 transects perpendicular to the coast. According to this survey, the reservoir hosted 4.5×10^{10} mussels, over 98 % of them along the coastal fringe between depths of ~1 and 10 m (Figs. 11 and 12). Deeper areas were invariably covered by a thick (up to over 23 m) layer of silt with practically no mussels. Mussel presence was closely associated with bottom type, where rocks yielded the highest mean values, and silt the lowest (Figs. 12 and 13). The mean density for the entire reservoir was 959 ind./m^2, or around 0.1 ind. per liter of reservoir water.

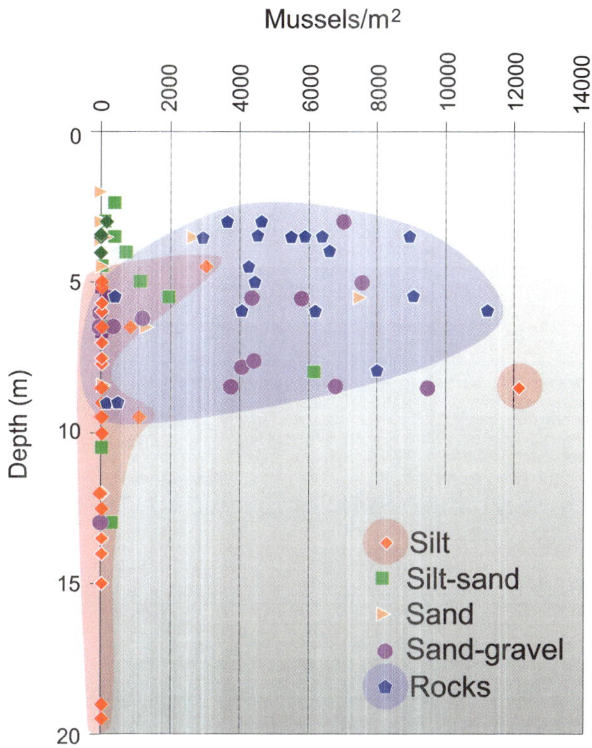

Fig. 12 *Limnoperna fortunei* densities on different substrata as a function of water depth in Embalse Río Tercero reservoir. The highest value (12,096 ind./m² on silty bottom) was recorded on an isolated hard object lying on the mud. (From Boltovskoy et al. 2009a)

In lakes and rivers, removal and resettlement of sediment particles decreases with depth (Bloesch 2004); this process is responsible for the fact that the deepest areas are usually covered by fine-grained sediments (clay and silt), whereas exposed rock, boulders and pebbles are restricted to shallower areas, normally along the coasts. Thus, the type of distribution of *L. fortunei* found in Embalse de Río Tercero reservoir, where colonization is restricted to the coastal fringe down to depths < 10 m, is probably characteristic of many other reservoirs, lakes and rivers. This constraint imposes a severe cost on the bivalve, especially in waterbodies where water-level fluctuations are large exposing extensive *L. fortunei* beds for periods long enough to produce massive kills. Such events have been observed in

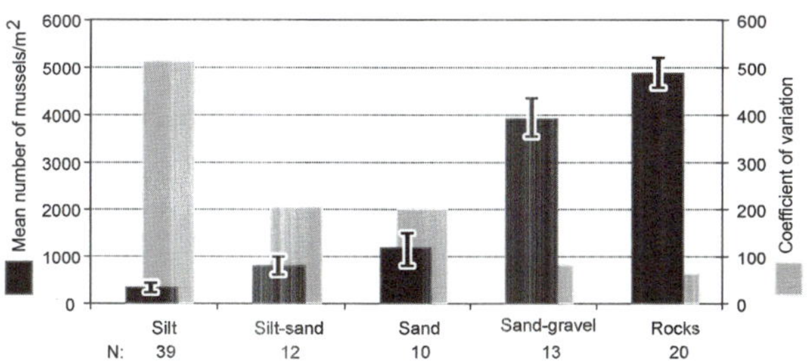

Fig. 13 Mean densities and variabilities of *L. fortunei* recorded on different substrate types (standard errors and coefficients of variation) in Embalse Río Tercero reservoir. (Modified after Boltovskoy et al. 2009a)

Fig. 14 Empty *Limnoperna fortunei* shells along Salto Grande Reservoir resulting from a massive mortality event due to extended exposure of coastal mussel beds during a period of low water levels. (From Boltovskoy et al. 2006)

several Argentine reservoirs, including Embalse de Río Tercero and Salto Grande (Figs. 5 and 14).

With the exception of the Uruguay River, where coastal stretches with rock outcrops and large boulders are common, the margins of most other rivers of the Río de la Plata basin are largely dominated by loose sediments, ranging from clay to sand (the main river channels are invariably soft bottom throughout). Hard substrata are therefore scant in most of the area colonized by the mussel, which may suggest that sessile populations are few, small and scattered. However, indirect evidence does not support this conclusion. According to data collected in 2005–2006, the mean annual density of *L. fortunei* planktonic larvae in Embalse de Río Tercero is 4168 ind./L, which are produced by a total population of 4.5×10^{10} animals spread over 47 km^2, or 959 mussels/m^2 (Boltovskoy et al. 2009a). Mean annual larval densities in the Paraná and Río de la Plata are around 6000–7000 larvae/m^3, (Boltovskoy et al. 2009b); assuming that the mussel's fertility is roughly similar in these water bodies (fertility is probably somewhat lower in Embalse de Río Tercero; Boltovskoy et al. 2009b), adult densities needed to produce the larval output recorded in the Paraná and Río de la Plata must be at least comparable to those of Embalse de Río Tercero. Thus, the significance of alternative substrata, in particular biological substrata (reed roots, emergent, submerged and floating plants, tree branches and trunks, Fig. 3), must be more important than the impression conveyed by a visual assessment of these environments.

In areas subject to tidal and/or wind-induced changes in water level, mussels have some tolerance to air exposure (see Chapter "Control of *Limnoperna fortunei* Fouling by Desiccation" in this volume), but densities drop sharply away from the permanently submerged sectors. Mudrock substrata along the coasts of the Río de la Plata estuary host extensive *L. fortunei* beds (Fig. 6); the tidal span in these areas is about 1–1.5 m, but the effects of wind can increase these values to over 5 m. Analysis of *L. fortunei* densities along transects perpendicular to the coastline shows that an increase in air exposure from 5 % (of the overall time) to 25 % results in a fourfold density drop (from ca. 8000–2000 ind./m^2, Fig. 7).

L. fortunei is clearly a ubiquitous species with very broad environmental tolerance. Most of the limiting factors that are important for *D. polymorpha*, including temperature, pollution, pH, nutrients and dissolved calcium (Ramcharan et al. 1992b, 1992a) seem to be well within the tolerance ranges of the golden mussel (Karatayev et al. 2007; Xu et al. 2013a). However, in some areas, such as the Pantanal wetlands associated with the Paraguay River, extreme conditions (e.g. very low calcium concentrations and carbonate mineral undersaturation, extensive anoxic events) may limit the distribution of *L. fortunei* or produce important seasonal die-offs (Oliveira et al. 2010a, 2010b, 2010c, 2011). Pollution is tolerated by the mussel (Contardo-Jara et al. 2009; Young et al. 2014), but at a cost: contaminated areas host lower densities and individuals have a lower length:width ratio, probably reflecting a slower growth rate (Bonel et al. 2013).

Depth-Related Colonization Trends

Some studies have noticed differences in the density and/or size structure of mussel beds in association with water depth. These variations have been tentatively ascribed to vertical gradients in the abundance of larvae (probably in response to environmental parameters, including light penetration, turbidity, food availability, temperature, dissolved oxygen), to behavioural responses of the recruits (Uryu et al. 1996), and/or to differences in predation pressure.

On artificial substrata deployed in Plover Cove Reservoir (Hong Kong) at five depths between 0 and 12 m, the highest densities of recruits were found between 6 and 9 m (Morton 1977). Brugnoli et al. (2011) also recorded higher densities of recruits at 10 m depth than at 0.5 m (Palmar Reservoir, Uruguay). In a study using artificial substrata deployed at 6, 12 and 18 m, Nakano et al. (2010) concluded that, after 105 days at 18 m densities of recruits were higher, but their sizes were lower than higher up in the water column.

In a series of laboratory experiments, Iwasaki (1997) noticed that mussels kept in a fish tank tend to climb up the glass walls, nearly 40 % of them attaching just beneath the water surface (see Chapter "Behavior and Taxis of Young and Adult *Limnoperna fortunei*" in this volume). A similar behaviour was also attributed to populations in aqueducts. He speculated that such behaviour may respond to several factors, including avoidance of deeper, brackish water in estuaries, avoidance of siltation, of deoxygenated water layers, and of benthic predators. Our own (unpublished) observations confirm that many mussels of variable size climb up the walls of the fish tank and re-attach next to the air–water interface, but the advantages of this behaviour are still unclear because most of these animals die when the water level in the tank drops from evaporation (as would presumably happen in nature).

While there is little doubt that recruits favour precolonized areas and crevices over open surfaces, it is not yet clear whether variations in mussel beds associated with depth are due to differences in the preferred living depth of the larvae, in their settling depths, or to postsettling effects associated with differential survival and predation.

Multiannual Changes in Adult Densities

The difficulties described above concerning estimates of adult densities that are unbiased by the patchy distribution of mussel beds are also responsible for the lack of reliable information on the evolution of *L. fortunei* populations over multiannual periods. Surveys aimed at investigating reproduction and population dynamics, both on natural substrata (Iwasaki and Uryu 1998; Belz et al. 2010; Spaccesi and Rodrigues Capitulo 2012) and on artificial ones (Morton 1977; Boltovskoy and Cataldo 1999; Pestana 2006; Sylvester 2006; Santos et al. 2008; Nakano et al. 2010, 2011; Bonel 2011; Brugnoli et al. 2011), yielding precise abundance estimates (albeit on small areas), are normally restricted to a single annual cycle.

Darrigran et al. (2003) reported densities of settled individuals at Bagliardi Beach (Río de la Plata estuary) between October 1991 and October 2001. However, interpretation of these data is difficult since only one or two monthly values were available for most years, and studies were not carried out in 1996, 1997, 1999 and 2000. Mata (2011) produced a similar series for Itaipú Reservoir between 2001 and 2010, based on 6–12 data points per year. A potential problem of comparing these abundance estimates is that they were performed collecting and counting all mussels from a known surface, which was obviously different on each new sampling date. Thus, the resulting series reflects two sources of variation (time and site), rather than time only. Despite these shortcomings, both studies concluded that *L. fortunei* reached peak densities 3–5 years after invading the corresponding water body, and decreased thereafter.

The longest multiannual record for the golden mussel is a 9-year series (2004–2013) of the abundance of its larvae in the reservoir Embalse de Salto Grande, based on weekly plankton samples (Boltovskoy et al. 2009b, 2013). *Limnoperna fortunei* was first detected in this reservoir around 2000, and by 2013 larval densities did not show signs of decreasing (Boltovskoy et al. 2013). It should be noted, however, that recruitment of the mussel in this water body is strongly affected by recurrent cyanobacterial blooms that kill *L. fortunei* larvae (Boltovskoy et al. 2013), and this may account for a unique long-term trend in these populations (see Chapter "Reproductive Output and Seasonality of *Limnoperna fortunei*" in this volume).

The fact that *L. fortunei*, as any other introduced species, needs some time to build up its population numbers after invading is obvious and has been observed repeatedly (Boltovskoy et al. 2009a, 2009b). On the other hand, subsequent density declines or significant interannual variations, like those described for *D. polymorpha* (Stanczykowska 1977; Ramcharan et al. 1992b; Burlakova et al. 2006; Strayer and Malcom 2006) have not yet been unequivocally demonstrated, although there are some hints that support the notion that populations of the golden mussel in South America have been waning in recent years. Observations in the lower delta of the Paraná River and Río de la Plata estuary seem to indicate that mussel beds are less dense and occur more sparsely than 10 years ago. This impression is confirmed by comments from local residents, who agree that mussel presence has decreased in recent years. Evidence from other waterbodies, such as the Embalse de Río Tercero reservoir, also point in the same direction. In particularly dry years, the water level in this reservoir can drop over 10 m exposing colonized areas. One such event occurred in November 2005, exposing hard substrata very densely covered by mussel growth. Eight years later the same substrata were exposed again by another strong drawdown, showing a dramatic decrease in *L. fortunei* densities (Fig. 5). Although none of this has been evaluated and quantified objectively, it agrees with the notion of a cycle characteristic of many invasive species, whereby the initial explosive population growth, shortly after introduction, is followed by a decline and subsequent stabilization (e.g. Stanczykowska 1977). Limitations in the carrying capacity, including availability of food and substrate, could explain a stabilization of population densities (although food shortage is unlikely, at least in the Paraná watershed; Sylvester et al. 2005), but not a decline. On the other hand, an increase in predation pressure by organisms that consume *L. fortunei*, due to growth of predators favoured by the availability of more high quality food, could account for lower survival rates. In the Great Lakes, *Dreissena* species have been observed to decline steadily after an initial density peak due to increasing predation pressure by several water fowl attracted to the area by the availability of mussels (Petrie and Knapton 1999). Many fish species have been reported to feed actively on both adults and larvae of the golden mussel (see Chapters "Trophic Relationships of *Limnoperna fortunei* with Larval Fishes" and "Trophic relationships of *Limnoperna fortunei* with adult fishes" in this volume); several of these take years to reach maturity (e.g. Sverlij et al. 1993), which could account for the lag between the mussel's peak population densities and their subsequent decline as predator populations increase.

Size Structure in Mussel Colonies

The size-frequency distribution in mussel beds depends chiefly on the time of year. During the reproductive season (typically between spring and autumn; see Chapter "Reproductive Output and Seasonality of *Limnoperna fortunei*" in this volume), tiny recruits 0.5–2 mm in length account for over 90 % of the population. In winter, their proportion is the lowest (5–20 %), but they rarely drop to zero, suggesting that reproduction never stops altogether (Fig. 15; Cataldo and Boltovskoy 2000). Sylvester et al. (2007) noticed that highest mortalities occur immediately after settlement, at sizes below 1 mm, when over 93 % of the juveniles are lost. For animals > 1 mm, mortality between successive size classes drops sharply, oscillating around 20 % for the interval between 2 and 20–23 mm. These data indicate that approximately 2 % of the animals that reach the settling stage survive until first reproduction (at about 7 mm, cf. Darrigran et al. 1999), and only 0.5 % survive the first year of life (approximately 20 mm, cf. Boltovskoy and Cataldo 1999). Overall densities decrease sharply during the winter, chiefly due to reduced recruitment and, probably, to enhanced mortality.

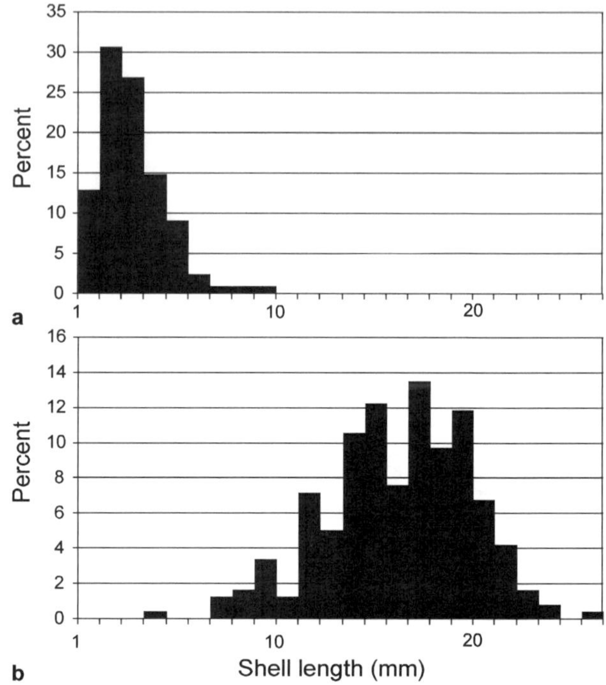

Fig. 15 Changes in the size–frequency distribution of *Limnoperna fortunei* shells **a** during the reproductive season (February) and **b** during the period of reproductive relaxation (October). (Based on data from Sylvester 2006)

Biomass

The biomass of mussels is closely associated with their size; the function that best describes the relationship between length and dry tissue weight is a power function (see Chapter "Population Dynamics and Growth of *Limnoperna fortunei*" and Fig. 6 in this volume). The strength of this association, however, changes with mussel size, with cutoff values at 10–15 mm (Sylvester 2006). While both size and weight increase with age, weight obviously increases faster, and growth from 2.5 to 30 mm in length involves a 12-fold length increase in size, but a 427-fold increase in tissue dry weight. Although this contrast may seem obvious, it underscores limitations of data reporting mussel densities alone, without information on the size structure of the population involved (Fig. 16; Young et al. 1996; Burlakova et al. 2006).

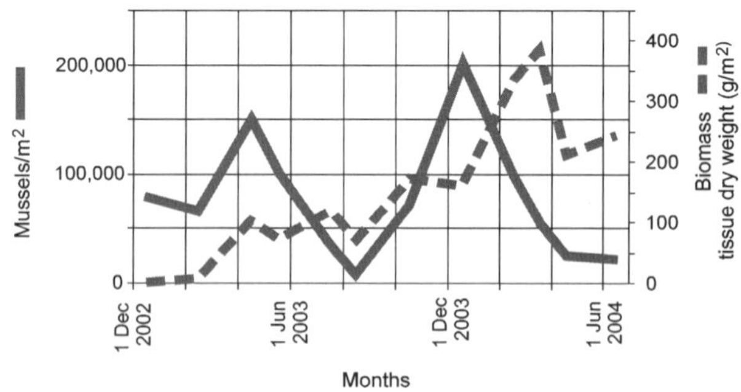

Fig. 16 Seasonal changes in *Limnoperna fortunei* density and biomass on artificial substrata deployed in the lower delta of the Paraná River between 6 November 2002 and 15 June 2004. (Based on data from Sylvester 2006)

The valve accounts for approximately 80% of the overall (dry) weight of the mussel. This proportion changes little throughout the life of the animal. Water represents around 93–94% of the weight of the mussel's tissue (excluding the shell), with slightly higher values in older specimens (Sylvester 2006).

Acknowledgements This work was partially financed by grants from the University of Buenos Aires, Argentina (UBA X-020 and 20020100100035) and from the Argentine Agencia Nacional de Promoción Científica y Tecnológica, Argentina (PICT 2007 1968) to DB.

References

Alvarenga C, Chalde T, Rosa J, Sánchez ML, Santoferrara L (2005) Fijación del bivalvo invasor *Limnoperna fortunei* sobre distintos sustratos en un estanque urbano. Facultad de Ciencias Exactas y Naturales, Universidad de Buenos Aires. Undergraduate course report. Buenos Aires (Argentina), pp 1–10

Balazote Oliver A (2011) Factores que inducen el establecimiento de larvas del mejillón dorado, *Limnoperna fortunei* (Bivalvia: Mytilidae). BS Thesis, Universidad de Buenos Aires (Argentina), pp 1–27

Belz CE (2006) Análise de risco de bioinvasão por *Limnoperna fortunei* (Dunker, 1857): um modelo para a bacia do rio Iguaçu, Paraná. PhD Thesis, Universidade Federal do Paraná (Brazil), pp 1–102

Belz CE, Darrigran GA, Bonel N, Mader Netto OS (2010) Density, recruitment, and shell growth of *Limnoperna fortunei* (Mytilidae), an invasive mussel in tropical South America. J Freshw Ecol 25:227–233

Bloesch J (2004) Sedimentation and lake sediment formation. In: O'Sullivan PE, Reynolds CS (eds) The lakes handbook. Volume 1. Limnology and limnetic ecology, Blackwell, Oxford, pp 197–229

Boltovskoy D, Cataldo D (1999) Population dynamics of *Limnoperna fortunei*, an invasive fouling mollusc, in the lower Paraná River (Argentina). Biofouling 14:255–263

Boltovskoy D, Correa N (2015) Ecosystem impacts of the invasive bivalve *Limnoperna fortunei* (golden mussel) in South America. Hydrobiologia 746:81–95.

Boltovskoy D, Correa N, Cataldo D, Sylvester F (2006) Dispersion and ecological impact of the invasive freshwater bivalve *Limnoperna fortunei* in the Río de la Plata watershed and beyond. Biol Invasions 8:947–963

Boltovskoy D, Karatayev A, Burlakova L, Cataldo D, Karatayev V, Sylvester F, Mariñelarena A (2009a) Significant ecosystem-wide effects of the swiftly spreading invasive freshwater bivalve *Limnoperna fortunei*. Hydrobiologia 636:271–284

Boltovskoy D, Sylvester F, Otaegui A, Leytes V, Cataldo D (2009b) Environmental modulation of the reproductive activity of the invasive mussel *Limnoperna fortunei* in South America. Austral Ecol 34:719–730

Boltovskoy D, Correa N, Bordet F, Leites V, Cataldo D (2013) Toxic *Microcystis* (cyanobacteria) inhibit recruitment of the bloom-enhancing invasive bivalve *Limnoperna fortunei*. Freshw Biol 58:1968–1981

Bonel N (2011) Demografía experimental de un bivalvo invasor, *Limnoperna fortunei* (Dunker, 1857), en la cuenca del Plata. PhD Thesis, Universidad Nacional de La Plata (Argentina), pp 1–180

Bonel N, Solari LC, Lorda J (2013) Differences in density, shell allometry and growth between two populations of *Limnoperna fortunei* (Mytilidae) from the Río de la Plata basin, Argentina. Malacologia 56:43–58

Bonetto AA (1998) Panorama sinóptico sobre la ictiofauna, la pesca y la piscicultura en los ríos de la cuenca del Plata con especial referencia al Paraná. Revista de Ictiología 6:3–15
Browne KA, Zimmer RK (2001) Controlled field release of a waterborne chemical signal stimulates planktonic larvae to settle. Biol Bull 200:87–91
Brugnoli E, Dabezies MJ, Clemente JM, Muniz P (2011) *Limnoperna fortunei* (Dunker 1857) en el sistema de embalses del Rio Negro, Uruguay. Oecologia Australis 15:576–592
Burke RD (1986) Pheromones and the gregarious settlement of marine invertebrate larvae. Bull Marine Sci 39:323–331
Burks RL, Tuchman NC, Call CA (2002) Colonial aggregates: effects of spatial position on zebra mussel responses to vertical gradients in interstitial water quality. J N Am Benthol Soc 21:64–75
Burlakova LE, Karatayev AY, Padilla DK (2006) Changes in the distribution and abundance of *Dreissena polymorpha* within lakes through time. Hydrobiologia 571:133–146
Callil CT, Mansur MCD, Marcelo SM (2006) Bivalves invasores no Pantanal. In: XVII Encontro Brasileiro de Malacologia, Rio De Janeiro (Brazil)
Cataldo D, Boltovskoy D (2000) Yearly reproductive activity of *Limnoperna fortunei* (Bivalvia) as inferred from the occurrence of its larvae in the plankton of the lower Paraná River and the Río de la Plata estuary (Argentina). Aquat Ecol 34:307–317
Cataldo D, Boltovskoy D, Hermosa JL, Canzi C (2005) Temperature-dependent larval development rates of *Limnoperna fortunei* (Bivalvia: Mytilidae). J Molluscan Stud 71:41–46
Cepero E (2003) Control de *Limnoperna fortunei*. Infestación de plantas potabilizadoras de agua. In: 13 Congreso Argentino de Saneamiento y Medio Ambiente, Buenos Aires (Argentina)
Contardo-Jara V, Galanti LN, Ame MV, Monferran MV, Wunderlin DA, Wiegand C (2009) Biotransformation and antioxidant enzymes of *Limnoperna fortunei* detect site impact in watercourses of Córdoba, Argentina. Ecotoxicol Environ Saf 72:1871–1880
Côté IM, Jelnikar E (1999) Predator-induced clumping behaviour in mussels (*Mytilus edulis* Linnaeus). J Exp Marine Biol Ecol 235:201–211
Chase ME, Bailey RC (1996) Recruitment of *Dreissena polymorpha* : Does the presence and density of conspecifics determine the recruitment density and pattern in a population? Malacologia 38:19–31
Cheung SG, Tong PY, Yip KM, Shin PKS (2004) Chemical cues from predators and damaged conspecifics affect byssus production in the green-lipped mussel *Perna viridis*. Mar Freshw Behav Physiol 37:127–135
Darrigran GA (2002) Potential impact of filter-feeding invaders on temperate inland freshwater environments. Biol Invasions 4:145–156
Darrigran GA, Ezcurra de Drago I (2000) Invasion of the exotic freshwater mussel *Limnoperna fortunei* (Dunker, 1857) (Bivalvia: Mytilidae) in South America. The Nautilus 114:69–73
Darrigran GA, Pastorino G (1993) Bivalvos invasores en el Rio de la Plata, Argentina. Comunicaciones de la Sociedad Malacológica del Uruguay 7:309–313
Darrigran GA, Penchaszadeh PE, Damborenea C (1999) The reproductive cycle of *Limnoperna fortunei* (Dunker, 1857) (Mytilidae) from a neotropical temperate locality. J Shellfish Res 18:361–365
Darrigran GA, Damborenea C, Penchaszadeh PE, Taraborelli C (2003) Reproductive stabilization of *Limnoperna fortunei* (Bivalvia, Mytilidae) after ten years of invasion in the Americas. J Shellfish Res 22:141–146
Ezcurra de Drago I Marchese M Wantzen KM (2004) Benthos of a large neotropical river: spatial patterns and species assemblages in the Lower Paraguay and its floodplains. Arc Hydrobiol 160:347–374
Freyre LR, Protogino LC, Iwaszkiw JM (1983) Demografía del pejerrey *Basilichtys bonaeriensis bonaeriensis* (Pisces, Atherinidae) en el Embalse Río Tercero, Córdoba. Descripción de las artes de pesca. Biología Acuática 4:1–25
Hadfield MG, Paul VJ (2001) Natural chemical cues for settlement and metamorphosis of marine invertebrate larvae. In: McClintock JB, Baker BJ (eds) Marine chemical ecology, CRC Press, Boca Raton, pp 431–461

Hebert PDN, Wilson CC, Murdoch MH, Lazar R (1991) Demography and ecological impacts of the invading mollusc *Dreissena polymorpha*. Can J Zool 69:405–409

Iwasaki K (1997) Climbing behaviour and tolerance to aerial exposure of a freshwater mussel, *Limnoperna fortunei*. Venus 56:15–25

Iwasaki K, Uryu Y (1998) Life cycle of a freshwater Mytilid mussel, *Limnoperna fortunei*, in Uji River, Kyoto. Venus 57:105–113

Kapusta SK, Fagondes de Freitas SM (2012) *Limnoperna fortunei* na bacia hidrográfica do baixo Rio Jacuí: relações com a comunidade de macroinvertebrados bentônicos. In: Mansur MCD, Santos CP, Pereira D, Padula Paz IC, Leite Zurita ML, Raya Rodriguez MT, Vilar Nehrke M, Aydos Bergonci PE (eds) Moluscos límnicos invasores no Brasil. Biologia, prevenção, controle. Redes Editora, Porto Alegre, pp 243–246

Karatayev AY, Boltovskoy D, Padilla DK, Burlakova LE (2007) The invasive bivalves *Dreissena polymorpha* and *Limnoperna fortunei* : parallels, contrasts, potential spread and invasion impacts. J Shellfish Res 26:205–213

Karatayev AY, Burlakova LE, Karatayev VA, Boltovskoy D (2010) *Limnoperna fortunei* versus *Dreissena polymorpha* : Population densities and benthic community impacts of two invasive freshwater bivalves. J Shellfish Res 29:975–984

Kavouras JH, Maki JS (2003) Effects of biofilms on zebra mussel postveliger attachment to artificial surfaces. Invertebr Biol 122:138–151

Kobak J, Kakareko T, Posnanska M (2010) Changes in attachment strength and aggregation of zebra mussel, *Dreissena polymorpha* in the presence of potential fish predators of various species and size. Hydrobiologia 644:195–206

Lopes M, Vieira J (2012) Space-time variation of the relative abundance of *Limnoperna fortunei* in deep zones of São Gonçalo Channel, Rio Grande do Sul, Brazil. Iheringia (Série Zoologia) 102:370–374

Lopes MN, Vieira JP, Burns MDM (2009) Biofouling of the golden mussel *Limnoperna fortunei* (Dunker, 1857) over the Anomura crab *Aegla platensis* Schmitt, 1942. Pan-American J Aquat Sci 4:222–225

López HL, Menni RC, Donato M, Miquelarena AM (2008) Biogeographical revision of Argentina (Andean and Neotropical Regions): an analysis using freshwater fishes. J Biogeogr 35:1564–1579

Mansur MCD, Santos CP, Darrigran G, Heydrich I, Callil CT, Cardoso FR (2003) Primeiros dados quali-quantitativos do mexilhão-dourado, *Limnoperna fortunei* (Dunker), no Delta do Jacuí, no Lago Guaíba e na Laguna dos Patos, Rio Grande do Sul, Brasil e alguns aspectos de sua invasão no novo ambiente. Revista Brasileira de Zoologia 20:75–84

Marçal SF, Callil CT (2008) Structure of invertebrates community associated with *Eichhornia crassipes* Mart. (Solms-Laubach) after the introduction of *Limnoperna fortunei* (Dunker, 1857) (Bivalvia, Mytilidae) in the Upper Paraguay River, MT, Brazil. Acta Limnologica Brasiliensia 20:359–371

Marçal SF, Callil CT (2012) *Limnoperna fortunei* associada a macrófitas aquáticas na bacia do Rio Paraguai, Mato Grosso. In: Mansur MCD, Santos CP, Pereira D, Padula Paz IC, Leite Zurita ML, Raya Rodriguez MT, Vilar Nehrke M, Aydos Bergonci PE (eds) Moluscos límnicos invasores no Brasil. Biologia, prevenção, controle. Redes Editora, Porto Alegre, pp 201–206

Mata FAR (2011) Abundância e distribuição temporal de *Limnoperna fortunei* Dunker, 1857 (Mollusca, Bivalvia) e os impactos da incrustação em usinas geradoras de energia elétrica. MSc Thesis, Universidade Federal de Ouro Preto (Brazil), pp 1–91

Michelan TS, Silveira MJ, Petsch DK, Pinha GD, Thomaz SM (2014) The invasive aquatic macrophyte *Hydrilla verticillata* facilitates the establishment of the invasive mussel *Limnoperna fortunei* in Neotropical reservoirs. J Limnol 73:598–602

Morton B (1977) The population dynamics of *Limnoperna fortunei* (Dunker 1857) (Bivalvia: Mytilacea) in Plover Cove Reservoir, Hong Kong. Malacologia 16:165–182

Nakano D, Kobayashi T, Sakaguchi I (2010) Predation and depth effects on abundance and size distribution of an invasive bivalve, the golden mussel *Limnoperna fortunei*, in a dam reservoir. Limnology 11:259–266

Nakano D, Kobayashi T, Endo N, Sakaguchi I (2011) Growth rate and settlement of *Limnoperna fortunei* in a temperate reservoir. J Molluscan Stud 77:142–148
Ohtaka A, Ryusei W, Sokrithy IM, Rachna C, Shinji T (2010) Spatial and seasonal changes of net plankton and zoobenthos in Lake Tonle Sap, Cambodia. Limnology 11:85–94
Ohtaka A, Tetsuya N, Takahiro K, Haruo K, Yuji A, Sokrithy IM, Rachna C, Shinji T (2011) Composition of aquatic invertebrates associated with macrophytes in Lake Tonle Sap, Cambodia. Limnology 12:137–144
Oliveira MD, Calheiros DF (2012) *Limnoperna fortunei* na bacia do Rio Paraguai, Mato Grosso do Sul. In: Mansur MCD, Santos CP, Pereira D, Padula Paz IC, Leite Zurita ML, Raya Rodriguez MT, Vilar Nehrke M, Aydos Bergonci PE (eds) Moluscos límnicos invasores no Brasil. Biologia, prevenção, controle. Redes Editora, Porto Alegre, pp 197–200
Oliveira MD, Hamilton SK, Calheiros DF, Jacobi CM (2010a) Oxygen depletion events control the invasive golden mussel (*Limnoperna fortunei*) in a tropical floodplain. Wetlands 30:705–716
Oliveira MD, Hamilton SK, Calheiros DF, Jacobi CM, Latini RO (2010b) Modeling the potential distribution of the invasive golden mussel *Limnoperna fortunei* in the Upper Paraguay River system using limnological variables. Braz J Biol 70:831–840
Oliveira MD, Hamilton SK, Jacobi CM (2010c) Forecasting the expansion of the invasive golden mussel *Limnoperna fortunei* in Brazilian and North American rivers based on its occurrence in the Paraguay River and Pantanal wetland of Brazil. Aquat Invasions 5:59–73
Oliveira MD, Calheiros DF, Jacobi CM, Hamilton SK (2011) Abiotic factors controlling the establishment and abundance of the invasive golden mussel *Limnoperna fortunei*. Biol Invasions 13:717–729
Pastorino G, Darrigran GA, Martín SM, Lunaschi L (1993) *Limnoperna fortunei* (Dunker, 1857) (Mytilidae), nuevo bivalvo invasor en aguas del Río de la Plata. Neotropica 39:101–102
Pestana D (2006) Aspetos bioecológicos do mexilhao dourado *Limnoperna fortunei* (Bivalvia, Mytilidae) (Dunker, 1857). PhD Thesis, Universidade Federal de Curitiba (Brazil), pp 1–123
Petrie SA, Knapton RW (1999) Rapid increase and subsequent decline of zebra and quagga mussels in Long Point Bay, Lake Erie: possible influence of waterfowl predation. J Great Lakes Res 25:772–782
Ramcharan CW, Padilla DK, Dodson SI (1992a) Models to predict potential occurrence and density of the zebra mussel, *Dreissena polymorpha*. Can J Fish Aquat Sci 49:2611–2620
Ramcharan CW, Padilla DK, Dodson SI (1992b) A multivariate model for predicting population fluctuations of *Dreissena polymorpha* in North American lakes. Can J Fish Aquat Sci 49:150–158
Rodriguez SR, Ojeda PF, Inestrosa NC (1993) Settlement of benthic marine invertebrates. Marine Ecol Prog Ser 97:193–207
Rojas Molina F (2010) Efectos del molusco invasor *Limnoperna fortunei* (Dunker) sobre el zooplancton del paraná medio. PhD Thesis, Universidad de Buenos Aires (Argentina), pp 1–205
Rojas Molina F, Williner V (2013) First record of the non-indigenous mussel *Limnoperna fortunei* (Bivalvia, Mytilidae) as an epibiont of the crab *Trichodactylus borellianus* (Decapoda, Trichodactylidae). Crustaceana 86:682–692
Rojas Molina F, Paggi JC, Devercelli M (2010) Zooplanktophagy in the natural diet and selectivity of the invasive mollusk *Limnoperna fortunei*. Biol Invasions 12:1647–1659
Santos CP, Mansur MCD, Wurdig NI (2008) Variaçoes no comprimento dos individuos de uma populacão do mexilhão dourado, *Limnoperna fortunei* (Mollusca: Bivalvia: Mytilidae), ao longo do ano, na Praia do Veludo, Lago Guaiba, Rio Grande do Sul, Brasil. Braz J Biol 25:389–396
Sardiña P, Cataldo D, Boltovskoy D (2009) Effects of conspecifics on settling juveniles of the invasive golden mussel, *Limnoperna fortunei*. Aquat Sci 71:479–486
Schloesser DW, Nalepa TF, Mackie GL (1996) Zebra mussel infestation of Unionid bivalves (Unionidae) in North America. Am Zool 36:300–310
Shu F-Y, Wu X (2005) Investigations of zebra mussel, *Limnoperna lacustris*, infestation of bivalves in Poyang Lake. Territory Nat Resour Stud 1:82–83 [In Chinese]

Spaccesi FG, Rodrigues Capitulo A (2012) Benthic communities on hard substrates covered by *Limnoperna fortunei* Dunker (Bivalvia, Mytilidae) at an estuarine beach (Río de la Plata, Argentina). J Limnol 71:144–153

Stanczykowska A (1977) Ecology of *Dreissena polymorpha* (Pall.) (Bivalvia) in lakes. Polskie Archiwum Hydrobiologii 24:461–530

Strayer DL, Malcom HM (2006) Long-term demography of a zebra mussel (*Dreissena polymorpha*) population. Freshw Biol 51:117–130

Strayer DL, Powell J, Ambrose P, Smith LC, Pace ML, Fischer DT (1996) Arrival, spread, and early dynamics of a zebra mussel (*Dreissena polymorpha*) population in the Hudson River estuary. Can J Fish Aquat Sci 53:1143–1149

Sverlij SB, Espinach Ros A, Orti G (1993) Sinopsis de los datos biológicos y pesqueros del sábalo *Prochilodus lineatus* (Valenciennes, 1847). Food and Agriculture Organization, Rome, pp 1–64

Sylvester F (2006) Biología alimentaria y ecología del molusco invasor *Limnoperna fortunei* (Mytilidae) en el Paraná inferior y Río de la Plata. PhD Thesis, Universidad de Buenos Aires (Argentina), pp 1–160

Sylvester F, Dorado J, Boltovskoy D, Juárez A, Cataldo D (2005) Filtration rates of the invasive pest bivalve *Limnoperna fortunei* as a function of size and temperature. Hydrobiologia 534:71–80

Sylvester F, Boltovskoy D, Cataldo D (2007) Fast response of freshwater consumers to a new trophic resource: Predation on the recently introduced Asian bivalve *Limnoperna fortunei* in the lower Parana River, South America. Austral Ecol 32:403–415

Takeda AM, Fujita DS (2012) Experiência da procura e monitoramento no início da invasão de mexilhão-dourado no Rio Paraná e no reservatório de Itaipu. In: Mansur MCD, Santos CP, Pereira D, Padula Paz IC, Leite Zurita ML, Raya Rodriguez MT, Vilar Nehrke M, Aydos Bergonci PE (eds) Moluscos límnicos invasores no Brasil. Biologia, prevenção, controle. Redes Editora, Porto Alegre, pp 207–217

Tamburri MN, Luckenbach MW, Breitburg DL, Bonniwell SM (2008) Settlement of *Crassostrea ariakensis* larvae: effects of substrate, biofilms, sediment and adult chemical cues. J Shellfish Res 27:601–608

Tuchman NC, Burks RL, Call CA, Smarrelli J (2004) Flow rate and vertical position influence ingestion rates of colonial zebra mussels (*Dreissena polymorpha*). Freshw Biol 49:191–198

Uryu K, Iwasaki K, Hinoue M (1996) Laboratory experiments on behaviour and movement of a freshwater mussel, *Limnoperna fortunei* (Dunker). J Molluscan Stud 62:327–341

Wainman BC, Hincks SS, Kaushik NK, Mackie GL (1996) Biofilm and substrate preference in the dreissenid larvae of Lake Erie. Can J Fish Aquat Sci 53:134–140

Walters LJ, Wethey DS (1996) Settlement and early post-settlement survival of sessile marine invertebrates on topographically complex surfaces: the importance of refuge dimensions and adult morphology. Marine Ecol Prog Ser 137:161–171

Wood H (2013) Post-settlement processes and early life history dynamics of zebra mussels (Dreissena polymorpha) in the Hudson River estuary. PhD Thesis, University of Illinois (USA), pp 1–122

Xu M, Wang Z, Duan X, Pan B (2013a) Effects of pollution on macroinvertebrates and water quality bio-assessment. Hydrobiologia, 729:247–259

Xu M, Wang ZY, Lin CC, Pan BZ, Zhao N (2013b) Experimental study of invasion and biofouling of freshwater mussel *Limnoperna fortunei*. Int J Geosci 4:1–7

Young BL, Padilla DK, Schneider DW, Hewett SW (1996) The importance of size-frequency relationships for predicting ecological impact of zebra mussel populations. Hydrobiologia 332:151–158

Young SS, Yang HN, Huang DJ, Liu SM, Huang YH, Chiang CT, Liu JW (2014) Using benthic macroinvertebrate and fish communities as bioindicators of the Tanshui River Basin around the greater Taipei area—multivariate analysis of spatial variation related to levels of water pollution. Int J Environ Res Public Health 11:7116–7143

Part II
Ecology and Environmental Impact

Ecology and Environmental Impact of *Limnoperna fortunei*: Introduction

Demetrio Boltovskoy

Abstract Comparisons between *Limnoperna fortunei* and much more thoroughly researched *Dreissena* species have been helpful in orienting work on the golden mussel, but they also encouraged unwarranted extrapolations to *L. fortunei* of ecological traits and effects of the zebra mussel on the systems invaded. A growing body of evidence indicates that while these mussels are functionally similar, intrinsic and environmental differences are responsible for the fact that their impacts on the waterbodies colonized often differ significantly. Interpretation of the impacts of the golden mussel on the ecosystems invaded is complicated by a priori judgments on the harm associated with this introduction, which often hamper objective analysis.

Keywords *Limnoperna fortunei* · Golden mussel · Impact · Ecology · Zebra mussel · *Dreissena*

The mechanisms by which *Limnoperna fortunei* modifies living conditions for other organisms are largely the same as those described for the zebra mussel (Karatayev et al. 1997; Ward and Ricciardi 2007; Kelly et al. 2010; Burlakova et al. 2012), but the final results of these interactions are not necessarily alike. Although in comparison with *Dreissena polymorpha*, which has been intensively studied for over a century (Karatayev et al. 2012), our knowledge of *L. fortunei* is still in its infancy, data at hand consistently show that intrinsic dissimilarities between the two species, as well as environmental differences between Europe-North America and South America (Karatayev et al. 2010), are responsible for significant differences in the impacts involved. Studies on the golden mussel have traditionally used *D. polymorpha* as a model, which resulted in useful guidelines for defining potential

D. Boltovskoy (✉)
Facultad de Ciencias Exactas y Naturales, Instituto de Ecología, Genética y Evolución de Buenos Aires (IEGEBA), Universidad de Buenos Aires-CONICET, Ciudad Universitaria, 1428 Buenos Aires, Argentina
e-mail: demetrio@ege.fcen.uba.ar

Museo Argentino de Ciencias Naturales 'Bernardino Rivadavia'-CONICET, Buenos Aires, Argentina

D. Boltovskoy (ed.), *Limnoperna fortunei*, Invading Nature - Springer Series in Invasion Ecology 10, DOI 10.1007/978-3-319-13494-9_8

interactions and fruitful research topics, but these similarities have often proved misleading when extrapolating to *L. fortunei* the effects of the zebra mussel on the systems invaded (Boltovskoy et al. 2006, 2013; Cataldo et al. 2012; Boltovskoy and Correa 2015).

The huge Paraguay-Paraná-Uruguay floodplain river system invaded by *L. fortunei* in South America has very marked differences with the colder, clearer and more oligotrophic North American waterbodies colonized by *Dreissena*. The mean transport of POC by the Paraná River has been calculated at 1 Tg/y, 20–40 % of it being labile and available for biologic consumption (Depetris and Pasquini 2007). This suggests that filtering organisms are not food-limited (Sylvester et al. 2005). Furthermore, because indigenous filter-feeding benthic animals in the Río de la Plata watershed are scarce, most of this organic matter is flushed out into the ocean through the Río de la Plata estuary (Boltovskoy et al. 2006). *L. fortunei*, the first and only abundant macrobenthic filter feeder in this ecosystem, is intercepting an important proportion of this particulate organic matter and retaining it in the system for use by a wide array of animals. This trophic shift involves not only *L. fortunei* larvae and adults, but also many other invertebrates whose abundances are enhanced by *L. fortunei* beds. In addition, the organic matter-enriched sediments derived from the "shunt" of suspended POC to the bottom as feces and pseudofeces further contributes to enhancing benthic invertebrate abundances (Sylvester et al. 2007, Sardiña et al. 2008). Although on local scales some effects of this mechanism have been explored, on the ecosystem scale we still have a very limited understanding of these potential influences and many others, including the biomagnification and transfer of contaminants (Villar et al. 1997), thermal shifts due to changes in the light environment (Yu and Culver 2000), the homogenization of faunal compositions across environments (Sardiña et al. 2011), "invasional meltdown" effects, (Ricciardi 2001), etc.

Key pieces of information for weighing the potential effects of *L. fortunei* in these lotic systems are reliable estimates of its abundances over reasonably large areas. These estimates, however, have not yet been achieved, largely because assessment of average densities over large areas is complicated by the fact that beds of *L. fortunei* have an extremely patchy distribution. Thus, practically all data available refer to very restricted areas.

Interpretation of the effects of *L. fortunei* on the ecosystem is further complicated by the fact that interactions are multiple, intricate and very dynamic, depending not only on the species and compartments considered, but also on regional conditions, season, interannual differences, etc. (Kelly et al. 2010; Boltovskoy et al. 2013; Boltovskoy and Correa 2015). Furthermore, through the action of multiple stressors, mussels can have opposite effects on the same variable (see Fig. 1 in Chapter "Nutrient recycling, phytoplankton grazing, and associated impacts of *Limnoperna fortunei*" in this volume). For example, mussel respiration and the decomposition of their feces and pseudofeces tend to decrease dissolved oxygen concentrations, whereas clarification of the water-column and the associated enhancement of macrophyte growth have the opposite effect (summarized in Boltovskoy and Correa 2015). Feedback effects have been described where the nutrient recycling activity of *L. fortunei* enhances the growth of toxic cyanobacteria, whose blooms in turn kill the mussel's larvae (Boltovskoy et al. 2013; see Chapter "Nutrient recycling, phytoplankton grazing, and associated impacts of *Limnoperna fortunei*" in this volume). The complexity of the relationships involved is illustrated by the fact that after a century of intensive studies on *D. polymorpha*, there is still no agreement on some of its fundamental effects on the environment, such as its impact on cyanobacterial blooms (Juhel et al. 2006a, b, Dionisio Pires et al. 2010).

Complications for interpreting the significance of these effects are even more critical when attempting to label the impacts as negative or positive. Unfortunately, the ecology of introduced species is too often associated with attempts to demonstrate that invasive organisms are fundamentally different from indigenous species, and particularly that their effects are detrimental to the ecosystem. This perspective has often hampered objective analysis and has accomplished little for advancing our understanding of how these species interact with their new environment (Davis et al. 2011, but see also Simberloff and signatories 2011). Just as not all nonindigenous species have large effects (Byers et al. 2002), different invaders can have different net effects, and the same or very similar species can have dissimilar effects in different areas.

The fact that most introduced species have had negative effects on the biota (Simberloff 2003) leaves little doubt about the potential harm involved in every new introduction. However, if eradication is not a viable option, assessment of its interactions with the local biota should be objective and untainted by the fact that other introductions have been harmful. Our results indicate that, after having established itself, *L. fortunei* interacts with other organisms like any other species and some of the outcomes of these relationships can be perceived as negative (e.g., enhancement of cyanobacterial blooms, grazing on some phyto- and zooplankton, introduction of new fish parasites), whereas others are probably positive (e.g., food for larval and adult fishes, enhancement of benthic abundance and diversity).

As far as we know, in South America the range of *L. fortunei* is still limited to the Río de la Plata and a few minor basins (see Chapter "Colonization and spread of *Limnoperna fortunei* in South America" in this volume). Infestation of the next large watershed—the Amazon, has not been reported to date, but is most probably inevitable. The Cuiabá River, a tributary of the Paraguay River, which has been colonized by *L. fortunei* at least since 2000 (Boltovskoy et al. 2006) is only 150 km from the Teles Pires River in the Tapajós River basin, within the Amazon watershed (Calazans et al. 2013). Both this proximity and the fact that the Amazon is navigable to ocean liners of virtually any tonnage, including ships with ballast water from infested ports along the Paraná-Uruguay-Río de la Plata waterways and the Guaíba basin (where compliance with international water ballast regulations is rather loosely enforced; Boltovskoy et al. 2011), suggests that sooner or later *L. fortunei* will invade this basin and, eventually, other South and North American freshwater bodies (Ricciardi 1998; Boltovskoy et al. 2006; Karatayev et al. 2007). An intriguing question is to what extent the lessons learned in South America will serve as a predictor of impacts elsewhere.

Acknowledgments This work was partially financed by grants from the University of Buenos Aires, Argentina (UBA X-020 and 20020100100035) and from the Argentine Agencia Nacional de Promoción Científica y Tecnológica, Argentina (PICT 2007 1968) to DB.

References

Boltovskoy D, Correa N (2015) Ecosystem impacts of the invasive bivalve *Limnoperna fortunei* (golden mussel) in South America. Hydrobiologia 746:81–95

Boltovskoy D, Correa N, Cataldo D, Sylvester F (2006) Dispersion and ecological impact of the invasive freshwater bivalve *Limnoperna fortunei* in the Río de la Plata watershed and beyond. Biol Invasions 8:947–963

Boltovskoy D, Almada P, Correa N (2011) Biological invasions: assessment of threat from ballast-water discharge in Patagonian (Argentina) ports. Environ Sci Policy 14:578–583

Boltovskoy D, Correa N, Bordet F, Leites V, Cataldo D (2013) Toxic *Microcystis* (cyanobacteria) inhibit recruitment of the bloom-enhancing invasive bivalve *Limnoperna fortunei*. Freshw Biol 58:1968–1981

Burlakova LE, Karatayev AY, Karatayev VA (2012) Invasive mussels induce community changes by increasing habitat complexity. Hydrobiologia 685:121–134

Byers JE, Reichard S, Randall JM, Parker IM, Smith CS, Lonsdale WM, Atkinson IAE, Seastedt TR, Williamson M, Chornesky E, Hayes D (2002) Directing research to reduce the impacts of nonindigenous species. Conserv Biol 16:630–640

Calazans SHC, Americo JA, Fernandes FD, Aldridge DC, Rebelo MD (2013) Assessment of toxicity of dissolved and microencapsulated biocides for control of the golden mussel *Limnoperna fortunei*. Marine Environ Res 91:104–108

Cataldo D, Vinocur A, O'Farrell I, Paolucci E, Leites V, Boltovskoy D (2012) The introduced bivalve *Limnoperna fortunei* boosts *Microcystis* growth in Salto Grande Reservoir (Argentina): evidence from mesocosm experiments. Hydrobiologia 680:25–38

Davis MA, Chew MK, Hobbs RJ, Lugo AE, Ewel JJ, Vermeij GJ, Brown JH, Rosenzweig ML, Gardener MR, Carroll SP, Thompson K, Pickett TA, Stromberg JC, Del Tredici P, Suding KN, Ehrenfeld JG, Grime JP, Mascaro J, Briggs JC (2011) Don't judge species on their origins. Nature 474:153–154

Depetris PJ, Pasquini AI (2007) The geochemistry of the Paraná River: an overview. In: Iriondo MH, Paggi JC, Parma MJ (eds) The Middle Paraná River. Limnology of a subtropical wetland. Springer Verlag, Berlin, pp 143–174

Dionisio Pires LM, Ibelings BW, van Donk E (2010) Zebra mussels as a potential tool in the restoration of eutrophic shallow lakes, dominated by toxic cyanobacteria. In: van der Velde G, Rajagopal S, Bij de Vaate A (eds) The zebra mussel in Europe. Backhuys Publishers, Leiden, pp 361–372

Juhel G, Davenport J, O' Halloran J, Culloty S, Ramsay R, James K, Furey A, Allis O (2006a) Pseudodiarrhoea in zebra mussels *Dreissena polymorpha* (Pallas) exposed to microcystins. J Exp Biol 209:810–816

Juhel G, Davenport J, O'Halloran J, Culloty SC, O'Riordan RM, James KF, Furey A, Allis O (2006b) Impacts of microcystins on the feeding behaviour and energy balance of zebra mussels, *Dreissena polymorpha*: a bioenergetics approach. Aquat Toxicol 79:391–400

Karatayev AY, Burlakova LE, Padilla DK (1997) The effects of *Dreissena polymorpha* (Pallas) invasion on aquatic communities in Eastern Europe. J Shellfish Res 16:187–203

Karatayev AY, Padilla DK, Minchin D, Boltovskoy D, Burlakova LE (2007) Changes in global economies and trade: the potential spread of exotic freshwater bivalves. Biol Invasions 9:161–180

Karatayev AY, Burlakova LE, Karatayev VA, Boltovskoy D (2010) *Limnoperna fortunei* versus *Dreissena polymorpha*: Population densities and benthic community impacts of two invasive freshwater bivalves. J Shellfish Res 29:975–984

Karatayev A, Claudi R, Lucy F (2012) History of *Dreissena* research and the ICAIS gateway to aquatic invasions research. Aquat Invasions 7:1–5

Kelly DW, Herborg L-M, MacIsaac HJ (2010) Ecosystem changes associated with *Dreissena* invasions: recent developments and emerging issues. In: van der Velde G, Rajagopal S, bij de Vaate A (eds) The zebra mussel in Europe. Backhuys Publishers, Leiden, pp 199–210

Ricciardi A (1998) Global range expansion of the Asian mussel *Limnoperna fortunei* (Mytilidae): another fouling threat to freshwater systems. Biofouling 13:97–106

Ricciardi A (2001) Facilitative interactions among aquatic invaders: is an "invasional meltdown" occurring in the Great Lakes? Can J Fish Aquat Sci 58:2513–2525

Sardiña P, Cataldo D, Boltovskoy D (2008) The effects of the invasive mussel, *Limnoperna fortunei*, on associated fauna in South American freshwaters: importance of physical structure and food supply. Fund Appl Limnol/Archiv für Hydrobiologie 173:135–144

Sardiña P, Chaves E, Marchese M (2011) Benthic community responses to invasion by the golden mussel, *Limnoperna fortunei* Dunker: biotic homogenization vs environmental driving forces. J N Am Benthol Soc 30:1009–1023

Simberloff D (2003) Confronting introduced species: a form of xenophobia? Biol Invasions 5:179–192

Simberloff D, signatories (2011) Non-natives: 141 scientists object. Nature 475:36

Sylvester F, Dorado J, Boltovskoy D, Juárez A, Cataldo D (2005) Filtration rates of the invasive pest bivalve *Limnoperna fortunei* as a function of size and temperature. Hydrobiologia 534:71–80

Sylvester F, Boltovskoy D, Cataldo D (2007) The invasive bivalve *Limnoperna fortunei* enhances benthic invertebrate densities in South American floodplain rivers. Hydrobiologia 589:15–27

Villar CA, Stripeikis J, Tudino M, D' Huicque L, Troccoli O, Bonetto C (1997) Use of invasive bivalves as biomonitor organisms for metal pollution at the Río de la Plata basin. In: 7° Conferencia Internacional sobre la Conservación y Gestión de Lagos (Lacar 97), San Martín de los Andes (Argentina)

Ward JM, Ricciardi A (2007) Impacts of *Dreissena* invasions on benthic macroinvertebrate communities: a meta-analysis. Divers Distrib 13:155–165

Yu N, Culver DA (2000) Can zebra mussels change stratification patterns in a small reservoir? Hydrobiologia 431:175–184

Nutrient Recycling, Phytoplankton Grazing, and Associated Impacts of *Limnoperna fortunei*

Demetrio Boltovskoy, Nancy Correa, Francisco Sylvester and Daniel Cataldo

Abstract Laboratory and field experiments indicate that the presence of *Limnoperna fortunei* decreases concentrations of particulate organic matter and increases ammonia, nitrate, and especially phosphate. Long-term series of field data partially confirm these results. After having been colonized by the mussel, a 47 km^2 reservoir developed higher concentrations of ammonia and phosphates, a higher P:N ratio, more transparency, less seston, and less phytoplankton and primary production. Phytoplankton clearance rates by the mussel vary widely, suggesting that "normal" values for adult organisms are around 100 mL/ind./h, or ca. 2–4 mL/mg DW/h. Data on grazing selectivity are inconclusive, but seem to indicate highest impacts on small (<1 mm) particles. Large plankton are negatively selected, but they may account for greater proportions of total biomass in the diet. Studies on consumption of toxic cyanobacteria yield conflicting results, but large golden mussel populations significantly enhance blooms of colonial *Microcystis* spp. through changes in nutrient availability, size-selective grazing, promotion of colony formation, and reduced grazing of toxic cells. These toxic blooms, in turn, suppress reproduction

D. Boltovskoy (✉) · F. Sylvester · D. Cataldo
Facultad de Ciencias Exactas y Naturales, Instituto de Ecología, Genética y Evolución de Buenos Aires (IEGEBA), Universidad de Buenos Aires-CONICET, Ciudad Universitaria, 1428 Buenos Aires, Argentina
e-mail: demetrio@ege.fcen.uba.ar

D. Cataldo
e-mail: daniel@ege.fcen.uba.ar

D. Boltovskoy · D. Cataldo
Museo Argentino de Ciencias Naturales 'Bernardino Rivadavia'-CONICET, Buenos Aires, Argentina

N. Correa
Servicio de Hidrografía Naval (Ministerio de Defensa), and Escuela de Ciencias del Mar (Instituto Universitario Naval), Av. Montes de Oca 2124, 1271 Buenos Aires, Argentina
e-mail: ncorrea@hidro.gov.ar

F. Sylvester
Facultad de Ciencias Naturales, Instituto para el Estudio de la Biodiversidad de Invertebrados (IEBI), Universidad Nacional de Salta, Av. Bolivia 5150, 4400 Salta, Argentina
e-mail: fsylvester@natura.unsa.edu.ar

D. Boltovskoy (ed.), *Limnoperna fortunei,* Invading Nature - Springer Series in Invasion Ecology 10, DOI 10.1007/978-3-319-13494-9_9

of the mussel, most probably killing the larvae. Growth of periphyton and aquatic macrophytes are enhanced significantly by the golden mussel.

Keywords *Limnoperna fortunei* · Golden mussel · Ecological impact · Nutrient recycling · Phytoplankton grazing · Grazing selectivity · Cyanobacterial blooms · *Microcystis*

Introduction

The influence of filter-feeding organisms on water-column properties, in particular the concentration of bioseston and modifications in the concentration and proportions of nutrients, has been investigated for decades in both marine and freshwater environments. Interest in this topic stems from the fact that filtration is among the most widespread feeding modes in a vast array of aquatic animals (Jørgensen 1966), and because it profoundly affects many water-column traits, as well as sediment characteristics. Filtration-related changes brought about by introduced species, in particular bivalves, have received special attention because they modify historical, preintroduction conditions, and because some nonindigenous species can attain very high densities and become invasive, thus enhancing their otherwise moderate impact on waterbodies. Figure 1 offers a visual overview of some of the most important modifications observed in association with the introduction of zebra and quagga mussels in Europe and North America (MacIsaac 1996; Karatayev et al. 2002; Kelly et al. 2010; Nalepa and Schloesser 2014), and of the golden mussel in Asia and South America (Mansur et al. 2012; Boltovskoy and Correa 2015). Even though this diagram includes but a fraction of the actual relationships that come into effect when one of these species colonizes a hitherto uninvaded waterbody, the maze of interactions is remarkable. It is noteworthy that the same ecosystem trait, process or component can be influenced by several different effects associated with the presence of mussels, and that the directions of these impacts can often be opposing. Further complications in pinpointing and quantifying impacts stem from the fact that many of these effects are site-dependent, which means that they can vary widely among waterbodies, or even in different areas of the same waterbody, and also as a function of time elapsed after initial introduction. Some shifts can be very strong during the initial postintroduction years and wane thereafter; others persist through time, and still others only become evident several years after introduction (Burlakova et al. 2005; Burlakova et al. 2006).

This section reviews current knowledge of the effects of the introduction of *Limnoperna fortunei* on nutrients and phytoplankton abundance and composition, as well as some consequences of the observed changes. As elsewhere in this volume, this chapter focuses on the golden mussel, rather than on invasive freshwater bivalves in general. As discussed below, while the mechanisms by which *L. fortunei* influences waterbodies are practically identical to those of the dreissenids, the final outcome of these interactions is often very dissimilar (Boltovskoy et al. 2006, 2013; Boltovskoy and Correa 2015).

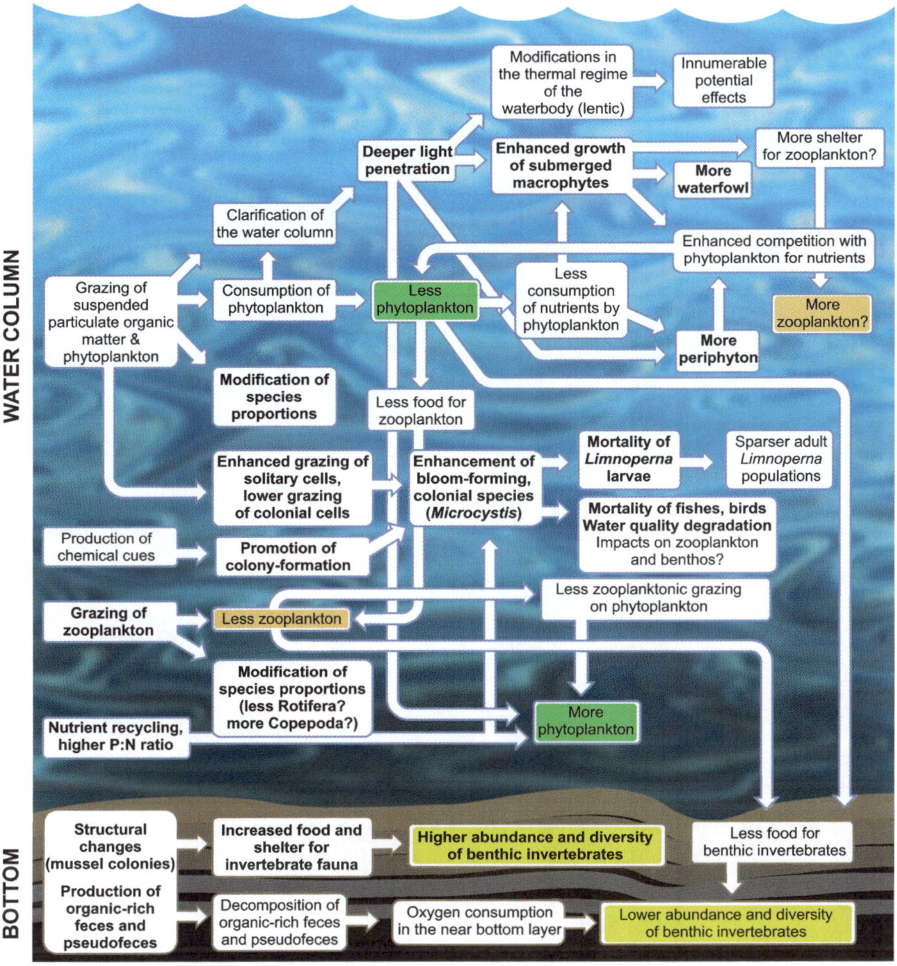

Fig. 1 Generalized schematic diagram of some salient effects of filter-feeding mussels on freshwater bodies (excluding relationships with fishes). Effects analyzed and demonstrated for *L. fortunei* are denoted in *bold* characters. Notice that several of the impacts shown have opposite effects on the same process or component; for example, clarification of the water and nutrient recycling can favor phytoplankton growth, but grazing and enhancement of periphyton and macrophytes can depress phytoplankton abundance (conflicting effects are denoted with the same color)

Nutrient Recycling

Short-term (24 h) experiments investigating effects of *L. fortunei* on nutrient concentrations and proportions have been carried out in laboratory settings and in field-deployed mesocosms (Cataldo et al. 2005; Cataldo et al. 2012a).

Laboratory experiments were conducted using plastic containers with 60 *L. fortunei* 18–27 mm in length in 15 L of water obtained in the Río de la Plata estuary (Cataldo et al. 2005). Nutrient concentrations were measured at 0, 3, 6, 12 and 24 h.

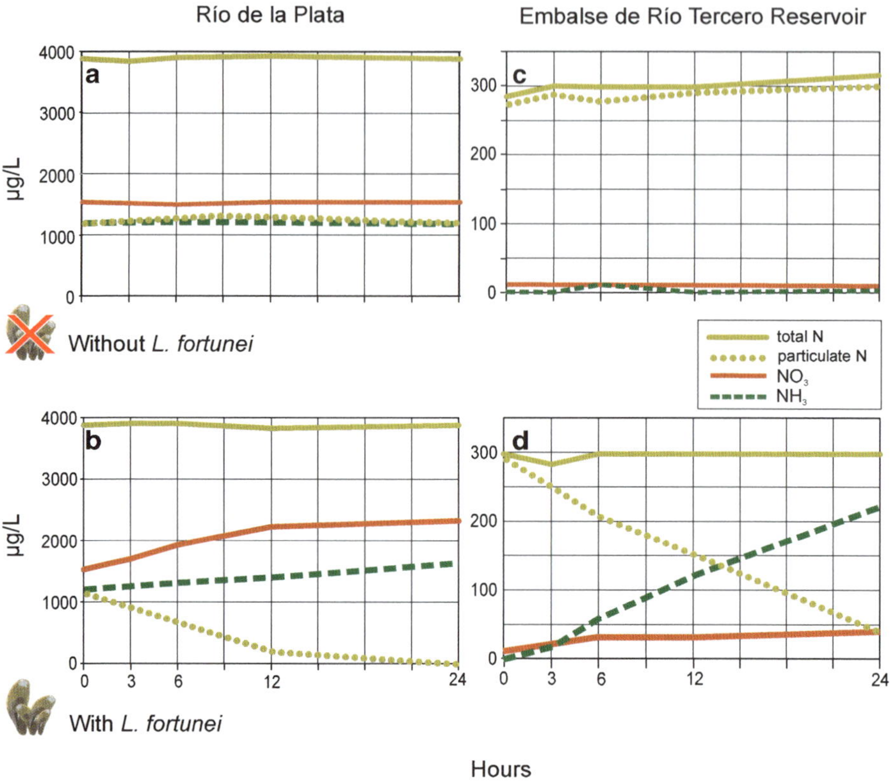

Fig. 2 Changes in the concentrations of nitrogen with and without *L. fortunei* in Río de la Plata waters (laboratory, **a, b**) and in the reservoir Embalse de Río Tercero (mesocosms, **c, d**) throughout a 24-h period. (**a** and **b** based on data from Cataldo et al. 2005; **c** and **d** from Cataldo et al. 2012a)

In controls without mussels, none of the measurements at 24 h differed significantly from initial (0 h) conditions. In containers with mussels, only total N and total P remained constant. Particulate N and P dropped to <25 % of their initial values (N from 1149 to 0 µg/L, and P from 122 to 31 µg/L). Ammonia, nitrate and phosphate, on the other hand, increased conspicuously (ammonia from 1190 to 1620 µg/L, nitrate from 1532 to 2333 µg/L, and phosphate from 121 to 212 µg/L) (Figs. 2a, b and 3a, b).

A similar experiment was performed in 400 l mesocosms deployed in a shallow, coastal area of the reservoir Embalse de Río Tercero, a medium-sized (47 km^2), meso-eutrophic waterbody (chlorophyll a around 3–6 µg/L; Boltovskoy et al. 2009a) located in central Argentina (32.37°S, 64.77°W) (Cataldo et al. 2012a). Four polyethylene terephthalate cylindrical mesocosms (75 cm in diameter, 105 cm high) with their bottoms sealed off with a polyethylene liner were filled with reservoir water to about 15 cm from the rim. Each of two of the mesocosms were stocked with ~1700 mussels 14–35 mm in length, collected nearby, whereas the other two

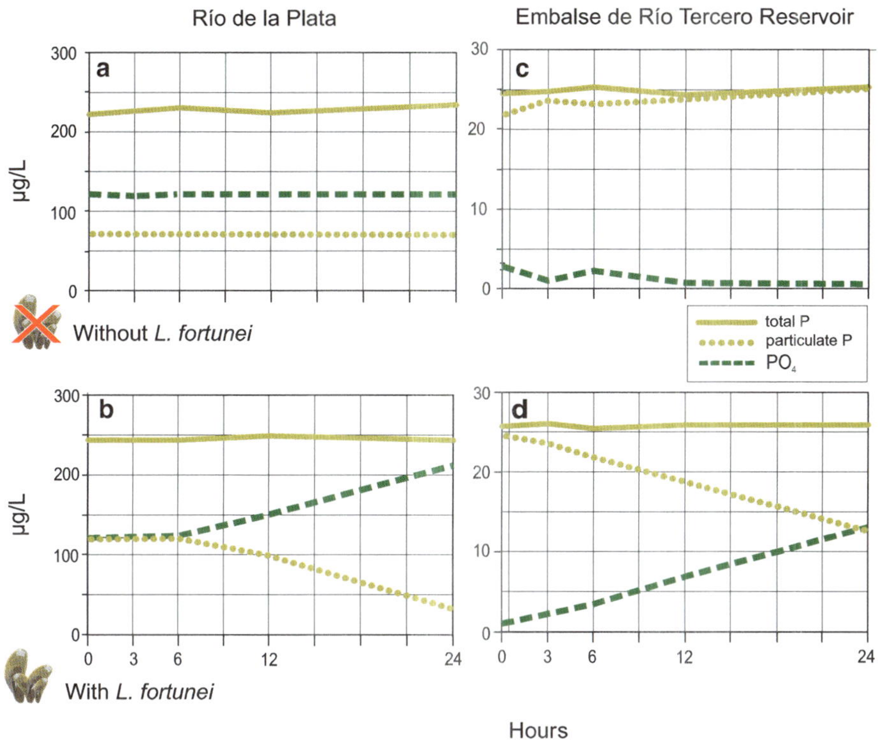

Fig. 3 Changes in the concentrations of phosphorus with and without *L. fortunei* in Río de la Plata waters (laboratory, **a, b**) and in the reservoir Embalse de Río Tercero (mesocosms, **c, d**) throughout a 24-h period. (**a** and **b** based on data from Cataldo et al. 2005;**c** and **d** from Cataldo et al. 2012a)

were used as controls (no mussels). As in the previous experiment, nutrient concentrations were measured at 0, 3, 6, 12, and 24 h. All variables remained practically constant in the controls (as in the lab experiments). In the mesocosms stocked with mussels, total N and P changed negligibly. Particulates dropped significantly (particulate N from 25 to 13 µg/L, particulate P from 122 to 31 µg/L), and ammonia, nitrate and phosphate increased strongly (ammonia from 0 to 220 µg/L, nitrate from 10 to 40 µg/L, and phosphate from 1 to 13 µg/L) (Fig. 2c and d; 3c and d).

It is noteworthy that, after 24 h, in both lab and mesocosm experiments, increases in the concentration of phosphate were much higher than those of nitrate (Figs. 2 and 3), thus modifying not only the total amount of major nutrients available for the autotrophs, but also their proportions. These results are generally similar to those obtained by Kawase (2011) in his 6-h filtration experiments with *L. fortunei*, where he recorded significant drops in turbidity and in the concentrations of particulate organic C and N.

Results obtained with longer incubations agreed with the above trends during the initial 24-h period, but subsequently the behavior of nutrients changed significantly. Cataldo et al. (2012b) assessed the effects of *L. fortunei* on water column properties

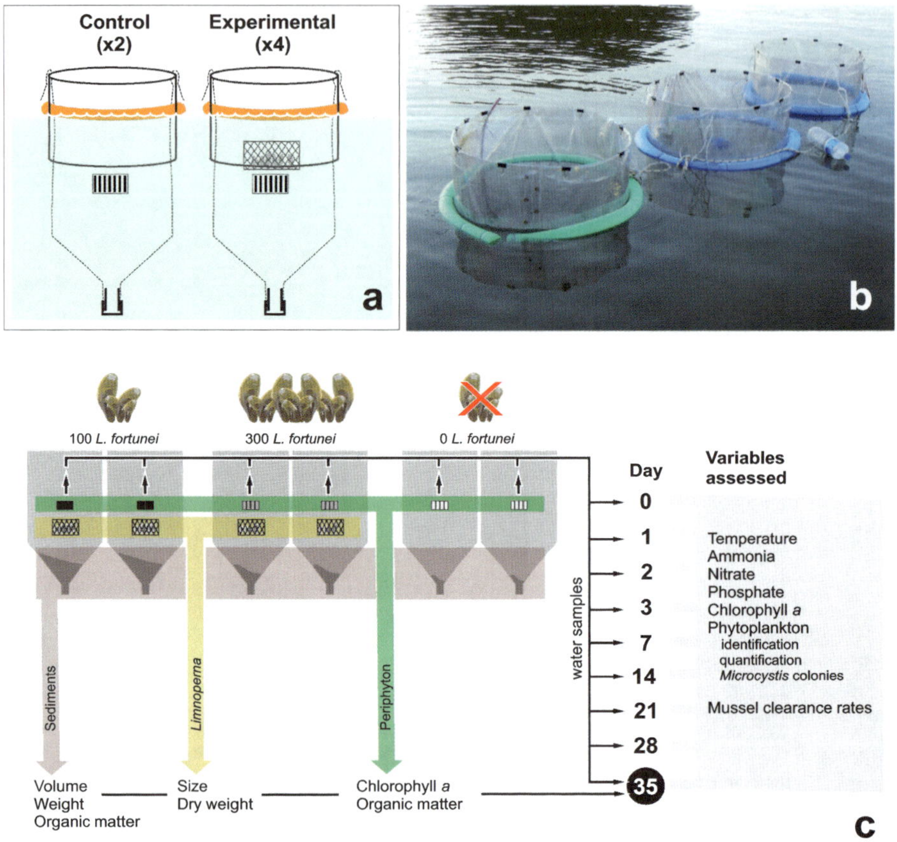

Fig. 4 Sketch (**a**) and photograph (**b**) of the mesocosms, and general scheme of the experimental setup and sampling design (**c**) used in experiments of phytoplankton grazing and nutrient regeneration in Salto Grande Reservoir by Cataldo et al. (2012b). (Modified from Cataldo et al. 2012b)

of Salto Grande Reservoir (Uruguay River, Argentina-Uruguay) using three pairs of floating 400-L mesocosms (Fig. 4a and b) stocked with either 100 or 300 mussels 16–20 mm in length, and without mussels (controls). Immediately after deployment (day 0), and on days 1, 2, 3, 7, 14, 21, 28, and 35 the following parameters were assessed (Fig. 4c): water temperature, concentrations of ammonia, nitrate, phosphate, and chlorophyll a, quantification and identification of phytoplanktonic algae, and evaluation of the size and density of *Microcystis* spp. colonies. Accumulated sediments were retrieved from each mesocosm at the end of the experiment to assess their wet and dry weight, and organic matter contents. All mesocosms were provided with four PVC plates suspended within the enclosure; these plates were retrieved on day 35 and periphytic organic matter and chlorophyll a were measured.

In the absence of *L. fortunei*, ammonia dropped from ca. 10 µg/L to below detection levels by the end of the incubation. In the mesocosms with mussels, on the other hand, ammonia increased from 10 µg/L (on day 0) to around 20–30 µg/L (on

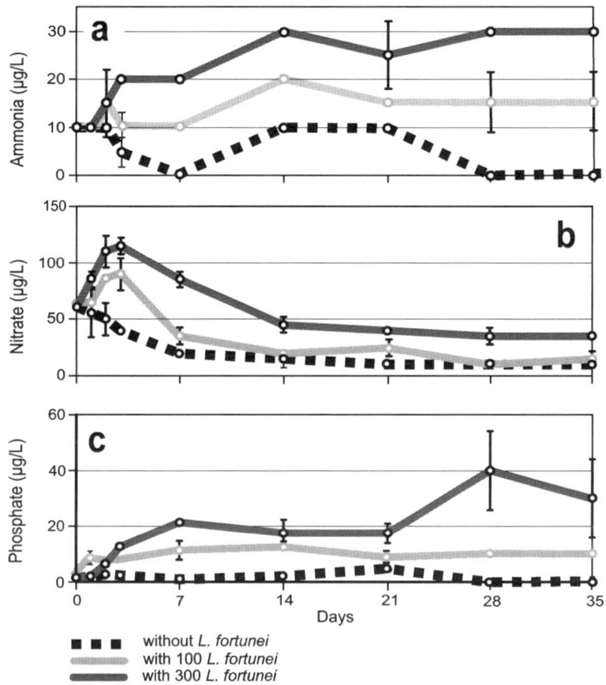

Fig. 5 Changes in the concentrations of ammonia (**a**), nitrate (**b**), and phosphate (**c**) in mesocosms with and without *L. fortunei* throughout a 35-day experimental period. (From Cataldo et al. 2012b)

day 35; Fig. 5a). For nitrate the pattern was different; in agreement with short-term studies, the presence of mussels enhanced nitrate concentrations until day 3, but from then on nitrate decreased gradually until termination of the experiment (day 35), when it was at around 40–60 % of initial values (Fig. 5b). In the controls, nitrate concentrations dropped from day 1, and by day 35 were at ~20 % of initial values. Phosphate concentrations increased on day 1 (in the mesocosms with 100 mussels; as in the short term surveys), or day 3 (in mesocosms with 300 mussels), and from there on showed moderate variations (with the exception of a second peak on day 28 in the mesocosms with 300 mussels). In the controls, phosphate values were very low throughout the entire experimental period (Fig. 5c).

While the usefulness of short and medium-term experiments for identifying the effects of invasive mussels on nutrients and the biota is beyond doubt, their ability to predict impacts over the long term (years to decades) is limited. Long-term impacts cannot be unequivocally assessed through laboratory tests or enclosure experiments, whereas field sampling programs very seldom cover periods long enough for such analyses. Furthermore, many of the variables that play fundamental roles in lakes, rivers and reservoirs (e.g., nearshore vs. offshore partitioning, external inputs of nutrients and organic matter, regional differences in substrate type, vertical mixing, and many others, see reviews in Kelly et al. 2010, Bootsma and Liao 2014) are not effective in experimental settings. For *L. fortunei*, the only survey available where the effects of this mussel were quantified on the basis of a long-term series of field data is the one carried out in Embalse de Río Tercero reservoir by Boltovskoy et al. (2009a). This reservoir, which was colonized by the golden mussel around 1998, has been monitored regularly since 1996. In 2006, average mussel densities over the entire reservoir were estimated at 960 ind./m^2 (see Fig. 11 in Chapter "*Limnoperna fortunei* Colonies: Structure, Distribution and Dynamics" in this volume), suggesting that these populations could potentially filter a volume equivalent to that of the entire water body every 2–8 days (Boltovskoy et al. 2009a). Comparison of data collected between 1996 and 2008 point at significant changes in several water column properties, especially at the station located in the area of highest mussel densities. Total N in the water increased 300 % (most probably largely on account of ammonia), ammonia increased ca. 400 %, and phosphate increased 200 %. Nitrate remained at pre-2000 levels. The phosphate:nitrate ratio increased from 0.061 (before 2000) to 0.112 (after 2000) (Fig. 6), probably as a result of high rates of phosphate release due to mobilization of iron-bound phosphorus in the anoxic guts of the mussels (Turner 2010). These shifts are particularly noticeable when contrasting the periods 1996–2000 versus 2002–2007, which suggests that it took the mussel around four years to build up a population large enough to start affecting the reservoir water (see Chapter "*Limnoperna fortunei* Colonies: Structure, Distribution and Dynamics" in this volume).

These results generally confirm previous information based on studies with *Dreissena* spp. (Karatayev et al. 2002; Turner 2010; Zhang et al. 2011), yet their effects on phytoplankton are different (see below).

Enhanced mineralization associated with the presence of mussels is not restricted to organic matter, but may also affect other substances present in the water, including pesticides. Di Fiori et al. (2012), concluded that concentrations of glyphosate, a phosphonate compound widely used as an herbicide for weed control of several genetically modified crops (soybean, maize, cotton, canola) decreased by 40 % in the presence of large mussels. The pathways responsible for this decrease, however, are still poorly understood.

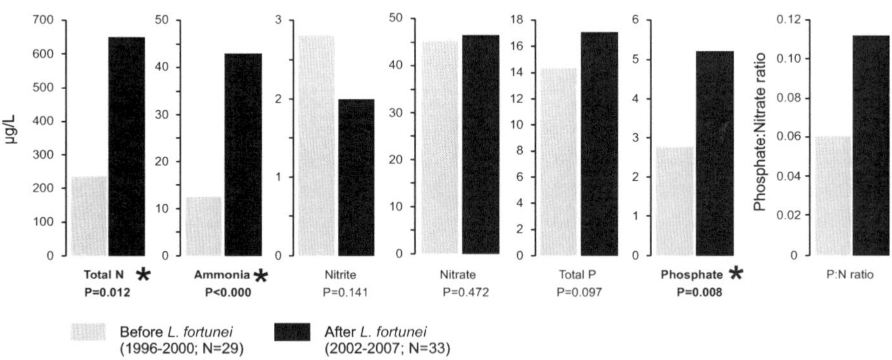

Fig. 6 Mean concentrations of N and P in the reservoir Embalse de Río Tercero in 1996–2000 (before the waterbody was influenced by the presence of *L. fortunei,* which invaded in 1998), and in 2002–2007 (with high *L. fortunei* population densities). (Based on data from Boltovskoy et al. 2009a)

Phytoplankton Grazing: Rates and Impact

Clearance rates of *L. fortunei* have been estimated in several studies, but the range of values reported is very large: 0.2 to 725 mL/ind./h, or 0.1 to 29.5 mL/mg dry tissue (DT)/h (Table 1). This spread is largely associated with differences in methodology. With very few exceptions (e.g., Gazulha et al. 2012a, b), studies did not differentiate actually ingested particles from those discarded as pseudofeces; the latter are embedded in mucus, rejected and settled on the bottom, thus disappearing from the water column and ending up being included in the estimates of consumption. Differences in food type and toxicity, pre-experiment starvation times, mussel size, experiment duration, amount of suspended solids, temperature, resuspension mechanisms used, pH, and water flow, among other factors, also strongly affect grazing estimates (Morton 1983). Despite their spread, these data suggest that normal filtration rates for adult individuals 15–25 mm in length range around 100 mL/ind./h, or ca. 2–4 mL/mg DW/h. Younger individuals and warmer temperatures yield higher specific filtration rates (Sylvester et al. 2005).

These figures are roughly within the range of those reported for many other freshwater mussels (Karatayev et al. 1997; Sylvester et al. 2005), very few of which, however, attain densities similar to those observed in *L. fortunei* beds. On suitable substrates golden mussel densities normally range around 5000 ind./m^2, and can occasionally exceed 200,000 ind./m^2 (Sylvester et al. 2007; Spaccesi and Rodrigues Capitulo 2012; see Chapter "*Limnoperna fortunei* Colonies: Structure, Distribution and Dynamics" in this volume). Thus, strong impacts on the water column are due to high mussel densities, rather than to exceptional individual filtration rates.

Results of laboratory and mesocosm experiments indicate strong drops in algal numbers over short periods. Studies carried out in a laboratory recirculating system and in 400-l mesocosms (Cataldo et al. 2005; 2012a; see above) showed dramatic drops in phytoplankton cell numbers over 24 h (Fig. 7a, d). Clearance and grazing rates were not monotonic, but changed over the course of the experiment. Clearance was found to be highest at 6 h, and decreased afterwards (Fig. 7b, e). Similar trends have been reported for *L. fortunei* by other authors (Pestana et al. 2009; Frau et al. 2013). However, because algal densities decrease with time, the numbers of algae eliminated from the water column (either ingested or rejected as pseudofeces) decrease very sharply (Fig. 7c, f). Decreased pumping (=clearance) rates are probably a response to satiation (Fig. 7a), low phytoplankton densities (Fig. 7d), or both. Longer-term experiments, however, indicate that after this initial decline algal numbers recover partially, most probably stimulated by higher nutrient availability and increasing light penetration (Cataldo et al. 2012b; Fig. 8). These results seem to mimic natural conditions; the few colonized waterbodies for which there are adequate historical records dropped in phytoplankton abundance and production after having been invaded by *L. fortunei*. The reservoir Embalse de Río Tercero lost about 30–40 % of its seston load, represented chiefly by algae (Fig. 9b), >40 % of its planktonic primary production (Fig. 9d), and became significantly clearer (Fig. 9a) (Boltovskoy et al. 2009a). Information on potential impacts in the main South American waterbodies colonized by the mussel—the large floodplain rivers

Table 1 Summary of results on filtration rates of *L. fortunei*

Filtration rate (mL/mussel/h)	Filtration rate (mL/mg tissue DW/h)	Food	Temperature (°C)	Size of mussels (mm)	Experiment duration (h)	Settings and method	Reference
92.5	17.2	*Microcystis viridis*	22–24	17–20	1	L-CC-N	Rückert et al. (2004)
133.8	24.5	*Pseudoanabaena* sp.	22–24	17–20	1	L-CC-N	Rückert et al. (2004)
89.2	11.9	*Selenastrum capricornutum*	22–24	17–20	1	L-CC-N	Rückert et al. (2004)
125–350	9.9–29.5	*Chlorella vulgaris*	15, 20, 25	15, 23	0.5	L-CC-U	Sylvester et al. (2005)
4–80[a] (max. 297)	ND	*Schizochytrium* sp.[b]	10, 15, 20, 25, 28, 30	>15	24	L-T	Pestana et al. (2009)
8-247[a](max. 725)	ND	*Scenedesmus* sp.	15, 20, 25	>15	24	L-T	Pestana et al. (2009)
8–53 (mean: 28)[c]	ND	Mixed plankton	24	30	1	L-CC-S	Fachini (2011)
78.3	ND	*Monoraphidium* sp. + toxic *Microcystis* extract[d]	24	30	1	L-CC-S	Fachini (2011)
56.6	ND	*Monoraphidium* sp. + nontoxic *Microcystis* extract	24	30	1	L-CC-S	Fachini (2011)
130.6	ND	*Monoraphidium* sp.	24	30	1	L-CC-S	Fachini (2011)
100–214	1.5–3.1	Mixed plankton	23	14–35	24	M-CC-U	Cataldo et al. (2012a)
ND	<0.1 to 2–4[e]	Mixed plankton	22–27	18.4±2.8	840 (35 d)	M-CC-U	Cataldo et al. (2012b)
24–30 µL/larva/h	ND	*Monoraphidium* sp. (nontoxic strain)	24	Larvae	24	L-CC-S	Gazulha (2010, 2012)
15–26 µL/larva/h	ND	*Microcystis* spp. (toxic strain)	24	Larvae	24	L-CC-S	Gazulha (2010, 2012)
17–32 µL/larva/h	ND	*Microcystis* spp. (non-toxic strain)	24	Larvae	24	L-CC-S	Gazulha (2010, 2012)

Table 1 (continued)

Filtration rate (mL/mussel/h)	Filtration rate (mL/mg tissue DW/h)	Food	Temperature (°C)	Size of mussels (mm)	Experiment duration (h)	Settings and method	Reference
36–63[c]	0.7–0.8[c]	*Microcystis aeruginosa* (nontoxic strain, solitary)	24	30	1 & 120	L-CC-S	Gazulha et al. (2012a)
32–55[c]	0.5–0.7[c]	*Microcystis aeruginosa* (toxic strain, solitary)	24	30	1 & 120	L-CC-S	Gazulha et al. (2012a)
0.2[c]	ND	*Nitzschia palea*	24	30	1 & 120	L-CC-S	Gazulha et al. (2012a)
4[c]	ND	*Microcystis aeruginosa* (toxic strain, solitary) + *Nitzschia palea* (50:50)	24	30	1 & 120	L-CC-S	Gazulha et al. (2012a)
11[c]	0.2[c]	*Microcystis aeruginosa* (nontoxic strain, colonial)	24	30	1	L-CC-S	Gazulha et al. (2012b)
503[c]	12.3[c]	*Microcystis aeruginosa* (nontoxic strain, solitary)	24	30	1	L-CC-S	Gazulha et al. (2012b)
136[c]	3.2[c]	*Planktothrix* sp. (nontoxic strain)	24	30	1	L-CC-S	Gazulha et al. (2012b)
19–357	ND	Mixed plankton	22–29	16.2 ± 2.7	72	M-CC-U	Frau et al. (2013)

Settings and method: *CC* cell counts, *L* laboratory, *M* mesocosms, *N* Neubauer chamber, *S* Sedgwick-Rafter chamber, *T* algal concentrations assessed from transmittance values, *U* Utermöhl technique, *ND* no data

[a] 25–75 % range of values obtained

[b] Dehydrated, supplied as aquarium food commercially known as Algamac-2000

[c] Ingestion rates (excludes pseudofeces)

[d] 9 µg of microcystin LR per liter

[e] 2–4 during the first 3 days of the experiment, dropping to <0.1 after 35 days

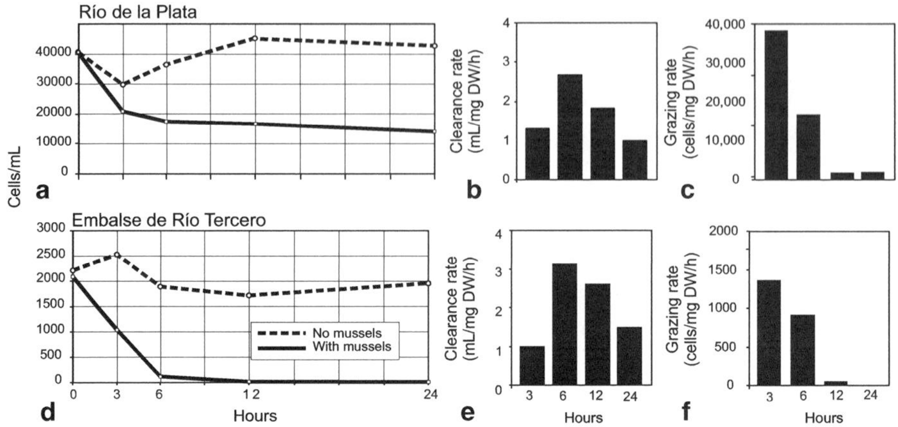

Fig. 7 Changes in algal densities (**a**, **d**), clearance rates (**b**, **e**), and grazing rates (**c**, **f**) by *L. fortunei* fed natural plankton along a 24-h experimental period in laboratory conditions (Río de la Plata, **a-c**) and in 400-l mesocosms (Embalse de Río Tercero reservoir, **d-e**). (Based on data from Cataldo et al. 2005; Cataldo et al. 2012a)

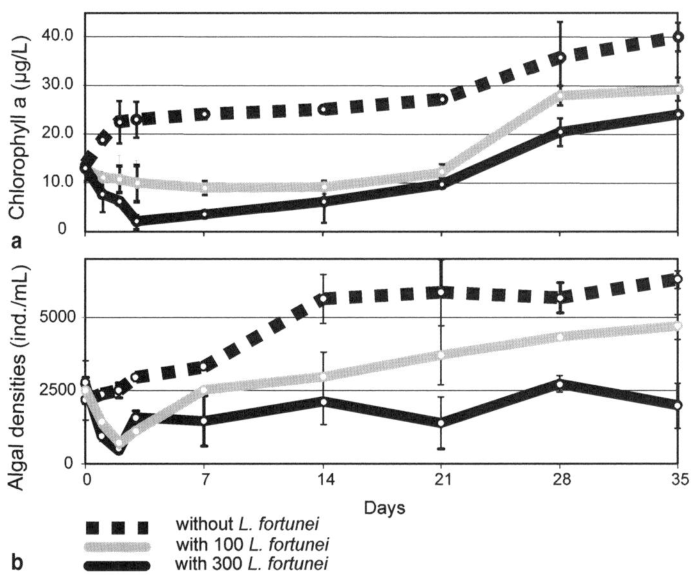

Fig. 8 Changes in the concentrations of chlorophyll a (**a**) and algal cells (**b**) in mesocosms with and without *L. fortunei* throughout a 35-day experimental period. (From Cataldo et al. 2012b)

(see Chapter "Colonization and Spread of *Limnoperna fortunei* in South America" in this volume) is still scant. Based on data from two tributaries of the Middle Paraná River collected before and after colonization by the mussel, Rojas Molina and José de Paggi (2008) concluded that zooplankton abundance (especially Rotifera) and chlorophyll a declined as a result of the invasion of *L. fortunei* (see

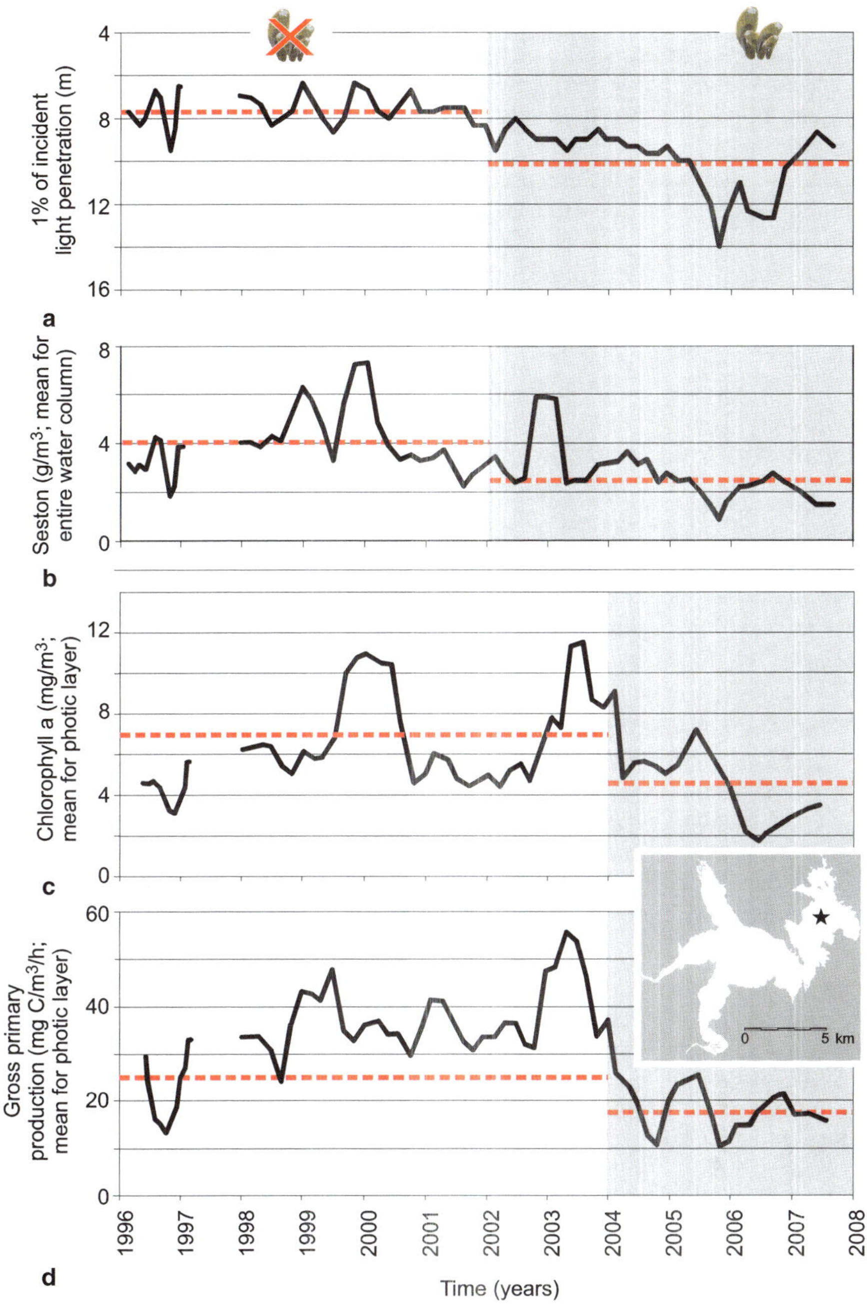

Fig. 9 Changes in several water-column properties in the reservoir Embalse de Río Tercero between 1996 and 2007. Curves (3-point running means; 5-point for **d**) are based on measurements at approximately 3-month intervals (ca. 70 data points). *Shaded areas* indicate periods with modified conditions (presumably due to colonization by the mussel). *Red broken lines* denote means for each period (significantly different at $p=0.05$ for all parameters except chlorophyll a, where $p=0.069$; Scheffé's post hoc tests). *Inset map* shows the position of the sampling station in the reservoir (*star*). (Modified from Boltovskoy et al. 2009a)

Chapter "Impacts of *Limnoperna fortunei* on Zooplankton" in this volume). Similar trends have also been suggested for a marginal lagoon and a tributary of the Middle Paraná River (Devercelli and Peruchet 2008). However, these rivers and associated marginal floodplains are relatively open systems strongly influenced by variable conditions in their upper reaches, and are subject to wide seasonal and interannual variations largely depending on precipitation and runoff regimes (see below), which complicates interpretation of causal relationships. Furthermore, in recent decades their catchment basins have been strongly modified by growing human populations and increasing land use for agricultural purposes, with the enhancement of the input of fertilizer-derived nutrients, pesticides and waste products, which further hinders pinpointing the concomitant effects of the invasive mussel.

Phytoplankton Grazing: Selectivity

Several surveys have addressed the issue of selectivity in *L. fortunei* grazing, using both natural, mixed plankton, and various combinations of cultured algae, with somewhat dissimilar results. In the 24-h mesocosm experiment performed in Embalse de Río Tercero described above, Cataldo et al. (2012a) found no association between prey cell size (across a range of 5–280,596 μm^3) and consumption rate. Laboratory experiments with *Microcystis viridis*, *Pseudoanabaena* sp. and *Selenastrum capricornutum* also yielded similar filtration rates for the three species (Rückert et al. 2004). On the other hand, strong selectivity was found when the mussel was fed a wider range of planktonic organisms, including zooplankton. Fachini et al (2012) reported negative selection for large (1–20 mm) filamentous algae and copepods, and positive selection for Rotifera, small (< 1 mm) filamentous algae, and several solitary algal cells, concluding that small to moderately sized particles and organisms with limited escape responses are favored by the mussel (see Chapter "Impacts of *Limnoperna fortunei* on Zooplankton" in this volume). Filtration experiments (72 h) performed in 200-L containers with natural plankton suggested that small flagellates are avoided by the mussel, whereas diatoms are positively selected (Frau et al. 2013).

Contrasting results may be partly explained by the fact that most studies did not differentiate particles actually ingested from those that are collected, embedded in mucus, and rejected as mucus-bound clumps that do not return to the water-column, but settle on the bottom. Gazulha et al. (2012b) differentiated between ingested particles and particles expelled as pseudofeces, and concluded that filtration rates of single-celled, colonial and filamentous cyanobacteria are similar, but while single cells are ingested, filamentous, and colonial forms are massively rejected as pseudofeces. A similar result was obtained when feeding *L. fortunei* with a mixture of cyanobacteria and diatoms. The diatom *Nitzschia palea* was found to disappear from the water faster than the cyanobacterium *Microcystis aeruginosa*, but ingestion rates were significantly higher for the latter, whereas *N. palea* was rejected (Gazulha et al. 2012a).

In summary, results available to date are still scarce and contradictory. Aside from the fact that the proportions of large organisms, especially those with well-developed avoidance abilities, are lower in the diet than in the water (although positive selection for several planktonic animals was also reported; see Chapter "Impacts of *Limnoperna fortunei* on Zooplankton" in this volume), data on grazing selectivity of the golden mussel are inconclusive. Furthermore, while impacts on larger plankton are probably lower than those on smaller particles, selectivity studies based on filtration experiments can be misleading with regard to the relative importance of the different items in the diet of *L. fortunei*, because the proportion of total biomass ingested may be dominated by large prey items (Rojas Molina et al. 2010; see Chapter "Impacts of *Limnoperna fortunei* on Zooplankton" in this volume).

Enhancement of Cyanobacterial Blooms

Blooms of toxic cyanobacteria, especially *Microcystis* spp., are usually associated with eutrophication and, in particular, with elevated P:N ratios (Smith 1983; Smith and Bennett 1999). Extensive river damming worldwide has created thousands of new waterbodies where stagnancy, enhanced vertical stratification, and growing nutrient input from agricultural land use boosts growth of cyanobacteria (Pizzolón et al. 1999; Jeong et al. 2003; Ruibal Conti et al. 2005; Relyea 2006). In recent decades, an additional bloom-enhancing effect has been described: some waterbodies have been observed to develop more frequent and stronger toxic cyanobacterial blooms after having been colonized by the zebra mussel (Bykova et al. 2006). Observational and experimental data on the effects of *L. fortunei* show that it also has a very significant impact on the abundance of Cyanobacteria.

In the experiment described above using 400-L mesocosms (Fig. 4), Cataldo et al. (2012a) found that *Microcystis* spp. densities increased from undetectable levels to ca. 1500–2000 cells/mL after 35 days in enclosures without mussels (Fig. 10a); however, in mesocosms with mussels, *Microcystis* spp. numbers soared to >200,000 cells/ml (Fig. 10b). Most significantly, colonial forms almost exclusively accounted for this increase, whereas solitary cells of *Microcystis* spp. remained at very low levels throughout the experiment (Fig. 10b). During the first week, only solitary cells of *Microcystis* spp. were found in all mesocosms. After 2 weeks, in the mesocosms without mussels the proportion of solitary *Microcystis* cells dropped to 34 %, and varied around 20–50 % until the end of the experiment. In contrast, in the mesocosms with mussels, solitary cells were dominant until week 3, being completely replaced thereafter by colonial individuals (Fig. 10b). Most significantly, this growth in the presence of mussels was accompanied by a strong increase in the size of *Microcystis* spp. colonies (Fig. 10c). In controls without mussels, the mean size of *Microcystis* spp. colonies remained around 50 μm (maximum dimension) throughout the entire experimental period. In the mesocosms with mussels, on the other hand, on week 4 all *Microcystis* cells were in colonies ~135 μm in size, and by week 5, they attained a mean size of ~179 μm

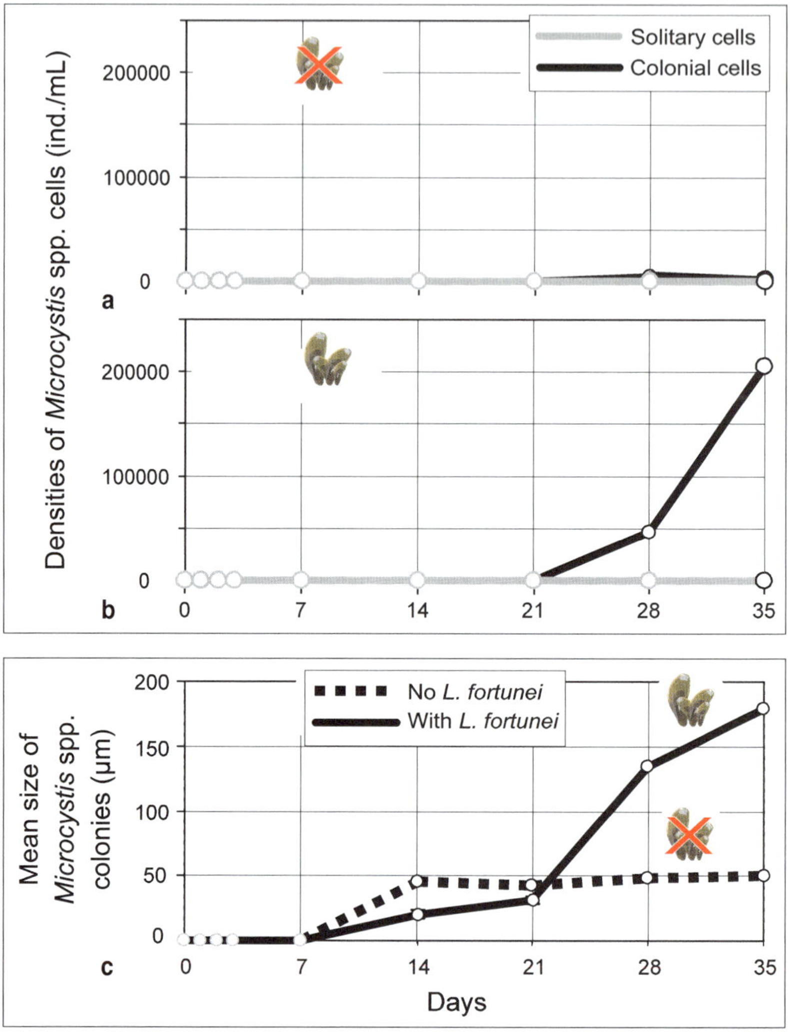

Fig. 10 Changes in the abundance of *Microcystis* spp. solitary cells and cells belonging to a colony in 400-L mesocosms without *L. fortunei* (**a**) and stocked with 300 mussels (**b**). Changes in mean size of *Microcystis* spp. colonies through time in mesocosms without mussels and in the presence of 300 mussels (**c**). (From Cataldo et al. 2012b)

(Fig. 10c). Other cyanobacterial species present in the enclosures (*Anabaena circularis, Aphanocapsa delicatissima, Chroococcus minutus, Pseudoanabaena mucicola*) also grew more in the presence of *L. fortunei* than in the controls, but their densities remained low.

Interpretation of these results suggests that there are several mechanisms converging to enhance *Microcystis* spp. densities in the presence of the mussels: (1) Changes in nutrient availability, (2) changes in the P:N ratio, (3) size-selective grazing, whereby small, solitary cells are eliminated more effectively than colonies,

(4) promotion of colony-formation by chemical signals that trigger aggregation of solitary cells in order to avoid grazing, and (5) Microcystin toxicity, deterring grazing as *Microcystis* spp. biomass builds up (Cataldo et al. 2012b).

Nutrient availability and higher P:N ratios have long been identified as cyanobacterial bloom-enhancing triggers (Smith 1983; Smith and Bennett 1999). Promotion of colony-formation in various autotrophs, including *Microcystis* spp., by predator-produced chemical signals has been described repeatedly (Yang et al. 2005). Selective grazing of solitary cells is supported by some previous results with *L. fortunei* (see above). Grazing of toxic *Microcystis* spp. by the golden mussel, on the other hand, is still a debatable issue. Some surveys have concluded that both toxic and nontoxic strains of *Microcystis* spp. are actively consumed by adult mussels (Rückert et al. 2004; Gazulha et al. 2012a). Fachini (2011) found that toxic and nontoxic strains of *Microcystis aeruginosa* are consumed alike (at microcystin LR concentrations in water around 6–9 ppb), but at lower ingestion rates than other algae (*Monorhaphidium* sp.). In contrast, preliminary tests performed in Salto Grande Reservoir, where very dense blooms of *Microcystis* spp. are a recurrent summer through autumn phenomenon (Chalar 2009; O' Farrell et al. 2012; Boltovskoy et al. 2013), indicate that at dissolved microcystin LR concentrations above 2 ppb the mussel ceases to filter, whereas at levels above 8 ppb significant mortality is observed (at 8 ppb 5 % of the mussels die after 2 h, whereas at 30 ppb mortality reaches 90 %) (Boltovskoy et al. 2009b). Interestingly, a similar controversy is also found in studies on the effects of *Microcystis* spp. on *Dreissena polymorpha.* Some surveys concluded that zebra mussels fed *Microcystis* spp. show significantly reduced grazing and acute irritant responses (Juhel et al. 2006a; Juhel et al. 2006b), or selectively reject them, especially large colonial aggregates of the unpalatable toxic strains (Vanderploeg et al. 2001), whereas others concluded that toxic *Microcystis* spp. are consumed as effectively as other algae, suggesting that the mussel could control cyanobacterial blooms (Dionisio Pires et al. 2010). This suggests that consumption of toxin-producing strains of these cyanobacteria varies depending on conditions that we still do not fully understand.

In any case, regardless of *L. fortunei*'s tolerance to microcystin, the results of Cataldo et al. (2012b) strongly suggest that the golden mussel boosts the growth of cyanobacteria. A major difference with the zebra mussel, however, is that while *D. polymorpha* enhances cyanobacterial numbers only in lakes with low to moderate P concentrations (< 25 µg total P/L; Nicholls et al. 2002; Raikow et al. 2004; Sarnelle et al. 2005; Knoll et al. 2008), *L. fortunei* in South America does so at very high total P levels (between 50 and > 100 µg total P/L in the reservoir where these experiments were carried out, Chalar 2006; Cataldo et al. 2012b; O' Farrell et al. 2012).

A remarkable consequence of mussel-induced toxic cyanobacterial growth is that these blooms suppress the bivalve's reproduction. This effect has been suggested by several laboratory and field studies (Boltovskoy et al. 2009b; Gazulha 2010; Gazulha et al. 2012b), and confirmed by the analysis of 9 years of observational data in Salto Grande reservoir. Recurrent blooms of *Microcystis* spp. in Salto Grande Reservoir interrupt production of larvae at a time when in all other waterbodies investigated (without cyanobacterial blooms) mussel reproduction is maximum (late spring–early autumn; Boltovskoy et al. 2013).

Periphyton and Aquatic Macrophytes

An indirect consequence of fast nutrient regeneration rates and clarification of the water column by elimination of suspended organic and inorganic matter (including nutrient-consuming phytoplankton) through filter feeding is enhanced growth of periphyton and aquatic macrophytes (Fig. 1; Pillsbury and Lowe 1994; Karatayev et al. 1997; Zhu et al. 2006; Karatayev et al. 2007; Kelly et al. 2010). Data for the golden mussel indicate that it has also had these effects in some of the reservoirs surveyed.

In the mesocosm survey in Salto Grande Reservoir (Fig. 4; Cataldo et al. 2012a), periphytic chlorophyll a in the enclosures with mussels was ca. 16 times higher than in those without *L. fortunei* 5 weeks after deployment. Periphyton biomass increased in both control and experimental enclosures, but the proportion of algal biomass was significantly higher in the latter, indicating a shift from heterotrophic to autotrophic dominance.

Since it was first filled in 1934, the reservoir Embalse de Río Tercero had no significant macrophyte populations. It was colonized by *L. fortunei* in 1998, and since around 2000 the macrophyte *Elodea callitrichoides* has been a dominant feature of this waterbody, forming large beds along several coastal stretches (Boltovskoy et al. 2009a). At approximately the same time, coot and grebe (*Fulica leucoptera, Fulica armillata, Podilymbus podiceps*) populations in the reservoir increased noticeably, most probably in response to the expansive growth of the beds of aquatic plants on which the birds feed. Coots and grebes have also been observed to retrieve clusters of *L. fortunei* from the bottom (M. Hechem, pers. comm.), suggesting that they also feed on the mussel too (as other coot species feed on *D. polymorpha* in North America; Molloy et al. 1997).

Concluding Remarks

Functional similarities between freshwater invasive byssate mussels, including *L. fortunei* and *Dreissena* spp., are responsible for similar forcing mechanisms, particularly in their effects on nutrient recycling and pelagic-benthic coupling (Karatayev et al. 1997; Boltovskoy et al. 2006; Ward and Ricciardi 2007; Kelly et al. 2010; Burlakova et al. 2012). However, intrinsic differences between *L. fortunei* and dreissenids, as well as environmental differences between the waterbodies invaded, are responsible for significant contrasts in ecosystem responses. For example, considerably higher calcium concentrations in European and North American waters than in South American waters (Karatayev et al. 2007) are presumably responsible for the fact that in the former empty mollusc shells represent an important source of substrate for *D. polymorpha* (Strayer et al. 1996; Burlakova et al. 2006; Strayer and Malcom 2006), whereas in South America dead mussels shells dissolve before they are colonized (Boltovskoy et al. 2006; Karatayev et al. 2007) (see Chapter "*Limnoperna fortunei* Colonies: Structure, Distribution and Dynamics" in this volume).

For reasons we still do not fully understand, in North American waterbodies *Dreissena* spp. enhance cyanobacterial blooms only when total P concentrations are below 20–25 µg/l (Nicholls et al. 2002; Raikow et al. 2004; Sarnelle et al. 2005; Knoll et al. 2008), but in South America *L. fortunei* boosts growth of *Microcystis* spp. even at levels as high as 50–100 µg P/L (Cataldo et al. 2012b).

Impacts of bivalve grazing on phytoplankton obviously depend on mussel densities and on the abundance of particulate organic matter (POM) in the water column. While mussel densities are generally comparable for these invaders (Karatayev et al. 2010), the attributes of the water bodies invaded are not. The Río de la Plata floodplain river system invaded by *L. fortunei* has very marked differences with the colder, clearer and more oligotrophic North American waterbodies colonized by *Dreissena*. One salient difference are the concentrations of POC, typically around 0.15–1 mg/L in the Great Lakes (Fanslow et al. 1995; Barbiero and Tuchman 2004; Johengen et al. 2008), but several times higher in the Río de la Plata watershed—about 3.5 mg/L (Depetris 1976; Depetris and Paolini 1991; Depetris and Pasquini 2007). Sylvester et al. (2005) estimated the energy that can be obtained by a filtering mussel from the phytoplankton and from the seston in general in the lower delta of the Paraná River. Their assessment indicates that phytoplankton alone cannot meet the energetic demands of *L. fortunei*, but when POM is considered the requirements of juvenile and adult mussels are exceeded 15–22 fold. Furthermore, although freshwater suspension feeding organisms are generally thought to be inefficient at using dissolved organic carbon (DOC) as a source of food (Lopez 1988), both veligers and adults of *D. polymorpha* have been shown to use DOC intensively, obtaining up to 50 % of their metabolic needs for carbon from this source (Roditi et al. 2000; Banard et al. 2006; Baines et al. 2007). No data are available for the golden mussel, but functional similarities with zebra mussels indicate that *L. fortunei* may also be able to use DOC. This suggests that filter-feeding organisms in these riverine systems are not food-limited (but they may be food-limited in some of the lentic bodies of water colonized: Boltovskoy et al. 2009b). This assumption contrasts sharply with some of the described impacts of *Dreissena* spp. in the northern hemisphere, where competition for food with the invader has been found to have strong effects on zooplankton and fish communities (Lozano et al. 2001; Bartsch et al. 2003; Strayer et al. 2004).

Indigenous filter-feeding benthic animals and fishes in the Paraná watershed are scarce, and the dominant feeding modes are associated with detrital and sedimentary organic matter (José de Paggi and Paggi 2007; Rossi et al. 2007). Thus, much of the organic matter carried downstream (around 1 Tg/y, Depetris and Pasquini 2007), is flushed out into the ocean through the Río de la Plata estuary. Since the 1990s, *L. fortunei*, the first abundant macrobenthic filter-feeder in this system, has been intercepting an important proportion of this POM and retaining it in the system for use by a wide array of animals. This trophic shift involves not only *L. fortunei* larvae and adults, but also many other invertebrates whose abundances are enhanced by *L. fortunei* beds, and the organic matter-enriched sediments derived from the "shunt" of suspended POM to the bottom as feces and pseudofeces (Sardiña et al. 2008; Kelly et al. 2010) (see Chapter "Relationships of *Limnoperna fortunei* with

Benthic Animals" in this volume). These organic matter-rich sediments are the main source of food for many organisms, including some of the most abundant fish species, like *Prochilodus lineatus*, whose biomass represents >60% of the overall fish biomass in the Paraná-Uruguay system (Bonetto 1998). On local scales some of the consequences of this invasion have been explored, but on the ecosystem scale we still know very little. Boltovskoy et al. (2006) suggested that enhanced feeding conditions for larval and adult fishes may be responsible for the three-fold increase in Argentine freshwater fish landings after introduction of *L. fortunei*. However, interpretation of these cause-effect relationships should be made with caution because several factors, including changes in fishing regulations, fishing pressure, fish export trends, etc. changed during the same period.

Acknowledgments This work was partially financed by grants from the University of Buenos Aires, Argentina (UBA X-020 and 20020100100035) and from the Argentine Agencia Nacional de Promoción Científica y Tecnológica, Argentina (PICT 2007 1968) to DB.

References

Baines SB, Fisher NS, Cole JJ (2007) Dissolved organic matter and persistence of the invasive zebra mussel (*Dreissena polymorpha*) under low food conditions. Limnol Oceanogr 52:70–78

Banard C, Martineau C, Frenette JJ, Dodson JJ, Vincent WF (2006) Trophic position of zebra mussel veligers and their use of dissolved organic carbon. Limnol Oceanogr 51:1473–1484

Barbiero RP, Tuchman ML (2004) The deep chlorophyll maximum in Lake Superior. J Great Lakes Res 30(Suppl 1):256–268

Bartsch LA, Richardson WB, Sandheinrich MB (2003) Zebra mussels (*Dreissena polymorpha*) limit food for larval fish (*Pimephales promelas*) in turbulent systems: a bioenergetics analysis. Hydrobiologia 495:59–72

Boltovskoy D, Correa N (2015) Ecosystem impacts of the invasive bivalve *Limnoperna fortunei* (golden mussel) in South America. Hydrobiologia 746:81–95

Boltovskoy D, Correa N, Cataldo D, Sylvester F (2006) Dispersion and ecological impact of the invasive freshwater bivalve *Limnoperna fortunei* in the Río de la Plata watershed and beyond. Biol Invasions 8:947–963

Boltovskoy D, Karatayev A, Burlakova L, Cataldo D, Karatayev V, Sylvester F, Mariñelarena A (2009a) Significant ecosystem-wide effects of the swiftly spreading invasive freshwater bivalve *Limnoperna fortunei*. Hydrobiologia 636:271–284

Boltovskoy D, Sylvester F, Otaegui A, Leytes V, Cataldo D (2009b) Environmental modulation of the reproductive activity of the invasive mussel *Limnoperna fortunei* in South America. Austral Ecol 34:719–730

Boltovskoy D, Correa N, Bordet F, Leites V, Cataldo D (2013) Toxic *Microcystis* (cyanobacteria) inhibit recruitment of the bloom-enhancing invasive bivalve *Limnoperna fortunei*. Freshw Biol 58:1968–1981

Bonetto AA (1998) Panorama sinóptico sobre la ictiofauna, la pesca y la piscicultura en los ríos de la cuenca del Plata con especial referencia al Paraná. Revista de Ictiología 6:3–15

Bootsma HA, Liao Q (2014) Nutrient cycling by dreissenid mussels. In: Nalepa TF, Schloesser DW (eds) Quagga and zebra mussels: biology, impacts, and control, 2nd edn. CRC Press, Boca Raton, pp 555–574

Burlakova LE, Karatayev AY, Padilla DK (2005) Functional changes in benthic freshwater communities after *Dreissena polymorpha* (Pallas) invasion and consequences for filtration. In:

Dame RF, Olenin S (eds) The comparative roles of suspension-feeders in ecosystems. Springer, Amsterdam, pp 263–275

Burlakova LE, Karatayev AY, Padilla DK (2006) Changes in the distribution and abundance of *Dreissena polymorpha* within lakes through time. Hydrobiologia 571:133–146

Burlakova LE, Karatayev AY, Karatayev VA (2012) Invasive mussels induce community changes by increasing habitat complexity. Hydrobiologia 685:121–134

Bykova O, Laursen A, Bostan V, Bautista J, Mc Carthy L (2006) Do zebra mussels (*Dreissena polymorpha*) alter lake water chemistry in a way that favours *Microcystis* growth? Sci Total Environ 371:362–372

Cataldo D, O'Farrell I, Paolucci EM, Sylvester F, Boltovskoy D (2005) Efectos de *Limnoperna fortunei* sobre el fitoplancton y los nutrientes. In: Tercer Congreso Argentino de Limnología, CAL III, Chascomús (Argentina)

Cataldo D, O'Farrell I, Paolucci E, Sylvester F, Boltovskoy D (2012a) Impact of the invasive golden mussel (*Limnoperna fortunei*) on phytoplankton and nutrient cycling. Aquat Invasions 7:91–100

Cataldo D, Vinocur A, O'Farrell I, Paolucci E, Leites V, Boltovskoy D (2012b) The introduced bivalve *Limnoperna fortunei* boosts *Microcystis* growth in Salto Grande Reservoir (Argentina): evidence from mesocosm experiments. Hydrobiologia 680:25–38

Chalar G (2006) Dinámica de la eutrofización a diferentes escalas temporales: Embalse Salto Grande (Argentina-Uruguay). In: Tundisi JG, Matsumura Tundisi T, Sidagis Galli C (eds) Eutrofização na América do Sul: causas, conseqüências e tecnologias de gerenciamento e controle. Instituto Internacional de Ecologia e Gerenciamento Ambiental, Academia Brasileira de Ciencias, Conselho Nacional de Desenvolvimento Científico e Tecnológico, InterAcademy Panel on International Issues, InterAmerican Network of Academies of Sciences, Sao Carlos, pp 87–101

Chalar G (2009) The use of phytoplankton patterns of diversity for algal bloom management. Limnol—Ecol Manage Inland Waters 39:200–208

Depetris PJ (1976) Hydrochemistry of the Paraná River. Limnol Oceanogr 21:736–739

Depetris PJ, Paolini JE (1991) Biogeochemical aspects of South American rivers: the Paraná and the Orinoco. In: Degens ET, Kempe S, Richey JE (eds) Biogeochemistry of major world rivers. Wiley, Chichester, pp 105–125

Depetris PJ, Pasquini AI (2007) The geochemistry of the Paraná River: an overview. In: Iriondo MH, Paggi JC, Parma MJ (eds) The Middle Paraná River. Limnology of a subtropical wetland. Springer Verlag, Berlin, pp 143–174

Devercelli M, Peruchet E (2008) Trends in chlorophyll-a concentration in urban water bodies within different man-used basins. Annales de Limnologie—Int J Limnol 44:75–84

Di Fiori E, Pizarro H, dos Santos Afonso M, Cataldo D (2012) Impact of the invasive mussel *Limnoperna fortunei* on glyphosate concentration in water. Ecotoxicol Environ Safety 81:106–113

Dionisio Pires LM, Ibelings BW, van Donk E (2010) Zebra mussels as a potential tool in the restoration of eutrophic shallow lakes, dominated by toxic cyanobacteria. In: van der Velde G, Rajagopal S, Bij de Vaate A (eds) The Zebra Mussel in Europe. Backhuys Publishers, Leiden, pp 361–372

Fachini A (2011) Filtração do bivalve invasor *Limnoperna fortunei* (Dunker, 1857), o mexilhão dourado, sobre a comunidade planctônica natural e na presença de cianobactéria tóxica. MSc Thesis, Universidade Federal do Rio Grande do Sul (Brazil), pp 1–68

Fachini A, Gazulha V, Pedrozo CS (2012) Os impactos do mexilhão-dourado sobre a comunidade planctônica. In: Mansur MCD, Santos CP, Pereira D, Padula Paz IC, Leite Zurita ML, Raya Rodriguez MT, Vilar Nehrke M, Aydos Bergonci PE (eds) Moluscos límnicos invasores no Brasil. Biologia, prevenção, controle. Redes Editora, Porto Alegre, pp 255–261

Fanslow DL, Nalepa TF, Lang GA (1995) Filtration rates of the zebra mussel (*Dreissena polymorpha*) on natural seston from Saginaw Bay, Lake Huron. J Great Lakes Res 21:489–500

Frau D, Rojas Molina F, Devercelli M, José de Paggi S (2013) The effect of an invading filter-feeding bivalve on a phytoplankton assemblage from the Paraná system: a mesocosm experiment. Mar Freshw Behav Physiol 45:303–316

Gazulha V (2010) O mexilhão dourado *Limnoperna fortunei* (Dunker, 1857) na presença de cianobactérias: taxas de filtração, comportamento alimentar e sobrevivência. PhD Thesis, Universidade Federal do Rio Grande do Sul (Brazil), pp 1–104
Gazulha V (2012) O impacto de *Limnoperna fortunei* sobre as cianobacterias. In: Mansur MCD, Santos CP, Pereira D, Padula Paz IC, Leite Zurita ML, Raya Rodriguez MT, Vilar Nehrke M, Aydos Bergonci PE (eds) Moluscos límnicos invasores no Brasil. Biologia, prevenção, controle. Redes Editora, Porto Alegre, pp 249–254
Gazulha V, Mansur MCD, Cybis LF, Azevedo SMFO (2012a) Feeding behavior of the invasive bivalve *Limnoperna fortunei* (Dunker, 1857) under exposure to toxic cyanobacteria *Microcystis aeruginosa* . Braz J Biol 72:41–49
Gazulha V, Mansur MCD, Cybis LF, Azevedo SMFO (2012b) Grazing impacts of the invasive bivalve *Limnoperna fortunei* (Dunker, 1857) on single-celled, colonial and filamentous cyanobacteria. Braz J Biol 72:33–39
Jeong KS, Kim DK, Wigham P, Joo GJ (2003) Modelling *Microcystis aeruginosa* bloom dynamics in the Nakdong River by means of evolutionary computation and statistic approach. Ecol Model 161:67–68
Johengen TH, Biddanda BA, Cotner JB (2008) Stimulation of Lake Michigan plankton metabolism by sediment resuspension and river runoff. J Great Lakes Res 34:213–227
Jørgensen CB (1966) Biology of suspension feeding. Pergamon Press, Oxford, pp 1–357
José de Paggi S, Paggi JC (2007) Zooplankton. In: Iriondo MH, Paggi JC, Parma MJ (eds) The middle Parana River. Limnology of a subtropical wetland. Springer, Berlin, pp 229–249
Juhel G, Davenport J, O' Halloran J, Culloty S, Ramsay R, James K, Furey A, Allis O (2006a) Pseudodiarrhoea in zebra mussels *Dreissena polymorpha* (Pallas) exposed to microcystins. J Exp Biol 209:810–816
Juhel G, Davenport J, O'Halloran J, Culloty SC, O'Riordan RM, James KF, Furey A, Allis O (2006b) Impacts of microcystins on the feeding behaviour and energy balance of zebra mussels, *Dreissena polymorpha*: a bioenergetics approach. Aquat Toxicol 79:391–400
Karatayev AY, Burlakova LE, Padilla DK (1997) The effects of *Dreissena polymorpha* (Pallas) invasion on aquatic communities in Eastern Europe. J Shellfish Res 16:187–203
Karatayev AY, Burlakova LE, Padilla DK (2002) Impacts of zebra mussels on aquatic communities and their role as ecosystem engineers. In: Leppäkoski E, Gollasch S, Olenin S (eds) Invasive aquatic species of Europe: distribution, impacts and management. Kluwer Academic Publishers, Dodrecht, pp 433–446
Karatayev AY, Boltovskoy D, Padilla DK, Burlakova LE (2007) The invasive bivalves *Dreissena polymorpha* and *Limnoperna fortunei*: parallels, contrasts, potential spread and invasion impacts. J Shellfish Res 26:205–213
Karatayev AY, Burlakova LE, Karatayev VA, Boltovskoy D (2010) *Limnoperna fortunei* versus *Dreissena polymorpha*: population densities and benthic community impacts of two invasive freshwater bivalves. J Shellfish Res 29:975–984
Kawase M (2011) Feeding and filtration in *Limnoperna lacustris* . http://www.mizuho-c.ac.jp/images/library/kiyo_05/amckiyo-no05-05.pdf. Accessed 11 Feb 2014 [In Japanese]
Kelly DW, Herborg L-M, MacIsaac HJ (2010) Ecosystem changes associated with *Dreissena* invasions: recent developments and emerging issues. In: van der Velde G, Rajagopal S, Bij de Vaate A (eds) The Zebra Mussel in Europe. Backhuys Publishers, Leiden, pp 199–210
Knoll LB, Sarnelle O, Hamilton SK, Kissman CEH, Wilson AE, Rose JB, Morgan MR (2008) Invasive zebra mussels (*Dreissena polymorpha*) increase cyanobacterial toxin concentrations in low-nutrient lakes. Can J Fish Aquat Sci 65:448–455
Lopez GR (1988) Comparative ecology of the macrofauna of freshwater and marine muds. Limnol Oceanogr 33:946–962
Lozano SJ, Scharold JV, Nalepa TF (2001) Recent declines in benthic macroinvertebrate densities in Lake Ontario. Can J Fish Aquat Sci 58:518–529
MacIsaac HJ (1996) Potential abiotic and biotic impacts of zebra mussels on the inland waters of North America. Am Zool 36:287–299

Mansur MCD, Santos CP, Pereira D, Padula Paz IC, Leite Zurita ML, Raya Rodriguez MT, Vilar Nehrke M, Aydos Bergonci PE (eds) (2012) Moluscos límnicos invasores no Brasil. Biologia, prevenção, controle. Redes Editora, Porto Alegre, pp 1–411

Molloy DP, Karatayev AY, Burlakova EB, Kurandina DP, Laruelle F (1997) Natural enemies of zebra mussels: predators, parasites, and ecological competitors. Rev Fish Sci 5:27–97

Morton B (1983) Feeding and digestion in Bivalvia. In: Wilbur KM, Saleuddin ASM (eds) The Mollusca, vol 5: Physiology, Part 2. Academic Press, New York, pp 65–147

Nalepa TF, Schloesser DW (eds) (2014) Quagga and zebra mussels: biology, impacts, and control. 2nd Edition. CRC Press, Boca Raton, pp 1–815

Nicholls KH, Heintsch L, Carney E (2002) Univariate step-trend and multivariate assessments of the apparent effects of P loading reductions and zebra mussels on the phytoplankton of the Bay of Quinte, Lake Ontario. J Great Lakes Res 28:15–31

O'Farrell I, Bordet F, Chaparro G (2012) Bloom forming cyanobacterial complexes co-occurring in a subtropical large reservoir: validation of dominant eco-strategies. Hydrobiologia 698:175–190

Pestana D, Ostrensky A, Pereira Boeger WA, Pie MR (2009) The effect of temperature and body size on filtration rates of *Limnoperna fortunei* (Bivalvia, Mytilidae) under laboratory conditions. Braz Arch Biol Technol 52:135–144

Pillsbury RW, Lowe RL (1994) The impact of zebra mussels on benthic algal communities in Saginaw Bay, Lake Huron. In: 4th International Zebra Mussel Conference, Madison (USA)

Pizzolón L, Tracanna B, Prosperi C, Guerrero JM (1999) Cyanobacterial blooms in Argentinean inland waters. Lakes & Reserv: Res Manage 4:101–105

Raikow DF, Sarnelle O, Wilson AE, Hamilton SK (2004) Dominance of the noxious cyanobacterium *Microcystis aeruginosa* in low-nutrient lakes is associated with exotic zebra mussels. Limnol Oceanogr 49:482–487

Relyea RA (2006) The impact of insecticides and herbicides on the biodiversity and productivity of aquatic communities. Ecol Appl 16:2027–2034

Roditi HA, Fisher NS, Sañudo-Wilhelmy SA (2000) Uptake of dissolved organic carbon and trace elements by zebra mussels. Nature 407:78–80

Rojas Molina F, Jose de Paggi SB (2008) Zooplankton in the Paraná River floodplain (South America) before and after the invasion of *Limnoperna fortunei* (Bivalvia). Wetlands 28:695–702

Rojas Molina F, Paggi JC, Devercelli M (2010) Zooplanktophagy in the natural diet and selectivity of the invasive mollusk *Limnoperna fortunei* . Biol Invasions 12:1647–1659

Rossi L, Cordiviola E, Parma MJ (2007) Fishes. In: Iriondo MH, Paggi JC, Parma MJ (eds) The middle Parana River. Limnology of a subtropical wetland. Springer, Berlin, pp 305–325

Rückert G, Campos MCS, Rolla ME (2004) Alimentação de *Limnoperna fortunei* (Dunker 1857): taxas de filtração com ênfase ao uso de Cyanobacteria. Acta Scientiarum 26:421–429

Ruibal Conti AL, Guerrero JM, Regueira JM (2005) Levels of microcystins in two Argentinean reservoirs used for water supply and recreation: differences in the implementation of safe levels. Environ Toxicol 20:263–269

Sardiña P, Cataldo D, Boltovskoy D (2008) The effects of the invasive mussel, *Limnoperna fortunei*, on associated fauna in South American freshwaters: importance of physical structure and food supply. Fund Appl Limnol/Archiv für Hydrobiologie 173:135–144

Sarnelle O, Wilson AE, Hamilton SK, Knoll LB, Raikow DF (2005) Complex interactions between the zebra mussel, *Dreissena polymorpha*, and the harmful phytoplankter, *Microcystis aeruginosa* . Limnol Oceanogr 50:896–904

Smith VH (1983) Low nitrogen to phosphorus ratios favor dominance by blue-green algae in lake phytoplankton. Science 221:669–671

Smith VH, Bennett SJ (1999) Nitrogen:phosphorus supply ratios and phytoplankton community structure in lakes. Archiv für Hydrobiologie 146:37–53

Spaccesi FG, Rodrigues Capitulo A (2012) Benthic communities on hard substrates covered by *Limnoperna fortunei* Dunker (Bivalvia, Mytilidae) at an estuarine beach (Río de la Plata, Argentina). J Limnol 71:144–153

Strayer DL, Malcom HM (2006) Long-term demography of a zebra mussel (*Dreissena polymorpha*) population. Freshw Biol 51:117–130

Strayer DL, Powell J, Ambrose P, Smith LC, Pace ML, Fischer DT (1996) Arrival, spread, and early dynamics of a zebra mussel (*Dreissena polymorpha*) population in the Hudson River estuary. Can J Fish Aquat Sci 53:1143–1149

Strayer DL, Hattala KA, Kahnle AW (2004) Effects of an invasive bivalve (*Dreissena polymorpha*) on fish in the Hudson River estuary. Can J Fish Aquat Sci 61:924–941

Sylvester F, Dorado J, Boltovskoy D, Juárez A, Cataldo D (2005) Filtration rates of the invasive pest bivalve *Limnoperna fortunei* as a function of size and temperature. Hydrobiologia 534:71–80

Sylvester F, Boltovskoy D, Cataldo D (2007) Fast response of freshwater consumers to a new trophic resource: Predation on the recently introduced Asian bivalve *Limnoperna fortunei* in the lower Parana River, South America. Austral Ecol 32:403–415

Turner CB (2010) Influence of zebra (*Dreissena polymorpha*) and quagga (*Dreissena rostriformis*) mussel invasions on benthic nutrient and oxygen dynamics. Can J Fish Aquat Sci 67:1899–1908

Vanderploeg HA, Liebig JR, Carmichael WW, Agy MA, Johengen TH, Fahnenstiel GL, Nalepa TF (2001) Zebra mussel (*Dreissena polymorpha*) selective filtration promoted toxic *Microcystis* blooms in Saginaw Bay (Lake Huron) and Lake Erie. Can J Fish Aquat Sci 58:1208–1221

Ward JM, Ricciardi A (2007) Impacts of *Dreissena* invasions on benthic macroinvertebrate communities: a meta-analysis. Divers Distrib 13:155–165

Yang Z, Kong FX, H.S. C, Shi XL (2005) Observation on colony formation of *Microcystis aeruginosa* induced by filtered lake water under laboratory conditions. Annales de Limnologie 41:169–173

Zhang H, Culver D, Boegman L (2011) Dreissenids in Lake Erie: an algal filter or a fertilizer? Aquat Invasions 6:175–194

Zhu B, Fitzgerald DG, Mayer CM, Rudstam LG, Mills EL (2006) Alteration of ecosystem function by zebra mussels in Oneida Lake: impacts on submerged macrophytes. Ecosystems 9:1017–1028

Impacts of *Limnoperna fortunei* on Zooplankton

Florencia Rojas Molina, Susana B. José de Paggi and Juan César Paggi

Abstract In addition to phytoplankton, adult *Limnoperna fortunei* feed actively on animal prey, consuming over 150 different organisms, including Protista, Rotifera, Nematoda, Cladocera, Copepoda, Ostracoda, and *L. fortunei* larvae. Animals are present in the stomachs of 96% of the individuals analyzed representing, on average, 67% of ingested biomass. Rotifers are the most frequent animal prey, followed by cladocerans and copepod nauplii. Comparison between the abundances of prey in plankton samples and the diet of *L. fortunei* indicates that all animals are positively selected, with the highest selectivity for the rotifers and small cladocerans. Selectivity is positively associated with prey size and negatively with avoidance capabilities. This selective grazing pressure is probably responsible for the fact that rotifer densities have dropped in several water bodies associated with the Middle Paraná River after *L. fortunei* colonized the area, while cladocerans and copepods remain at pre-invasion levels. In addition, grazing pressure probably accounts for significant post-invasion decreases in zooplankton density differences during low- and high-water periods. Densities of *L. fortunei* veligers normally exceed those of rotifer+crustacean zooplankton for 8–9 months of the year, underscoring their potential significance as competitors of other zooplankton for food, and as food for various animals.

Keywords *Limnoperna fortunei* · Golden mussel · Ecological impact · Zooplankton grazing · Grazing selectivity

---

F. Rojas Molina (✉) · S. B. José de Paggi
Instituto Nacional de Limnología
(INALI, Consejo Nacional de Investigaciones Científicas y Técnicas-Universidad Nacional del Litoral), Facultad de Bioquímica y Ciencias Biológicas (Universidad Nacional del Litoral), Ciudad Universitaria, 3000 Santa Fe, Argentina
e-mail: frojasmolina@inali.unl.edu.ar

S. B. José de Paggi
e-mail: sjdepaggi@gmail.com

J. C. Paggi
Instituto Nacional de Limnología (INALI, Consejo Nacional de Investigaciones Científicas y Técnicas-Universidad Nacional del Litoral),
Ciudad Universitaria, 3000 Santa Fe, Argentina
e-mail: juanpaggi@gmail.com

D. Boltovskoy (ed.), *Limnoperna fortunei,* Invading Nature - Springer Series in Invasion Ecology 10, DOI 10.1007/978-3-319-13494-9_10

Introduction

Benthic–pelagic coupling mechanisms may involve strong effects of suspension feeding bivalves on the plankton community, dramatically affecting planktonic abundance and structure by selective feeding (Wong and Levinton 2006). In order to understand the effects of mussel grazing on the plankton, it is important to determine both the amount and the type of organisms retained, ingested, and rejected (Cohen et al. 1984; Caraco et al. 1997; Pace et al. 1998; Rojas Molina and José de Paggi 2008). Plankton that is withdrawn from the water column and digested by the mussels or trapped in their pseudofeces may constitute a significant loss of biomass from the pelagial (Horgan and Mills 1997). Although phytoplankton is usually the major source of nutrition for bivalves (Cohen et al. 1984; Bastviken et al. 1998; Prins et al. 1998), several studies have shown that they can also feed on zooplankton (MacIsaac et al. 1995; Wong et al. 2003; Wong and Twining 2003; Wong and Levinton 2005).

Animals in the Diet of *Limnoperna fortunei*: Field Data and Laboratory Experiments

Although most data on the feeding of *Limnoperna fortunei* are centered on the phytoplankton (see Chapter "Nutrient Recycling, Phytoplankton Grazing, and Associated Impacts of *Limnoperna fortunei*" in this volume), the zooplankton represents a significant fraction of the diet of the mussel. In a survey carried out in several floodplain environments of the Middle Paraná River, Rojas Molina et al. (2010) recorded a total of 156 taxa in the stomach contents of the golden mussel, covering a wide spectrum of sizes ranging from 4 to slightly over 1000 μm in length. Algae included 81 taxa (Cyanobacteria, Chlorophyceae, Xanthophyta, Bacillariophyceae, Euglenophyta, and Dinophyta), whereas animals were represented by 46 species of Rotifera, 17 of Cladocera, 4 of Copepoda, several kinds of Protista, Ostracoda, and Nematoda, as well as larvae of *L. fortunei* (Table 1).

Despite the fact that animals were represented by much lower numbers of individuals (18 ind./stomach) than phytoplankton (1825 ind./stomach; Fig. 1), they were present in 96 % of the individuals analyzed. Rotifers, especially *Keratella* spp. and *Lecane* spp., were the most frequent animal prey items accounting for 81 % of all animals, with a mean abundance of 15 ind./stomach (maximum: 233 ind./stomach). Cladocera, mainly represented by Chydoridae and Bosminidae, were less frequent (present in 64 % of the stomachs analyzed), accounting for 10 % of the animals in the diet. Copepoda, represented by nauplii and copepodites only, made up <2 % of the animals and were present in 24 % of the stomachs examined. Proportions of the other animal groups present in the diet (protists, nematodes) were negligible and only occurred in a few stomachs (Rojas Molina 2010; Rojas Molina et al. 2010).

Table 1 Zooplanktonic organisms recorded in *Limnoperna fortunei* stomachs in the Middle Paraná River, Argentina. (From Rojas Molina 2010)

Protozoa	Rotifera	Crustacea	Others
Difflugia spp.	*Ascomorpha* sp.	*Alona dentifera* (Cl.)	Unid. nematodes
Arcella spp.	*Brachionus calicyflorus*	*Bosminopsis deitersi* (Cl.)	*Limnoperna fortunei* larvae
Trinema spp.	*Cephalodella gibba*	*Camptocercus dadayi* (Cl.)	Unid. insect remains
Tecamoebae	*Colurella adriatica*	*Ceriodaphnia cornuta* (Cl.)	
	Euchlanis sp.	*Chydorus eurynotus* (Cl.)	
	Epiphanes sp.	*Chydorus pubescens* (Cl.)	
	Keratella cochlearis	*Chydorus* sp. (Cl.)	
	Keratella tropica	*Euryalona* sp. (Cl.)	
	Lecane aculeata	*Macrothrix elegans* (Cl.)	
	Lecane amazonica	*Moinodaphnia macleayi* (Cl.)	
	Lecane bulla	*Notoalona sculpta* (Cl.)	
	Lecane cornuta	*Picripleuroxus quasidenticulatus* (Cl.)	
	Lecane curvicornis	*Picripleuroxus* sp. (Cl.)	
	Lecane decipiens	*Chidoridae* (Cl.)	
	Lecane halyclista	Unidentified remains (Cl.)	
	Lecane hamata	*Tropocyclops* sp. (Co.)	
	Lecane herzigi	Calanoida (Co.)	
	Lecane leontina	Cyclopida (Co.)	
	Lecane ludwigi	Nauplii (Co.)	
	Lecane luna	Unidentified remains (Co.)	
	Lecane lunaris	Unid. Ostracoda	
	Lecane monostyla		
	Lecane quadridentata		

Table 1 (continued)

Protozoa	Rotifera	Crustacea	Others
	Lepadella acuminata		
	Lepadella ovalis		
	Lepadella patella		
	Mytilina mucronata		
	Mytilina ventralis		
	Platyias quadricornis		
	Plationus patulus		
	Polyarthra sp.		
	Scaridium longicaudum		
	Testudinella cf. *alhstromi*		
	Testudinella patina		
	Trichocerca bicristata		

Cl cladocera, *Co* Copepoda

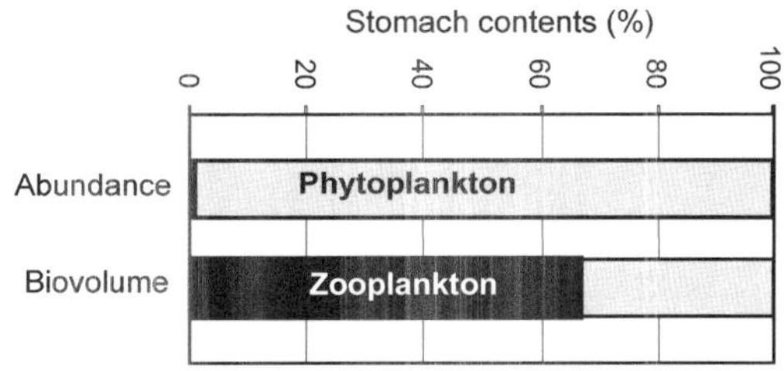

Fig. 1 Mean relative density and biovolume of the plankton recorded in *Limnoperna fortunei* stomachs. Data are based on 140 individuals. (Modified from Rojas Molina 2010)

As opposed to abundance, biomass of animal food was more important than that of the algae: on average, 67 % of the biomass in stomach contents was represented by animals, chiefly cladocerans and copepod larvae (Fig. 1).

Comparison between the yields of plankton samples and the diet of the mollusc indicates that *L. fortunei* may exert strong selective pressure on the zooplankton. All animal groups are positively selected, albeit with different degrees of preference (Fig. 2a). Furthermore, in terms of biovolume, zooplankton is preferred over phytoplankton (Fig. 2b). Among the animals, rotifers and small cladocerans, especially from the families Bosminidae and Chydoridae, are favored by *L. fortunei* (Fig. 2a). These conclusions align with the results of a

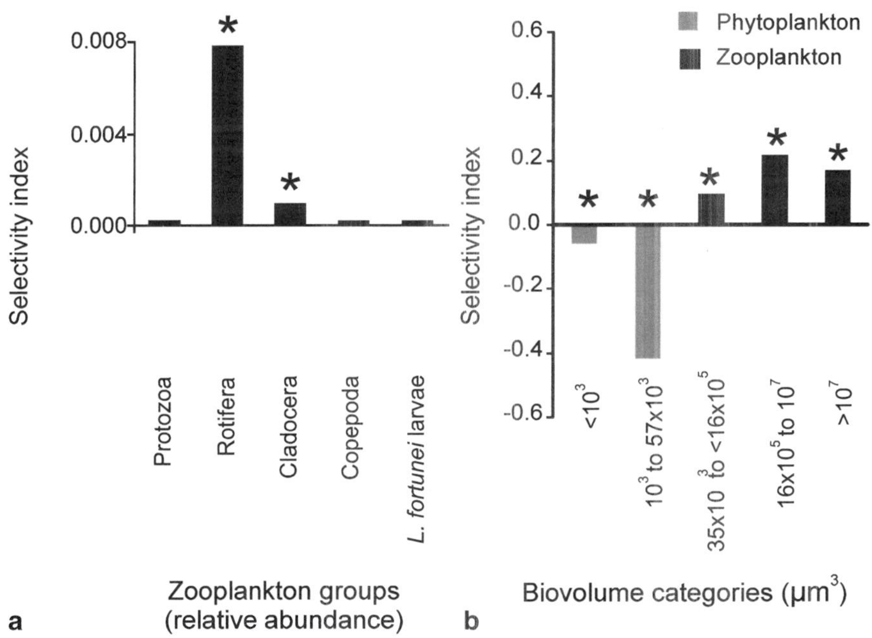

Fig. 2 Feeding selectivity of *Limnoperna fortunei* adults in the Middle Paraná River as indicated by comparisons between the relative abundance (**a**) and biovolume (**b**) of the various planktonic food items in the water column and in the mussel's stomach contents (Strauss' linear index of selectivity; <0: negative selection, 0: no selection, >0: positive selection). *Asterisks* denote selectivity different from 0 at $p<0.05$, t-test. (Modified from Rojas Molina 2010)

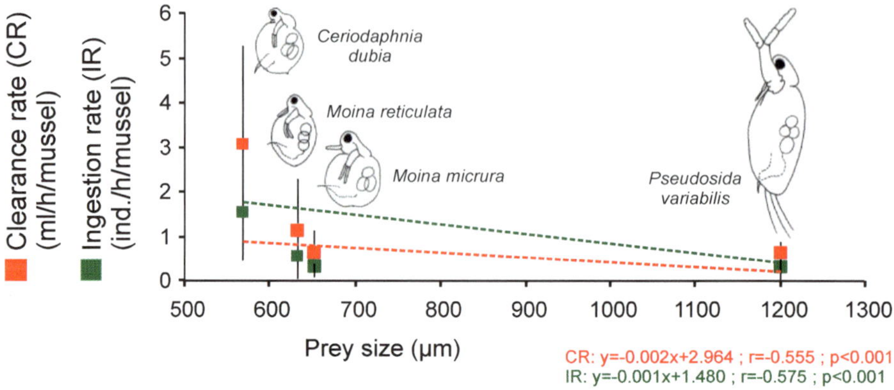

Fig. 3 Relationships between cladoceran body size and clearance (*CR*) and ingestion (*IR*) rates of *Limnoperna fortunei.* Cladoceran sketches are to scale

short-term microcosm feeding experiment on natural plankton where rotifers had the highest positive selection by the mussel (unpublished data). This selective feeding was positively associated with small to moderate size, low mobility, and limited escape responses, which characterize most Rotifera (Rojas Molina et al. 2010; Rojas Molina et al. 2011). Nevertheless, some rotifers, like *Polyarthra* spp., *Synchaeta* spp., and *Filinia* spp., have powerful swimming appendages and may exhibit fast escape responses; accordingly, these species represented higher proportions of the zooplankton in the water column than in stomach contents.

Among the microcrustaceans, small cladocerans with poor evasive behavior, such as *Ceriodaphnia dubia*, bosminids, and chydorids, are clearly favored over medium and large-sized species like *Moina* spp., sidids, and copepods. The latter are comparatively less vulnerable to predation by the mussel due to their large, strong, and active swimming appendages capable of performing rapid evasive maneuvers (Rojas Molina et al. 2010; Rojas Molina et al. 2011). Laboratory microcosm experiments show that cladoceran survival rates are higher for the larger forms, and that body size is negatively associated with clearance and ingestion rates by *L. fortunei* (Fig. 3; Rojas Molina et al. 2011). These results suggest that selective predation pressure can be important in altering species composition and zooplankton size-structure.

Mesocosm (72 h) and microcosm (12 h) experiments suggest strong impacts of *L. fortunei* on zooplankton, especially rotifers (Rojas Molina et al. 2012; Frau et al. 2013), with filtration rates between 60 and 128 ml/ind./h (unpublished data). Grazing affects not only the abundance, but also the composition and the size-structure of the zooplankton community. In mesocosm trials, some rotifers, like *Keratella* spp. and *Polyarthra* spp., decreased by 90 % in 48 h, and disappeared from the mesocosms altogether by the end of the experiment (72 h; Fig. 4), whereas microcrustaceans like *Bosmina* spp. and copepod nauplii also decreased in abundance, but only slightly (Rojas Molina et al. 2012). While other antagonistic relationships, like the competition between *L. fortunei* and zooplankton for the same food resources,

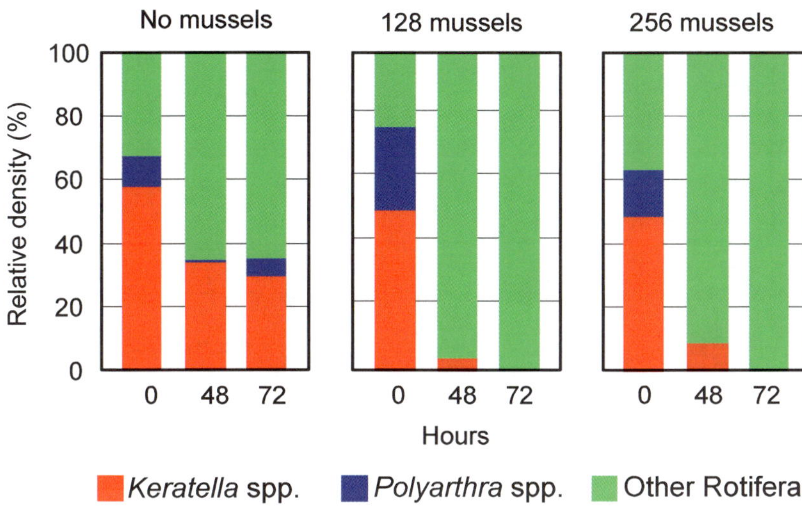

Fig. 4 Relative densities of dominant rotifers in 200 L experimental mesocosms over 72 h grazing experimental times. (Modified from Rojas Molina et al. 2012)

cannot be ruled out, strong competition-related effects over the short experimental times used are unlikely.

The wide spectrum of food items utilized by *L. fortunei* indicates that the mussel can consume phytoplankton, zooplankton, and detritus. In terms of biomass, zooplankton seems to be more important than phytoplankton. The importance of detrital matter, while probably high, has not been assessed. High consumption rates of rotifers are particularly significant because both marine and freshwater bivalves can assimilate rotifer carbon very efficiently (up to 70 %, according to Wong et al. 2003; Wong and Twining 2003). Furthermore, the mixture of phytoplankton and zooplankton in the diet yields a better growth performance than either of them separately (Wong and Levinton 2004; Safi and Hayden 2010). This trophic elasticity, associated with an ample environmental tolerance (Karatayev et al. 2007a; Karatayev et al. 2007b), confers important adaptive advantages to *L. fortunei*, such as an extended reproduction period (Boltovskoy et al. 2009b), and ample distribution in areas where phytoplankton is insufficient for supporting mussel populations, such as many large South American rivers (Sylvester et al. 2005).

In comparison with *Dreissena polymorpha*, *L. fortunei* can consume larger zooplankton. Laboratory experiments showed that the zebra mussel can consume small organisms like rotifers and small cladocerans (<400 μm), but larger microcrustaceans are not impacted (Shevtsova et al. 1986; MacIsaac et al. 1991). It is likely that structural dissimilarities of the siphonal region and the lamellae of the demibranchs (Morton 1973; 1993) play a role in the differences between the type and size of particles available to the two mussel species (Rojas Molina et al. 2011).

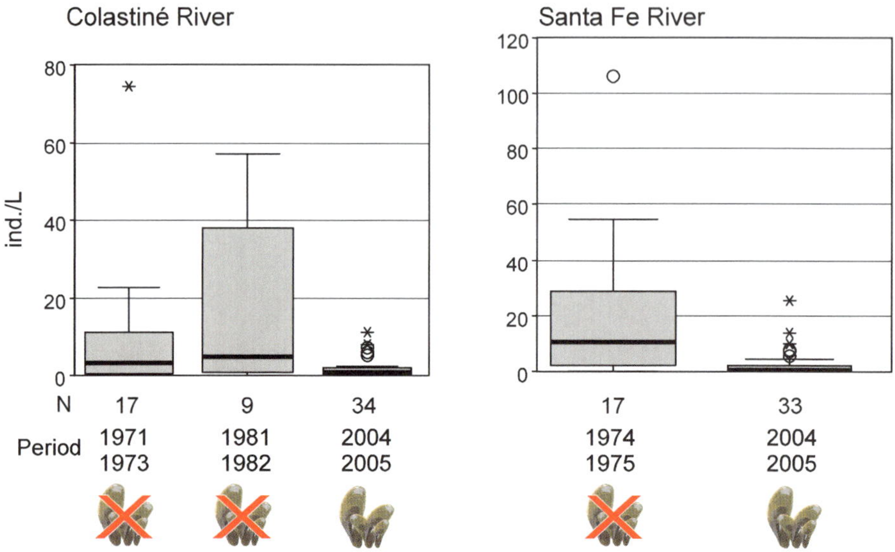

Fig. 5 Changes in the density of *Keratella* spp. before (1971–1972, 1974–1975, 1981–1982) and after invasion by *Limnoperna fortunei* (2004–2005) in Colastiné and Santa Fe rivers. *Outer lines:* 95th percentiles; *box boundaries:* 90th percentiles; *horizontal line in the box:* median value. *Circles* and *asterisks* are atypical and extreme values, respectively. (Modified from Rojas Molina 2010)

Impacts on the Ecosystem

The filtration rates of bivalve populations may be significant, occasionally exceeding those of other filter feeders in the ecosystem, including pelagic grazers (Strayer et al. 1999). Studies carried out in rivers, lakes, and reservoirs have shown that the filtering activity of *L. fortunei* can modify the characteristics of the water column, including nutrient concentrations and proportions, phytoplankton abundance and production, water transparency, suspended matter, etc. (Boltovskoy et al. 2009a; Cataldo et al. 2012a, b; Rojas Molina et al. 2012; see Chapter "Nutrient Recycling, Phytoplankton Grazing, and Associated Impacts of *Limnoperna fortunei*" in this volume). Consumption of animal plankton, especially rotifers, chydorids, and bosminids, and to a lesser extent larger microcrustaceans, can also have an impact on the populations of these organisms, as shown by field and experimental studies in the Middle Paraná River (Rojas Molina and José de Paggi 2008; Rojas Molina et al. 2011; Rojas Molina et al. 2012).

In the Middle Paraná River, adults and planktonic larvae of *L. fortunei* were first recorded in the late 1990s (Darrigran and Ezcurra de Drago 2000; José de Paggi and Paggi 2008). Comparison of pre-invasion (1971–1982) data with data collected after establishment of the golden mussel (2004–2005) in two secondary channels of the Middle Paraná River (Santa Fe and Colastiné rivers) showed important changes

Table 2 Zooplankton densities (ind./L) recorded in two secondary waterways associated with the Middle Paraná River before (1971–1973, 1974–1975, 1981–1982), and after colonization by *Limnoperna fortunei* (2004–2005). (From Rojas Molina and José de Paggi 2008, Rojas Molina 2010)

Site	Colastiné River			Santa Fe River	
Period	1971–1972	1981–1982	2004–2005	1974–1975	2004–2005
Rotifera, mean [maximum]	27.2 [166.4]	24.9 [69.2]	8.5 [42.4]	30.9 [140.5]	9.5 [56.7]
% of Rotifera	≥60	≥60	≥45	≥70	≥40
Cladocera, mean	2.0	1.4	0.4	0.7	0.7
Copepoda, mean	2.0	2.7	1.7	2.2	3.2

in the zooplankton, with rotifer abundance drops of up to 66 % (especially *Keratella* spp., the dominant genus among the Rotifera; Fig. 5), whereas abundances of Cladocera and Copepoda remained relatively unchanged (Rojas Molina and José de Paggi 2008; Table 2).

In many floodplain rivers, including the Middle Paraná, zooplankton densities are higher during low-water periods, partly because dilution of particles is lower, and partly because lower flow rates and longer water residence times favor buildup of numbers through in situ reproduction (Pace et al. 1992; Thorp et al. 1994; Basu and Pick 1996; Kobayashi et al. 1998; José de Paggi and Paggi 2007; Zalocar de Domitrovic et al. 2007). In agreement with this pattern, Santa Fe and Colastiné rivers had noticeably higher zooplankton abundances during low water (and lower flow velocities) than during high water (and higher flow velocities) before introduction of *L. fortunei*, (Fig. 6). However, after the invasion of *L. fortunei* this situation disappeared and zooplankton densities during low- and high-water periods became much more similar, presumably because of enhanced filtration rates by benthic mussels during low flow conditions (Descy et al. 2003). A similar effect was observed on concentrations of chlorophyll a, which after the invasion dropped 53–80 %, despite the growing trend in nutrient levels from increasing land use and urbanization (Rojas Molina and José de Paggi 2008). Significantly, these effects have recently been observed not only in the secondary waterways, but also in the main channel of the Paraná River (José de Paggi et al. 2014).

These conclusions are supported by observations based on northern hemisphere rivers invaded by the zebra mussel. Significant reductions in the abundance of zooplankton, mainly rotifers, and chlorophyll a were observed in the Hudson, St. Lawrence, Rideau, Moselle, and Spree rivers after having been invaded by *D. polymorpha* (MacIsaac et al. 1995; Basu and Pick 1997; Viroux 1997; Pace et al. 1998; Welker and Walz 1998; Strayer et al. 1999; Descy et al. 2003; Higgins and Vander Zanden 2010). Interestingly, zooplankton abundance and chlorophyll a concentrations ceased to correlate with flow, water level, and/or flow velocity after the invasion of *D. polymorpha* into the Hudson and Rideau rivers (Basu and Pick 1997; Pace et al. 1998).

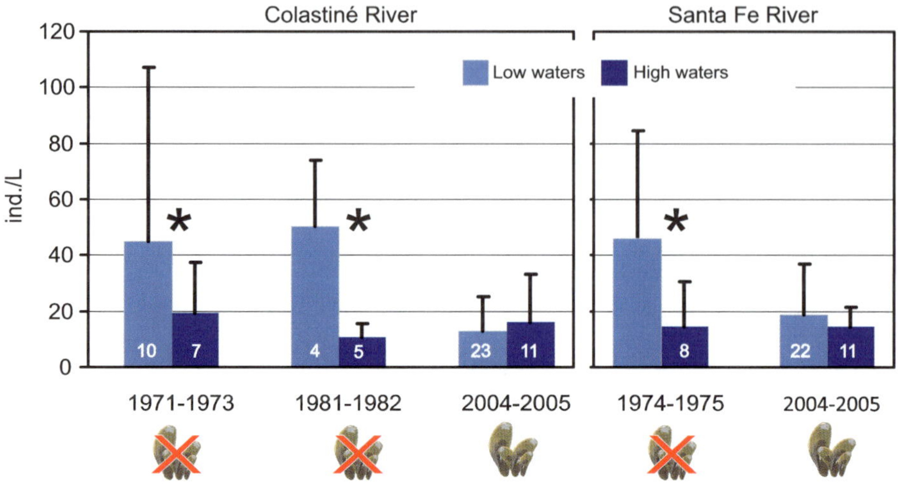

Fig. 6 Mean density and SD (*bars*) of zooplankton recorded during low- and high-water periods in Colastiné and Santa Fe rivers during pre- (1971–1973, 1974–1975, 1981–1982) and post-invasion (2004–2005) periods. *Numbers in bars* indicate the number of samples; *asterisks* denote significant ($p<0.05$) differences between high- and low-water periods. (Modified from Rojas Molina 2010)

Table 3 Mean densities [and ranges] of rotifers+microcrustaceans (R+M) and *Limnoperna fortunei* larvae (ind./L) in the Middle Paraná River. (From Rojas Molina and José de Paggi 2008; José de Paggi et al. 2014)

River (sampling period)	R+M	*Limnoperna fortunei* larvae
Colastiné (2004–2005)	10.7 [0.7–44.6]	17.2 [0.1–62.2]
Santa Fe (2004–2005)	13.5 [1.4–62.3]	2.9 [0.4–26.1]
Paraná (2007–2008)	9.0 [0.2–98.5]	22.7 [0.1–87.7]

Limnoperna fortunei Larvae and Native Zooplankton

Studies on the reproduction of *L. fortunei* (see Chapter "Reproductive Output and Seasonality of *Limnoperna fortunei*" in this volume) show that larvae are present in the water column in significant numbers for up to 8–9 months of the year (Boltovskoy et al. 2009b), thus representing an almost permanent component of the zooplankton. In the Middle Paraná River, average densities of *L. fortunei* larvae can be several times higher than those of the rest of the zooplankton (Table 3). During peak mussel reproductive events veligers can account for over 90% of overall zooplankton numbers (Rojas Molina 2010). Similar situations have been recorded in other South American water bodies, including the Paraguay River, and the reservoirs Embalse de Río Tercero and Salto Grande (Paolucci et al. 2007; Paolucci et al. 2010b; Boltovskoy et al. 2013). These extremely high densities suggest that veligers may successfully compete for food with native zooplakton, but very little is known about the feeding behavior of golden mussel larvae. The transition from trochophore to veliger stage is accompanied by a change in the feeding mode, from lecithotrophy to planktotrophy (Ezcurra de Drago et al. 2006); in the straight-hinged stage animals already feed on external sources (Cataldo et al. 2005). Gazulha (2010, 2012) reported clearance rates between 15 and 32 µL/ind./h (see Chapter "Nutrient Recycling, Phytoplankton Grazing, and Associated Impacts of *Limnoperna fortunei*" in this volume), but data at hand are too scarce to allow reasonable estimates of the grazing impact.

Dreissena polymorpha veligers feed on particles 1–13 µm in diameter, including bacteria, cyanobacteria, small green algae, very fine detritus, and dissolved organic matter (Gliwicz 1969; Sprung 1989; Wacker et al. 2002; Dionisio Pires et al. 2004; Banard et al. 2006). Although individual clearance rates are low (62–420 µL/ind./day, MacIsaac et al. 1992), high larval concentrations can have significant impacts on the plankton. In Lake Erie, at larval densities around 630 ind./L (Leach 1993), larval grazing can clear up to 20% of the water column per day (MacIsaac et al. 1992). Stable isotope studies of the trophic structure of the St. Lawrence River estuarine zone, where larvae of *D. polymorpha* are the dominant component of the zooplankton in summer, indicate that, despite these high overall grazing rates, dietary overlap with native zooplankton is limited (Banard et al. 2006). Extrapolation of these results to *L. fortunei* in South America, however, is questionable, because the two bivalves differ in several important aspects, and because the large, floodplain rivers invaded by *L. fortunei* contrast sharply with the clearer and colder waters invaded by the zebra mussel in the northern hemisphere (Boltovskoy et al. 2006). Zooplanktonic organisms also seem to differ in that South American inland waters are dominated by small-sized zooplankton (Fernando 1994; Iglesias et al. 2011).

In contrast to the effects of adult *L. fortunei* on zooplankton populations, mussel larvae may have a different impact. Larvae of this mussel represent an important food item for several fish larvae in the Paraná, Paraguay, and Uruguay Rivers (Paolucci et al. 2007, 2010a, b; see Chapter "Trophic Relationships of *Limnoperna fortunei* with Larval Fishes" in this volume). This could represent a positive impact on zooplankton because of the reduction of predation pressure upon this community; however, these effects have not yet been evaluated.

Acknowledgments This work was supported by CAI+D PI 6–44 (UNL), CAI+D PJ 1482 (UNL), CAI+D 14–78 (UNL), PICT- 2010–2350, and SECTeI N°2010-042-13 Res 117/13 projects.

References

Banard C, Martineau C, Frenette JJ, Dodson JJ, Vincent WF (2006) Trophic position of zebra mussel veligers and their use of dissolved organic carbon. Limnol Oceanogr 51:1473–1484

Bastviken DTE, Caraco NF, Cole JJ (1998) Experimental measurements of zebra mussel (*Dreissena polymorpha*) impacts on phytoplankton community composition. Freshw Biol 39:375–386

Basu BK, Pick FR (1996) Factors regulating phytoplankton and zooplankton development in temperate rivers. Can J Fish Aquat Sci 52:804–815

Basu BK, Pick FR (1997) Phytoplankton and zooplankton development in a lowland, temperate river. J Plankton Res 19:237–253

Boltovskoy D, Correa N, Cataldo D, Sylvester F (2006) Dispersion and ecological impact of the invasive freshwater bivalve *Limnoperna fortunei* in the Río de la Plata watershed and beyond. Biol Invasions 8:947–963

Boltovskoy D, Karatayev A, Burlakova L, Cataldo D, Karatayev V, Sylvester F, Mariñelarena A (2009a) Significant ecosystem-wide effects of the swiftly spreading invasive freshwater bivalve *Limnoperna fortunei*. Hydrobiologia 636:271–284

Boltovskoy D, Sylvester F, Otaegui A, Leytes V, Cataldo D (2009b) Environmental modulation of the reproductive activity of the invasive mussel *Limnoperna fortunei* in South America. Austral Ecol 34:719–730

Boltovskoy D, Correa N, Bordet F, Leites V, Cataldo D (2013) Toxic *Microcystis* (cyanobacteria) inhibit recruitment of the bloom-enhancing invasive bivalve *Limnoperna fortunei*. Freshw Biol 58:1968–1981

Caraco NF, Cole JJ, Raymond PA, Strayer DL, Pace ML, Findlay SEG, Fischer DT (1997) Zebra mussel invasion in a large, turbid river: phytoplanton response to increased grazing. Ecology 78:588–602

Cataldo D, Boltovskoy D, Hermosa JL, Canzi C (2005) Temperature-dependent larval development rates of *Limnoperna fortunei* (Bivalvia: Mytilidae). J Molluscan Stud 71:41–46

Cataldo D, O'Farrell I, Paolucci E, Sylvester F, Boltovskoy D (2012a) Impact of the invasive golden mussel (*Limnoperna fortunei*) on phytoplankton and nutrient cycling. Aquat Invasions 7:91–100

Cataldo D, Vinocur A, O'Farrell I, Paolucci E, Leites V, Boltovskoy D (2012b) The introduced bivalve *Limnoperna fortunei* boosts *Microcystis* growth in Salto Grande Reservoir (Argentina): evidence from mesocosm experiments. Hydrobiologia 680:25–38

Cohen RRH, Dresler PV, Phillips EJP, Cory RL (1984) The effect of the Asiatic clam, *Corbicula fluminea*, on phytoplankton of the Potomac River, Maryland. Limnol Oceanogr 29:170–180

Darrigran GA, Ezcurra de Drago I (2000) Limnoperna fortunei (Dunker, 1857) (Mytilidae) distribution in the Plate Basin, Neotropical Region. Medio Ambiente 13:75–79

Descy JP, Everbecq E, Gosselain V, Viroux L, Smitz JS (2003) Modelling the impact of benthic filter-feeders on the composition and biomass of river plankton. Freshw Biol 48:404–417

Dionisio Pires LM, Jonker RR, Van Donk E, Laanbroek HJ, (2004) Selective grazing by adults and larvae of the zebra mussel (*Dreissena polymorpha*): application of flow cytometry to natural seston. Freshw Biol 49:116–126

Ezcurra de Drago I Montalto L Oliveros OB (2006) Desarrollo y ecología larval de *Limnoperna fortunei*. In: Darrigran G, Damborenea C (eds) Bio-invasión del mejillón dorado en el continente americano. Editorial de la Universidad de La Plata, La Plata, pp 85–93

Fernando CH (1994) Zooplankton, fish and fisheries in tropical freshwaters. Hydrobiologia 272:105–123

Frau D, Rojas Molina F, Devercelli M, José de Paggi S (2013) The effect of an invading filter-feeding bivalve on a phytoplankton assemblage from the Paraná system: a mesocosm experiment. Marine and Freshwater Behaviour and Physiology 45:303–316

Gazulha V (2010) O mexilhão dourado *Limnoperna fortunei* (Dunker, 1857) na presença de cianobactérias: taxas de filtração, comportamento alimentar e sobrevivência. PhD Thesis, Universidade Federal do Rio Grande do Sul (Brazil), pp 1–104

Gazulha V (2012) O impacto de Limnoperna fortunei sobre as cianobacterias. In: Mansur MCD, Santos CP, Pereira D, Padula Paz IC, Leite Zurita ML, Raya Rodriguez MT, Vilar Nehrke M, Aydos Bergonci PE (eds) Moluscos límnicos invasores no Brasil. Biologia, prevenção, controle. Redes Editora, Porto Alegre, pp 249–254

Gliwicz ZM (1969) Studies on the feeding of pelagic zooplankton in lakes with varying trophy. Ekologia Polska, Series A 17:663–706

Higgins SN, Vander Zanden MJ (2010) What a difference a species makes: a meta–analysis of dreissenid mussel impacts on freshwater ecosystems. Ecol Monographs 80:179–196

Horgan MJ, Mills EL (1997) Clearance rates and filtering activity of Zebra Mussel (*Dreissena polymorpha*): Implications for freshwater Lakes. Can J Fish Aquat Sci 54:249–255

Iglesias C, Mazzeo N, Meerhoff M, Lacerot G, Clemente JM, Scasso F, Kruk C, Goyenola G, Garcia Alonso J, Amsinck SL, Paggi JC, José de Paggi S, Jeppesen E (2011) High predation is of key importance for dominance of small-bodied zooplankton in warm shallow lakes: evidence from lakes, fish exclosures and surface sediments. Hydrobiologia 667:133–147

José de Paggi S, Paggi JC (2007) Zooplankton. In: Iriondo MH, Paggi JC, Parma MJ (eds) The middle Parana River. Limnology of a subtropical wetland. Springer, Berlin, pp 229–249

José de Paggi SB, Paggi JC (2008) Hydrological connectivity as a shaping force in the zooplankton community of two lakes in the Paraná river floodplain. Int Rev Hydrobiol 93:659–678

José de Paggi SB, Devercelli M, Rojas Molina F (2014) Zooplankton and their driving factors in a large subtropical river during low water periods. Fund Appl Limnol/Archiv Hydrobiologie 184:125–139

Karatayev AY, Boltovskoy D, Padilla DK, Burlakova LE (2007a) The invasive bivalves *Dreissena polymorpha* and *Limnoperna fortunei* : parallels, contrasts, potential spread and invasion impacts. J Shellfish Res 26:205–213

Karatayev AY, Padilla DK, Minchin D, Boltovskoy D, Burlakova LE (2007b) Changes in global economies and trade: the potential spread of exotic freshwater bivalves. Biol Invasions 9:161–180

Kobayashi T, Shiel RJ, Gibbs P, Dixon PI (1998) Freshwater zooplankton in the Hawkesbury-Nepean river: comparison of community structure with other rivers. Hydrobiologia 377:133–145

Leach JH (1993) Impacts of the zebra mussel (*Dreissena polymorpha*) on water quality and fish spawning reefs in western Lake Erie. In: Nalepa TF, Schloesser D (eds) Zebra mussels: biology, impacts, and control. Lewis Publishers, Boca Raton, pp 381–397

MacIsaac HJ, Sprules WG, Leach JH (1991) Ingestion of small-bodied zooplankton by zebra mussels (*Dreissena polymorpha*): can cannibalism on larvae influence population dynamics? Can J Fish Aquat Sci 48:2051–2060.

MacIsaac HJ, Sprules WG, Johannsson OE, Leach JH (1992) Filtering impacts of larval and sessile zebra mussels (*Dreissena polymorpha*) in western Lake Erie. Oecologia 92:30–39

MacIsaac HJ, Lonnee CJ, Leach JH (1995) Suppression of microzooplankton by zebra mussels: importance of mussel size. Freshw Biol 34:379–387

Morton B (1973) Some aspects of the biology and functional morphology of the organs of feeding and digestion of *Limnoperna fortunei* (Dunker) (Bivalvia: Mytilacea). Malacologia 12:265–281

Morton B (1993) The anatomy of *Dreissena polymorpha* and the evolution and success of the heteromyarian form in the Dreissenoidea. In: Nalepa TF, Schloesser D (eds) Zebra mussels: biology, impacts, and control. Lewis Publishers, Boca Raton, pp 185–215

Pace ML, Findlay SEG, Lints D (1992) Zooplankton in advective environments: the Hudson River community and a comparative analysis. Can J Fish Aquat Sci 49:1060–1069

Pace ML, Findlay SEG, Fischer DT (1998) Effects of an invasive bivalve on the zooplankton community of the Hudson River. Freshw Biol 39:103–116

Paolucci EM, Cataldo DH, Fuentes CM, Boltovskoy D (2007) Larvae of the invasive species *Limnoperna fortunei* (Bivalvia) in the diet of fish larvae in the Paraná River, Argentina. Hydrobiologia 589:219–233

Paolucci EM, Cataldo DH, Boltovskoy D (2010a) Prey selection by larvae of *Prochilodus lineatus* (Pisces): indigenous zooplankton versus larvae of the introduced bivalve *Limnoperna fortunei.* Aquat Ecol 44:255–267

Paolucci EM, Thuesen EV, Cataldo DH, Boltovskoy D (2010b) Veligers of an introduced bivalve (*Limnoperna fortunei*) are a new food resource that enhances growth of larval fish in the Paraná River (South America). Freshw Biology 55:1831–1844

Prins TC, Smaal AC, Dame RF (1998) A review of the feedbacks between bivalve grazing and ecosystem processes. Aquat Ecol 31:349–359

Rojas Molina F (2010) Efectos del molusco invasor *Limnoperna fortunei* (Dunker) sobre el zooplancton del paraná medio. PhD Thesis, Universidad de Buenos Aires (Argentina), pp 1–205

Rojas Molina F, Jose de Paggi SB (2008) Zooplankton in the Paraná River floodplain (South America) before and after the invasion of *Limnoperna fortunei* (Bivalvia). Wetlands 28:695–702

Rojas Molina F, Paggi JC, Devercelli M (2010) Zooplanktophagy in the natural diet and selectivity of the invasive mollusk *Limnoperna fortunei.* Biol Invasions 12:1647–1659

Rojas Molina F, José de Paggi S, Boltovskoy D (2011) Vulnerability of microcrustaceans to predation by the invasive filter-feeding mussel *Limnoperna fortunei* (Dunker). Marine and Freshwater Behaviour and Physiology 44:329–338

Rojas Molina F, Jose de Paggi SB, Frau D (2012) Impacts of the invading golden mussel *Limnoperna fortunei* on zooplankton: a mesocosm experiment. Zool Stud 51:733–744

Safi KA, Hayden B (2010) Differential grazing on natural planktonic populations by the mussel *Perna canaliculus*. Aquat Biol 11:113–125

Shevtsova LV, Zhadanova GA, Movchan VA, Primak AB (1986) Experimental interrelationship between *Dreissena* and planktic invertebrates. Hydrobiol J 22:36–39

Sprung M (1989) Field and laboratory observations of Dreissena polymorpha larvae: abundance, growth, mortality and food demands. Archiv Hydrobiologie 115:537–561

Strayer DL, Caraco NF, Cole JJ, Findlay S, Pace ML (1999) Transformation of freshwater ecosystems by bivalves. A case study of zebra mussels in the Hudson River. Bioscience 49:19–27

Sylvester F, Dorado J, Boltovskoy D, Juárez A, Cataldo D (2005) Filtration rates of the invasive pest bivalve *Limnoperna fortunei* as a function of size and temperature. Hydrobiologia 534:71–80

Thorp JH, Black AR, Haags KH (1994) Zooplankton assemblages in the Ohio River: seasonal, tributary and navigation dam effects. Can J Fish Aquat Sci 51:1634–1643

Viroux L (1997) Zooplankton development in two large lowland rivers, the Moselle (France) and the Meuse (Belgium), in 1993. J Plankton Res 19:1743–1762

Wacker A, Becher P, von Elert E (2002) Food quality effects of unsaturated fatty acids on larvae of the zebra mussel *Dreissena polymorpha*. Limnol Oceanogr 47:1242–1248

Welker M, Walz N (1998) Can mussels control the plankton in rivers?—a planktological approach applying a Langrangian sampling strategy. Limnol Oceanogr 43:753–762

Wong WH, Levinton JS (2004) Culture of the blue mussel *Mytilus edulis* (Linnaeus, 1758) fed both phytoplankton and zooplankton: a microcosm experiment. Aquac Res 35:965–969

Wong WH, Levinton JS (2005) Consumption rates of two rotifer species by zebra mussels *Dreissena polymorpha*. Marine and Freshwater Behaviour and Physiology 38:149–157

Wong WH, Levinton JS (2006) The trophic linkage between zooplankton and benthic suspension feeders: direct evidence from analyses of bivalve fecal pellets. Marine Biol 148:799–805.

Wong WH, Twining BS (2003) Assimilation of micro- and mesozooplankton by zebra mussels: a demonstration of the food web link between zooplankton and benthic suspension feeders. Limnol Oceanogr 48:308–312

Wong WH, Levinton JS, Twining BS, Fisher NS, Kelaher BP, Alt AK (2003) Assimilation of carbon from a rotifer by the mussels, *Mytilus edulis* and *Perna viridis* : a potential food web link between zooplankton and benthic suspension feeders. Marine Ecol Prog Ser 253:175–182

Zalocar de Domitrovic Y, Devercelli M, García de Emiliani O (2007) Phytoplankton. In: Iriondo MH, Paggi JC, Parma MJ (eds) The middle Parana River. Limnology of a subtropical wetland. Springer, Berlin, pp 175–203

Relationships of *Limnoperna fortunei* with Benthic Animals

Francisco Sylvester and Paula Sardiña

Abstract Similar to other invasive bivalves, *Limnoperna fortunei* has a variety of effects on other benthic animals. These effects have been studied in the Paraná-Paraguay-Uruguay river system, Río de la Plata estuary, and in the reservoir Embalse Río Tercero since the invasion of South America by the bivalve. The bulk of information accumulated indicates that *L. fortunei* has predominantly positive effects on meiofaunal groups. Increases in the abundance, biomass, and richness of many groups are attributed to substrate enrichment from the bivalve's feces and pseudofeces as well as refugia provision amid the valves. Nonetheless, negative impacts on some groups (gastropods) and the homogenization of benthic fauna following colonization by the mussel have also been reported. Large-sized invertebrates can also be detrimentally affected by this mussel's biofouling, as severe cases of epifaunal growth have been reported for native crabs and mussels, including the invasive clam *Corbicula fluminea*. However, consequences to affected individuals and impacts at the population level have not yet been assessed. A variety of animals, including fish, crabs, turtles, waterfowl, and some mammals, may benefit from predation on this new abundant prey item, although the consequences to predator populations remain unstudied. Despite marked similarities with *Dreissena polymorpha,* there are a number of differences regarding the effects of the two bivalves arising from differences in their biology and ambient dissimilarities between their respective environments. The extrapolation of results obtained for *Dreissena* species, abundant in the *L. fortunei* literature, can be misleading due to these differences.

F. Sylvester (✉)
Facultad de Ciencias Exactas y Naturales, Instituto de Ecología, Genética y Evolución de Buenos Aires (IEGEBA), Universidad de Buenos Aires-CONICET, Ciudad Universitaria, 1428 Buenos Aires, Argentina
e-mail: fsylvester@natura.unsa.edu.ar

Facultad de Ciencias Naturales, Instituto para el Estudio de la Biodiversidad de Invertebrados (IEBI),, Universidad Nacional de Salta, Av. Bolivia 5150, 4400 Salta, Argentina

P. Sardiña
Departamento de Biología, Bioquímica y Farmacia, Instituto de Investigaciones Biológicas y Biomédicas del Sur (INBIOSUR), Universidad Nacional del Sur-CONICET, San Juan 670, 8000 Bahía Blanca, Argentina
e-mail: psardina@gmail.com

D. Boltovskoy (ed.), *Limnoperna fortunei,* Invading Nature - Springer Series in Invasion Ecology 10, DOI 10.1007/978-3-319-13494-9_11

Keywords *Limnoperna fortunei* · Golden mussel · Ecological impact · Ecosystem engineer · Facilitation · Zoobenthos · Biodiversity · Meiofauna

Introduction

The golden mussel establishes a variety of interactions with other animals after invasion. Over the last 15 years, coincidental with the widespread colonization of Japanese and South American aquatic ecosystems by *Limnoperna fortunei,* a wealth of scientific knowledge has built up describing these interactions. A substantial part of this knowledge has been produced in the area of the Río de la Plata basin, while the rest of the available information comes from other basins in South America. In Japan, investigations have centered on its population dynamics and particularly on biofouling control measures, while research on its relationships with native fauna has lagged behind. Conclusions on the effects of *L. fortunei* are, therefore, necessarily biased towards the characteristics of South American environments, typically dominated by large floodplain rivers with high organic-mater contents and sediment load, hosting a diverse community of fishes and low abundance of native filter-feeders (Boltovskoy and Correa 2015). While some of the effects will most probably differ in other environmental settings and continents (see Chapter "Ecology and Environmental Impact of *Limnoperna fortunei*: Introduction" in this volume), the literature reviewed here comes from three countries—Argentina, Uruguay, and Brazil, comprising most of the bivalve's invasive range outside Asia.

This chapter reviews the consequences of *L. fortunei* invasions for other animals, chiefly meiofaunal, benthic invertebrates (relationships with the zooplankton and with fishes are dealt with in other chapters of this volume). These interactions are comparatively well studied and give rise to a large number of effects that are transmitted to other compartments of aquatic ecosystems. The information available on epizoic macrofouling by *L. fortunei* and its consumption by nonfish animals is also reviewed. We examine similarities and differences with the zebra mussel, the better-studied, northern-hemisphere counterpart of *L. fortunei*. However, care has been taken to discriminate between knowledge based on actual observations and assumptions derived from the extrapolation of information reported for the zebra mussel. In view of the information compiled, future research steps are suggested.

Effects on Accompanying Benthic Invertebrates

Types of Studies

Several studies have assessed the effects of *L. fortunei* on benthic invertebrate communities. Studies dealing with the facilitation and, to a lesser extent, inhibition of fauna living on or amid valves of *L. fortunei* can be broadly classified in two types:

(1) Studies based on the comparison of natural or preexisting artificial underwater substrata colonized by mussels with substrata barren of them and (2) studies based on the evaluation of artificial hard substrata deliberately deployed for colonization by mussels and other invertebrates. Most surveys belong to the first type (Mansur et al. 2008; Fagondes de Freitas and Kapusta 2010; Karatayev et al. 2010; Sardiña et al. 2011; Burlakova et al. 2012; Kapusta and Fagondes de Freitas 2012; Spaccesi and Rodrigues Capitulo 2012). Only one study has attempted to compare benthic communities before and after invasion by *L. fortunei* (Darrigran et al. 1998), but its results are limited by the fact that pre-*L. fortunei* data were derived from a variety of sources, time periods, and areas whose direct comparability with post-*L. fortunei* information is uncertain.

In the second type of study, artificial hard substrata deployed for colonization were used by Sylvester et al. (2007) and Sardiña et al. (2008). Comparisons in surveys using artificial substrata are more straightforward because they contrast invertebrate communities on areas with and without *L. fortunei* established on the same type of substrate and right next to each other. In contrast, natural substrata normally involve comparison of communities on *L. fortunei* beds (hard substrata) with nearby soft-bottom communities on silt or sand. Another advantage is that for artificial substrata precise knowledge of the age of the community is available, which allows for analyses of time-related changes. A limitation of this method resides in the extrapolation of results to natural conditions, as assemblages developed on artificial substrata over limited periods may differ from natural ones, or correspond only to early successional stages. Sampling of mature assemblages evolved on natural substrata over long periods can yield more realistic information on the long-term impacts of the mussel.

Population and Community Effects

Available studies indicate that *L. fortunei* has a strong effect on benthic invertebrate communities. An exhaustive revision of the literature yielded at least 200 species reported to be directly affected by the presence of *L. fortunei* in invaded freshwater ecosystems across South America (Table 1). These organisms belong to a wide range of taxonomic groups including Mollusca (at least 51 taxa), Annelida (42), Insecta (38), Crustacea (32), Rotifera (7), Turbellaria (4), Collembola (3), Bryozoa (3), Chelicerata (2), Tardigrada (2), Nematoda, Hydrozoa, Kamptozoa, and Porifera (for the latter four no further taxonomic information was provided; Table 1). It should be noted that the number of taxa for which an effect has been quantitatively assessed and statistically established is considerably lower (see below). Despite this limitation, the bulk of evidence at hand very clearly indicates that *L. fortunei* is a significant ecosystem component having a marked influence on benthic invertebrate communities.

Although negative impacts on invertebrate populations have been reported, positive effects predominate in the literature likely reflecting their prevalence also in nature. A comparison of invertebrate communities associated to artificial substrata

Table 1 Summary of conclusions reporting effects of *L. fortunei* on benthic invertebrates

Phylum/ subphylum	Class/order	Functional feeding group	Prevailing effect	Total records	Taxa most affected	Trait affected	Source
Cnidaria	Hydrozoa	P	+	2 (+)	Hydridae	A	9,10
Platyhelminthes	Turbellaria	P	+	6 (5+, 1±)		A B	7,9,10,11,18,21
Rotifera	Rotifera	CG P	+	7 (+)		A B	20,21
Nematoda	Nematoda	CG P	+	5 (4+, 1−)		A B	7,17,18,20,21
Tardigrada	Tardigrada	MP	+	3 (+)		A B	17,20,21
Mollusca	Bivalvia	CF	−	14 (12−, 2+)	*Corbicula fluminea*, Anodontites, Diplodon	A R	10,11,13,14
	Gastropoda	CG (Planorbidae), S	+/−	30 (17+, 12−, 1±)	Positive mainly Cochliopidae Negative mainly Planorbidae and Chilinidae	A B	5,7,9,10,11,13,14,18,20,21
Annelida	Aphanoneura	CF	+	1 (+)	Aelosoma	A	7
	Hirudinea	P	+	20 (16+, 3−, 1±)	Helobdela	A B R	7,9,10,11,17,20,21
	Oligochaeta	CG	+	39 (32+, 6−, 1±)	Naididae and Tubificidae	A B R	1,7,9,10,11,16,17,18,20,21
	Polychaeta	CG	+	2 (+)		A	9,10
Crustacea	Amphipoda	CG	+	7 (5+, 1−, 1±)	Hyalella	A B R	7,9,10,14,18,20
	Cladocera	CF CG S (Alona, Bosmina)	+	9 (6+, 3−)	Chyrodirae (Alona, Camptocercus), Bosminidae, Moinidae	A B	15,17,20,21
	Copepoda	CF	+	11 (+)	Cyclopoida and Harpacticoida	A B	17,18,20,21
	Decapoda	CG	+/−	4 (2+, 2−)	Aegla	A	6,11,12,21
	Isopoda	CG Sh	+	4 (3+, 1−)		A B R	7,17,18
	Ostracoda	CF	+	4 (3+, 1−)	Cyprididae	A B	17,18,20,21

Table 1 (continued)

Phylum/subphylum	Class/order	Functional feeding group	Prevailing effect	Total records	Taxa most affected	Trait affected	Source
	Tanaidacea	CG	+	3 (+)	Sinelobus	A B R	7,18,20
Chelicerata	Arachnida	P (Hydrachnidiae), S Sh (Orbitidae)	+	4 (4+, 1±)	Hydrachnidiae and Orbatidae	A B	9,10,17,20,21
Entognatha	Collembola	CG	+	4 (3+, 1−)		A	9,10,20
Insecta	Coleoptera	S (larvae), Sh (adult)	+	2 (+)	Elmidae	A	9,10
	Diptera (Chironomidae)	CF CG (Chironominae), P (Tanypodinae)	+/−	16 (9+, 5 −, 2±)	Positive mainly Chironominae and Tanypodinae Negative mainly Chironominae	A B	7,9,10,11,18,20,21
	Diptera (others)	CG P S Sh	+	12 (8 +, 3 −, 1 ±)	Ceratopogonidae, Psychodidae, Tabanidae, Tipulidae	A	9,10,11,20
	Ephemeroptera	CG S Sh	+	11 (9+, 2−)		A	2,9,10,11,20,21
	Odonata	P	+	2 (+)	Coenagrionidae	A	9,10
	Trichoptera	CF CG P S Sh	+	9 (7+, 1−, 1±)		A B	2,9,10,11,18

Functional feeding group: *CF* collector-filterer, *CG* collector-gatherer, *MP* macrophyte piercer, *P* predator, *S* scraper, *Sh* shredder. Prevailing effect: + positive, − negative, ± neutral. Total records: number of publications reporting interaction events between *L. fortunei* and the specified taxon. Trait affected: *A* abundance, *B* biomass, *R* taxonomic richness. Sources: *1* Armendáriz et al. (2011), *2* Brugnoli et al. (2005), *3* Burlakova et al. (2012), *4* Carvalho Torgan et al. (2009), *5* César et al. (2012), *6* Darrigran (2002), *7* Darrigran et al. (1998), *8* Darrigran et al. (2000), *9* Fagondes de Freitas and Kapusta (2010), *10* Kapusta and Fagondes de Freitas (2012), *11* Karatayev et al. (2010), *12* Lopes et al. (2009), *13* Mansur et al. (2003), *14* Mansur et al. (2008), *15* Marçal and Callil (2008), *16* Ramseyer and Marchese (2009), *17* Sardiña et al. (2008), *18* Sardiña et al. (2011), *19* Scarabino (2004), *20* Spaccesi and Rodrigues Capitulo (2012), *21* Sylvester et al. (2007), *22* Uhde et al. (2012)

covered by *L. fortunei* and substrata barren of the mussel conducted by Sylvester et al. (2007) showed that the establishment of *L. fortunei* beds increases benthic invertebrate abundance and biomass. Temporal profiles of invertebrate occupation over colonized and barren substrata were sharply different with barren areas saturating earlier (Fig. 1). A positive correlation between mussel and accompanying invertebrate biomass (Fig. 1) suggests that mussel beds increase the carrying capacity of benthic habitats.

The establishment of *L. fortunei* also enhances invertebrate richness. A survey conducted in the Jacuí River (southeastern Brazil) observed a considerably higher taxonomic richness (23 vs. 15 families) associated with *L. fortunei* colonies than with barren sediments (Kapusta and Fagondes de Freitas 2012). Similar results were obtained by Karatayev et al. (2010) in a reservoir in central Argentina, where significantly more invertebrate species were found in samples from *L. fortunei* druses (clumps or aggregations of mussels around either a stone or other hard object, held together by byssal threads), than in samples from the surrounding sediment. This effect is likely a result of the fact that *L. fortunei* beds represent isolated and highly populated islands of hard, biologically modified substrate in a sea of sparsely populated mud (Boltovskoy and Correa 2015). As noticed by Burlakova et al. (2012), mussel aggregates create habitat for species that would otherwise be infrequent in the environment, providing them with shelter and food.

Local increases in species richness, however, do not necessarily translate into a higher overall diversity (as measured by numbers of species and their proportions). Armendáriz et al. (2011) found that the presence of *L. fortunei* increased oligochaete richness but not evenness. The study by Karatayev et al. (2010) found that while species' richness increased in *L. fortunei* druses, similarity between samples with *L. fortunei* was significantly higher than that between sediment samples, suggesting a reduction of β-diversity This result agrees with that of Sardiña et al. (2011), who concluded that the spread of *L. fortunei* can promote faunal homogenization across benthic communities even though overall invertebrate abundance and biomass are enhanced. The bottoms of South American rivers are dominated by soft muddy and sandy sediments offering limited habitat opportunities for epifaunal groups. These studies suggest that while *L. fortunei* valves create islands of complex hard substrata that facilitate benthic invertebrate communities, these islands are more similar between themselves than were the original communities (Fig. 2). Thus, while *L. fortunei* promotes benthic invertebrate communities locally, a likely concomitant effect is the homogenization of bottom faunas across habitats. A potentially important caveat to these results is the fact that druses and mussel beds host significantly higher diversities and abundances, which may yield inventories less biased by omissions of the rare species, and thus artificially enhance bed-to-bed or druse-to-druse similarities when compared with similarities between bare sediment samples.

An important factor regulating the magnitude of facilitation and inhibition by *L. fortunei* is the amount of hard substrate available for colonization by the mussel. Partially due to the lack of information on this trait, most assessments of *L. fortunei* densities are anecdotal and normally restricted to local density maxima (Darrigran 2002; Boltovskoy et al. 2006; but see Boltovskoy et al. 2009). The dominance of

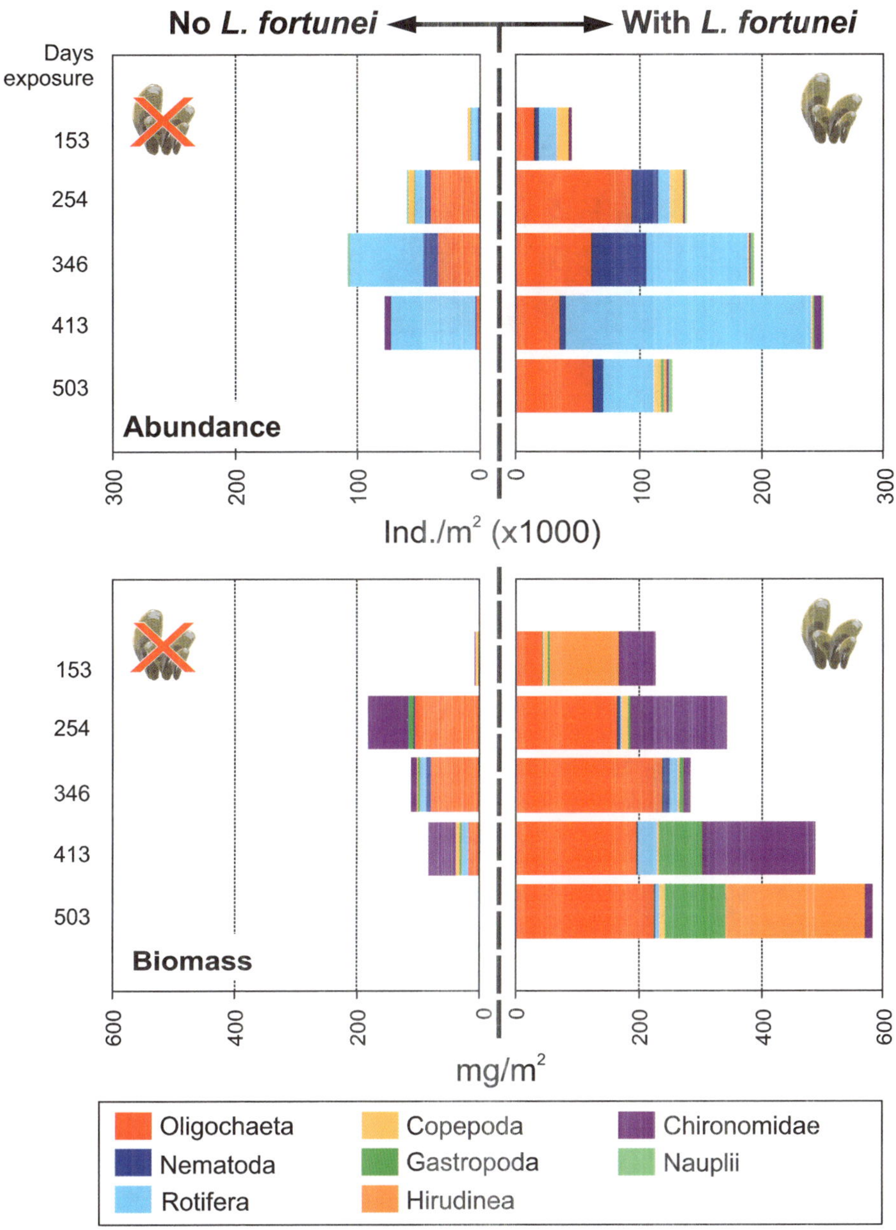

Fig. 1 Abundance and biomass of the dominant invertebrate groups in areas with and without *Limnoperna fortunei* mussels throughout a 503-day study period. Experimental substrata were deployed in the Lower Paraná River delta, South America, on 6 Nov. 2002 and successively retrieved on five occasions 153–503 days after the deployment. The accumulated overgrowth was calculated, and invertebrates were evaluated separately for substrate areas covered by *L. fortunei* and those barren of the mussel. The positive correlation between mussel and associated invertebrate biomass suggests that mussel beds increase the carrying capacity of benthic habitats. (Based on data from Sylvester et al. 2007)

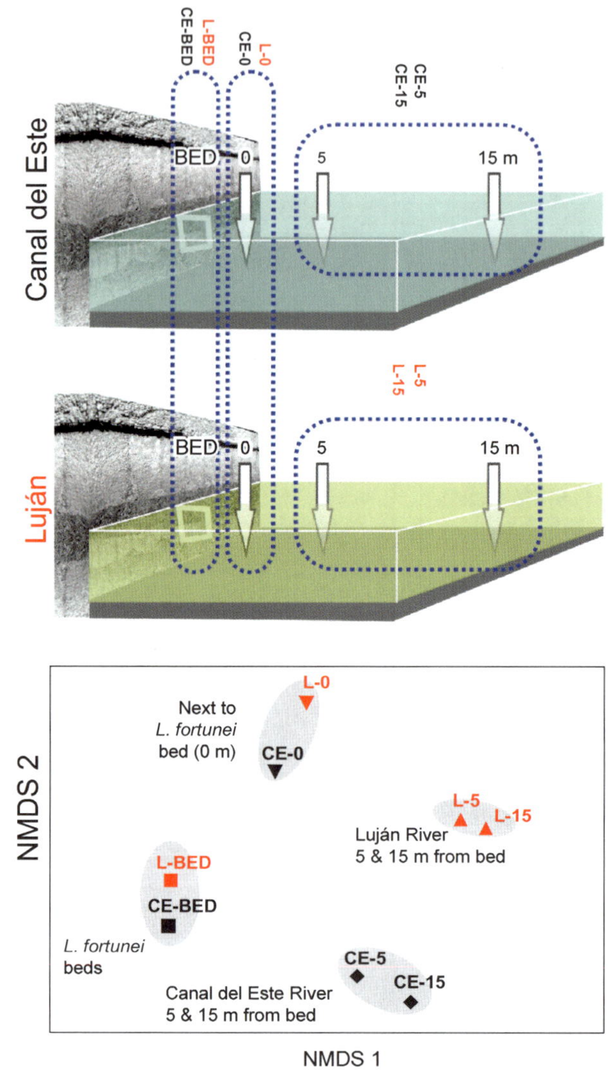

Fig. 2 Experimental setting (*upper panel*) and nonmetric multidimensional scaling (*NMDS*) analysis (*lower panel*) of a study evaluating the interaction between *Limnoperna fortunei* and benthic invertebrate communities at different spatial scales and under different environmental conditions: highly polluted Luján River (*L*) versus the less polluted Canal del Este River (*CE*; Lower Paraná River delta, South America). At each river, benthic samples were obtained from mussel beds (*L-BED* and *CE-BED*), and from sediments next to the mussel beds (0 m; *L-0* and *CE-0*), 5 m downstream (*L-5* and *CE-5*), and 15 m downstream (*L-15* and *CE-15*) from the mussel beds. Sampling location in the mussel beds is represented by the white sampling frame on the revetment. Invertebrate communities did not differ significantly between rivers in *L. fortunei* beds and at sites immediately downstream (0 m), but differed strongly at sites 5 and 15 m downstream from the mussel beds; *dotted frames* (*upper panel*) and *shaded areas* (*NMDS* analysis, *lower panel*) denote similar sample pairs. (Based on data from Sardiña et al. 2011)

silt and sandy bottoms unsuitable for the establishment of *L. fortunei* colonies likely curbs the species' ecological impacts over large areas in South America.

Facilitation of invertebrate communities by *L. fortunei* also occurs in the silt-bottom-habitats around mussel beds, thus extending the influence of *L. fortunei* beyond the range of colonizable substrata (Sardiña et al. 2011). This effect is presumably due to the downstream dispersal of feces and pseudofeces that settle out on the bottom.

Effects on benthic invertebrate communities also cascade along food webs into other compartments of aquatic ecosystems. For example, Sardiña et al. (2011) observed significantly higher abundance and biomass of predatory isopods and turbellarians in mussel beds than elsewhere, which they associated with an increase in prey availability. Benthivorous fish, birds, and mammals may also benefit from these increases. Thus, the spread of positive interactions to neighboring habitats and along food webs can counterbalance the facilitating effects of *L. fortunei* on invertebrates.

Data at hand are still too scarce and too fragmentary to judge whether the changes observed to date are definitive, or will invertebrate diversities and abundances change in the future, as noticed for the zebra mussel. Long-term invertebrate declines due to competition for food (Lozano et al. 2001) are unlikely in South American waterbodies rich in particulate organic carbon (POC) (Sylvester et al. 2005; see Chapter "Ecology and Environmental Impact of *Limnoperna fortunei*: Introduction" in this volume). On the other hand, consumption of *L. fortunei* by predators has likely increased due to a rise in fish populations resulting from the introduction of a new abundant food item, and their acclimation to the new prey after over two decades of coexistence (Boltovskoy et al. 2006; see Chapter "Trophic Relationships of *Limnoperna fortunei* with Adult Fishes" in this volume). Higher predation pressure might reduce *L. fortunei* abundances to a lower level compared to early invasion values. If *L. fortunei* populations decline after initially peaking, ecological effects might wane (see Chapter "*Limnoperna fortunei* Colonies: Structure, Distribution and Dynamics" in this volume; Haynes et al. 1999). However, these effects may wax globally as *L. fortunei* keeps invading new watersheds.

Taxonomic and Functional Patterns

It is not easy to find general trends in the response of different taxonomic groups to the presence of *L. fortunei,* but some overall patterns may be extracted. In general, deposit-feeders, collector-gatherers, and scrapers appear to benefit from the presence of *L. fortunei*. Some predators are also positively impacted, while some small-sized groups (e.g., small scrapers) can be negatively impacted (Sardiña et al. 2011; Burlakova et al. 2012). Although evidence is mixed, hirudineans appear to be more positively than negatively impacted. The same applies to oligochaetes. All reports for water mites and sessile taxa, such as hydroids, suggest positive effects, although these groups are still largely understudied and conclusions are preliminary. Most reports for mobile meiofaunal crustaceans (including amphipods, cladocerans, copepods, isopods, and tanaidaceans) and aquatic insects (including water beetles, mayflies, dragonflies, caddisflies, and springtails) indicate positive effects. Both positive and negative effects on dipterans can be found in the literature (Table 1). Negative impacts have been observed for a number of groups, among which bivalves and gastropods have been the most frequently cited (Darrigran et al. 1998; Darrigran et al. 2000; Mansur et al. 2003; Scarabino 2004). However, negative responses have only been reliably demonstrated for a few species. These include some gastropod molluscs (*Heleobia parchappei, Biomphalaria* sp., *Potamolithus* sp., and some *Planorbidae*), chironomids (*Cryptochironomus* sp., *Harnischia* sp., and *Procladius* sp.), the oligochaete *Branchiura sowerbyi,* the ostracod *Cyprideis hartmanni,* and a ceratopogonid midge, *Culicoides* sp. (Sylvester et al. 2007; Sardiña et al. 2008, 2011; Karatayev et al. 2010; Table 1).

Although the general trends of the effects of *L. fortunei* on benthic invertebrates are reasonably clear, literature on the subject is not without disagreements. A vivid example is provided by gastropods, whose densities were reported to have decreased after the colonization by *L. fortunei* (Darrigran et al. 1998), whereas other field (Karatayev et al. 2010; Sardiña et al. 2011) and experimental data (Sylvester et al. 2007) suggest otherwise. Similarly, the diversity and abundance of Oligochaeta were found to be significantly higher in *L. fortunei* beds as compared with nearby substrate without the mussel (Darrigran et al. 1998; Sylvester et al. 2007; Sardiña et al. 2008), but the opposite trend was recorded when studying *L. fortunei* druses in comparison with nearby soft sediments (Karatayev et al. 2010). Stable mussel beds on natural and artificial hard substrata accumulate large amounts of silt between the mussels (Sardiña et al. 2008), whereas druses are highly mobile and less efficient in retaining the particulate material derived from the golden mussel's feces and pseudofeces. Thus, the comparative scarcity of Oligochaeta—typically soft bottom, burrowing organisms—in druses is probably associated with the fact that loose, soft sediments are scarce in these aggregates.

While some of the disagreements encountered in the literature may be attributable to species–specific effects or differences in regional or seasonal settings (Radziejewska et al. 2009; Karatayev et al. 2010; Sardiña et al. 2011), it is also conceivable that the lower precision of abundance estimates associated with the less abundant taxa and methodological, sampling-related disparities play an important role. For example, of the ten taxa found in higher numbers in the sediments than in *L. fortunei* druses by Karatayev et al. (2010), five were recorded in only one or two out of ten samples. In the comparison of pre- and post-*L. fortunei* conditions by Darrigran et al. (1998), only gastropods, hirudineans, and isopods were included in the pre-*L. fortunei* dataset, while other organisms, such as oligochaetes, flatworms, nematodes, crustaceans, and chironomids, were considered only in post-*L. fortunei* samples. Hirudineans were found to be less abundant in *L. fortunei* beds than on bare hard substrate (Sardiña et al. 2008); yet another study found some leeches (*Gloiobdella michaelseni* and *Helobdella stagnalis*) to be dominant in golden mussel-associated communities when analyzed at the species level (Karatayev et al. 2010)

Mechanisms for Positive Effects

Positive effects on invertebrate fauna are commonly attributed to the enhancement of food supply, provision of substrate, and refugia in the new habitat created among the valves (Fig. 3 and 4). The increase in food for benthic organisms chiefly occurs through biodeposition of organic-matter-rich feces and pseudofeces. *L. fortunei* is a bottom filter-feeding organism that essentially removes matter and energy from the water column and transfers it to the benthos in the form of feces, pseudofeces, and its own body mass, a function generally referred to as benthic-pelagic coupling (Karatayev et al. 2007). The combination of high filtration rates and elevated densities can make this process very effective in some areas. After colonization by *L. fortunei,* in the Río de la Plata basin a large amount of particulate organic matter that was previously flushed into the ocean became redirected to the benthos (Sylvester et al. 2005). The deposition of organic matter over the riverbed is linked to an increase in biomass of heterotrophs. Among the mussels' valves, a food-rich environment is generated that is profited by several organisms in the guilds of the deposit feeders and scrappers (Sardiña et al. 2011; Fig. 4).

Besides this biotic effect, there is also a physical effect. The sole presence of mussel valves transforms a flat surface into a rough, three-dimensional structure which offers a larger surface area for sessile organisms, as well as refuge from predators and the physical stress of wave and current action (Fig. 3 and 4). In studies conducted in several Argentine waterbodies, the local abundances of at least three gastropod species, *Heleobia piscium, Gundlachia moricandi,* and *Stenophysa marmorata,* were found to have increased substantially due to the protection supplied by the mussels' valves (Darrigran et al. 1998; Karatayev et al. 2010). Even

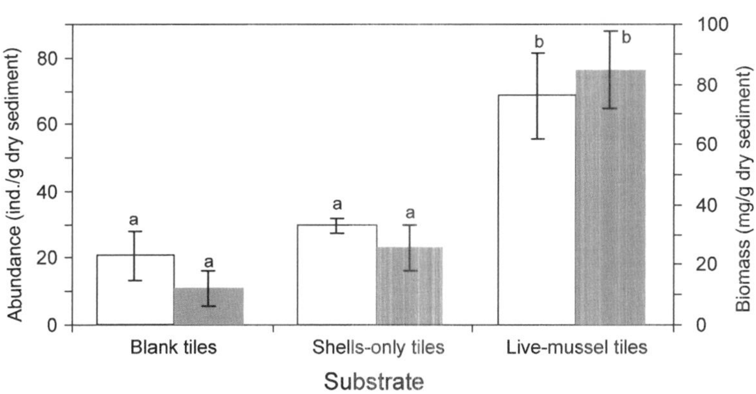

Fig. 3 Mean abundance and biomass (±SD) of invertebrates settled on artificial colonization tiles in three treatments: blank tiles (barren of *Limnoperna fortunei*), shells-only tiles (empty shells of *L. fortunei* glued together and to the upper and lateral surfaces of the tiles), and live-mussel tiles (living *L. fortunei* settled on the tiles). Different letters indicate significant differences among treatments. (Based on data from Sardiña et al. 2008)

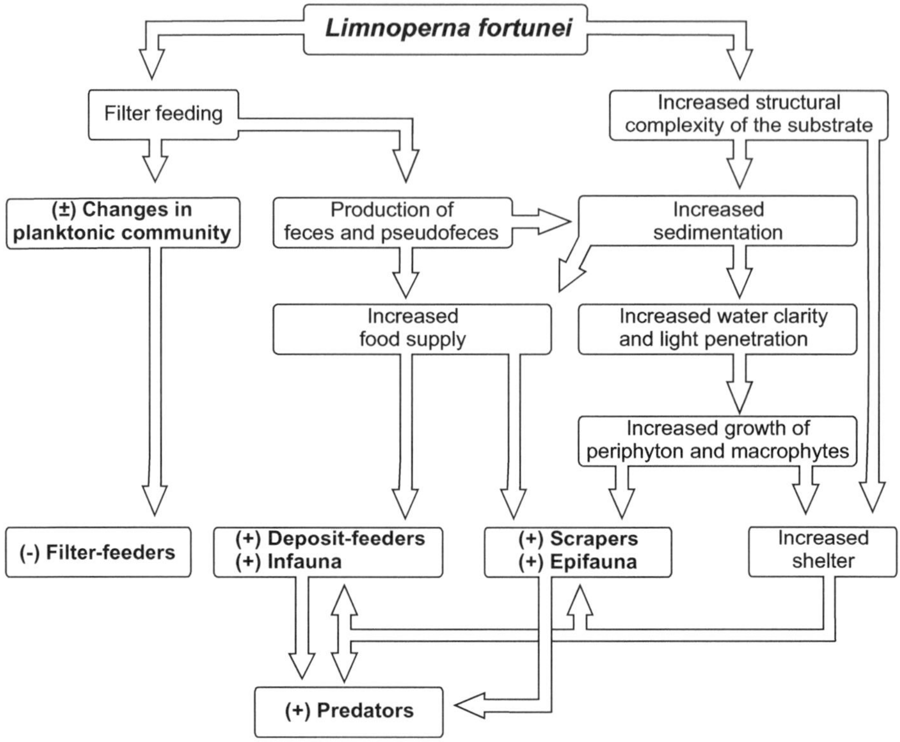

Fig. 4 Mechanisms for positive (+) and negative (−) effects of *Limnoperna fortunei* on benthic invertebrate functional feeding groups

some bivalves (*Sphaerium* sp.), a group commonly considered to be negatively impacted by *L. fortunei* (see below), have been found at higher abundances within *L. fortunei* beds (Karatayev et al. 2010). Sphaerid bivalves are tiny and mobile, and they might be able to find refuge from predators within mussel beds. While chiefly filter-feeders, sphaerids are also capable of deposit feeding and it is conceivable that some species find a suitable habitat among the valves of *L. fortunei* due to both abundant food and protection.

It is difficult to establish through descriptive studies whether physical or biotic factors play a greater role in the promotion of bottom invertebrates. Moreover, the two effects are not independent, as valve complexity also increases sedimentation of organic particles through the buffering of current and wave action (Sylvester et al. 2007; Sardiña et al. 2008; Fig. 4). In an attempt to distinguish between the two, Sardiña et al. (2008) deployed bare hard substrates (tiles), tiles with empty mussel shells glued to them, and tiles colonized by live mussels. Upon retrieval, they found that both live mussels and empty shells had promoted significantly higher invertebrate densities than blank tiles. While empty shells actually promoted higher sedimentation rates (in the absence of the cleansing currents generated by the mussels' siphons), live mussels supported a greater invertebrate biomass due to a richer organic matter content of their deposits (Fig. 3). The authors concluded that invertebrate communities in the Paraná River are enhanced by the presence of *L. fortunei* beds, and this is due more to the biotic effect of sediment enrichment than to increased substrate complexity.

Mechanisms for Negative Effects

Mechanisms proposed for population declines observed in South American aquatic habitats following the invasion of *L. fortunei* include substrate eutrophication (chironomids), biomagnification of contaminants (ostracods and nematodes), oxygen depletion (oligochaetes), and competition (gastropods) (Sylvester et al. 2007; Karatayev et al. 2010; Sardiña et al. 2011). Plankton depletion, direct predation of eggs and gameta, and competition for space have also been proposed as potential mechanisms underlying adverse effects on invertebrates (Scarabino 2004; Mansur et al. 2008). However, in most if not all cases, the mechanisms underlying negative effects were untested but assumed based on the extrapolation from the literature on North American *Dreissena* spp. invasions.

Available evidence suggests that both positive and negative effects of *L. fortunei* are habitat dependent. For example, oxygen depletion may be associated with *L. fortunei* beds as a result of the decomposition of feces and pseudofeces (Karatayev et al. 2010). While this effect is possible in deep, poorly mixed zones of lakes and ponds, it is unlikely in well-mixed rivers. Similarly, in sediments of stagnant habitats, burrowing species may be more exposed to anoxia than epifaunal species. As discussed below, indirect effects on benthic communities may also be strongly influenced by other environmental factors and ecosystem characteristics, such as the load of POC and the fauna species complement (see Chapter "Parallels and contrasts between *Limnoperna fortunei* and species of *Dreissena*" in this volume). The effects of *L. fortunei* on invertebrate fauna are also species dependent, as physiological tolerances may increase or decrease vulnerability to these effects.

Effects on Other Benthic Animals

Macrofouling

The most conspicuous and frequently cited negative impact by *L. fortunei* is epifaunal macrofouling, particularly on bivalves and gastropods. An extreme case illustrative of such impact was the observation of a crab, *Aegla platensis,* weighing barely one tenth the weight of the colony of *L. fortunei* attached to its back (Lopes et al. 2009). Specimens of the gastropod *Pomacea canaliculata,* and many bivalve species (such as *Anodontites trapesialis, A. trapezeus, Diplodon deceptus, D. koseritzi,* and *Leila blainvilliana*) have also been found fouled in aquatic habitats in

Argentina and Brazil (Darrigran et al. 2000; Mansur et al. 2003; Mansur et al. 2008; Karatayev et al. 2010). *L. fortunei* can also form colonies on top of other invasive bivalves such as the Asian clam *Corbicula fluminea* (Darrigran et al. 2000; Mansur et al. 2003). Shu and Wu (2005) reported that 35 % of the bivalves (*Arconaia lanceolata, Larnprotula leai, Larnprotula caveata,* and *Larnprotula rochechouarti*) of Lake Poyang (China) are fouled by *L. fortunei*. An interesting case is the macrofouling of *Trichodactylus borellianus*. This native South American crab can be subjected to severe macrofouling by *L. fortunei,* but mature specimens have the ability to feed on the mussel (Rojas Molina and Williner 2013). The net effect of this interaction remains a question. Unfortunately, the same uncertainties discussed above for benthic meiofauna apply to much of the information on macrofaunal fouling by *L. fortunei*. While some of the cases reported are very dramatic, they constitute isolated observations and their effects on the population level remain unclear.

Predation by Benthic Animals

The colonization of South American waterbodies by *L. fortunei* has not only offered shelter and food resources to benthic animals, but the mussel itself has become an attractive food item for a variety of resident groups. Besides fishes, which are well known to prey on larval and adult forms (see Chapters "Trophic Relationships of *Limnoperna fortunei* with Larval Fishes" and "Trophic Relationships of *Limnoperna fortunei* with Adult Fishes" in this volume), several invertebrates and vertebrates can prey on settled *L. fortunei* mussels. For example, in laboratory trials the native crabs *Zilchiopsis collastinensis* and *T. borellianus* have been observed to consume several sizes of mussels (Torres et al. 2012; Carvalho et al. 2013). While *L. fortunei* was not a preferred prey item for *T. borellianus,* its consumption by crabs ranging ~4–11 cm indicates that the mussel might constitute a new, potentially important, alternative prey item in times of shortage of other food (Carvalho et al. 2013).

Turtles also benefit from the new prey. The Brazilian slider, *Trachemys dorbigni,* feeds on *L. fortunei* mussels attached to dock pilings and other structures (Bujes et al. 2007). Even though a study by Hahn et al. (2014) did not find *L. fortunei* in the stomachs of 73 individuals of this species along the shores of São Gonçalo Canal and Mirim Lake in 2002–2003, this result should not be taken as indicative of a lack of consumption capabilities because this area was not colonized by the mussel until 2005 (Burns et al. 2006; Capítoli et al. 2008; Colling et al. 2012). The presence of shell debris in Brazilian slider feces confirms consumption of *L. fortunei* by this turtle (Bujes et al. 2007).

Other animals have been suggested as potential predators of golden mussels, including the crab-eating raccoon *Procyon cancrivorus,* the giant otter *Pteronura brasiliensis,* and the neotropical river otter *Lontra longicaudis*. Waterfowl such as coots, cormorants, grebes, gulls, ducks, and swans are also likely to consume this new food resource, although this assumption is largely based on the extrapolation of observations made on waterfowl feeding on zebra mussels in North America (Sylvester et al. 2007). It has been suggested that small invertebrates (crustaceans such as predatory isopods, decapods, and copepods, as well as other small invertebrates including leeches, gastropods, and insect larvae) may consume early settled stages of *L. fortunei* causing strong impacts on its populations (Sylvester et al. 2007; Nakano et al. 2010), although this type of predation has never been verified nor quantified.

Similarities and Differences with *Dreissena* spp.

The effects of *L. fortunei* on associated fauna are remarkably similar to those of *Dreissena* species, in particular *D. polymorpha*. Both are strong ecosystem engineers that increase the structural complexity of the substrate and provide shelter and food for other benthic invertebrates. Although negative impacts have been described for some species and habitats, both mussels have predominantly positive influences on their accompanying fauna. The ultimate impact of these changes has been associated with increased benthic invertebrate density, biomass, and taxonomic richness, and with decreased community diversity (Ward and Ricciardi 2007; Karatayev et al. 2010; Sardiña et al. 2011). Like *Dreissena* spp., *L. fortunei* positively affects predators and scrapers, particularly leeches (*Hirudinea*), flatworms (*Turbellaria*), and mayflies (*Ephemeroptera*), and negatively affects other bivalves (Ward and Ricciardi 2007). Also similar to *Dreissena* spp., *L. fortunei* exerts a mixture of positive and negative effects on gastropods. Unlike the zebra mussel, however, which has been linked to declines of large-bodied snails, particularly in the family Pleuroceridae (Ward and Ricciardi 2007), the golden mussel has been associated with a decline of small-bodied snails such as *Potamolithus* sp. and Planorbids (Sylvester et al. 2007; Sardiña et al. 2011). Competitive exclusion by larger snails within *L. fortunei* beds, and size-limiting interstitial spaces created amongst *D. polymorpha* shells have been proposed as the possible ecological venues in each case (Ricciardi et al. 1997; Sardiña et al. 2011).

Probably the most important difference between *L. fortunei* and *Dreissena* spp. derives from the fact that *Dreissena* spp. (in particular *Dreissena rostriformis bugensis*) are able to colonize soft sediments, while *L. fortunei* is rarely found on soft substrata (see Chapter "Parallels and Contrasts Between *Limnoperna fortunei* and Species of *Dreissena*" in this volume). Moreover, the strength and direction of the interactions between *Dreissena* spp. and other macroinvertebrates is correlated with sediment particle size. For example, infaunal (burrowing) organisms such as nematodes are positively effected in the presence of dreissenid mussels on hard substrata but negatively so on fine sediments (Ward and Ricciardi 2007). In contrast, the golden mussel favors nematodes in almost every case (Table 1). Similarly, strong positive interactions prevail between *L. fortunei* and epifaunal organisms such as gammarid amphipods and isopod crustaceans, whereas the positive effects of *Dreissena* spp. on these groups decline with decreasing particle size (Ward and Ricciardi 2007).

Another noteworthy difference between *L. fortunei* and *Dreissena* spp. relates to their effects on collector-gatherer organisms. The presence of *L. fortunei* is almost invariably associated with enhanced densities and biomass of deposit-feeders, mainly oligochaetes (82 % of the interactions reported are positive; Table 1). Conversely, a meta-analysis of 47 study sites conducted by Ward and Ricciardi (2007) found that the overall effect of *Dreissena* spp. on these organisms was neutral. It should be borne in mind, however, that this represents an average trend, and site-specific records for strong positive interactions abound in the literature on *Dreissena* spp. (Botts et al. 1996; Ricciardi et al. 1997; Bially and MacIsaac 2000).

Other differences between these mussels may arise from ambient differences between the ranges invaded by each species. There is, for instance, a sharp difference between the faunal composition of South American and North American waterbodies. While the North American lakes hosting a large proportion of *Dreissena* spp. populations are dominated by pelagic fish, most of the South American populations of *L. fortunei* occur in rivers hosting a wealth of benthivorous/detritivorous fish species that can benefit from substrate enrichment by the mussel. As a result, the benthic-pelagic coupling enhanced by the mussel is likely more significant in South American rivers colonized by *L. fortunei* than in the invasive range of the dreissenids (but see Karatayev et al. 2007). Another important environmental difference between both continents is the POC load of lotic habitats. In North American rivers, depletion by *Dreissena* spp. of the relatively scant bioseston can indirectly depress macroinvertebrates in sites away from mussel beds (Strayer and Smith 2001). In contrast, bioseston is probably not limiting in South American POC-rich rivers and this depression is less likely.

One can envision that marked differences such as long-term declines of mussel populations due to competition for food, as those documented for *Dreissena* invasions (Lozano et al. 2001; Ratti and Barton 2003) are unlikely in South America (Boltovskoy and Correa 2015). There still is, however, a need for long-term studies on the effects of *L. fortunei* on benthic communities in order to judge whether the changes described above are transient or permanent. In general, and despite rapid progress made in the last few years, particularly since the invasion of South America, scientific knowledge on *L. fortunei* runs far behind that of *Dreissena* spp.

Future Research Lines and Concluding Remarks

While notable progress has been made in the study of the relationships between *L. fortunei* and other animals, there is still a long way to go before we can understand the nature and magnitude of these relationships, and the ecosystem changes brought about by this invasion. Our major gaps in scientific information include the following:

Effects on Microfauna and Smaller Organisms Most of the studies on fauna associated with *L. fortunei* are restricted to meiofauna and macroinvertebrates, whereas studies on smaller organisms are almost completely lacking. Carvalho Torgan et al.

(2009) have documented at least 18 diatom species living on the valves of *L. fortunei*. It is likely that small animals are also affected and the onset of *L. fortunei* is having profound yet unexplored effects on benthic microfaunal and microbial communities.

Indirect and Long-Term Effects While most studies have looked into local, direct effects, ecosystem responses may vary at broader time and spatial scales as a consequence of indirect and feedback effects. For example, increases in water clarity and benthophagous fishes promoted by invasive mussels can subsequently impact benthic invertebrate populations (Ward and Ricciardi 2007; Fig. 4). These impacts may differ between mussel beds and distant habitats (Strayer and Smith 2001).

Synergistic and Antagonistic Effects In addition to golden mussels, in South American water bodies, there are other native and nonnative ecosystem engineers, such as rushes. The invasive macrophyte *Hydrilla verticillata* has been reported to host more *L. fortunei* on its surface than other (native) macrophytes (Michelan et al. 2014). The comparison of faunal facilitation by *L. fortunei* and other ecosystem engineers, as well as potential synergistic effects between them, remain largely unexplored.

Interactions Between Faunal Facilitation and Pollution Contrasts between abundances of *L. fortunei*-associated invertebrates at polluted and comparatively clean sites indicated that facilitation is the highest at low levels of environmental pollution in the Paraná River delta (Sardiña et al. 2011). While based on only two sites, this result suggests that some of the benefits produced to the benthos by *L. fortunei*, and from there to other compartments of the food web, may be offset by environmental pollution. This interaction and the potential rerouting and bio-magnification of contaminants through bottom food webs by *L. fortunei* are important research priorities because introduction gateways and some of the most densely invaded areas are estuaries and ports heavily impacted by human activities.

These information gaps are mere examples of potential future research. Actual blanks in the available information span a wide range of topics, many of which are key for understanding changes brought about by the invasion of *L. fortunei*.

We would like to conclude this chapter summarizing some of the problems that, in our view, have significantly slowed down the buildup of knowledge in this field of study. In the first place, extrapolation of conclusions from the much better studied zebra and quagga mussels is widespread in the literature on *L. fortunei*, particularly in the area of its impacts on native organisms. Some of the effects on other animals attributed to *L. fortunei* and most of the mechanisms underlying these effects have been drawn from literature on *D. polymorpha*; yet we have seen that species-specific and environment-specific differences may result in very different outcomes. While reviews highlighting likely impacts and native taxonomic groups potentially at risk that are based on the observations of *Dreissena* spp. were important during the early invasional stages of *L. fortunei* (e.g., Scarabino 2004; Brugnoli et al. 2005), current research needs to move forward with firsthand experimental work. The repetition time and again of untested conclusions can be both misleading and discouraging of research over matters for which we have developed a false perception of understanding.

In the second place, the dissemination of anecdotal, uninformative and nonquantitative information has become a problem in the literature on interactions between *L. fortunei* and benthic invertebrates. Circumstantial reports of macrofouling by the mussel or its incorporation in the diet of native species have shed little light on its effects on native ecosystems. These interactions need to be quantified at the individual and population levels.

In the third place, quantitative assessments need adequate statistical support. Out of about 280 cases reviewed in the literature, less than 45 (15 %) based their conclusions on statistical evidence or overwhelming differences (an order of magnitude or higher).

References

Armendáriz LC, Rodrigues CA, Ambrosio ES (2011) Relationships between the spatial distribution of oligochaetes (Annelida, Clitellata) and environmental variables in a temperate estuary system of South America (Río de la Plata, Argentina). N Z J Mar Freshw Res 45:263–279

Bially A, MacIsaac HJ (2000) Fouling mussels (*Dreissena* spp.) colonize soft sediments in Lake Erie and facilitate benthic invertebrates. Freshw Biol 43:85–97

Boltovskoy D, Correa N (2015) Ecosystem impacts of the invasive bivalve *Limnoperna fortunei* (golden mussel) in South America. Hydrobiologia. 746:81–95

Boltovskoy D, Correa N, Cataldo D, Sylvester F (2006) Dispersion and ecological impact of the invasive freshwater bivalve *Limnoperna fortunei* in the Río de la Plata watershed and beyond. Biol Invasions 8:947–963

Boltovskoy D, Karatayev A, Burlakova L, Cataldo D, Karatayev V, Sylvester F, Mariñelarena A (2009) Significant ecosystem-wide effects of the swiftly spreading invasive freshwater bivalve *Limnoperna fortunei*. Hydrobiologia 636:271–284

Botts PS, Patterson BA, Schloesser DW (1996) Zebra Mussel effects on benthic invertebrates: physical or biotic? J N Am Benthol Soc 15:179–184

Brugnoli E, Clemente J, Boccardi L, Borthagaray A, Scarabino F (2005) Golden mussel *Limnoperna fortunei* (Bivalvia: Mytilidae) distribution in the main hydrographical basins of Uruguay: update and predictions. Anais da Academia Brasileira de Ciências 77:235–244

Bujes CS, Ely I, Verrastro L (2007) *Trachemys dorbigni* (Brazilian Slider) diet. Herpetol Rev 38:335

Burlakova LE, Karatayev AY, Karatayev VA (2012) Invasive mussels induce community changes by increasing habitat complexity. Hydrobiologia 685:121–134

Burns MDdM, Geraldi RM, Garcia AM, Bemvenuti CE, Capitoli RR, Vieira JP (2006) Primeiro registro de eocorrencia do mexilhao dourado *Limnoperna fortunei* na bacia de drenagem da Lagoa Mirim, RS, Brasil. Biociências 14:83–84

Capítoli RR, Colling LA, Bemvenuti CE (2008) Cenários de distribuição do Mexilhão Dourado *Limnoperna fortunei* (Mollusca—Bivalvia) sob distintas condições de salinidade no complexo lagunar Patos-Mirim, RS—Brasil. Atlântica 30:35–44

Carvalho Torgan L, Salomoni SE, Brugalli Bicca A (2009) Diatomáceas sobre *Limnoperna fortunei* (Dunker), molusco introduzido no Lago Guaíba, Sul do Brasil. Revista Brasileira de Botanica 32:23–31

Carvalho DA, Collins PA, De Bonis CJ (2013) Predation ability of freshwater crabs: age and prey-specific differences in *Trichodactylus borellianus* (Brachyura: Trichodactylidae). J Freshw Ecol 28:573–584

César II, Martín SM, Rumi A, Tassara M (2012) Mollusks (Gastropoda and Bivalvia) of the multiple-use reserve Martín García Island, Río de la plata river: biodiversity and ecology. Braz J Biol 72:121–130

Colling LA, Pinotti RM, Bemvenuti CE (2012) *Limnoperna fortunei* na Bacia da Lagoa dos Patos e Lagoa Mirim. In: Mansur MCD, Santos CP, Pereira D, Padula PIC, Leite Zurita ML, Raya RMT, Vilar NM, Aydos BPE (eds) Moluscos límnicos invasores no Brasil. Biologia, prevenção, controle, Redes Editora, Porto Alegre, pp 187–191

Darrigran GA (2002) Potential impact of filter-feeding invaders on temperate inland freshwater environments. Biol Invasions 4:145–156

Darrigran GA, Martin SM, Gullo B, Armendariz L (1998) Macroinvertebrates associated with *Limnoperna fortunei* (Dunker, 1857) (Bivalvia, Mytilidae) in Río de la Plata, Argentina. Hydrobiologia 367:223–230

Darrigran GA, Penchaszadeh PE, Damborenea C (2000) An invasion tale: *Limnoperna fortunei* (Dunker, 1857) (Mytilidae) in the Neotropics. In: 10th International Aquatic Nuisance Species and Zebra Mussels Conference, Toronto (Canada)

Fagondes de Freitas SM, Kapusta SK (2010) O mexilhão dourado como agente transformador da fauna bentônica (trans. 1° Workshop de Invertebrados Límnicos Invasores na América do Sul: Prevenção, Monitoramento e Controle, vol, edn. Centro de Ecologia (CENECO), Universidade Federal do Rio Grande do Sul, Porto Alegre (Brazil)

Hahn AT, Rosa CA, Bager A, Krause L (2014) Dietary variation and overlap in D'Orbigny's slider turtles *Trachemys dorbigni* (Duméril and Bibron 1835) (Testudines: Emydidae). J Nat Hist 48:721–728

Haynes JM, Stewart TW, Cook GE (1999) Benthic macroinvertebrate communities in southwestern Lake Ontario following invasion of *Dreissena*: continuing change. J Gt Lakes Res 25:828–838

Kapusta SK, Fagondes de Freitas SM (2012) *Limnoperna fortunei* na bacia hidrográfica do baixo Rio Jacuí: relações com a comunidade de macroinvertebrados bentônicos. In: Mansur MCD, Santos CP, Pereira D, Padula PIC, Leite Zurita ML, Raya RMT, Vilar NM, Aydos BPE (eds) Moluscos límnicos invasores no Brasil. Biologia, prevenção, controle, Redes Editora, Porto Alegre, pp 243–246

Karatayev AY, Boltovskoy D, Padilla DK, Burlakova LE (2007) The invasive bivalves *Dreissena polymorpha* and *Limnoperna fortunei*: parallels, contrasts, potential spread and invasion impacts. J Shellfish Res 26:205–213

Karatayev AY, Burlakova LE, Karatayev VA, Boltovskoy D (2010) *Limnoperna fortunei* versus *Dreissena polymorpha*: Population densities and benthic community impacts of two invasive freshwater bivalves. J Shellfish Res 29:975–984

Lopes MN, Vieira JP, Burns MDM (2009) Biofouling of the golden mussel *Limnoperna fortunei* (Dunker, 1857) over the Anomura crab *Aegla platensis* Schmitt, 1942. Panam J Aquat Sci 4:222–225

Lozano SJ, Scharold JV, Nalepa TF (2001) Recent declines in benthic macroinvertebrate densities in Lake Ontario. Can J Fish Aquat Sci 58:518–529

Mansur MCD, Santos CP, Darrigran G, Heydrich I, Callil CT, Cardoso FR (2003) Primeiros dados quali-quantitativos do mexilhão-dourado, *Limnoperna fortunei* (Dunker), no Delta do Jacuí, no Lago Guaíba e na Laguna dos Patos, Rio Grande do Sul, Brasil e alguns aspectos de sua invasão no novo ambiente. Revista Brasileira de Zoologia 20:75–84

Mansur MCD, Santos CP, Richinitti LMZ, Pereira D, Batista CB, da Silveira MB, de Fraga A, Regina M, Pons MC (2008) Ocorrência de moluscos límnicos e crustáceo em macroaglomerados do mexilhão dourado, *Limnoperna fortunei* (Dunker, 1857) sobre sarandi no lago Guaíba (RS, Brasil). Biotemas 21:179–182

Marçal SF, Callil CT (2008) Structure of invertebrates community associated with *Eichhornia crassipes* Mart. (Solms-Laubach) after the introduction of *Limnoperna fortunei* (Dunker, 1857) (Bivalvia, Mytilidae) in the Upper Paraguay River, MT, Brazil. Acta Limnologica Brasiliensia 20:359–371

Michelan TS, Silveira MJ, Petsch DK, Pinha GD, Thomaz SM (2014) The invasive aquatic macrophyte *Hydrilla verticillata* facilitates the establishment of the invasive mussel *Limnoperna fortunei* in Neotropical reservoirs. J Limnol 73:598–602

Nakano D, Kobayashi T, Sakaguchi I (2010) Predation and depth effects on abundance and size distribution of an invasive bivalve, the golden mussel *Limnoperna fortunei*, in a dam reservoir. Limnology 11:259–266

Radziejewska T, Fenske C, Wawrzyniak-Wydrowska B, Riel P, Woźniczka A, Gruszka P (2009) The zebra mussel (*Dreissena polymorpha*) and the benthic community in a coastal Baltic lagoon: another example of enhancement? Mar Ecol 30:138–150

Ramseyer U, Marchese M (2009) Leaf litter of *Erythrina crista-galli* L. (ceibo): trophic and substratum resources for benthic invertebrates in a secondary channel of the middle Paraná river. Limnetica 28:1–10

Ratti C, Barton DR (2003) Decline in the diversity of benthic invertebrates in the wave-zone of eastern Lake Erie, 1974–2001. J Gt Lakes Res 29:608–615

Ricciardi A, Whoriskey FG, Rasmussen JB (1997) The role of the zebra mussel (*Dreissena polymorpha*) in structuring macroinvertebrate communities on hard substrata. Can J Fish Aquat Sci 54:1–13

Rojas Molina F, Williner V (2013) First record of the non-indigenous mussel *Limnoperna fortunei* (Bivalvia, Mytilidae) as an epibiont of the crab *Trichodactylus borellianus* (Decapoda, Trichodactylidae). Crustaceana 86:682–692

Sardiña P, Cataldo D, Boltovskoy D (2008) The effects of the invasive mussel, *Limnoperna fortunei*, on associated fauna in South American freshwaters: importance of physical structure and food supply. Fundam Appl Limnol/Archiv Hydrobiologie 173:135–144

Sardiña P, Chaves E, Marchese M (2011) Benthic community responses to invasion by the golden mussel, *Limnoperna fortunei* Dunker: biotic homogenization vs environmental driving forces. J N Am Benthol Soc 30:1009–1023

Scarabino F (2004) Conservación de la malacofauna uruguaya. Comunicaciones de la Sociedad Malacológica del Uruguay 8:267–273

Shu F-Y, Wu X (2005) Investigations of zebra mussel, *Limnoperna lacustris*, infestation of bivalves in Poyang Lake. Territ Nat Resour Study 1:82–83 [In Chinese]

Spaccesi FG, Rodrigues Capitulo A (2012) Benthic communities on hard substrates covered by *Limnoperna fortunei* Dunker (Bivalvia, Mytilidae) at an estuarine beach (Río de la Plata, Argentina). J Limnol 71:144–153

Strayer DL, Smith LC (2001) The zoobenthos of the freshwater tidal Hudson river and its response to the zebra mussel (*Dreissena polymorpha*) invasion. Archiv für Hydrobiologie 139 (Supplementband, Monographische Beiträge):1–52

Sylvester F, Dorado J, Boltovskoy D, Juárez A, Cataldo D (2005) Filtration rates of the invasive pest bivalve *Limnoperna fortunei* as a function of size and temperature. Hydrobiologia 534:71–80

Sylvester F, Boltovskoy D, Cataldo D (2007) Fast response of freshwater consumers to a new trophic resource: Predation on the recently introduced Asian bivalve *Limnoperna fortunei* in the lower Parana River, South America. Austral Ecol 32:403–415

Torres MV, Williner V, Giri F (2012) Size selective predation on an invasive bivalve, *Limnoperna fortunei* (Mytilidae), by a freshwater crab, *Zilchiopsis collastinensis* (Trichodactylidae). J Crustac Biol 32:698–710

Uhde V, Massoli EV Jr, Callil CT (2012) Efeito do macrofouling sobre a comunidade de invertebrados aquáticos. In: Mansur MCD, Santos CP, Pereira D, Padula PIC, Leite Zurita ML, Raya RMT, Vilar NM, Aydos BPE (eds) Moluscos límnicos invasores no Brasil. Biologia, prevenção, controle, Redes Editora, Porto Alegre, pp 235–241

Ward JM, Ricciardi A (2007) Impacts of *Dreissena* invasions on benthic macroinvertebrate communities: a meta-analysis. Divers Distrib 13:155–165

Trophic Relationships of *Limnoperna fortunei* with Larval Fishes

Esteban M. Paolucci and Erik V. Thuesen

Abstract This chapter reviews investigations concerning the importance of veligers of the exotic bivalve *Limnoperna fortunei* in the diets of larval fish in the Río de la Plata basin. These studies have shown that of the 25 fish taxa studied, 18 consumed veligers of *L. fortunei*. These species included the most abundant members of Characiformes and Siluriformes. The relative frequency and biomass contribution of *L. fortunei* larvae differed strongly in pimelodid and *Prochilodus lineatus* larvae at different developmental stages and in different environments. Thus, as fish larvae grew, their diets shifted from veligers to other prey items. The fact that the earliest fish larvae are the most active consumers of veligers is particularly significant because these early larvae usually represent the most vulnerable life stage when mortality rates are the highest. In addition, field data and laboratory experiments indicate that small crustaceans have been largely replaced by *L. fortunei* veligers in diets of fish larvae, especially when veligers are abundant. Selectivity for feeding on veligers was recorded in the field and in laboratory experiments by manipulating prey density. Experiments also demonstrated that *P. lineatus* larvae grew to a significantly larger size with a high concentration of veligers in the diet. This new and abundant food resource appears to have a very important impact on the survival and growth of *P. lineatus* and probably other fish species as well.

Keywords *Limnoperna fortunei* · Golden mussel · Predation by fish · Ecological impact · Trophic interactions · Fish diet · Fish larvae · Veligers

---

E. M. Paolucci (✉)
Facultad de Ciencias Exactas y Naturales,
Instituto de Ecología, Genética y Evolución de Buenos Aires (IEGEBA),
Universidad de Buenos Aires-CONICET, Ciudad Universitaria, 1428 Buenos Aires, Argentina
e-mail: estebanmpaolucci@gmail.com

Museo Argentino de Ciencias Naturales 'Bernardino Rivadavia'-CONICET,
Buenos Aires, Argentina

E. V. Thuesen
Laboratory 1, The Evergreen State College, Olympia, WA 98505, USA
e-mail: thuesene@evergreen.edu

D. Boltovskoy (ed.), *Limnoperna fortunei*, Invading Nature - Springer Series in Invasion Ecology 10, DOI 10.1007/978-3-319-13494-9_12

Introduction: Bivalve Invaders as Prey in Aquatic Ecosystems

One of the most important and extensively studied effects of introduced bivalves is their impact on trophic relationships and food web structures (Karatayev et al. 2007b; Bulté and Blouin-Demers 2008; García-Ortega et al. 2010). For example, several species of adult fish have incorporated the zebra mussel into their diets, but not always with beneficial results (French and Bur 1996; Molloy et al. 1997; García-Ortega et al. 2010). In South America, at least 50 fish species have been recorded to feed on adult *Limnoperna fortunei,* and this mussel has become an important food item for native fish of ecological and economic importance (see Chapter "Trophic Relationships of *Limnoperna fortunei* with Adult Fishes" in this volume; Boltovskoy and Correa 2015). Furthermore, the planktivorous larvae and juveniles of several fish species benefit from very high densities of the planktonic larval stages of *L. fortunei* (Paolucci et al. 2007).

Many planktivorous fish larvae in South American rivers are the product of reproductive migrations, in which mature adults migrate upstream to spawn, after which the larvae drift passively downstream until they reach a marginal wetland (Carolsfield et al. 2004). Even though these species migrate upstream for spawning, other aspects of these movements have species-specific traits, such as the extent and timing of migrations (Welcomme 1979; Fuentes and Espinach Ros 1998). This results in an ichthyoplankton characterized by a mix of different species and different developmental stages. In addition, larval feeding behaviors of these species are also different, depending on their developmental stage and environmental conditions (Rossi 2008). For example, larvae of the sábalo, *Prochilodus lineatus,* migrate from the main channel toward marginal lagoons while depending mostly on their yolk reserves for energy (Rossi 1992; Fuentes and Espinach Ros 1998; Paolucci 2002). Active feeding begins once they have reached marginal lagoons, which serve as nursery areas for several key fish species (Rossi 2008). In contrast, catfish larvae (Pimelodidae) and the larvae and juveniles of several ichthyophagous species start feeding in the channel shortly after hatching (Merigoux and Ponton 1998; Rossi 2001; Makrakis et al. 2008).

Importance of Veligers in Larval Fish Diets

Using samples collected in 1996–1997, Rossi (2008) studied the trophic behavior of larval fish in the main and secondary channels of the Middle Paraná River (Fig. 1a) and found that veligers of *L. fortunei* were consumed by ten fish taxa (Table 1). Of these fishes, pimelodid larvae were the heaviest consumers of this new prey item. On a larger scale study of the impact of these veligers, Paolucci et al. (2007) analyzed fish larvae in the Middle and Lower Paraná during 2000–2001 and in subsidiary marginal lagoons in 2004 (Fig. 1a, b). They reported that *L. fortunei*

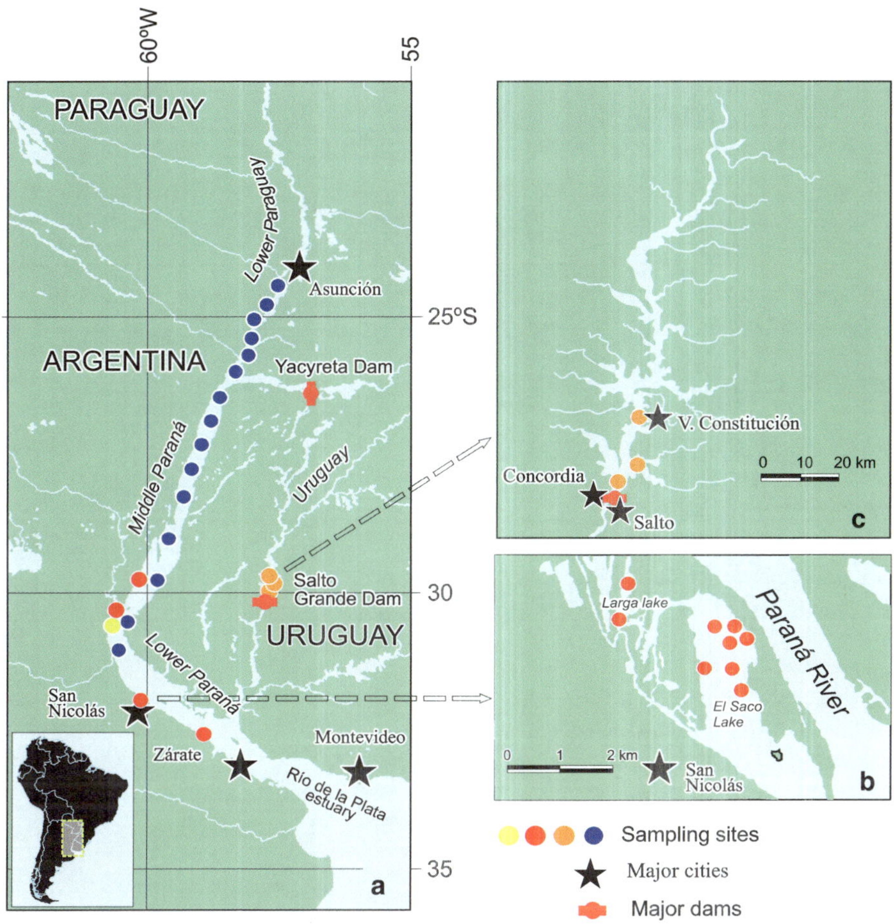

Fig. 1 Location of sampling sites for studies investigating larval fish predation on veligers of *Limnoperna fortunei*. *Red circles*: data from Paolucci et al. (2007). (**a** Lower and Middle Paraná River. **b** Marginal environments); *yellow circle:* data from Rossi (2008); *orange* and *blue circles:* data from Paolucci (2010) (**a** Paraná and Paraguay rivers. **c** Salto Grande Reservoir)

veligers were actively consumed by 11 of 15 larval fish taxa surveyed (Table 1). At that time, *L. fortunei* was recorded in the guts of *P. lineatus, Iheringichthys labrosus, Luciopimelodus pati, Sorubim lima, Pimelodus* spp., and *Parapimelodus valenciennis,* as well as the larvae of other unidentified species of Anostomidae, Doradidae, Characiformes, and Pimelodidae (Fig. 2a–g). These first studies showed that veligers of *L. fortunei* were also present, although at low importance, in the gut contents of piscivorous species, such as *Pseudoplatystoma* sp. and *Rhaphiodon vulpinus*. In a subsequent analysis of the local ichthyoplankton, which included the Lower Paraguay River, Middle Paraná River, and the Salto Grande Reservoir (Fig. 1a, c), the list of consumers of veligers was extended to 18 out of a total of 25 analyzed taxa (Table 1; Paolucci 2010). In the Paraguay River and its marginal

Table 1 Frequency of occurrence (FO) and total biomass (dry weight) of veligers recorded in the gut contents of larval fishes collected in different environments in Argentina

Species/taxa	Total FO (%) [N]	Maximum FO (%)	Total biomass (μg)	Main channel, Middle Paraná River[a]	Main channel, Middle and Lower Paraná River[b]	Marginal lagoons, San Nicolás[b]	Paraná and Paraguay rivers[c]	Salto Grande Reservoir[c]
Sampling period				1996–1997	2000–2001	2004	2005	2005–2009
Siluriformes								
Fam. Pimelodidae								
Pimelodidae NI	64.4 [114]	100.0	321.66	x	x	**X**	x	x
Sorubim lima	85.7 [6]	85.7	14.10	x	**X**			
Parapimelodus valenciennis	47.0 [16]	47.1	46.20		**X**			
Pimelodus spp.	44.9 [40]	50.7	181.67	x	**X**	x	x	
Iheringichthys labrosus	42.1 [32]	68.2	150.10	x	**X**		x	
Pseudoplatystoma spp.	27.6 [13]	29.7	18.80	x	**X**	x		
Luciopimelodus pati	18.7 [3]	100.0	7.38		**X**		x	
Fam. Doradidae	50.0 [1]	50.0	0.70	x		**X**		
Siluriformes NI	10.0 [1]	25.0	0.18				x	**X**
Characiformes								
Fam. Anostomidae	42.8 [6]	50.0	6.07			**X**	x	
Fam. Characidae	25.0 [3]	25.0	2.00	x		**X**		
Aphyocharax sp.	100.0 [1]	100.0	0.67				x	
Fam. Prochilodontidae								
Prochilodus lineatus	22.6 [31]	100.0	42.15	x	**X**	x	x	
Fam. Parodontidae								
Apareiodon affinis	13.7 [8]	46.1	0.19					**X**

Table 1 (continued)

Species/taxa	Total FO (%) [N]	Maximum FO (%)	Total biomass (µg)	Main channel, Middle Paraná River[a]	Main channel, Middle and Lower Paraná River[b]	Marginal lagoons, San Nicolás[b]	Paraná and Paraguay rivers[c]	Salto Grande Reservoir[c]
Sampling period				1996–1997	2000–2001	2004	2005	2005–2009
Fam. Cynodontidae								
Rhapiodon vulpinus	-	-	-	**X**				
Characiformes NI	25.0 [2]	20.0	5.46				**X**	x
Clupeiformes								
Fam. Engraulidae								
Lycengraulis grossidens	4.2 [3]	11.1	0.04					**X**
Pleuronectiformes								
Fam. Achiridae								
Catathyridium jenynsii	73.9 [17]	85.7	3.52				**X**	x
Perciformes								
Fam. Sciaenidae	5.2 [1]	11.1	0.05	x			x	**X**
Total				10	8	7	11	7

Maximum FO was calculated per environment and its corresponding environment is marked in capital bold font
N total number of guts containing veligers of *L. fortunei,* NI not identified
[a] Data of veliger predators are from: Rossi 2008
[b] Data of veliger predators are from: Paolucci et al. 2007
[c] Data of veliger predators are from: Paolucci 2010

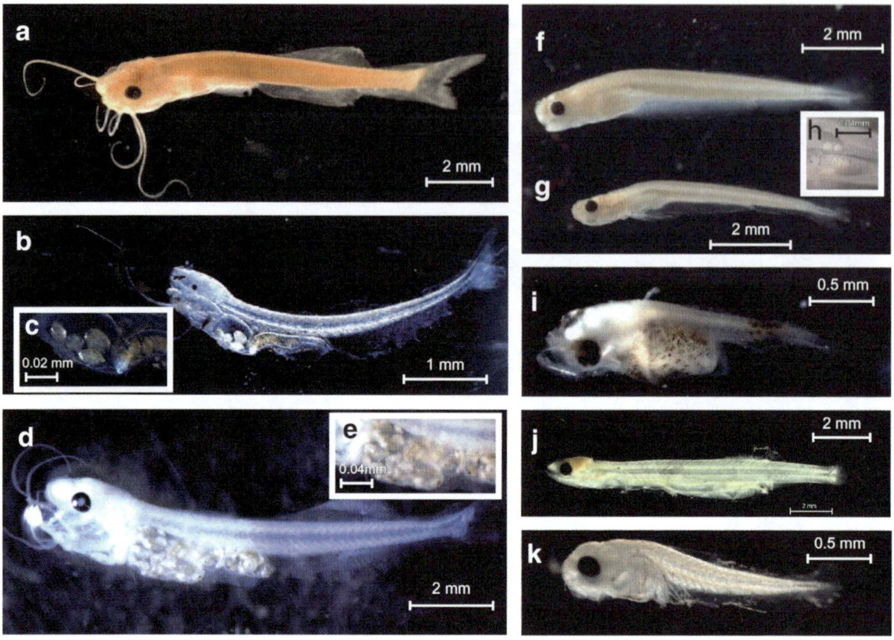

Fig. 2 Some of the larval fish species that feed on *Limnoperna fortunei* veligers. **a** Mesolarva of catfish, *Iheringichthys labrosus*. (**b**, **d**) Pimelodid protolarva. (**c**, **e**, **h**) Veligers of *L. fortunei* in gut contents. **f** *Prochilodus lineatus* protolarva. **g** Anostomid protolarva. **i** Flatfish, *Catathyridium jenynsii*. **j** Anchovy larva, *Lycengraulis grossidens*. **k** Sciaenid larva. (Modified from Paolucci 2010)

and lentic environments, several species of Clupeiformes, Pleuronectiformes, and Perciformes (Fig. 2i–k) were added to the list of predators. The temporal overlap between fish and mussel reproductive periods results in a stable food supply for the larval fishes and is a key factor in this relationship. In comparison, the zebra mussel, *D. polymorpha,* often has a shorter reproductive period (Karatayev et al. 2007a; see Chapter "Parallels and Contrasts between *Limnoperna fortunei* and Species of *Dreissena*" in this volume), and this may be one of the reasons that zebra mussel veligers are much less important in the diets of North American fish larvae (Banard et al. 2006).

Feeding on Veligers by Larval Fish in the Main Channel Versus Marginal Environments

Some larval fish start feeding in the main river channel, while others forage in marginal water-bodies of the alluvial plain, and data from both environments were compared by Paolucci et al. (2007). These authors found that in the main channel of the Paraná River, eight taxa had *L. fortunei* larvae in their guts. Among these, pimelodid larvae such as *I. labrosus*, *L. pati,* and *S. lima* were by far the most

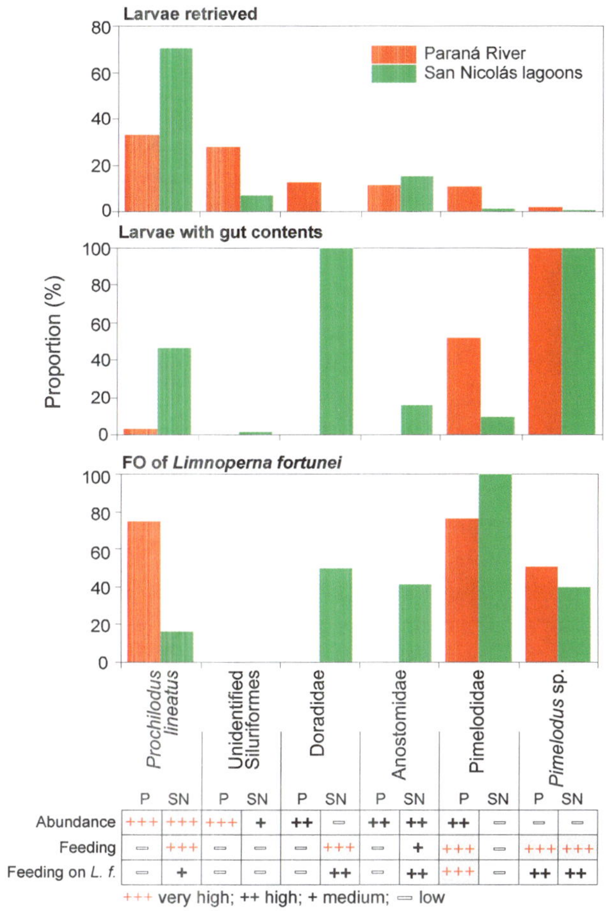

Fig. 3 Taxonomic composition, proportions of larvae with gut contents, and frequency of occurrence (FO) of *Limnoperna fortunei* for the dominant fish taxa (those comprising >94 % of all the larvae retrieved) in the Paraná River and San Nicolás lagoons. (Based on data from Paolucci et al. 2007)

active consumers of veligers (Fig. 3). Among the Characiformes in their study, only *P. lineatus* consumed *L. fortunei* in the main channel; however, veligers were the only identifiable food item recorded for all of these fish larvae. In the San Nicolás lagoons (a marginal environment connected to the river; Fig. 1a), seven out of ten taxa they investigated had *L. fortunei* in their guts. Approximately 20 % of the Characiformes consumed veligers. For the Siluriformes, the proportion was similar (23 %), but these fishes were comparatively scarce in these lagoons (ca. 10 % of all larvae collected).

Differences in taxonomic compositions and feeding activities of larval fish assemblages between the main channel and marginal environments seem to be the main factors affecting the frequency of occurrence (FO) of *L. fortunei* in diets found by Paolucci et al. (2007; Fig. 3). The proportions of feeding larvae that had *L. fortunei* in their gut varied between 20 % (San Nicolás lagoons) and 56 % (Paraná River). In terms of biomass, however, the mean contributions of veligers in larval fish diets was smaller, ranging from 8 % (Paraná) to 12 % (San Nicolás) (mean weighted values), but contributions as high as 100 % were recorded for some taxa in the Paraná River. Larvae of the sábalo, *P. lineatus*, which is the most abundant species in the Paraná-Uruguay watershed (Sverlij et al. 1993; Bonetto 1998), were dominant in both environments (Fig. 3). In the Paraná River, only 3 % had gut contents, while in San Nicolás almost half of the specimens had food in their stomachs. On the other hand, for Siluriformes, proportions of nonempty guts in the Paraná were comparable to those in San Nicolás. Similar observations of differences between the main channel and secondary rivers have been made by Rossi (1992).

Ontogenetic Dietary Shift and Veliger Impact

It is well known that interactions between larval fish and their prey change over time (Lazzaro 1987; Merigoux and Ponton 1998), and field studies on the importance of *L. fortunei* veligers in the diets of larval fishes have also demonstrated ontogenetic changes (Paolucci et al. 2007; Rossi 2008; Paolucci 2010). In absolute terms, the FO and biomass of *L. fortunei* larvae in the guts of proto-, meso-, and metalarvae of mostly pimelodid species collected in the Lower and Middle Paraná and Paraguay rivers had more or less similar values (Paolucci et al. 2007; Paolucci 2010). However, as a percentage of total occurrence or biomass, size-related patterns in the diets of fish larvae were evident (Fig. 4). Protolarvae fed chiefly on *L. fortunei* veligers and cladocerans with fewer copepods and fish larvae (Fig. 4). Mesolarvae consumed veligers, cladocerans, and copepods in similar proportions, but exhibited increased frequencies of fish and insect larvae. Finally, metalarvae consumed veligers and cladocerans less often, but exhibited an increased amount of copepods and fish larvae in their diets. Because the biomass of copepods and fish larvae is 5–10 and 50 times greater, respectively, than that of veligers and cladocerans, the relative importance of veliger biomass dropped from 15 to 45 % in protolarvae and mesolarvae, to only 3 % in the metalarvae (Fig. 4). Thus, as larvae grew, their diet shifted from *L. fortunei* veligers to other larger prey items. Similar trends in the importance of veligers during the development of larval fish were observed by Rossi (2008) for pimelodid species such as *S. lima*, *Pimelodus* sp., and particularly *Pseudoplatystoma cf. corruscans*. These results clearly highlight the importance of *L. fortunei* veligers as prey during the earliest developmental stages of larval fishes. The fact that the earliest fish larvae are the most active consumers of veligers is particularly significant because they usually represent the most vulnerable life history stage where mortality rates are the highest (Elliott and Persson 1978).

Fig. 4 Average frequency of occurrence (FO) and total biomass (% contribution) of major prey items found in protolarvae, mesolarvae, and metalarvae, pooled data for the Paraguay River and Middle Paraná River (*upper panels*), and the Middle and Lower Paraná River and San Nicolás lagoons (*lower panels*). *Letters* denote significant differences between developmental stages at $p<0.01$ (a) or $p<0.05$ (b) (ANOVA, Duncan post-hoc test). (Based on data from Paolucci et al. 2007 and Paolucci 2010)

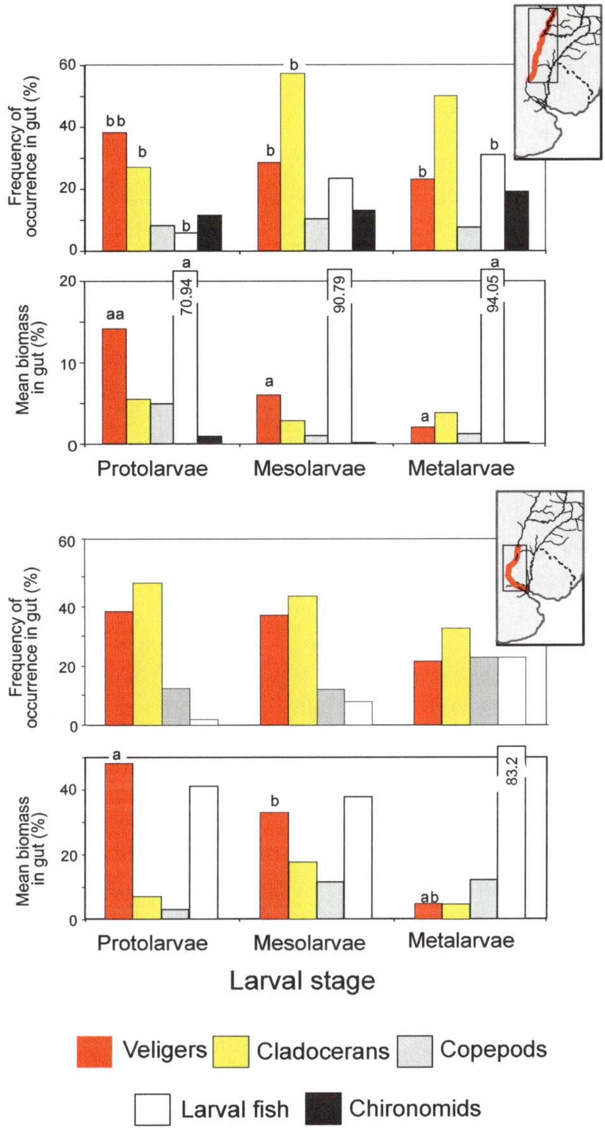

Selectivity for Veligers by Fish Larvae as a Function of Prey Density: Field Evidence

Preliminary analyses of the feeding preferences of native larval fishes, such as *P. lineatus* and anostomid larvae, in marginal environments (Fig. 5a), showed that the veligers of *L. fortunei* were selected positively over cladocerans and copepods (Paolucci et al. 2007). The importance of *L. fortunei* in the diets of larval fishes was up to seven times higher than that expected based on the relative abundance of veligers in the zooplankton. A subsequent study of selectivity and feeding behavior of larval fish collected in the main channel recorded higher impact and preference values for veligers of *L. fortunei* in the Paraná River than in the Paraguay River (Fig. 5b; Paolucci et al. 2015). Comparison of the larval fish diet between the Paraguay and Paraná rivers indicates that average veliger FO was ca. five times more common in fishes caught in the Paraná River (68.4 % of which consumed *L. fortunei* vs.14.2 %; Fig. 6). In the Paraguay River, in contrast, cladocerans were usually favored over other prey, often accounting for significantly higher proportions of all items in the stomachs than in the water-column (Fig. 5b). Selectivity behaviors were not evenly distributed among taxa, but restricted to Characiformes, *C. jenynsii, I. labrosus,* and pimelodid larvae (Fig. 5b). In addition, selectivity toward veligers was positively correlated with the absolute abundance of veligers in the water column (Fig. 6); when more veligers were available, fish larvae relied on them more as prey. As seen in other studies (Deudero and Morales-Nin 2001; Graeb et al. 2004; Fulford et al. 2006), veliger density plays a central role in selectivity by the predators, mainly due to an increase in the rate of predator-prey encounters.

Effects of Developmental Stage and Veliger Density on Selectivity: Experimental Evidence

As in field studies for other fish species, comparisons of gut contents versus available food determined in laboratory experiments demonstrated that selectivity behavior was a function of prey density and the developmental stage of *P. lineatus* (Paolucci 2010; Paolucci et al. 2010a). For protolarvae, average proportions of veligers, small cladocerans, and nauplii were always higher in the gut contents than in the prey offered indicating selectivity for these items (positive values in Fig. 7) (small cladocerans and nauplii were the main prey of *P. lineatus* larvae before *L. fortunei* was present in South America; Rossi 1992). Mesolarvae preyed on veligers selectively only when these were very abundant in the experimental tanks (enriched veliger concentrations in Fig. 7); however, when veliger concentrations dropped, they were consumed less selectively, with gut contents yielding lower proportions of veligers. Mesolarvae were especially efficient at consuming small and medium-sized cladocerans, whose selectivity indices were almost invariably posi-

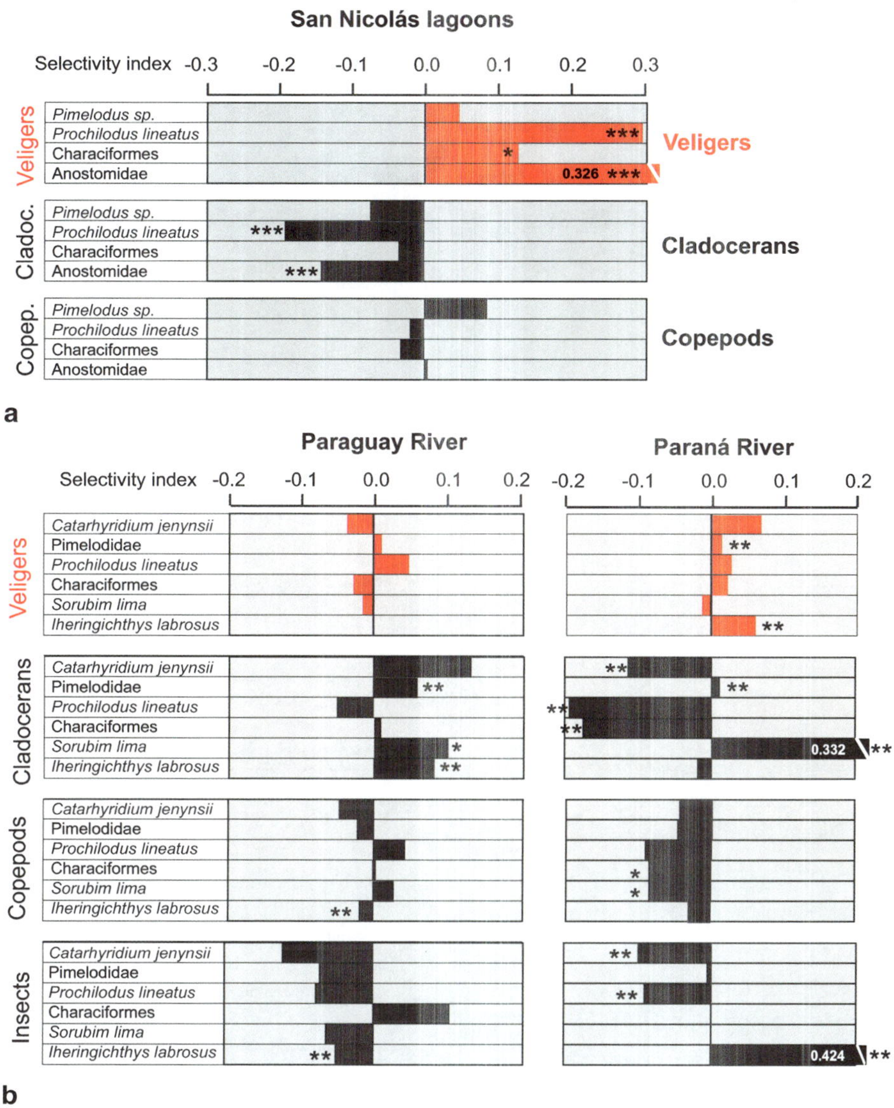

Fig. 5 Mean selectivity index for the prey items assessed for the most abundant fish species present in San Nicolás lagoons (**a**) and in the Paraguay and Paraná rivers (**b**). *Asterisks* denote significant differences between proportions of the corresponding prey in the water column and in the gut contents at $p<0.05$ (*), $p<0.01$ (**), or $p<0.001$ (***) (Chi-square tests). (Based on data from Paolucci et al. 2007 and Paolucci et al. 2015)

tive and statistically significant (Fig. 7). In contrast with younger larvae, metalarvae never selected veligers, regardless of their concentration in the experimental tanks, and they clearly favored small and medium-sized cladocerans. For *P. lineatus*, as

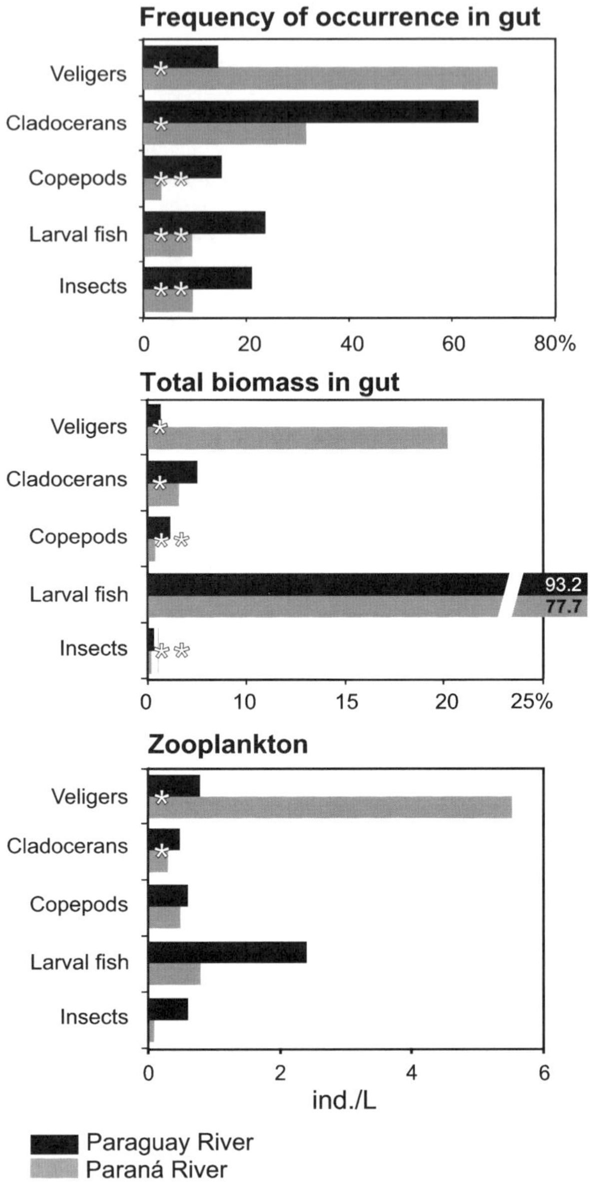

Fig. 6 Contribution of the five main food items to the diet of fish larvae and zooplankton composition recorded in the Paraguay and Paraná Rivers. Statistically significant differences between the two rivers are denoted with *asterisks* (* $p<0.05$; ** $p<0.01$, Kruskal-Wallis test). (Modified from Paolucci et al. 2015)

well as for pimelodid larvae, the highest positive selectivity values were also associated with the highest proportions of veligers in the zooplankton.

Laboratory results and available field data for these fish species (Paolucci et al. 2007; Rossi 2008; Paolucci et al. 2010a) indicate that small crustaceans have been largely replaced by veligers, especially when veligers are abundant. This dietary switch, however, is restricted to the earliest larvae. As fishes grow larger and de-

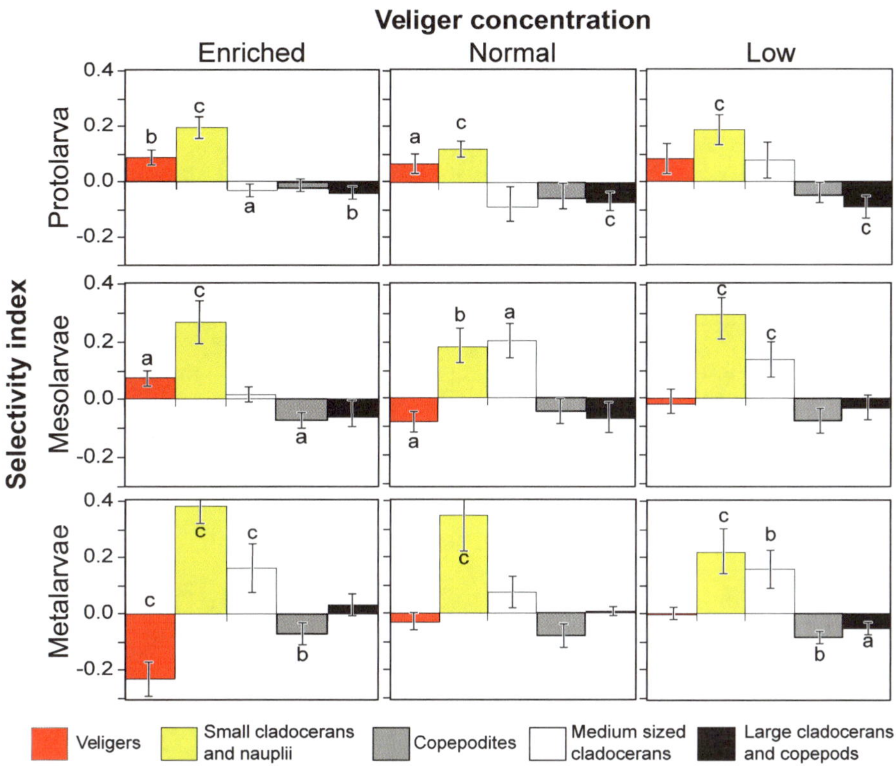

Fig. 7 Average selectivity index per prey item assessed for protolarvae ($n = 15$), mesolarvae ($n = 15$), and metalarvae ($n = 15$) in three experimental settings (*enriched, normal,* and *low concentrations* of veligers). Error bars denote confidence intervals at $p < 0.05$. *Letters* denote significant differences between available prey and gut contents at $p < 0.05$ (**a**), $p < 0.01$ (**b**), or $p < 0.001$ (**c**) (Chi-square tests) (Modified from Paolucci et al. 2010a)

velop a larger mouth gape (Fig. 8a), the relative importance of large crustaceans increases. The replacement of crustaceans by veligers of *L. fortunei* in the diet of larval fishes is chiefly due to the fact that veligers are poor swimmers, with limited neuromuscular coordination and poor predator-avoidance behavior compared to crustacean zooplankton. Cladocerans, in turn, are slower and less agile than copepods. This gradient in predator-avoidance capabilities seems to be the main prey-selection factor during the earliest life stages of a fish. Indeed, for many fish species mollusc larvae have been reported to be preferred over crustaceans (Pepin and Penney 1997; Lehtiniemi et al. 2007), and cladocerans are generally preferred over copepods (Cooper and Goldman 1980; Vanderploeg et al. 1982; Clarke et al. 2004), sometimes regardless of prey size (Werner 1974). However, it is worth noting that several of these fish species also prey on the juvenile and adult stages of *L. fortunei* when they reach a larger developmental stage (see Chapter "Trophic Relationships of *Limnoperna fortunei* with Adult Fishes" in this volume).

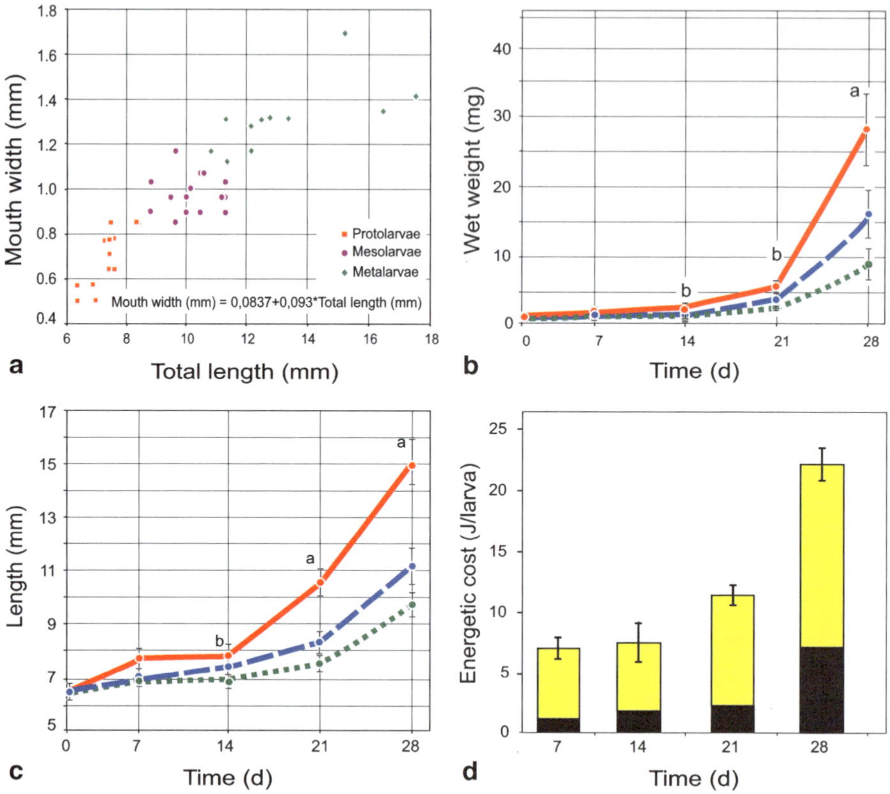

Fig. 8 **a** Correlation between total length and mouth width of the *Prochilodus lineatus* larvae used in the selectivity experiments ($R=0.917$; $p<0.01$; $n=45$; modified from Paolucci et al. 2010a). Growth as **b** wet weight (mg) and **c** total length (mm) of larval fish fed three different diets: veliger-enriched (*solid red line*), natural (*dashed blue line*), and low-veliger (*dotted green line*). *Vertical bars* denote 95 % confidence intervals. *Letters* denote significant differences. (ANOVA, $p<0.01$) as **a** between all diets; or (**b**) between at least one pair of diets. **d** weekly energetic costs of growth (*black bars*) and standard metabolic rate (*yellow bars*) according to the mean weight of the larval fish used in respiration experiments. Error bars denote 95 % confidence intervals of the combined energetic costs. (**b**, **c**, and **d** modified from Paolucci et al. 2010b)

Effects on Growth of Larval Fishes

Even in fish that are important predators of bivalves, some negative effects have been reported (French and Bur 1996; Nagelkerke and Sibbing 1996; Magoulick and Lewis 2002). These have been explained by the low caloric content of bivalve larvae compared to traditional food items, mainly due to the presence of shells that are not assimilated. However, growth experiments carried out using newly hatched *P. lineatus* larvae (with yolk-sac absorption completed) and different veliger concentrations showed positive rather than negative effects (Paolucci et al. 2010b). This experimental approach not only supported the idea that this new and abundant resource is selectively preyed upon by this larval fish, but also demonstrates that veligers of *L. fortunei* can significantly enhance the growth of *P. lineatus* larvae. Different veliger concentrations had significant effects on growth, for both total length and wet weight, of *P. lineatus* larvae; those fed a veliger-enriched diet had the highest growth performance, followed by those fed natural and low-veliger diets (Fig. 8b, c).

Similar to that observed in other studies with larval fishes (Halver 2001; Teshima et al. 2004), the enhanced growth rates observed by Paolucci et al. (2010b) may depend on the biochemical composition and caloric content of the veligers, as well as the energy costs involved in prey capture. Chemical composition analyses showed high protein and lipid contents for veligers with lipid contents being higher than those of cladocerans and copepods. This combination of high protein and fat contents, like that found in veligers of *L. fortunei,* has been highlighted by several authors as important in the diet of larval fish (Sargent et al. 1999; Lazo 2000; Rønnestad et al. 2007). While protein is the most important body component and accounts for over 50 % of the ash free dry weight (AFDW) in these organisms, lipids provide necessary energy during the fast-paced larval fish development period. Mostly as a consequence of high lipid content (17 % of the AFDW), a significantly higher specific caloric content was found in veligers also (24.88 ± 1.81 kJ/g dry weight) followed by cladocerans and copepods (Paolucci et al. 2010b). In addition to high energy density, veligers had a higher dry biomass than crustacean prey of the same or greater total length, and consequently veligers had comparatively higher total energy content sufficient to support the costs of growth and standard metabolic rate (Fig. 8d).

The energy density of veligers of *L. fortunei* is slightly higher than that recorded for adults of the invasive bivalve *Dreissena polymorpha* and other bivalve larvae (between 17.3 and 22.7 kJ/g) (Blaber 1979). In addition to the biochemical composition, the reduced energetic costs associated with the capture of slower prey, such as veligers, in comparison with faster prey, such as cladocerans, and especially copepods, may also have had an effect on larval growth. The results of these physiological studies combined with observations made during experimental and field investigations imply that selective feeding on slow and easy-to-capture prey results in a lower energetic cost of feeding and can result in a positive energetic impact that could enhance growth rates of larval fishes (Lazzaro 1987; Lankford and Targett 1997).

Impacts at Population and Community Levels

All these results suggest that fish species whose larvae have been observed to feed on veligers of *L. fortunei* have greatly benefited from its presence, and the impact of this new resource on fish populations is most likely very important. It is noteworthy that fish species whose diets rely heavily on *L. fortunei* are among the most abundant and ecologically important in the Paraná-Paraguay river system (Sverlij et al.

1993; Espinach Ros and Fuentes 2001). For example, deposit-feeding adults of *P. lineatus* constitute the main food item of larger ichthyophagous species (Sverlij et al. 1993). Thus, feeding conditions for *P. lineatus* may strongly affect abundances of many other fish species. Consequently, the effects of these shifts in the feeding behavior of larval fishes are conceivably not restricted to the organisms directly involved in the interactions, but may have cascading effects both up and down trophic webs (MacIsaac et al. 1999; Yan et al. 2001; Clarke et al. 2004). Insofar as the new interactions modify established grazing pressures, they can strongly affect species composition and size structure of the zooplankton community, which in turn may change phytoplankton abundance and composition (Strecker and Arnott 2008). Indirect impacts on other fishes may also derive from these ecological rearrangements through the direct consumption of veligers (Paolucci et al. 2007), or from changes in the availability of other food items.

Larval Fish Predation as Biological Control

Whether or not grazing on veligers by larval fish is able to curtail the growth of mussel populations is a question of major interest. Sylvester et al. (2007) suggested that adult mussels will not be controlled by fish predation, but a similar estimate for veligers is more complicated. Several key elements remain unknown; in particular, the reproductive output of *L. fortunei* on a basin-wide scale is unclear. Assuming conservative densities of 1000 ind/m^3 for veligers (Boltovskoy et al. 2009) and around 3 ind/m^3 for fish larvae (Fuentes and Espinach Ros 1998), and an ingestion rate of 2 veligers/h (Paolucci et al. 2010a), one could speculate that on a steady-state basis *P. lineatus* consumes daily between 10 and 20% of the standing stock of veligers. This figure could probably be doubled to include other species of fish that consume veligers (Paolucci et al. 2007; Paolucci et al. 2010a). However, the reproductive period of *L. fortunei* (around September–April) is much longer than that of most fishes (between November–December and February–March (cf. Rossi et al. 2007), and this must decrease the long-term impact significantly. Thus, although these estimates are very rough, they agree with the conclusion of Sylvester et al. (2007) who concluded that predation impact on the geographical spread of *L. fortunei* is probably minor, and the potential for predation to control the spread of the bivalve seems limited.

References

Banard C, Martineau C, Frenette JJ, Dodson JJ, Vincent WF (2006) Trophic position of zebra mussel veligers and their use of dissolved organic carbon. Limnol Oceanogr 51:1473–1484

Blaber SJM (1979) The biology of filter feeding teleosts in Lake St Lucia, Zululand. J Fish Biol 15:37–59

Boltovskoy D, Correa N (2015) Ecosystem impacts of the invasive bivalve *Limnoperna fortunei* (golden mussel) in South America 746:81–95

Boltovskoy D, Sylvester F, Otaegui A, Leytes V, Cataldo D (2009) Environmental modulation of the reproductive activity of the invasive mussel *Limnoperna fortunei* in South America. Austral Ecol 34:719–730

Bonetto AA (1998) Panorama sinóptico sobre la ictiofauna, la pesca y la piscicultura en los ríos de la cuenca del Plata con especial referencia al Paraná. Revista de Ictiología 6:3–15

Bulté G, Blouin-Demers G (2008) Northern map turtles (*Graptemys geographica*) derive energy from the pelagic pathway through predation on zebra mussels (*Dreissena polymorpha*). Freshw Biol 53:497–508

Carolsfield J, Harvey B, Ross C, Baer A (2004) Migratory fish of South America: biology, fisheries and conservation status. World Bank, Victoria, pp 1–380

Clarke LR, Letizia PS, Bennett DH (2004) Autumn-to-Spring energetic and diet changes among Kokanee from North Idaho lakes with and without *Mysis relicta*. N Am J Fish Manag 24:597–608

Cooper SD, Goldman CR (1980) Opossum shrimp (*Mysis relicta*) predation on zooplankton. Can J Fish Aquat Sci 37:909–919

Deudero S, Morales-Nin B (2001) Prey selectivity in planktivorous juvenile fishes associated with floating objects in the western Mediterranean. Aquac Res 32:481–490

Elliott JM, Persson L (1978) The estimation of daily rates of food consumption for fish. J Animal Ecol 47:977–991

Espinach Ros A, Fuentes CM (2001) Recursos pesqueros y pesquerías de la Cuenca del Plata. In: Bezzi S, Akselman R, Boschi EE (eds) Síntesis del estado de las pesquerías marítimas argentinas y de la Cuenca del Plata. Años 1997–1998. Instituto Nacional de Investigación y Desarrollo Pesquero, Mar del Plata, pp 353–358

French JRP, Bur MT (1996) The effect of zebra mussel consumption on growth of freshwater drum in Lage Erie. J Freshw Ecol 11:283–289

Fuentes CM, Espinach Ros A (1998) Variación de la actividad reproductiva del sábalo, *Prochilodus lineatus* (Valenciennes, 1847), estimada por el flujo de larvas en el rıo Paraná Inferior. Nat Neotrop 29:25–32

Fulford RS, Rice JA, Miller TJ, Binkowski FP, Dettmers JM, Belonger B (2006) Foraging selectivity by larval yellow perch (*Perca flavescens*): implications for understanding recruitment in small and large lakes. Can J Fish Aquat Sci 63:28–42

García-Ortega A, Verreth J, Vermis K, Nelis HJ, Sorgeloos P, Verstegen M (2010) Laboratory investigation of daily food intake and gut evacuation in larvae of African catfish *Clarias gariepinus* under different feeding conditions. Aquac Int 18:119–134

Graeb BDS, Dettmers JM, Wahl DH, Cáceres CE (2004) Fish size and prey availability affect growth, survival, prey selection, and foraging behavior of larval yellow perch. Transactions Am Fish Soc 133:504–514

Halver JE (2001) My 50 years in fish nutrition, 1949–99. Aquac Res 32:615–622

Karatayev AY, Boltovskoy D, Padilla DK, Burlakova LE (2007a) The invasive bivalves *Dreissena polymorpha* and *Limnoperna fortunei*: parallels, contrasts, potential spread and invasion impacts. J Shellfish Res 26:205–213

Karatayev AY, Padilla DK, Minchin D, Boltovskoy D, Burlakova LE (2007b) Changes in global economies and trade: the potential spread of exotic freshwater bivalves. Biol Invasions 9:161–180

Lankford TE, Targett TE (1997) Selective predation by juvenile weakfish: post-consumptive constraints on energy maximization and growth. Ecology 78:1049–1061

Lazo J (2000) Conocimiento actual y nuevas perspectivas en el desarrollo de dietas para larvas de peces marinos. In: Avances en Nutrición Acuícola. V Simposium Internacional de Nutrición Acuícola, Mérida (Venezuela)

Lazzaro X (1987) A review of planktivorous fishes: their evolution, feeding behaviours, selectivities, and impacts. Hydrobiologia 146:97–167

Lehtiniemi M, Hakala T, Saesmaa S, Viitasalo M (2007) Prey selection by the larvae of three species of littoral fishes on natural zooplankton assemblages. Aquat Ecol 41:85–94

MacIsaac HJ, Grigorovich IA, Hoyle JA, Yan ND, Panov VE (1999) Invasion of Lake Ontario by the Ponto–Caspian predatory cladoceran *Cercopagis pengoi*. Can J Fish Aquat Sci 56:1–5
Magoulick DD, Lewis LC (2002) Predation on exotic zebra mussels by native fishes: effects on predator and prey. Freshw Biol 47:1908–1918
Makrakis MC, Nakatani K, Bialetzki A, Gomes LC, Sanches PV, Baumgartner G (2008) Relationship between gape size and feeding selectivity of fish larvae from a neotropical reservoir. J Fish Biol 72:1690–1707
Merigoux S, Ponton D (1998) Body shape, diet and ontogenetic diet shifts in young fish of the Sinnamary River, French Guiana, South America. J Fish Biol 52:556–569
Molloy DP, Karatayev AY, Burlakova EB, Kurandina DP, Laruelle F (1997) Natural enemies of zebra mussels: predators, parasites, and ecological competitors. Rev Fish Sci 5:27–97
Nagelkerke LAJ, Sibbing FA (1996) Efficiency of feeding on zebra mussel (*Dreissena polymorpha*) by common bream (*Abramis brama*), white bream (*Blicca bjoerkna*), and roach (*Rutilus rutilulus*): the effects of morphology and behavior. Can J Fish Aquat Sci 53:2847–2861
Paolucci EM (2002) Actividad reproductiva del sábalo *Prochilodus lineatus* (Valenciennes, 1847) y otras especies de interés comercial y deportivo en el río Uruguay inferior, estimada por la abundancia de estadios larvales en la deriva. MSc Thesis, Universidad de Buenos Aires (Argentina), pp 1–74
Paolucci EM (2010) Impacto del molusco invasor *Limnoperna fortunei* sobre el ecosistema: interacción trófica y efectos sobre las larvas de peces nativos. PhD Thesis, Universidad de Buenos Aires (Argentina), pp 1–162
Paolucci EM, Cataldo DH, Fuentes CM, Boltovskoy D (2007) Larvae of the invasive species *Limnoperna fortunei* (Bivalvia) in the diet of fish larvae in the Paraná River, Argentina. Hydrobiologia 589:219–233
Paolucci EM, Cataldo DH, Boltovskoy D (2010a) Prey selection by larvae of *Prochilodus lineatus* (Pisces): indigenous zooplankton versus larvae of the introduced bivalve *Limnoperna fortunei*. Aquat Ecol 44:255–267
Paolucci EM, Thuesen EV, Cataldo DH, Boltovskoy D (2010b) Veligers of an introduced bivalve (*Limnoperna fortunei*) are a new food resource that enhances growth of larval fish in the Paraná River (South America). Freshw Biol 55:1831–1844
Paolucci EM, Almada P, Cataldo D, Boltovskoy D (2015) Native fish larvae take advantage of introduced mussel larvae: field evidence of feeding preferences on veligers of the introduced freshwater bivalve *Limnoperna fortunei*. Hydrobiologia 745:211–224
Pepin P, Penney RW (1997) Patterns of prey size and taxonomic composition in larval fish: are there general size-dependent models? J Fish Biol 51:84–100
Rønnestad I, Kamisaka Y, Conceição LEC, Morais S, Tonheim SK (2007) Digestive physiology of marine fish larvae: hormonal control and processing capacity for proteins, peptides and amino acids. Aquaculture 268:82–97
Rossi LM (1992) Evolucion morfológica del aparato digestivo de postlarvas y prejuveniles de *Prochilodus lineatus* (Val., 1847) (Pisces, Curimatidae) y su relacion con la dieta. Revue d'Hydrobiologie Tropicale 25:159–107
Rossi LM (2001) Ontogenetic diet shifts in a neotropical catfish, *Sorubim lima* (Schneider) from the River Paraná System. Fish Manag Ecol 8:141–152
Rossi LM (2008) Ecologia trófica de peces de importancia económica del Río Parana durante su ontogenia temprana. PhD Thesis, Universidad Nacional de La Plata (Argentina), pp 1–174
Rossi L, Cordiviola E, Parma MJ (2007) Fishes. In: Iriondo MH, Paggi JC, Parma MJ (eds) The middle Parana River. Limnology of a subtropical wetland. Springer, Berlin, pp 305–325
Sargent J, Bell G, McEvoy L, Tocher D, Estevez A (1999) Recent developments in the essential fatty acid nutrition of fish. Aquaculture 177:191–199
Strecker AL, Arnott SE (2008) Invasive predator, *Bythotrephes*, has varied effects on ecosystem function in freshwater lakes. Ecosystems 11:490–503
Sverlij SB, Espinach Ros A, Orti G (1993) Sinopsis de los datos biológicos y pesqueros del sábalo *Prochilodus lineatus* (Valenciennes, 1847). Food and Agriculture Organization, Rome, pp 1–64

Sylvester F, Boltovskoy D, Cataldo D (2007) Fast response of freshwater consumers to a new trophic resource: Predation on the recently introduced Asian bivalve *Limnoperna fortunei* in the lower Parana River, South America. Austral Ecology 32:403–415

Teshima SI, Koshio S, Ishikawa M, Alam MS, Hernandez LHH (2004) Effects of protein and lipid sources on the growth and survival of red sea bream *Pagrus major* and Japanese flounder *Paralichthys olivaceus* receiving micro-bound diets during larval and early juvenile stage. Aquac Nutr 10:279–287

Vanderploeg HA, Bowers JA, Chapelski O, Soo HK (1982) Measuring in situ predation by *Mysis relicta* and observations on underdispersed microdistributions of zooplankton. Hydrobiologia 93:109–119

Welcomme RL (1979) Fisheries ecology of floodplain rivers. Longman, London, pp 1–317

Werner EE (1974) The fish size, prey size, handling time relation in several sunfishes and some implications. J Fish Res Board Can 31:1531–1536

Yan ND, Blukacz A, Sprules WG, Kindy PK, Hackett D, Girard RE, Clark BJ (2001) Changes in zooplankton and the phenology of the spiny water flea, *Bythotrephes*, following its invasion of Harp Lake, Ontario, Canada. Can J Fish Aquat Sci 58:2341–2350

Trophic Relationships of *Limnoperna fortunei* with Adult Fishes

Daniel Cataldo

Abstract In South America, the inventory of species that feed on *Limnoperna fortunei* has increased steadily; in 2006, 18 fish species had been identified as predators of *L. fortunei*, whereas 7 years later, the list had grown to almost 50 species. In some areas, fishes that consume *L. fortunei* represent >50 % of the species regularly present in commercial fisheries, including traditionally omnivorous, iliophagous, and ichthyophagous forms. Several economically important species have significantly changed their feeding habits since the mussel's introduction, shifting from a diet based on plants and detritus to one dominated by adult mussels. Consumption of golden mussels is not restricted to fishes provided with teeth that can crush and grind the shells; many toothless species swallow whole specimens or nibble on the exposed siphons and mantle edges of the bivalve. Golden mussels can account for up to 100 % of the gut contents of some fish species, especially during the summer. Feeding of fishes on *L. fortunei* often involves the selective consumption of the smaller mussel size classes. Fish predation pressure on the mussel is likely high and it probably represents the most significant mechanism that modulates *L. fortunei* populations, but it is very unlikely to eradicate the mussel altogether. No comprehensive, large-scale studies are yet available on the effects of this new food supply on local fish stocks, but ancillary information suggests that these effects are likely very significant. Impacts are not restricted to species that consume the mollusc, but also affect species that benefit from this new food resource indirectly, including large ichthyophagous species feeding on molluscivorous forms, as well as on those that consume the organic matter-enriched sediments by the mussel's feces and pseudofeces.

Keywords *Limnoperna fortunei* · Golden mussel · Predation by fish · Ecological impact · Trophic interactions · Fish diet

D. Cataldo (✉)
Facultad de Ciencias Exactas y Naturales, Instituto de Ecología, Genética y Evolución de Buenos Aires (IEGEBA), Universidad de Buenos Aires-CONICET, Ciudad Universitaria, 1428 Buenos Aires, Argentina
e-mail: daniel@ege.fcen.uba.ar

Museo Argentino de Ciencias Naturales 'Bernardino Rivadavia'-CONICET, Buenos Aires, Argentina

D. Boltovskoy (ed.), *Limnoperna fortunei*, Invading Nature - Springer Series in Invasion Ecology 10, DOI 10.1007/978-3-319-13494-9_13

Introduction

The fish fauna of the Río de la Plata basin comprises around 500 species (Bonetto 1986), including many large species with significant economic value. From the point of view of their mode of feeding, these species have historically been attributed to a few classical categories: omnivorous, iliophagous/detritivorous, planktophagous, and ichthyophagous, but the boundaries between these categories are diffuse and change with fish age, habitat, and time of the year, among others. These trophic shifts suggest that feeding preferences are flexible and often depend on the most profitable resource available.

Before the introduction of the Asian clam *Corbicula fluminea* (probably in the 1970s: Ituarte 1994), and the golden mussel *Limnoperna fortunei*, around 1990 (Pastorino et al. 1993), molluscs were generally a minor component of fish diets (Gneri and Angelescu 1951; Bonetto et al. 1963; Ringuelet et al. 1967; Alonso 1978; Oliva et al. 1981). The invasion of these bivalves, and particularly of *L. fortunei*, in the Río de la Plata watershed confirmed the behavioral and trophic adaptability of South American fishes: many species originally included in either of the trophic categories above are presently known to actively feed on these invasive bivalves, especially *L. fortunei*. The feeding modes used by different species to prey on *L. fortunei* are diverse, as most probably are the energetic benefits involved.

The inventory of species that feed on *L. fortunei* has increased steadily, in part due to new surveys, and in part because of the mussel's geographic expansion. In 2006, 18 fish species had been identified as predators of *L. fortunei* in South American inland waters (García and Montalto 2006), whereas 7 years later, the list had grown to almost 50 species (Table 1). In some areas, fishes that consume *L. fortunei* represent >50 % of the species regularly present in commercial fisheries.

This chapter reviews our current knowledge of the consumption of adult *L. fortunei* by fishes. It furnishes an overview of the species known to feed on the mussel, as well as an account of the techniques employed by species with dissimilar anatomical adaptations to feed on this prey. The potential impact of this resource on local fish assemblages is discussed. The chapter is restricted to information from Argentine and Brazilian waterbodies; no data on dietary changes have been reported from the other countries invaded by the mussel.

Omnivorous Species

Omnivorous fishes are characterized by their high trophic plasticity, consuming a variety of items and often favoring widely available resources. In the Río de la Plata watershed, one of the typical and abundant representatives of this group is *Leporinus obtusidens* (locally known as "boga," in Argentina, and "piapara" or "piava," in Brazil). Originally, the diet of this species chiefly included aquatic plants, seeds, and to a lesser extent smaller fishes and molluscs (Mastrarrigo 1950;

Table 1 Fish species known to consume *L. fortunei*. Geographic areas are *GW* Guaíba watershed (Patos and Mirim lagoons, São Gonçalo Channel), *LP* Lower Paraná River, *MP* Middle Paraná River, *RP* Río de la Plata estuary, *UP* Upper Paraná River, *UP(I)* Upper Paraná, Itaipu Reservoir, *UR* Uruguay River. Bold species names denote fishes that consistently use *L. fortunei* in their diet in sizable numbers

Species	Area	References
Astyanax aff. fasciatus	GW	Lopes and Vieira (2012)
Auchenipterus osteomystax	UP(I)	Oliveira et al. (2010)
Brochiloricaria chauliodon	RP	García and Protogino (2005)
Crenicichla punctata	GW	Lopes and Vieira (2012)
Cyphocharax voga	GW	Lopes and Vieira (2012)
Cyprinus carpio	LP-RP	Cataldo et al. (2002)
Geophagus brasiliensis	GW	Lopes and Vieira (2012)
Hoplias malabaricus	UP(I)-GW	Lopes and Vieira (2012); Oliveira et al. (2010)
Hypostomus cf. laplatae	GW	Montalto et al. (1999)
Hypostomus commersoni	UP(I)-GW	Lopes and Vieira (2012); Oliveira et al. (2010)
Hypostomus regani	UP(I)	Oliveira et al. (2010)
Hypostomus ternetzi	UP(I)	Oliveira et al. (2010)
Hypostomus uruguayensis	LP	Boltovskoy and Cataldo (1999)
Iheringichthys labrosus	UP(I)-UR	Oliveira et al. (2010); Masdeu et al. (2011); Belz et al. (2012)
Leporinus friderici	UP(I)	Oliveira et al. (2010)
Leporinus macrocephalus	UP(I)	Oliveira et al. (2010)
Leporinus obtusidens	UP-MP-LP-RP	Montalto et al. (1999); Boltovskoy and Cataldo (1999)
Loricaria Loricaria nudiventris	LP-RP	Cataldo et al. (2002)
Loricaria Loricaria vetula	LP-RP	Cataldo et al. (2002)
Loricariichthys anus	GW	Lopes and Vieira (2012)
Megalancistrus aculeatus	UP	Belz et al. (2012)
Megalancistrus parananus	UP(I)	Oliveira et al. (2010)
Metynnis lippincotianus	UP(I)	Oliveira et al. (2010)
Micropogonias furnieri	GW-RP	López Armengol and Casciotta (1998); Lopes and Vieira (2012)

Table 1 (continued)

Species	Area	References
Oxydoras kneri	LP-RP	Cataldo et al. (2002)
Paraloricaria cf. vetula	LP-RP	Boltovskoy and Cataldo (1999); García and Protogino (2005)
Parauchenipterus galeatus	UP(I)	Oliveira et al. (2010)
Piaractus mesopotamicus	UP(I)	Lösch et al. (2009); Oliveira et al. (2010)
Pimelodus albicans	LP-MP-RP	Montalto et al. (1999); Boltovskoy and Cataldo (1999)
Pimelodus maculatus	UP-MP-LP-RP-GW	Montalto et al. (1999); Cataldo et al. (2002); Baptista and Zibetti (2006)
Pimelodus pintado	GW	Lopes and Vieira (2012); Vieira and Lopes (2013)
Pimelodus sp.	MP	Montalto et al. (1999)
Pirinampus pirinampu	UP(I)	Oliveira et al. (2010)
Plagioscion squamosissimus	UP(I)	Oliveira et al. (2010)
Potamotrygon cf. brachiurus	MP	Montalto et al. (1999)
Potamotrygon motoro	UP(I)	Oliveira et al. (2010)
Prochilodus lineatus	UP(I)	Oliveira et al. (2010)
Pterodoras granulosus	UP-MP-LP-UR- RP-GW	Boltovskoy and Cataldo (1999); Montalto et al. (1999); Ferriz et al. (2000); Cataldo et al. (2002); García and Protogino (2005); Cantanhêde et al. (2008); Vermulm and Giamas (2008); González-Bergonzoni et al. (2010); Oliveira et al. (2010); Belz et al. (2012)
Pterygoplichthys anisitsi	UP(I)	Oliveira et al. (2010)
Rhamdia quelen	UP(I)-GW	Lösch et al. (2009); Lopes and Vieira (2012)
Rhinodoras dorbignyi	MP-RP	Montalto et al. (1999) García and Protogino (2005)
Rineloricaria microlepidogaster	GW	Lopes and Vieira (2012)
Rineloricaria strigilata	GW	Lopes and Vieira (2012)
Satanoperca pappaterra	UP(I)	Oliveira et al. (2010)
Schizodon borellii	UP(I)-MP	Montalto et al. (1999); Oliveira et al. (2010)
Serrasalmus maculatus	UP(I)	Oliveira et al. (2010)
Serrasalmus marginatus	UP(I)	Oliveira et al. (2010)

Ringuelet et al. 1967). However, after *L. fortunei* became available, *L. obtusidens* has largely switched its diet to mussels that now represent 64–100 % of its gut contents (Montalto et al. 1999; Penchaszadeh et al. 2000; Cataldo et al. 2002). This species has strong teeth that allow it to bite, tear off, and grind mussel valves (Braga 1993). Other related species that have also been reported to consume *L. fortunei* are *Leporinus macrocephalus, Leporinus friderici,* and *Schizodon borellii* (Table 1).

Piaractus mesopotamicus is a large characid native to the Paraná-Paraguay river system that is widely cultured in Brazil, Argentina, and Paraguay. It has strong molar teeth adapted to crushing and fragmenting its food, which allows it to widely utilize *L. fortunei* in its diet. When raised in fishnet cages in lakes and reservoirs colonized by *L. fortunei*, the mussel can represent a major nuisance because it grows on the nets and clogs them rapidly. On the other hand, these mussels may represent a major source of food for the enclosed fishes that have been observed to feed on them exclusively (Lösch et al. 2009).

Another group of very abundant typically omnivorous fishes, known by the vernacular name of "armado" (Argentina) or "armal" (Brazil), have also benefited from the new food resource represented by *L. fortunei*. *Pterodoras granulosus,* one of the most widely distributed representatives of this group, is a euryphagous species which feeds on most widely available items (Panatieri and Del Barco 1981). After the invasion of *L. fortunei*, this species has changed its diet significantly, and mussels now represent up to 82–100 % of the biomass of its gut contents. Up to 2830 shells of *L. fortunei* have been recorded in a single digestive tract of *P. granulosus* (Montalto et al. 1999; Cataldo et al. 2002; García and Protogino 2005). *P. granulosus* lacks strong dentition, and therefore shells are swallowed whole. Other related species, including *Rhinodoras dorbignyi* and *Oxidoras kneri*, have also been reported to consume golden mussels (Montalto et al. 1999; Cataldo et al. 2002).

Among the benthic, euryphagous species, catfishes (*Pimelodus maculatus, Pimelodus albicans, Pimelodus argenteus*) are the most abundant in the Río de la Plata watershed. Their small oral villiform teeth cannot crush mussel shells, which are thus ingested whole (Montalto et al. 1999; Cataldo et al. 2002). These species are important in riverine food webs, as they represent a major component of the diet of most large, ichthyophagous species.

The carp (*Cyprinus carpio*), introduced to Argentina around the turn of the nineteenth century (Baigún and Quirós 1985), is present in large numbers in most South American lentic and lotic waterbodies. This typically omnivorous fish feeds on insect larvae, crustaceans, plants, detrital material, etc. (Colautti 1997, 2001; Menni 2004), and since the introduction of *L. fortunei*, it has been recorded to feed on the mussel as well (Cataldo et al. 2002).

Iliophagous Species

Iliophagous species typically feed on organic matter-rich sediments, but they also consume small particulate periphytic material scraping the surface of objects covered by an organic film. Organic films on hard substrata often encompass mussels, and these bivalves have become an occasionally important component of the diet of iliophagous fishes. Among the species that depict this feeding behavior, the members of the family Loricariidae are very important because of their abundance and diversity (*Hypostomus uruguayensis, Hypostomus laplatae, Hypostomus commersoni, Hypostomus regani, Hypostomus ternetzi, Paraloricaria vetula, Megalancystrus parananus, Pseudoheminodon laticeps*). The sucking, ventrally located mouths of these species are adapted to scraping the surface of leaves, rocks, branches, and other objects collecting adhering material, including small (usually <5 mm) *L. fortunei*. The mussel has often been observed to represent up to 100 % of the diet of various Loricariidae (Montalto et al. 1999; Cataldo et al. 2002; García and Protogino 2005; Oliveira et al. 2010; Belz et al. 2012; Lopes and Vieira 2012).

Prochilodus lineatus (locally known as "sábalo" in Argentina, or "curimbatá" in Brazil) deserves special attention. This medium-sized fish (adult individuals weight about 2–3 kg; Sverlij et al. 1993) represents >60 % of the fish biomass in the Paraná-Uruguay river system (Bonetto 1998). Economically, the sábalo is the most important exploitable species in the Río de la Plata watershed. In Argentina, it accounts for ca. 90 % of the freshwater fish exports (Iwaszkiw 2005; Iwaszkiw and Lacoste 2011), with landings varying around 15,000–30,000 t per year in the last decade (Food and Agriculture Organisation, FAO data). Surveys carried out in the middle and upper Paraná River, including the Itaipu Reservoir, show that adult golden mussels are seldom present in the gut of *P. lineatus* (Montalto et al. 1999; Lösch et al. 2009; Oliveira et al. 2010). On the other hand, specimens recovered in Salto Grande Reservoir (Uruguay River, Argentina-Uruguay), have been observed to contain large numbers of adult *L. fortunei* in their stomach contents (José Venzal, pers. comm.).

Observation of medium-sized sábalo specimens kept in captivity in fish tanks stocked with *L. fortunei* shows that fishes hover over the mussel clusters and bite off pieces of soft tissue protruding from the partly open valves. Although this species is provided with only small incisiform teeth, its protractile mouth allows it to efficiently grasp chunks of siphons or mantle edge, tearing them off the mussel. After a few days, only the empty valves of exposed mussels remain on the substrate, the soft tissue having been totally consumed by the fishes. Admittedly, no in situ observations have been made to provide proof that this feeding behavior takes place in the field, but there does not seem to be any reason to assume otherwise. Assessment of consumption of golden mussels by fishes is normally based on records of the mussels' shells, or pieces thereof, in the guts of the predators (Montalto et al. 1999; Penchaszadeh et al. 2000; Cataldo et al. 2002). The fact that *P. lineatus* can tear off the soft tissue leaving the valves intact suggests that the examination of stomach contents can significantly underestimate the importance of mussels as a trophic resource of the sábalo, as well as for any other fish species with a similar behavior.

In addition to direct consumption of mussels, iliophagous fishes, including *P. lineatus*, can benefit from the organic matter-enriched sediments derived from the "shunt" of suspended particulate organic matter to the bottom as mussel feces and pseudofeces (Sardiña et al. 2008; Cataldo et al. 2012; Boltovskoy and Correa 2015; see Chapter "Nutrient Recycling, Phytoplankton Grazing, and Associated Impacts of *Limnoperna fortunei*" in this volume).

Ichthyophagous Species

Ichthyophagous fishes comprise mostly large, actively swimming species provided with canine and villiform teeth used for piercing and holding the prey. These species do not normally consume mussels, but exceptions have been reported. *Hoplias malabaricus* ("tararira" in Argentina, "traíra" in Brazil) is a typical fish predator common in most South American freshwater bodies. In Brazil, where it represents an important fisheries resource, it has been reported to feed actively on small (ca. 1 cm) *L. fortunei*, which can account for up to 20% of its gut contents (Oliveira et al. 2010; Lopes and Vieira 2012). Even fishes of the family Serrasalmidae, which include several species of "palometa" and "piraña", have been recorded with *L. fortunei* in their stomachs (Oliveira et al. 2010).

Marine Species

Dietary changes associated with the invasion of the golden mussel in South America are not restricted to freshwater fishes, but have also been recorded in marine species that regularly enter the freshwater zone of the Río de la Plata estuary. The whitemouth croaker (*Micropogonias furnieri*) is a marine demersal species widely distributed from the Gulf of Mexico (around 24°N) to the Gulf of San Matías (Argentina, 41°S), which supports important fisheries in Brazil, Uruguay, and Argentina (Sardiña and Lopez Cazorla 2005; Acha et al. 2008). During the spring and summer, the whitemouth croaker enters the estuary to spawn (Acha et al. 2008), at which time adults come in contact with *L. fortunei* beds. Small fish do not consume golden mussels, but most of those > 200 mm do. Mussel shells are crushed by the croaker's strong molariform teeth (López Armengol and Casciotta 1998).

Predator–Prey Size Relationships

Several studies have shown that feeding of fishes on *L. fortunei* involves the selective consumption of the smaller size classes. For example, the size of mussels recorded in the gut contents of 12 fish species from São Gonçalo Channel (Brazil) (3–15 mm) was substantially lower than that of the local mussel populations (4–32 mm) (Lopes and Vieira 2012; Table 1; Fig. 1a). Similar results have been reported for *Pimelodus pintado* from Mirim Lake (Vieira and Lopes 2013; Fig. 1b), and *Rhamdia quelen* from Itaipu Reservoir (Brazil) (Lösch et al. 2009).

The strength of this relationship between the size of the predator and that of its prey, as well as the relative importance of mussels as a food item, is largely modulated by the feeding mode of the fishes. Two major groups can be identified: one comprised of fishes that cannot break the shell and therefore ingest whole organisms, and the other comprised of animals with dentition that allows them to bite off shell fragments and crush the valves before ingesting them.

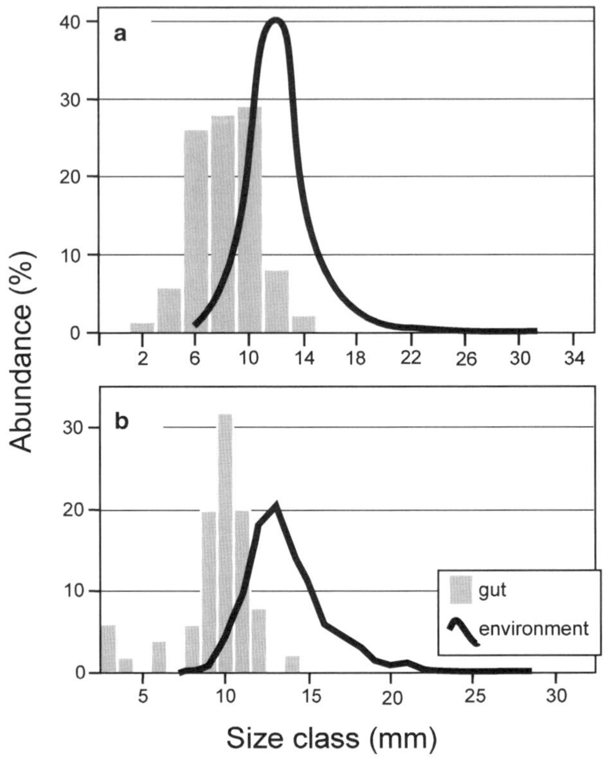

Fig. 1 Comparison of shell size distributions of *L. fortunei* collected in the environment (*line*) and retrieved from fish guts (*bars*). **a** Eight fish species (*Rineloricaria strigilata, Pimelodus pintado, Rhamdia* aff. *quelen, Hoplias malabaricus, Rineloricaria microlepdogaster, Astyanax fasciatus, Hypostomus commersoni,* and *Crenicichla punctata*) from the Patos-Mirim lagoon system (Brazil; from Lopes and Vieira 2012). **b** *Pimelodus pintado* from São Gonçalo Channel (Brazil). (From Vieira and Lopes 2013)

Most opportunistic species ingest whole shells, whereby the ability to tear off animals from the mussel bed and the size of the predator's mouth play a major role in the feeding process. Vieira and Lopes (2013) noticed that *P. pintado* below 10 cm in length do not consume *L. fortunei*, but as the fish grows in size the mussel becomes an increasingly more important food item. At 10–15 cm, ca. 5 % of the fishes consume *L. fortunei*, whereas at 25–30 mm around 50 % do (Fig. 2a). Additionally, the size of the mussels consumed changes little with fish size (Fig. 2b). Montalto et al. (1999), in their study in the Middle Paraná River encompassing nine fish species, also noticed a clear association between the size of the predator and that of the mussels consumed: small fishes selected *L. fortunei* below 6 mm in length, whereas larger species fed on mussels up to over 15 mm.

These results suggest that for fishes that ingest whole mussels, a major limitation is the size of their mouth. In contrast, fishes with teeth capable of crushing the shell

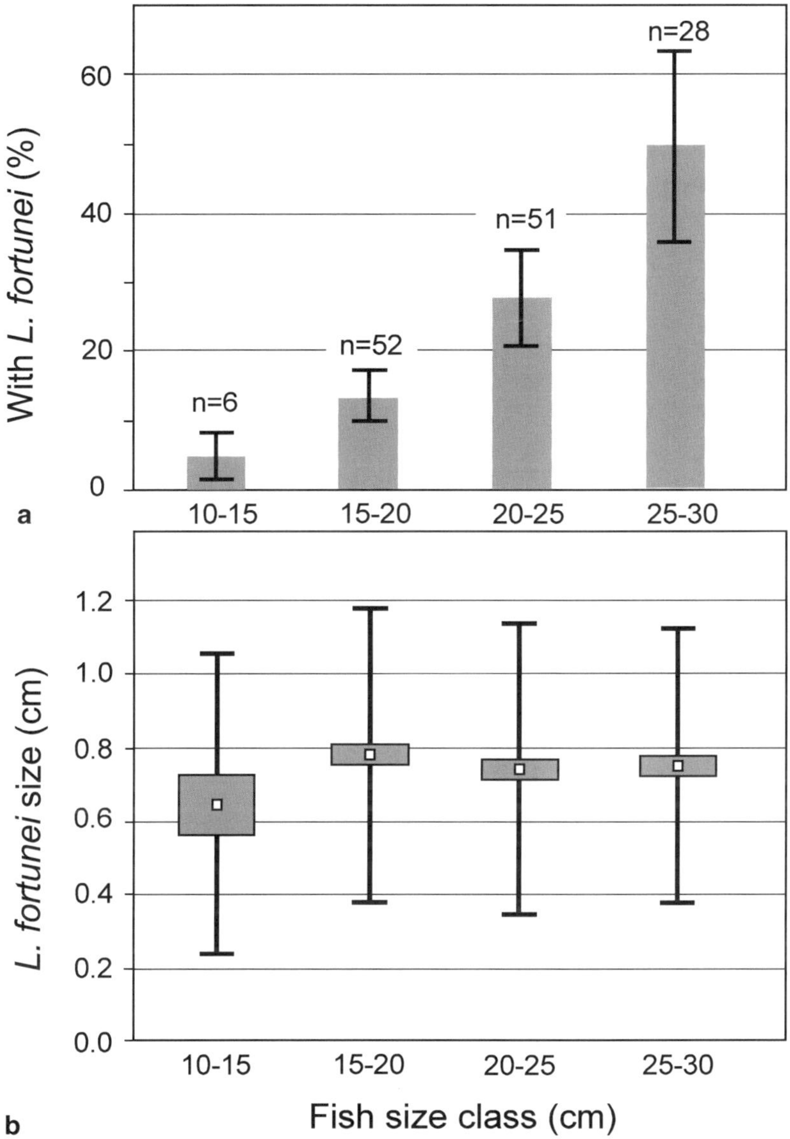

Fig. 2 **a** Proportions of *Pimelodus pintado* of different size with *L. fortunei* in their guts (mean and standard deviation; *n:* number of fishes). **b** Box plot of lengths of *L. fortunei* shells consumed by *P. pintado* of different size (median value, first and third quartiles, and range of values). Based on data from São Gonçalo Channel (Brazil) collected in spring 2005. (Modified from Vieira and Lopes 2013)

can benefit from a larger size-range of prey. Among the latter, a salient example is *L. obtusidens.* This species is a medium-sized fish (up to 80 cm in length), widely used for human consumption and production of fishmeal in the Río de la Plata

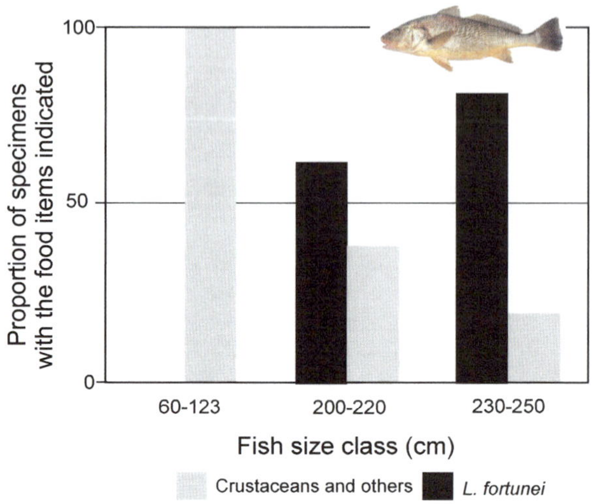

Fig. 3 Proportions of whitemouth croakers (*Micropogonias furnieri*) with *L. fortunei* and with other prey in their digestive tracts as a function of fish size. Data from the Río de la Plata estuary (Argentina) collected in Oct–Nov 1996; total number of fishes analyzed: 17. (From López Armengol and Casciotta 1998)

watershed countries. The range of sizes consumed by this species is very wide; specimens of *L. fortunei* below 6 mm are mostly found intact in stomach contents, whereas larger mussels are crushed. Despite this ability to break shells, mussels over 15 mm in length are chiefly consumed by large *L. obtusidens* (Montalto et al. 1999; Penchaszadeh et al. 2000). Lopes and Vieira (2012) suggested that enhanced predation of smaller mussels might be due to the fact that they tend to break free from the substrate and wander about more often than larger individuals (Uryu et al. 1996), thus spending more time away from a mussel agglomerate, where they are less vulnerable to predation. Direct evidence of this effect, however, is lacking.

Working on the whitemouth croaker, *Micropogonias furnieri,* López Armengol and Casciotta (1998) noticed that fishes below 123 mm in length fed chiefly on crustaceans, whereas in larger size classes the importance of crustaceans decreased and that of *L. fortunei* increased (Fig. 3).

It should be noticed that estimates of *L. fortunei* consumption by fishes that crush the shells is complicated and quite probably often biased, particularly with respect to the number and size of the mussels ingested. In a survey carried out in fish culture net cages deployed in Itaipu Reservoir (Upper Paraná River), all examined specimens of *P. mesopotamicus* ("pacú") had their gut contents filled with *L. fortunei,* but they were destroyed to such a degree that neither the number of mussels nor their sizes could be determined (Lösch et al. 2009). Other fish species have posed the same problem for gut analyses, even when their stomachs are totally occupied by mussel remains (e.g., *L. obtusidens*; Cataldo et al. 2002). Observations made in captivity indicate that *L. obtusidens* does not always tear off whole mussels, often biting off parts of the shell, while the remainder is left attached to the substratum. Large mussels often show bite scars on their periostracum (Penchaszadeh et al. 2000). In order to circumvent this problem to estimate the number of mussels consumed by fish species that destroy the shell, authors have resorted to using various proxies, such as the "beaks" (the umbonal region of the shell) (Penchaszadeh et al. 2000), or the ligament (López Armengol and Casciotta 1998) as an indicator of mussel numbers. However, the feeding mode described above, whereby only the distal part of the shell is torn off and ingested by the fish (while the "beak" and the ligament are left behind) may introduce a significant bias in these methods.

Seasonal Trends in Mussel Consumption by Fishes

A salient aspect of *L. fortunei* as a food item is the fact that, unlike many other organisms with shorter life cycles (including most planktonic resources), its adults are available for consumption throughout the year. Unsurprisingly, time-series studies of its use by fishes indicate that *L. fortunei* is preyed upon almost uninterruptedly (Penchaszadeh et al. 2000). Nevertheless, regardless of the availability of mussel prey, fish's feeding activity differs among seasons. In the summer, 80–100% of the specimens of *L. obtusidens* retrieved had fed on mussels, whereas during the winter (July–August), specimens with *L. fortunei* in their stomachs dropped to 0% (Fig. 4; Penchaszadeh et al. 2000). Ancillary observations on seasonal differences in the feeding of several fishes of the Río de la Plata estuary also suggest that feeding activity is lowest during the winter (Cataldo et al. 2002).

This seasonal trend, however, does not seem to hold elsewhere in the Río de la Plata watershed, where in addition to seasonal changes, interannual differences have been described. In a survey carried out in Mirim Lake (Brazil) in 2005 and 2008, Vieira and Lopes (2013) found no *L. fortunei* in the stomachs of *P. pintado* in the summer, autumn, and winter of 2005. The mussel was first recorded in the diet of this species in the spring of 2005 (September–November), when it was consumed by 22% of the 180 specimens surveyed. Three years later, in the spring of 2008, proportions of *P. pintado* with *L. fortunei* in their guts increased to 61%, and the mussel had become the most important item in the diet of this predator (Vieira and Lopes 2013). Since *L. fortunei* started invading Mirim Lake around 2005 (Langone 2005), this increase is likely associated with the spread of the mussel in the system.

In some cases, a delay in the use of this new prey by its potential consumers may conceivably be associated with the time required by predators to get used to the novel trophic resource. However, an increase in the use of *L. fortunei* by fish over time is more likely to stem from the growth of the predators' populations in response to better feeding and survival conditions.

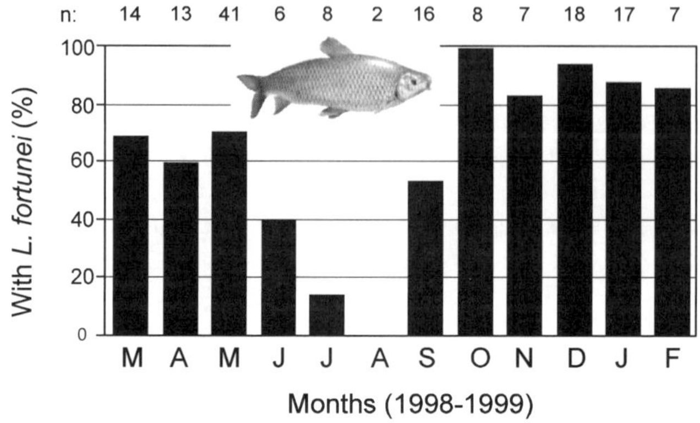

Fig. 4 Proportions of *Leporinus obtusidens* from the Río de la Plata estuary (Argentina) with *L. fortunei* prey in their digestive tract throughout an annual cycle. (Based on data from Penchaszadeh et al. 2000)

Degree of Digestion of the Bivalves

As reviewed above, consumption of *L. fortunei* is not limited to those fish species that are anatomically best adapted to obtain and ingest mussels, but also involves typically omnivorous, iliophagous, and ichthyophagous species. However, regardless of the importance of mussels in their diet, the ability to digest this prey may differ between species.

Fishes possessing strong teeth that can crush the shells gain full access to the soft tissues. In the stomach contents of *L. obtusidens* and *L. macrocephalus*, 80–100 % of the mussels are finely fragmented and the soft tissue is partially or totally digested (Montalto et al. 1999; Cataldo et al. 2002; Oliveira et al. 2010). Crushed shells and digested soft tissue have also been recorded in several other species, including *P. mesopotamicus, M. furnieri, Astyanax fasciatum*, *Geophagus brasilensis,* and *Crenicichla punctata* (López Armengol and Casciotta 1998; Lösch et al. 2009). Some other species, including several not armed with strong crushing or grinding teeth, have also been recorded with their stomachs full of broken and fragmented *L. fortunei* shells (e.g., *Megalancistrus parananus*, *Pimelodus maculatus*, *P. albicans*; Montalto et al. 1999; Oliveira et al. 2010).

For fishes that swallow whole shells, the degree to which mussels are effectively digested is probably slightly, but not significantly lower. Oliveira et al. (2010) noticed that ~80–90 % of *L. fortunei* specimens recorded in the digestive tract of *P. granulosus* and *Serrasalmus marginatus* are intact. However, in *P. granulosus* from the Middle Paraná River whole *L. fortunei* shells had their valves open and their soft tissues partly digested (Montalto et al. 1999). Belz et al. (2012) investigated the feasibility of long distance transport of *L. fortunei* in the stomachs of fish. They analyzed the diet of five of the most likely candidates to disperse the mussel: *P. granulosus, Megalancistrus aculeatus, Satanoperca papaterra, Potamotrygon motoro,* and *Iheringichthys labrosus*. Live *L. fortunei* were only recorded in three specimens of only one of the five species analyzed, *P. granulosus*. Of the 2198 individuals of *L. fortunei* present in the digestive tract of this species, 70 were found alive in the stomach, but only four were alive in the distal section of the intestine. These results suggest that crushing the shells upon ingestion may favor digestion, but breakage is not required for fishes to benefit from this trophic resource.

It has been suggested that the sharp edges of crushed *L. fortunei* shells may lacerate the anal area of their predators (e.g., *Rineloricaria microlepidogaster, R. srtrigilata*; Lopes and Vieira 2012). However, wounds could have been the result of the animals attempting to free themselves from the net. Ad hoc observations of many specimens in a wide range of species failed to reveal evidence of lacerations caused by ingested shell fragments (López Armengol and Casciotta 1998; Montalto et al. 1999; Penchaszadeh et al. 2000).

Effects of *L. fortunei* on Local Fish Populations

The number of fish species that feed on *L. fortunei* has increased steadily, largely because of the mussel's northward geographic expansion into areas with increasingly higher fish diversity. Predation pressure on the mussel is likely high and, together with consumption of veligers by larval fishes (see Chapter "Trophic Relationships of *Limnoperna fortunei* with Larval Fishes" in this volume), is probably the most significant mechanism that modulates *L. fortunei* populations, but it is very unlikely to eradicate the mussel altogether.

As useful as they are, the studies reviewed above are limited in scope and fall short of providing a comprehensive assessment of the effects of this new food supply on fish stocks. These effects are likely very significant. In the delta of the Lower Paraná River, predators (presumably mostly represented by fishes) consume ca. 6 kg of whole mussel mass per square meter, eliminating up to over 90% of the yearly production of *L. fortunei* (Sylvester et al. 2007). Nakano et al. (2010) estimated that predators eliminate ca. 97% of the mussels in Lake Ohshio, a Japanese reservoir, affecting not only biomass but also the size-structure of the populations. In the Itaipu Reservoir (Upper Paraná River), 24 of the 36 species (3752 specimens) surveyed in 2005–2006 were found to prey on *L. fortunei* (Oliveira et al. 2010).

Impacts are not restricted to species that consume the mollusc, but also affect species that benefit from this new food resource indirectly, such as the large and economically most valuable ichthyophagous species that feed on other fishes (e.g., *Pseudoplatystoma fasciatum, Pseudoplatystoma corruscans, Salminus maxillosus, H. malabaricus, Paulicea luetkeni, Luciopimelodus pati*). Furthermore, *L. fortunei* transfers large amounts of organic matter from the water column to the sediments through filtration and the formation of feces and pseudofeces (Sardiña et al. 2008; Cataldo et al. 2012), which boosts invertebrate densities (Sylvester et al. 2007; Sardiña et al. 2008, 2011; see Chapter "Relationships of *Limnoperna fortunei* with Benthic Animals" in this volume). This is important for deposit-feeding fish species, some of which, like *P. lineatus*, are very abundant, represent important fishing resources, and are the main food items of most ichthyophagous species (Bonetto 1998). Adult *L. fortunei* represent not only an additional food item but also one energetically more profitable than the plant- and detritus-based foods which characterized the diet of these fish species before the introduction (Ferriz et al. 2000).

The impact of these trophic shifts on local fish stocks is probably high, but has not yet been quantified. Argentine freshwater fish landings increased three-fold after the introduction of *L. fortunei* (Boltovskoy et al. 2006), which may suggest better recruitment and survival conditions, but interpretation of this trend is complicated by several factors, including changes in fishing regulations, fishing pressure, fish export trends, and profitability of the industry during the time span involved. In addition, exploitation of freshwater fish resources in the countries colonized by the mussel is largely artisanal and statistical information is scarce, fragmentary, and most probably very incomplete (Iwaszkiw 2001, 2005).

Dreissena polymorpha, the zebra mussel, has been shown to increase the abundance of littoral fish species (through enrichment of coastal bottom areas with organic matter), and decrease the abundance of pelagic fishes (due to depletion of zooplankton forage species through grazing) (Strayer et al. 2004). This trend, however, subsequently changed and open-water species returned to preinvasion levels (Strayer et al. 2014). The effects of *L. fortunei* on South American fish stocks are probably different because filter-feeding fishes are less abundant (the most abundant species in these large floodplain rivers are iliophagous and detritivorous; Bonetto 1998) and particulate organic carbon (POC) loads are very high. The mean concentration of POC in the Paraná River (about 3.5 mg/L: Depetris 1976; Depetris and Paolini 1991; Depetris and Pasquini 2007) is much higher than in most of the waterbodies invaded by *Dreissena* species (typically around 0.15–1 mg/L in the Great Lakes; Fanslow et al. 1995; Barbiero and Tuchman 2004; Johengen et al. 2008), which suggests that filtering organisms are not food-limited in most South American waterbodies invaded by *L. fortunei* (Sylvester et al. 2005).

Indigenous filter-feeding benthic animals in the Paraná watershed are scarce, and therefore most of the POC is flushed out into the ocean through the Río de la Plata estuary. *L. fortunei*, the first and only abundant macrobenthic filter-feeder, is intercepting an important proportion of this particulate organic matter and retaining it in the system for use by a wide array of animals (Boltovskoy et al. 2006; Boltovskoy and Correa 2015), which most probably represents an important energetic subsidy for the entire system.

While *L. fortunei* as a new food resource has likely had positive effects on fish populations, trophic relationships are but one of many possible interactions between fishes and the mussel. In Japan, trematode parasites introduced with the golden mussel (as an intermediate host) have been found to inhibit gamete production in some fish species (Tanaka et al. 2004; see Chapter "Parasites of *Limnoperna fortunei*" in this volume). Indirect interactions are numerous and can operate through various ecosystem compartments. For example, cyanobacterial blooms promoted by the mussel (Cataldo et al. 2012) can trigger massive fish kills. Clarification of the water-column can facilitate visual predation of zooplankton by fish. Enhancement of macrophyte growth (Boltovskoy et al. 2009) can provide shelter for adult and larval fishes, etc. Thus, it is conceivable that the positive effects of *L. fortunei* as food may be offset by its negative impacts through any of these, or other as yet unknown mechanisms.

References

Acha ME, Mianzan H, Guerrero R, Carreto J, Giberto D, Montoya N, Carignan M (2008) An overview of physical and ecological processes in the Rio de la Plata Estuary. Cont Shelf Res 28:1579–1588

Alonso C (1978) Estudio del contenido gástrico de *Pimelodus clarias maculatus* (Lacepede 1803) (Pices, Pimelodidae). Iheringia (Série Zoologia) 51:45–61

Baigún CM, Quirós R (1985) Itroducción de peces exóticos en la República Argentina. INIDEP (Instituto Nacional de Investigación y Desarrollo Pesquero, Argentina), Departamento de Aguas Continentales, Informe Técnico. INIDEP (Instituto Nacional de Investigación y Desarrollo Pesquero, Argentina), Departamento de Aguas Continentales, Informe Técnico, pp 1–90

Baptista CT, Zibetti V (2006) Mexilhão dourado (*Limnoperna fortunei*) como recurso alimentar de peixes do arroio Pelotas, Pelotas, RS: resultados preliminares. XV Congresso de Iniciação Cientifica, VIII Encontro de Pós-Graduação, Universidade Federal de Pelotas (Brazil)

Barbiero RP, Tuchman ML (2004) The deep chlorophyll maximum in Lake Superior. J Gt Lakes Res 30(Supplement 1):256–268

Belz CE, Darrigran GA, Netto OSM, Boeger WA, Ribeiro PJ (2012) Analysis of four dispersion vectors in inland waters: the case of the invading bivalves in South America. J Shellfish Res 31:777–784

Boltovskoy D, Cataldo D (1999) Population dynamics of *Limnoperna fortunei*, an invasive fouling mollusc, in the lower Paraná River (Argentina). Biofouling 14:255–263

Boltovskoy D, Correa N (2015) Ecosystem impacts of the invasive bivalve *Limnoperna fortunei* (golden mussel) in South America. Hydrobiologia 746:81–95

Boltovskoy D, Correa N, Cataldo D, Sylvester F (2006) Dispersion and ecological impact of the invasive freshwater bivalve *Limnoperna fortunei* in the Río de la Plata watershed and beyond. Biol Invasions 8:947–963

Boltovskoy D, Karatayev A, Burlakova L, Cataldo D, Karatayev V, Sylvester F, Mariñelarena A (2009) Significant ecosystem-wide effects of the swiftly spreading invasive freshwater bivalve *Limnoperna fortunei*. Hydrobiologia 636:271–284

Bonetto AA (1986) Fish of the Paraná system. In: Davies BR, Walker KF (eds) The ecology of river systems. Dr. W. Junk Publishers, Dordrecht, pp 573–588

Bonetto AA (1998) Panorama sinóptico sobre la ictiofauna, la pesca y la piscicultura en los ríos de la cuenca del Plata con especial referencia al Paraná. Rev Ictiol 6:3–15

Bonetto AA, Pignalberi C, Cordiviola E (1963) Ecología alimentaria del "amarillo" y "moncholo" *Pimelodus calrias* (Bloch) y *Pimelodus albicans* (Valenciennes) (Pisces, Pimelodidae). Physis 24:87–94

Braga L (1993) Los Anostomidae (Pisces, Characiformes) de Argentina. Fauna de Agua Dulce de la República Argentina 40:1–61

Cantanhêde G, Hahn NS, Gubiani ÉA, Fugi R (2008) Invasive molluscs in the diet of *Pterodoras granulosus* (Valenciennes, 1821) (Pisces, Doradidae) in the Upper Paraná River floodplain, Brazil. Ecol Freshw Fish 17:47–53

Cataldo D, Boltovskoy D, Marini V, Correa N (2002) Limitantes de la expansión de *Limnoperna fortunei* en la Cuenca del Plata: la predación por peces. Terceras Jornadas para la Conservación de la Fauna Íctica del Río Uruguay, Paysandú (Uruguay)

Cataldo D, Vinocur A, O'Farrell I, Paolucci E, Leites V, Boltovskoy D (2012) The introduced bivalve *Limnoperna fortunei* boosts *Microcystis* growth in Salto Grande Reservoir (Argentina): evidence from mesocosm experiments. Hydrobiologia 680:25–38

Colautti DC (1997) Ecología de la carpa *Cyprinus carpio* en la cuenca del río Salado, provincia de Buenos Aires. PhD Thesis, Universidad Nacional de La Plata (Argentina), pp 1–260

Colautti DC (2001) La carpa y el pejerrey, ¿enemigos? In: Grosman F (ed) Fundamentos biológicos, económicos y sociales para una correcta gestión del recurso pejerrey. Editorial Astyanax, Azul (Argentina), pp 85–94

Depetris PJ (1976) Hydrochemistry of the Paraná River. Limnol Oceanogr 21:736–739

Depetris PJ, Paolini JE (1991) Biogeochemical aspects of South American rivers: the Paraná and the Orinoco. In: Degens ET, Kempe S, Richey JE (eds) Biogeochemistry of major world rivers. Wiley, Chichester, pp 105–125

Depetris PJ, Pasquini AI (2007) The geochemistry of the Paraná River: an overview. In: Iriondo MH, Paggi JC, Parma MJ (eds) The middle Paraná River. Limnology of a subtropical wetland. Springer, Berlin, pp 143–174

Fanslow DL, Nalepa TF, Lang GA (1995) Filtration rates of the zebra mussel (*Dreissena polymorpha*) on natural seston from Saginaw Bay, Lake Huron. J Gt Lakes Res 21:489–500

Ferriz RA, Villar CA, Colautti D, Bonetto C (2000) Alimentacion de *Pterodoras granulosus* (Valenciennes) (Pisces, Doradidae) en la baja Cuenca del Plata. Revista del Museo Argentino de Ciencias Naturales "Bernardino Rivadavia" 2:151–156

García M, Montalto L (2006) Los peces depredadores de *Limnoperna fortunei* en los ambientes colonizados. In: Darrigran GA, Damborenea C (eds) Bio-invasión del mejillón dorado en el continente americano. Editorial Universitaria de La Plata, La Plata, pp 113–129

García ML, Protogino LC (2005) Invasive freshwater molluscs are consumed by native fishes in South America. J Appl Ichthyol 21:34–38

Gneri FS, Angelescu V (1951) La nutrición de los peces iliófagos en relación con el metabolismo general del ambiente acuático. Rev Insti Investig Cienc Nat (anexo al Museo Argentino de Ciencias Naturales "Bernardino Rivadavia") 2:1–44

González-Bergonzoni I, Teixeira-de Mello F, Vidal N, D'Anatro A, Masdeu M (2010) Reappearance and diet of juvenile armado catfish (*Pterodoras granulosus*) in Lower Uruguay River, (Río Negro, Uruguay). Bol Soc Zool Urug 19:42–46

Ituarte CF (1994) *Corbicula* and *Neocorbicula* (Bivalvia: Corbiculidae) in the Paraná, Uruguay and Río de la Plata basins. Naut 107:129–135

Iwaszkiw JM (2001) Pesqueras continentales del tramo argentino de la Cuenca del Plata. Consejo Federal de Inversiones. Consejo Federal de Inversiones, Buenos Aires (Argentina), pp 1–279

Iwaszkiw JM (2005) Estadisticas de exportaciones pesqueras de la Cuenca del Plata, Argentina 1994–2005. IV Taller Internacional sobre enfoques regionales para el desarrollo y gestión de embalses en la Cuenca del Plata, Salto Grande, Argentina-Uruguay

Iwaszkiw JM, Lacoste FF (2011) La pesca artesanal en la Cuenca del Plata (Argentina) y sus implicancias en la conservación de la biodiversidad. Revista del Museo Argentino de Ciencias Naturales "Bernardino Rivadavia". Nueva Serie 13:21–25

Johengen TH, Biddanda BA, Cotner JB (2008) Stimulation of Lake Michigan plankton metabolism by sediment resuspension and river runoff. J Gt Lakes Res 34:213–227

Langone JA (2005) Notas sobre el mejillón dorado *Limnoperna fortunei* (Dunker, 1857) (Bivalvia, Mytilidae) en Uruguay. Publicacion Extra del Museo Nacional De Historia Natural y Antropologia (Montevideo, Uruguay) 1:1–18

Lopes M, Vieira J (2012) Predadores potenciais para o controle do mexilhão-dourado. In: Mansur MCD, Santos CP, Pereira D, Padula Paz IC, Leite Zurita ML, Raya Rodriguez MT, Vilar Nehrke M, Aydos Bergonci PE (eds) Moluscos límnicos invasores no Brasil. Biologia, prevenção, controle. Redes Editora, Porto Alegre, pp 357–363

López Armengol MF, Casciotta JR (1998) First record of the predation of the introduced freshwater bivalve *Limnoperna fortunei* (Mytilidae) by the native fish *Micropogonias furnieri* (Sciaenidae). Iberus 16:105–108

Lösch JA, Boscolo WG, Feiden A, Kleber Lorenz E, Bittencourt F (2009) Presença de mexilhão dourado no trato gastrointestinal de três espécies nativas de peixes cultivadas em tanques rede no reservatório de Itaipu. Anais do I Seminário Internacional de Ciência, Tecnologia e Ambiente, Cascavel (Brazil)

Masdeu M, Teixeira-de Mello F, Loureiro M, Arim M (2011) Feeding habits and morphometry of *Iheringichthys labrosus* (Lütken, 1874) in the Uruguay River (Uruguay). Neotrop Ichthyol 9:657–664

Mastrarrigo V (1950) La Boga. Contribución a su conocimiento biológico. Almanaque del Ministerio de Agricultura y Ganadería (Argentina) 25:417–426

Menni RC (2004) Peces y ambientes en la Argentina continental. Monografía del Museo Argentino de Ciencias Naturales "Bernardino Rivadavia" 5:1–316

Montalto L, Oliveros OB, Ezcurra De Drago I, Demonte LD (1999) Peces del Río Parana medio predadores de una especie invasora: *Limnoperna fortunei* (Bivalvia, Mytilidae). Rev Fac Bioquud Cienc Biol (Universidad Nacional del Litoral) 3:85–101

Nakano D, Kobayashi T, Sakaguchi I (2010) Predation and depth effects on abundance and size distribution of an invasive bivalve, the golden mussel *Limnoperna fortunei*, in a dam reservoir. Limnology 11:259–266

Oliva A, Ubeda CA, Vignes IE, Uriondo A (1981) Contribución al conocimiento de la ecología alimentaria del bagre amarillo (*Pimelodus maculatus* Lacepede 1803) del Río de La Plata (Pisces, Pimelodidae). Comuncaciones del Museo Argentino de Ciencias Naturales "Bernardino Rivadavia". Ecologia 1:31–50

Oliveira CR, Fugi R, Brancalhão KP (2010) Fish as potential controllers of invasive mollusks in a neotropical reservoir. Nat Conserv 8:140–144

Panatieri AE, Del Barco D (1981) Peces de la provincia de Santa Fe. 9. Peces omnívoros preferentemente de fondo, de verano. Armado gallego (*Pterodoras granulosus*). CYTA, Ministerio de Agricultura y Ganadería de Santa Fé. Argentina 25:21–23

Pastorino G, Darrigran GA, Martín SM, Lunaschi L (1993) *Limnoperna fortunei* (Dunker, 1857) (Mytilidae), nuevo bivalvo invasor en aguas del Río de la Plata. Neotropica 39:101–102

Penchaszadeh PE, Darrigran GA, Angulo C, Averbuj A, Brögger M, Dogliotti A, Pírez N (2000) Predation of the invasive freshwater mussel *Limnoperna fortunei* (Dunker, 1857) (Mytilidae) by the fish *Leporinus obtusidens* Valenciennes, 1846 (Anostomidae) in the Rio de la Plata, Argentina. J Shellfish Res 19:229–231

Ringuelet RA, Aramburu RH, Alonso de Aramburu A (1967) Los peces argentinos de agua dulce. Comisión de Investigación Científica, Gobernación de la Provincia de Buenos Aires, La Plata (Argentina), pp 1–602

Sardiña P, Lopez Cazorla AC (2005) Trophic ecology of the whitemouth croaker, *Micropogonias furnieri* (Pisces: Sciaenidae), in south-western Atlantic waters. J Marin Bio Assoc U K 85:405–413

Sardiña P, Cataldo D, Boltovskoy D (2008) The effects of the invasive mussel, *Limnoperna fortunei*, on associated fauna in South American freshwaters: importance of physical structure and food supply. Fundam Appl Limnol/Archiv Hydrobiol 173:135–144

Sardiña P, Chaves E, Marchese M (2011) Benthic community responses to invasion by the golden mussel, *Limnoperna fortunei* Dunker: biotic homogenization vs environmental driving forces. J N Am Benthol Soc 30:1009–1023

Strayer DL, Hattala KA, Kahnle AW (2004) Effects of an invasive bivalve (*Dreissena polymorpha*) on fish in the Hudson River estuary. Can J Fish Aquat Sci 61:924–941

Strayer DL, Hattala KA, Kahnle AW, Adams RD (2014) Has the Hudson River fish community recovered from the zebra mussel invasion along with its forage base? Can J Fish Aquat Sci. 71:1146–1157

Sverlij SB, Espinach Ros A, Orti G (1993) Sinopsis de los datos biológicos y pesqueros del sábalo *Prochilodus lineatus* (Valenciennes, 1847). Food and Agriculture Organization, Rome (Italy), pp 1–64

Sylvester F, Dorado J, Boltovskoy D, Juárez A, Cataldo D (2005) Filtration rates of the invasive pest bivalve *Limnoperna fortunei* as a function of size and temperature. Hydrobiologia 534:71–80

Sylvester F, Boltovskoy D, Cataldo D (2007) Fast response of freshwater consumers to a new trophic resource: predation on the recently introduced Asian bivalve *Limnoperna fortunei* in the lower Parana River, South America. Austral Ecol 32:403–415

Tanaka H, Deno T, Kondo T (2004) Effect of gasterostome trematode infection on growth and reproduction in *Limnoperna fortunei* (Bivalvia: Mytilidae). Memoirs of Osaka Kyoiku University, Series III, Nat Sci Appl Sci 53:1–7

Uryu K, Iwasaki K, Hinoue M (1996) Laboratory experiments on behaviour and movement of a freshwater mussel, *Limnoperna fortunei* (Dunker). J Molluscan Stud 62:327–341

Vermulm JH, Giamas MTD (2008) Ocorrência do mexilhão dourado *Limnoperna fortunei* (Dunker, 1857), (Mollusca; Bivalvia; Mytilidae), no trato digestivo do "Armal" *Pterodoras granulosus* (Valenciennes, 1821) (Siluriformes; Doradidae), do Rio Paraná, São Paulo, Brasil. Boletim do Instituto de Pesca de São Paulo 34:175–179

Vieira JP, Lopes MN (2013) Size-selective predation of the catfish *Pimelodus pintado* (Siluriformes: Pimelodidae) on the golden mussel *Limnoperna fortunei* (Bivalvia: Mytilidae). Zoologia 30:43–48

Behavior and Taxis of Young and Adult *Limnoperna fortunei*

Keiji Iwasaki

Abstract Young and adult specimens of *Limnoperna fortunei* exhibit crawling, climbing, and aggregating behavior, and they are able to resecrete byssal threads when they become detached from the substratum or from other mussels. Smaller mussels have greater motility and produce more byssal threads than the larger ones. Negative phototaxis, positive thigmotaxis, and negative geotaxis are involved in these behaviors, which differ in intensity between small and large mussels. These behaviors likely represent adaptive responses enhancing mussel survival.

Keywords *Limnoperna fortunei* · Golden mussel · Behavior · Taxis · Translocation

Introduction

After completing their larval development, mytilid bivalves, including *Limnoperna fortunei,* are sessile and attach to the substratum by means of byssal threads secreted by the byssal gland. Most mytilid species form clumps, making their movements and behavior much less conspicuous than those of the free-living sedentary bivalves (Morton 1964; Ansell 1969; Stanley 1970; Brand 2006; Takada et al. 2013). However, nonlocomotive behaviors exhibited in mussel clumps, such as feeding (Riisgård et al. 2011 and references therein), gaping (Robson et al. 2010; Nicastro et al. 2012; Dowd and Somero 2013), and byssal thread secretion (Petraitis 1987; Côté 1995; Ishida and Iwasaki 1999), have been well documented and their adaptive significance in aquatic habitats discussed. When detached from their congeners, young and adult mussels crawl and reaggregate in clumps. The implications of such behavior have been discussed with respect to vulnerability to predation, desiccation, and dislodgement (Senawong 1970; Tan 1975; Uryu et al. 1996; Iwasaki 1997; Côté and Jelnikar 1999; de Vooys 2003; Schneider et al. 2005; Nicastro et al. 2007).

Studies on the behavior of *L. fortunei* have been carried out with free-swimming larvae, and with sessile juveniles and adults. Field studies on the small-scale distribution of recently settled larvae suggest that they preferentially attach to the

K. Iwasaki (✉)
Institute for Natural Science, Nara University, 1500 Misasagicho, Nara 631-8502, Japan
e-mail: iwasaki@daibutsu.nara-u.ac.jp

D. Boltovskoy (ed.), *Limnoperna fortunei,* Invading Nature - Springer Series in Invasion Ecology 10, DOI 10.1007/978-3-319-13494-9_14

shaded underside of boulders, the upper side of tunnels, and into cracks, angles, and crevices of hard substrata, usually selecting sites already colonized by conspecifics (Morton 1975, 1977; Sardiña et al. 2009; Akehoshi 2011; Uchida 2011).

Uryu et al. (1996) and Iwasaki (1997) studied the movements and behavior of *L. fortunei* postlarvae in the laboratory. They reported crawling, climbing, and aggregating behavior, in which negative phototaxis, negative geotaxis, and positive thigmotaxis are involved.

This section describes the results of a series of laboratory experiments on the behavior of the golden mussel and discusses its adaptive significance. Understanding the significance is important not only for gaining insight into the behavioral ecology of this species but also for the development of precautionary measures to prevent further spread and population expansion by the postlarvae.

Methods of Laboratory Experiments

Mussels used in the laboratory experiments (Uryu et al. 1996; Iwasaki 1997) were collected from the Lake Biwa–Uji River system in central Japan. They were maintained for 1 or 2 days in an aerated aquarium where pond water with abundant *Chlorella* spp. and *Euglena* spp. was renewed every 3–4 days. The mussels were detached from the substratum by cutting their byssal threads using a scalpel, and then they were individually placed at the center of the bottom of either transparent rectangular (18 cm long, 10 cm wide, 2.5 cm high) or cylindrical (9.5 cm in diameter, 16 cm high) plastic vessels filled with water, without aeration. The vessels were placed on a flat desk with a black surface near a window facing north. Room and water temperatures were maintained at 25 and 22 ℃, respectively. All the experiments were conducted for 23 h from 15:00 to 14:00 h of the next day under natural illumination conditions. In order to allow video recording in the dark (at 30 s intervals), a weak fluorescent light was used during the night (from 18:00 to 6:00 h the next day).

Mussels used in experiments ranged from 3.9 to 34.2 mm in length. For statistical analyses, they were divided into two size groups: small (<15 mm) and large (≧15 mm). Each mussel was used only once in an experiment. Mussels not straying from their initial position or initial direction at the end of the experiments were regarded as weaklings and thus were excluded from the study.

Crawling Behavior

Small mussels crawl actively with their muscular foot extending forward onto the substratum and by contracting their feet they pull themselves forward (Uryu et al. 1996). Mussels placed in the rectangular vessels moved randomly at first, but after reaching the vessels' sidewalls, they moved only along the angles between the sidewall and the bottom. The total distance traveled during the 23-h experiment differed with shell length, from approximately 1 cm for larger mussels (29–34 mm), to approximately 300 cm for small mussels (12 mm; Fig. 1a). Total distances decreased exponentially with increasing shell lengths (Fig. 1a).

Crawling activity was not consistent throughout the experiment, but exhibited three peaks (measured as the overall distance covered by all mussels during each hour interval as a percentage of the distance covered by all mussels during the entire 23-h experiment). These occurred approximately 2–5, 10, and 17 h after the start of the experiment (Fig. 1b). Mussel activity was highest during the first 5 h. A decrease in activity 4 h after the experiment started (at 18:00 h) seems to have been caused by a sudden change in light conditions when a light was activated.

Displacement of small (<15 mm) mussels placed on the bottom of rectangular vessels showed that, after 23 h, significantly higher numbers of individuals moved to the angles between the bottom and the walls, than to other locations on the flat bottom (Fig. 2). No such trend was observed with large mussels (>15 mm). Furthermore, angles seemed to be more frequently perceived as the definitive, adequate location, as opposed to the flat bottom, since a significantly higher proportion of the animals secreted byssal threads in the angles (Fig. 2).

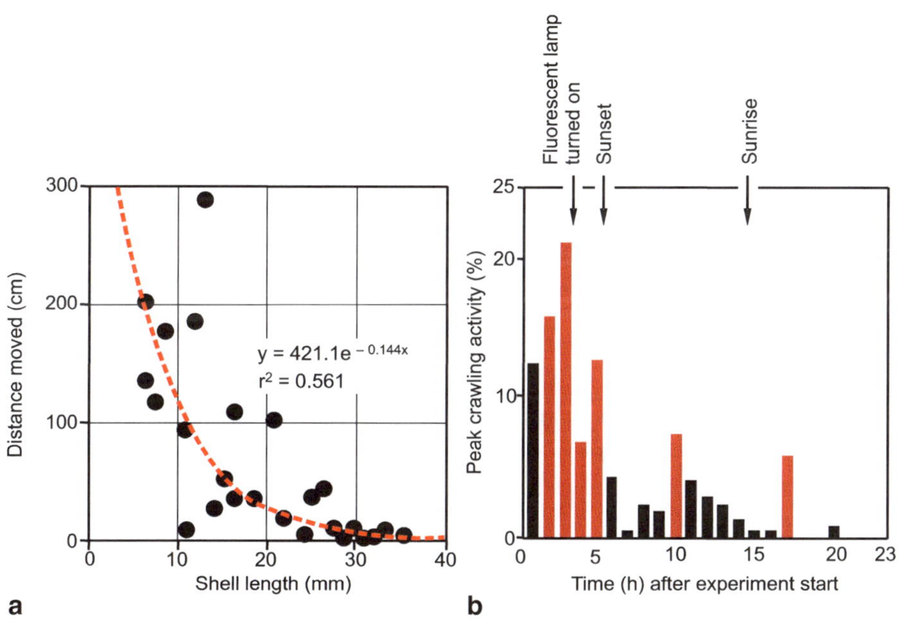

Fig. 1 Results of laboratory experiments investigating crawling behavior of *Limnoperna fortunei* within rectangular vessels filled with water. **a** Overall distance covered by 25 mussels during the duration of the experiment (23 h) as a function of shell length. **b** Periods of peak crawling activity (*red bars*), measured as the overall distance covered by all mussels during each hour interval as a percentage of the distance covered by all mussels during the entire 23-h experiment. (Redrawn from Uryu et al. 1996)

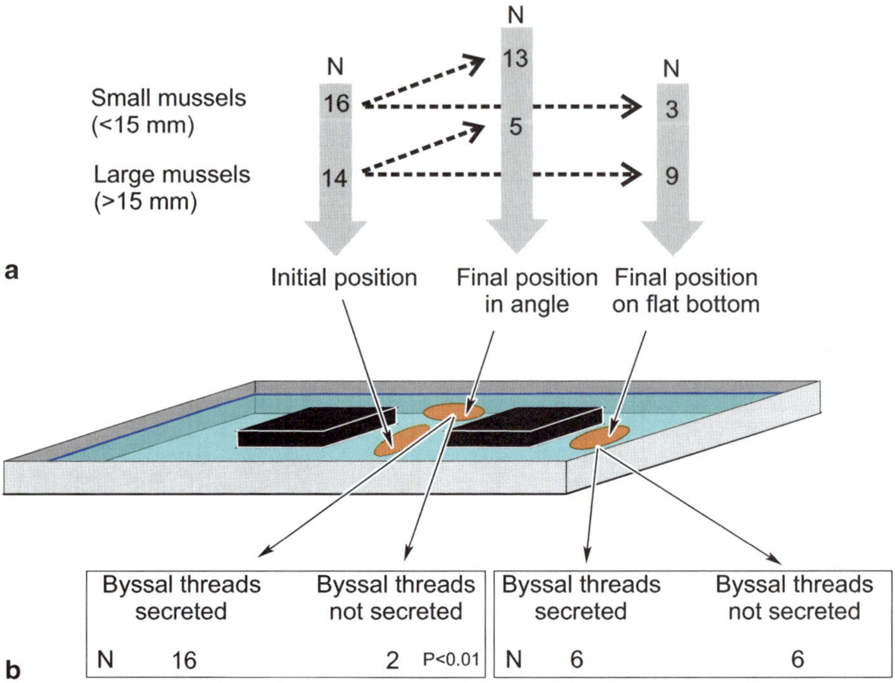

Fig. 2 Displacement of small and large *Limnoperna fortunei* after 23 h in rectangular containers from an initially central position on the flat bottom between two rectangular, black, plastic blocks (**a**), and production of byssal threads as a function of final position (**b**). *G*-test with Williams' correction for differences in behavior between small and large mussels: $P<0.05$. For small mussels (but not for large ones), selection of an angle between the bottom and the wall as the final location was significantly higher than the flat bottom (test of binomial proportions, $P<0.05$). Probability of byssus secretion was significantly higher for mussels whose final location was an angle between the bottom and the wall, than for those located on the flat bottom, away from a wall (test of binomial proportions, $P < 0.01$). (Based on data from Uryu et al. 1996)

In order to investigate phototropic behavior, mussels were placed individually in the center of rectangular vessels either half of which was shaded by a black vinyl sheet. After 23 h, significantly higher numbers of large mussels had moved to the shaded section of the container than the illuminated one (Fig. 3). Small mussels followed the same pattern, but differences for shaded versus unshaded sectors were not significant. These results indicate that detached large individuals prefer to resettle in shaded areas.

In summary, smaller mussels actively crawl on the substratum immediately after detachment and resettle in the angles, thus depicting positive thigmotaxis, whereas larger mussels move shorter distances but have a marked tendency to relocate in shaded areas. These results agree with Morton's (1977) observations based on colonization of experimental substrata. This behavior results in higher colonization rates of the underside of boulders or overhanging rocks, crevices, cracks, and

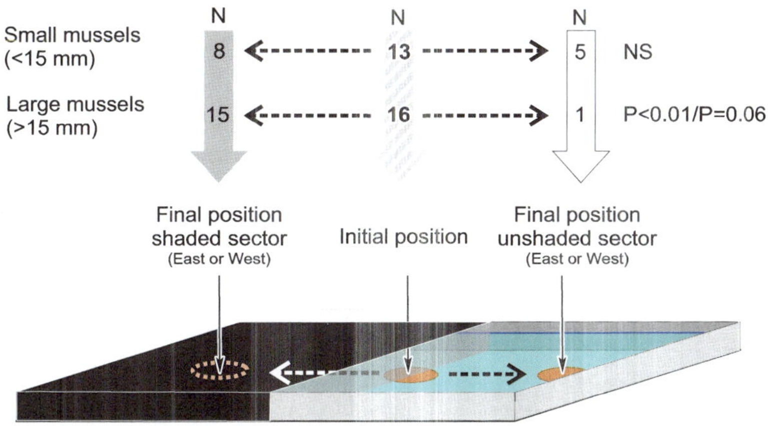

Fig. 3 Displacement of small and large *Limnoperna fortunei* from a central position after 23 h in rectangular containers one half of which was shaded with a black vinyl sheet. Significantly more large (but not small) mussels moved toward the shaded half of the container (test of binomial proportions). Data shown summarize results obtained in two consecutive trials shading either the western or the eastern half of the container. (Based on data from Uryu et al. 1996)

other protected sites, which presumably helps to prevent predation, desiccation, and dislodgement.

Climbing Behavior

Uryu et al. (1996) and Iwasaki (1997) described the "climbing behavior" of *L. fortunei* as an upward crawling movement along the inner walls of experimental vessels. When mussels around 5–25 mm in length were individually placed in horizontally oriented plastic cylinders in an aerated aquarium, they crawled up and down along the inner wall of the cylinder. At the end of the 23-h experiment, more mussels remained on the lower half of the cylinder than elsewhere, especially when this lower half was shaded (Fig. 4a). However, when the upper half or the entire cylinder was shaded, the preference for the lower half disappeared. These results suggest positive geotaxis in illuminated conditions, which was reinforced when the lower half was shaded.

In order to investigate how far upward from the bottom mussels move, a total of 200 animals were placed in cylindrical vessels filled with water to a height of 10 cm. The vessels stood upright in a dark room throughout the experiment. After 23 h, 29 % of the mussels climbed up the walls and attached at different distances from the bottom (Table 1). This climbing behavior was strongest in the smallest size class (>50 % of the mussels below 10 mm in length), and tended to decrease with increasing shell length (Table 1). Of the 56 climbing mussels, 21 (37.5 %) attached to the vessel walls 9–10 cm above the bottom, i.e., just beneath the water surface

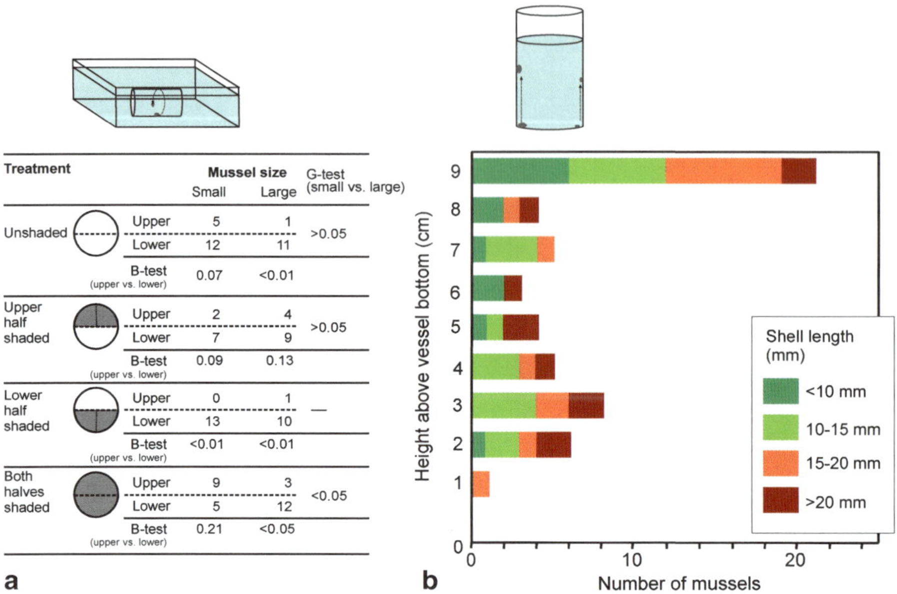

Fig. 4 Results of laboratory experiments of the climbing behavior of *Limnoperna fortunei*. **a** Location of small and large mussels 23 h after having been placed on the lower wall of horizontally oriented plastic cylinders in an aerated aquarium. *G*-test with Williams' correction for differences in the final location of the mussels; *B*-test of binomial proportions for differences between small and large mussels. (Modified from Uryu et al. 1996). **b** Maximum height reached by mussels of different size climbing along the internal wall of cylindrical tanks. (Modified from Iwasaki 1997)

Table 1 Climbing experiments with *Limnoperna fortunei*. Numbers of mussels climbing walls of cylindrical vessels filled with water to a height of 10 cm and mean height of attachment sites. After 23 h, water was removed, and mussels attached to the walls were exposed to air for 10 days; last column is the number of mussels that returned to the bottom during this period. (After Iwasaki 1997)

Shell length (mm)	No. of mussels	Mussels attaching above the bottom		Mussels returning to the bottom; no. (% of climbers)
		No. (%)	Height, cm (mean ± SD)	
< 10	24	13 (54.2)	8.1 ± 2.3	1 (7.7)
10–15	56	18 (32.1)	6.4 ± 3.0	2 (11.1)
15–20	62	14 (22.6)	7.2 ± 3.2	2 (14.3)
20–25	35	9 (25.7)	6.4 ± 2.8	1 (12.5)
25	12	2 (14.3)	3.7 ± 1.5	1 (33.3)
Total	191	56 (29.0)	6.8 ± 2.9	7 (12.5)

(Fig. 4b). The height reached by the climbing mussels tended to decrease with increasing shell length.

These results indicate that a sizable proportion of the animals display climbing behavior and negative geotaxis in the dark, eventually resettling in a new location, often near the air–water interface. Thus, the uneven colonization of various substrata, including pipelines and tunnels, may not only be influenced by settling selectivity of pediveliger larvae (Morton 1977; Sardiña et al. 2009) but also be due to migration and reattachment of individuals that have previously settled elsewhere.

Iwasaki (1997) suggested that the negative geotaxis and climbing behavior of juveniles and adults may have several adaptive advantages, including (1) avoidance of brackish deep waters in estuaries, (2) avoidance of bottom conditions where siltation and concentration of contaminants is highest, (3) avoidance of hypoxic bottom waters, and (4) avoidance of benthic predators, such as crabs and crayfishes (Torres et al. 2012; Carvalho et al. 2013).

Admittedly, this behavior also has negative implications for survival in inland waters, where water levels can fluctuate widely, because mussels attaching near the water surface are vulnerable to desiccation from aerial exposure. The opposite behavior was tested by Iwasaki (1997), that is—descent back towards the bottom after water has descended below the site of attachment, thus causing the mussel to be exposed to air. For this experiment, water was removed from the cylindrical vessel where 56 mussels had climbed up the walls and animals were left exposed to air for 10 days. Only seven mussels (12.5 %) returned to the bottom within 2 days following the start of the air exposure, and all of them survived. Although the mode of descent (downward crawling or detachment and falling) was not observed, no byssal threads were left at the attachment sites. The remaining 49 mussels stayed above the bottom during these 10 days and eventually died there. Thus, most mussels did not detach their byssal threads from the wall despite these hostile conditions. Iwasaki (1997) suggested that this climbing behavior towards the water surface has probably evolved in large lakes or the lower reaches of large rivers, where water levels can be relatively stable and climbing toward the surface has adaptive advantages.

Aggregating Behavior

Uryu et al. (1996) found that artificially detached and isolated mussels tend to aggregate in small clumps in the laboratory. They positioned 25 isolated, small (3.9–14.8 mm) mussels at 25 mm intervals in a petri dish (18.6 cm diameter, 3.5 cm depth) filled with water (Fig. 5a, inset). Mussel displacement and reattachment were analyzed in five replicates of the experiment. On average, 7.9 % of the mussels stopped crawling when they came into contact with other conspecifics. The number of solitary mussels decreased and clump size (i.e., number of mussels in a clump) increased with time. After 23 h, clumps with 2 and up to 15 mussels were formed (Fig. 5a), but 44 % of the animals remained isolated.

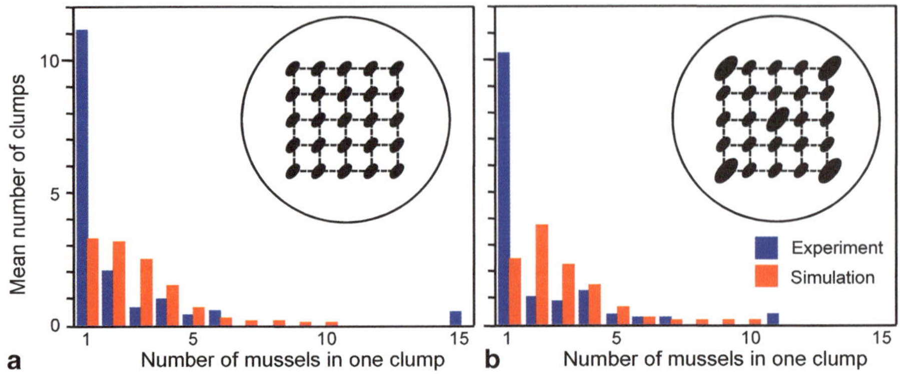

Fig. 5 Results of laboratory experiments and computer simulations (see text for details) testing the aggregating behavior of *L. fortunei* within a glass petri dish 18.6 cm in diameter. **a** Schematic of the 25 small mussels positioned at regular 25 mm intervals in the dish (*inset*), and size of clumps formed after 23 h (*bars*). **b** Schematic of the 25 small and large mussels positioned at regular 25 mm intervals in the dish (*inset*), and size of clumps formed after 23 h (*bars*). Data shown are averages for five replicate trials (*blue*) and results of numerical simulations (*red*). (Modified from Uryu et al. (1996)

A similar experiment with 20 small (7.0–14.9 mm) mussels and five large (19.2–26.1 mm) mussels was conducted to examine the effect of large mussels on aggregating behavior (i.e., clump size) of small mussels (Fig. 5b, inset). Again, the experiment was conducted in quintuplicate. On average, 15.7% of the small mussels stopped crawling and remained stationary after they came into contact with others. However, by the end of the experiment (23 h), mean clump size did not differ significantly from the previous trial (Fig. 5b). The presence of large mussels tended to suppress further movement of small mussels, but did not affect aggregating intensity.

Uryu et al. (1996) concluded that mussel clumps and the presence of large mussels promote byssal thread secretion by small mussels. In the experiments described above, small mussels in contact with conspecifics secreted byssal threads at significantly higher rates than small solitary mussels in both experiments (Table 2). Proportions of aggregated mussels that did secrete byssal threads were significantly higher when large mussels were present.

The presence of clumps including large mussels probably indicates that the site is stable and therefore fit for mussel growth and survival. Gregariousness reduces the risk of dislodgement due to water turbulence and predation (Bertness and Grosholz 1985). However, large mussel beds with high densities hinder growth and survival of small individuals (Bertness and Grosholz 1985; Okamura 1986). Thus, behavioral traits shown in the above experiments, such as cessation of crawling when small clumps that include large mussels are encountered, thus precluding the buildup of larger clumps, may have adaptive significance for the juveniles (Uryu et al. 1996).

Table 2 Byssal thread secretion by mussels in contact with other mussels after moving and aggregating, and by solitary mussels (not in contact with conspecifics; see Fig. 5 for experimental details). *G*-test with Williams' correction for differences between aggregated and solitary mussels; *B*-test of binomial proportions for differences between mussels secreting and not secreting byssal threads. (After Uryu et al. 1996)

	Small mussels only; *P* (*G*-test) < 0.05			Small and large mussels; *P* (*G*-test) < 0.01		
	Byssal threads secreted	Byssal threads not secreted	*P* (*B*-test)	Byssal threads secreted	Byssal threads not secreted	*P* (*B*-test)
Aggregated	42	11	<0.01	44	2	< 0.01
Solitary	24	23	0.5	17	17	0.5

Several studies suggested that chemotaxis plays an important role in the aggregation of mytilids (Senawong 1970; Tan 1975; de Vooys 2003; Nicastro et al. 2007). Uryu et al. (1996) examined this hypothesis for *L. fortunei* by comparing the above results with those from computer simulations, and applying the following assumptions: (1) all model mussels move randomly, both in distance and in direction, (2) model displacement speeds equal those of mussels of the same size as derived from the results of experiments of crawling behavior, and (3) when mussels come into contact with conspecifics, 7.9% (experiments with small mussels only) or 15.7% (experiments with small and large mussels) cease further displacement. For each group, 100 simulations were conducted in order to calculate final clump size. Simulations yielded much lower numbers of solitary mussels compared with actual experimental data, but simulations had higher numbers of clumps with 2–4 mussels (Fig. 5). In other words, aggregation intensity was higher in the simulation than that in the actual experiments. This result casts a doubt on the importance of chemotaxis as a driver of mussel aggregation (Uryu et al. 1996).

Byssal Thread Secretion

For mytilid mussels, byssus secretion is an important trait that ensures firm attachment to sites suitable for settlement. As described above, even postlarval *L. fortunei* secrete byssal threads that enable resettling after crawling, climbing, and aggregating in the laboratory.

Uryu et al. (1996) estimated byssal thread production in *L. fortunei* as a function of size, concluding that small mussels secrete significantly higher numbers of threads than large mussels. They placed small (4.6–14.6 mm) and large (15.9–26.8 mm) mussels in plastic containers filled with water and checked byssal thread production after 1, 2, and 3 days. Their results showed that small mussels produced significantly more byssal threads (~4–10) than large ones (~2–5).

Comparison with *Dreissena polymorpha*

The zebra mussel, *D. polymorpha,* is another invasive freshwater bivalve. It belongs to the family Dreissenidae (order Veneroida), whereas *L. fortunei* belongs to the Mytilidae (order Mytiloida). Native to the fresh and brackish waters of the Caspian and Black Sea drainage basins (Mordukhai-Boltovskoi 1960), *D. polymorpha* is currently one of the world's most invasive freshwater bivalves, spreading actively in Europe and North America (Nalepa and Schloesser 2014) (see chapter "Parallels and Contrasts Between *Limnoperna fortunei* and Species of *Dreissena*" in this volume).

In laboratory experiments, detached zebra mussels also actively move around. The distance moved is inversely proportional to mussel size, and *D. polymorpha* demonstrates strong negative phototaxis (Toomey et al. 2002). Aggregating and climbing behavior and reattachment to vertical walls of aquaria have also been described (Eckroat et al. 1993; Toomey et al. 2002). The zebra mussel has been shown to detach byssal threads and move along the surface of conspecific mussel clumps, both in the field and in laboratory settings (Eckroat et al. 1993; Toomey et al. 2002). Two types of byssal threads have been described for this species: permanent and temporary (Eckroat et al. 1993). Eckroat et al. (1993) concluded that the number of permanent threads secreted during 1–4 weeks is higher in large mussels (16–26 mm in length) than in small ones (8–12 mm), suggesting that smaller mussels can secrete only temporary threads that allow for temporary (rather than permanent) attachment.

This type of voluntary detachment and reattachment behavior has not been adequately investigated in *L. fortunei,* but it also seems to occur, albeit probably not as commonly as in *D. polymorpha.* The ability to detach from the glass walls of the fish tank when exposed to air indicates that at least some individuals can free themselves and relocate. Preliminary results based on laboratory experiments have also shown that firmly attached individuals can detach, crawl to a new position, and reattach (D. Duchini, pers. comm.).

Concluding Remarks

Observations on the behavior of juvenile and adult golden mussels indicate that the formation of mussel beds is partly due to preferential settling of larvae on sites already occupied by conspecifics, but detachment, displacement, and reattachment of juveniles and adults also play a large role in the fine-scale distribution. Selection of reattachment sites, presumably driven by adaptive advantages, has important implications for mussel fouling. Further studies on the mechanisms and stimuli involved in the detachment of individuals, as well as the drivers that control selection of reattachment sites, may greatly contribute to designing more efficacious control strategies to control fouling of *L. fortunei* in industrial installations.

References

Akehoshi Y (2011) Study of ecological features and prevention of damage of *Limnoperna fortunei* in the irrigation institution. Rep Yahagi River Inst 15:55–63 [In Japanese]

Ansell AD (1969) Leaping movements in the Bivalvia. Proc Malacol Soc Lond 38:387–399

Bertness MD, Grosholz E (1985) Population dynamics of the ribbed mussel, *Geukensia demissa* : the costs and benefits of an aggregated distribution. Oecologia 67:192–204

Brand AR (2006) Scallop ecology: distributions and behaviour. In: Shumway SE, Parsons GJ (eds) Scallops: biology, ecology and aquaculture, Elsevier, Amsterdam, pp 651–744

Carvalho DA, Collins PA, De Bonis CJ (2013) Predation ability of freshwater crabs: age and prey-specific differences in Trichodactylus borellianus (Brachyura: Trichodactylidae). J Freshwater Ecol 28:573–584

Côté IM (1995) Effects of predatory crab effluent on byssus production in mussels. J Exp Mar Biol Ecol 188:233–241

Côté IM, Jelnikar E (1999) Predator-induced clumping behaviour in mussels (*Mytilus edulis* Linnaeus). J Exp Mar Biol Ecol 235:201–211

de Vooys CGN (2003) Effect of a tripeptide on the aggregation behaviour of the blue mussel *Mytilus edulis*. Mar Biol 142:1119–1123

Dowd WW, Somero GN (2013) Behavior and survival of *Mytilus* congeners following episodes of elevated body temperature in air and seawater. J Exp Biol 216:502–514

Eckroat LR, Masteller EC, Shaffer JC, Steele LM (1993) The byssus of the zebra mussel (*Dreissena polymorpha*): morphology, byssal thread formation, and detachment. In: Nalepa TF, Schloesser D (eds) Zebra mussels: biology, impacts, and control, Lewis Publishers, Boca Raton, pp 239–263

Ishida S, Iwasaki K (1999) Immobilization of muricid whelks by byssal threads of an intertidal mussel, Hormomya mutabilis. Venus (Japanese Journal of Malacology) 58:55–59

Iwasaki K (1997) Climbing behaviour and tolerance to aerial exposure of a freshwater mussel, *Limnoperna fortunei*. Venus (Japanese Journal of Malacology), 56:15–25

Mordukhai-Boltovskoi FD (1960) Caspian fauna in the Azov and Black Sea basins. Akademia Nauk SSSR, Moscow, pp 1-286 [In Russian]

Morton B (1975) The colonization of Hong Kong's raw water supply system by *Limnoperna fortunei* (Dunker 1857) (Bivalvia: Mytilacea) from China. Malacol Rev 8:91–105

Morton B (1977) The population dynamics of *Limnoperna fortunei* (Dunker 1857) (Bivalvia: Mytilacea) in Plover Cove Reservoir, Hong Kong. Malacologia 16:165–182

Morton JE (1964) Locomotion. In: Wilbur KM, Yonge CM (eds) Physiology of Mollusca, Vol. 1, Academic Press, New York, pp 383–423

Nalepa TF, Schloesser DW (eds) (2014) Quagga and zebra mussels: biology, impacts, and control, 2nd ed. CRC Press, Boca Raton, pp 1–815

Nicastro KR, Zardi GI, McQuaid CD (2007) Behavioural response of invasive *Mytilus galllopro-vincialis* and indigenous *Perna perna* mussels exposed to risk of predation. Mar Ecol Prog Ser 336:169–175

Nicastro KR, Zardi GI, McQuaid CD, Pearson GA, Serrano EA (2012) Love thy neighbour: group properties of gaping behaviour in mussel aggregations. PLoS ONE 7(10):e47382. doi:10.13710047382.

Okamura B (1986) Group living and the effects of spatial position in aggregations of *Mytilus edulis*. Oecologia 69:341–347

Petraitis PS (1987) Immobilization of the predatory gastropod, *Nucella lapillus*, by its prey, *Mytilus edulis*. Biol Bull 172:307–314

Riisgård HU, Egede PP, Saavedra IB (2011) Feeding behaviour of the mussel, *Mytilus edulis* : new observations, with a minireview of current knowledge. J Mar Bio 2011:1–13. doi:10.1155/2011/312459

Robson AA, García de Leaniz C, Wilson RP, Halsey LG (2010) Behavioural adaptations of mussels to varying levels of food availability and predation risk. J Molluscan Stud 76:348–353

Sardiña P, Cataldo D, Boltovskoy D (2009) Effects of conspecifics on settling juveniles of the invasive golden mussel, *Limnoperna fortunei*. Aquat Sci 71:479–486

Schneider KR, Wethey DS, Helmuth BST, Hilbish TJ (2005) Implications of movement behavior on mussel dislodgement: exogenous selection in a *Mytilus* spp. hybrid zone. Mar Biol 146:333–343

Senawong C (1970) Biological studies of a littoral mussel, *Hormomya mutabilis* (Gould) I. Preliminary observations on the moving ability. Publ Set Mar Biol Lab 18:233–242

Stanley SM (1970) Relation of shell form to life habits of the Bivalvia (Mollusca). Geol Soc Am, Mem 125:1–296

Takada Y, Ito Y, Hayashi I (2013) Swimming ability of juvenile *Mactra chinensis* by pedal flapping. Venus (Japanese Journal of Malacology) 71:199–207

Tan WH (1975) The effects of exposure and crawling behaviour on the survival of recently settled green mussels (*Mytilus viridis* L.). Aquac 6:357–368

Toomey MB, McCabe D, Marsden JE (2002) Factors affecting the movement of adult zebra mussels (*Dreissena polymorpha*). J N Am Benthol Soc 21:468–475

Torres MV, Williner V, Giri F (2012) Size selective predation on an invasive bivalve, *Limnoperna fortunei* (Mytilidae), by a freshwater crab, *Zilchiopsis collastinensis* (Trichodactylidae). J Crustac Biol 32:698–710

Uchida A (2011) Occurrence of larvae and adults of the invasive mussel, *Limnoperna fortunei*, in the Yahagi River from 2007 to 2009. Rep Yahagi River Inst 15:65–70 [In Japanese]

Uryu K, Iwasaki K, Hinoue M (1996) Laboratory experiments on behaviour and movement of a freshwater mussel, *Limnoperna fortunei* (Dunker). J Molluscan Stud 62:327–341

Parallels and Contrasts Between *Limnoperna fortunei* and Species of *Dreissena*

Alexander Y. Karatayev, Demetrio Boltovskoy, Lyubov E. Burlakova and Dianna K. Padilla

Abstract *Limnoperna fortunei* (the golden mussel), *Dreissena polymorpha* (the zebra mussel), and *Dreissena rostriformis bugensis* (the quagga mussel) are considered among the most aggressive freshwater invaders. All three species share several biological traits, such as their sessile mode of life attached to hard substrata by a byssus (although quagga mussels can also dwell on muddy bottoms), similar sizes, similar longevity, and similar time to sexual maturity. The spawning period, however, is usually longer for *L. fortunei*. Ecologically, they also share similarities (e.g., suspension feeding mode), but the dreissenids thrive and reproduce in colder waters (especially *D. r. bugensis*), and are significantly less tolerant to low pH and calcium concentrations, hypoxic conditions, and pollution. Rates of intrabasin spread of *L. fortunei* in South America are roughly similar to those of *D. polymorpha* in North America, but interbasin spread is generally faster for the zebra mussel, probably partly due to cultural and economic differences between their respective invasive ranges. Geographic spread of quagga mussels has been much slower than that of

A. Y. Karatayev (✉)
Great Lakes Center, SUNY Buffalo State, 1300 Elmwood Avenue, Buffalo, NY 14222, USA
e-mail: karataay@buffalostate.edu

D. Boltovskoy
Facultad de Ciencias Exactas y Naturales, Instituto de Ecología, Genética y Evolución de Buenos Aires (IEGEBA), Universidad de Buenos Aires-CONICET, Ciudad Universitaria, 1428 Buenos Aires, Argentina
e-mail: demetrio@ege.fcen.uba.ar

Museo Argentino de Ciencias Naturales 'Bernardino Rivadavia'-CONICET, Buenos Aires, Argentina

L. E. Burlakova
The Research Foundation of The State University of New York, SUNY Buffalo State, Buffalo, NY 14222, USA
e-mail: burlakle@buffalostate.edu

D. K. Padilla
Department of Ecology and Evolution, Stony Brook University, Stony Brook, NY 11794-5245, USA
e-mail: padilla@life.bio.sunysb.edu

D. Boltovskoy (ed.), *Limnoperna fortunei,* Invading Nature - Springer Series in Invasion Ecology 10, DOI 10.1007/978-3-319-13494-9_15

zebra mussels, but once the former colonize waterbodies already populated by zebra mussels, they usually become dominant, both spatially and numerically. Judging from their respective environmental tolerance limits, in particular calcium concentrations, it is expected that both species of *Dreissena* may eventually colonize much of Europe, Asia, and North America, but colonization of South America, Africa, and Australia is less likely. In contrast, *L. fortunei,* which tolerates much lower calcium concentrations, could spread to areas presently occupied by the dreissenids as well as Africa and Australia. Should the three species overlap, it seems likely that *L. fortunei* will outcompete the dreissenids in warmer, more polluted, less oxygenated, and more acidic waters as well as in waters with lower calcium concentrations. However, the outcome of their competitive interactions when conditions are suitable for all three species is unclear. *L. fortunei* and both species of *Dreissena* are functionally similar, and as a consequence, many of their impacts on the systems they invade are also similar, yet the magnitude of these effects, and in some cases even their sign, can differ widely depending on the invasive species and environmental constraints. Future research on the golden mussel should focus on shedding light on the many unknown aspects of its biology and ecology, which are particularly critical for a comprehensive assessment of its interactions with local biota.

Keywords *Limnoperna fortunei* · *Dreissena polymorpha* · *Dreissena rostriformis bugensis* · Ecological impact · Distribution · Environmental tolerance · Geographic spread

Introduction

Although *Limnoperna fortunei* (the golden mussel) is taxonomically unrelated to *Dreissena polymorpha* (the zebra mussel) and *Dreissena rostriformis bugensis* (the quagga mussel), they have similar life histories, share many ecological traits, and are functionally similar. Therefore, their ecological and economic impacts on waterbodies they invade are often similar as well. Due to their high rates of spread, large numbers of colonized waterbodies, and the extent of their ecological and economic impacts, both species of *Dreissena* and *L. fortunei* are considered among the most aggressive freshwater invaders (Karatayev et al. 2007a, 2010a). All three are spreading at virtually all spatial scales and are expected to continue doing so (Karatayev et al. 2007a, 2007b, 2011, 2015; Pollux et al. 2010; Benson 2014; Boltovskoy and Correa 2015).

The overall impact of an invader depends on many factors, including, among others, the number of waterbodies colonized, its total population density in a given waterbody, its population dynamics, and distribution within a waterbody (Karatayev et al. 2010b, 2011). The number of waterbodies colonized will depend on the invader's ability to use different transport vectors, propagule pressure, environmental limits, and its life history and biological parameters (e.g., fecundity, growth, and survival), which ultimately determine total population size, population dynamics, and distribution within an invaded waterbody. Therefore, to accurately predict the potential spread and ecological impacts of invaders, it is essential to know their environmental limits and their biology. Although *D. polymorpha* is among the best studied freshwater invertebrates, less data are available for *D. r. bugensis* and *L. fortunei* (Karatayev et al. 2007a, 2007b, 2015, 2014a; Nalepa 2010; Boltovskoy and Correa 2015).

The aims of this chapter are to review similarities and differences among *L. fortunei* and both species of *Dreissena* in their biological traits, environmental limits, rates of spread, population dynamics, and ecological impacts, and to identify the essential information needed to better understand their geographic spread and their effects on ecosystems.

Life History

L. fortunei belongs to the largely marine bivalve family Mytilidae, while zebra and quagga mussels belong to the Dreissenidae, which is of brackish water origin. *L. fortunei* and *Dreissena* spp. represent an unusual ecological type in freshwaters and have traits typical of marine mussels, including free-swimming larvae and a sessile, attached adult stage.

Extensive research has been conducted on the biology, reproduction, growth and other life history traits of *D. polymorpha,* but relatively fewer studies have focused on *D. r. bugensis* and *L. fortunei* (Karatayev et al. 2007a, 2007b, 2014a, 2015; Nalepa 2010). We know that all three species mature at approximately the same age, and have a similar body size (Table 1).

In *L. fortunei* and *D. polymorpha* gonads are fully developed by the spring, and spawning typically occurs in spring–summer (Lvova and Makarova 1994; Boltovskoy et al. 2009b; see Chapter "Reproductive Output and Seasonality of *Limnoperna fortunei*" in this volume). In cold deep waters, gonads of *D. polymorpha* and *D. r. bugensis* may be ripe for a longer period of time during the year and spawning extends over more months, producing smaller recruitment events over a longer period (Bacchetta et al. 2010; Nalepa 2010). The duration of the reproductive period depends on the temperature regime and is longer in warmer regions. In the northern part of its range, spawning of *Dreissena* spp. lasts for about 3–5 months (Lvova and Makarova 1994). In Lake Mead (Arizona-Nevada, USA), however, quagga mussel veligers are present in the plankton year round, suggesting a much longer spawning season (Wong et al. 2012), similar to the spawning season of golden mussels in the tropics and subtropics (up to 10 months, Boltovskoy et al. 2009b). In contrast, in temperate and cold-temperate areas (Japan, Korea) *L. fortunei* produces larvae for only 1 month or less each year (Choi and Shin 1985; Nakano et al. 2010a) (see Figs. 2 and 3 in Chapter "Reproductive Output and Seasonality of *Limnoperna fortunei*", this volume).

For *L. fortunei,* most studies from Asia and South America concur that reproduction starts when water temperatures reach around 15–18 °C (Morton 1977; Choi

Table 1 Size and life history parameters for *Limnoperna fortunei, Dreissena polymorpha,* and *Dreissena rostriformis bugensis*

Parameter	*L. fortunei*	*D. polymorpha*	*D. r. bugensis*
Typical [maximum] length (mm)	20–30 (Boltovskoy et al. 2009a), [50.5] (Karatayev et al. 2010a)	20–30 (Karatayev et al. 2007a), [49] (Son 2007)	20–30 (Karatayev et al. 2014c)
Longevity (years)	2–3 (Boltovskoy and Cataldo 1999)	4–5 (Lvova et al. 1994b, Karatayev et al. 2006)	4–5 (Mills et al. 1996; Orlova et al. 2004)
Time to sexual maturity (months)	3–4 (Boltovskoy and Cataldo 1999; Darrigran et al. 1999)	3–11 (Lvova and Makarova 1994); 8–10 (McMahon and Bogan 2001)	No data
Typical spawning period (months/year)	<1 (temperate areas) to 10 (tropical and subtropical areas) (Choi and Shin 1985; Boltovskoy et al. 2009b; Nakano et al. 2010a)	3–5 (Lvova and Makarova 1994)	3–10 (Nalepa et al. 2010; Wong et al. 2012)
Fecundity (eggs per reproductive season)	No data	275,000–300,000 (Lvova 1977); up to 1,000,000 (Sprung 1991)	No data

and Shin 1985; Cataldo and Boltovskoy 2000; Nakano et al. 2010a; Brugnoli et al. 2011). So far, the golden mussel has not been reported from waterbodies where year round temperatures are below 15–18 °C, although there are many records from areas where water temperature is always above 15–18 °C (Mata 2011; Oliveira et al. 2011). Interestingly, in these tropical waterbodies the reproductive cycle is less regular, and slows noticeably in the winter (July–August in the southern hemisphere). Even when water temperatures are well above the threshold for reproduction (see Chapter "Reproductive Output and Seasonality of *Limnoperna fortunei*" in this volume), larval production typically decreases or even ceases in the winter. Zebra mussels usually initiate spawning when water temperatures reach 12–15 °C, typically in the late spring (May to June in the northern hemisphere), and continue to spawn until the end of summer (August or September) (Table 1; Sprung 1987; Borcherding 1991; Lvova et al. 1994a; Karatayev et al. 1998, 2010b; Pollux et al. 2010). Quagga mussels, which usually live deeper, can spawn at water temperatures as low as 4.5–6.0 °C (Nalepa 2010). However, in areas where they co-occur with zebra mussels, both dreissenid species may initiate spawning at the same time (e.g., 18–20°C; Claxton and Mackie 1998).

Thus, the golden mussel requires higher temperatures for reproduction (15–18 °C), followed by the zebra mussel (12–15 °C), and the quagga mussel can reproduce in much colder waters (variable, but occasionally as low as 5–6 °C).

Fecundity data are only available for *D. polymorpha* (Table 1). Female zebra mussels can spawn up to 10^6 eggs, and males up to nearly 10^{10} sperm, comprising more than 30 % of their body weight prior to spawning (Sprung 1991). In the absence of similar data for quagga mussels, their fecundity is often assumed to be the same as for zebra mussels (e.g., Keller et al. 2007). However, this may be not the case. Zebra and quagga mussels have very different population dynamics in the waterbodies they invade. The time lag between when a species is first detected in a waterbody and when it reaches its maximum population size being much shorter for zebra mussels (2.5 ± 0.2 years) than for quagga mussels (12.2 ± 1.5 years) (Karatayev et al. 2011). The shorter lag time for zebra mussels may reflect their higher reproductive potential. Information for *L. fortunei* is still too scant and fragmentary for comparison, but the few data at hand seem to indicate that the lag time is closer to that of the quagga than the zebra mussel (see Chapter "*Limnoperna fortunei* Colonies: Structure, Distribution and Dynamics" in this volume). This, however, does not necessarily imply comparatively lower fecundity because carrying capacity depends on many intrinsic (e.g., fecundity), and extrinsic traits (e.g., predation pressure, competition, etc.).

Longevity of zebra mussels (up to 4–5 years, reviewed in Lvova et al. 1994b; Mills et al. 1996; Orlova et al. 2004; Karatayev et al. 2007b), seems somewhat greater than that of *L. fortunei* (around 2–3 years; Morton 1977; Boltovskoy and Cataldo 1999) (Table 1).

Environmental Limits

Temperature

The lower temperature limit for both species of *Dreissena* is close to 0 °C. The upper temperature limit for *D. polymorpha,* determined from field observations in both Europe and North America, is around 32–33 °C (Aldridge et al. 1995; Karatayev et al. 1998; Allen et al. 1999; Table 2). Field observations indicate that quagga mussels are likely somewhat less tolerant of high temperatures than zebra mussels (reviewed in Mills et al. 1996; Karatayev et al. 1998; Garton et al. 2014). Data from the Zaporozhskoe Reservoir (Ukraine) show that quagga mussels survive in waters ≤ 30.5 °C, while zebra mussels tolerate waters ≤ 33 °C (Dyga and Zolotareva 1976).

For *L. fortunei,* the upper thermal limit is around 35 °C, which is somewhat higher than that of both dreissenids (Table 2). In South America, minimum winter temperatures of the waterbodies colonized by *L. fortunei* are around 10 °C, but in Japan golden mussels survive at water temperatures of 5–6 °C (Magara et al. 2001), and in Korea *L. fortunei* populations have been reported from the Paldang Reservoir, which freezes for 1–2 months every winter (Choi and Kim 1985; Choi and Shin 1985; Park et al. 2013; Hae-Kyung Park, pers. comm.).

While on the basis of these data, it is tempting to speculate that low winter temperatures are unlikely to be a deterrent for the spread of *L. fortunei* into cooler waterbodies, minimum survival temperature may not be a good indicator of the

Table 2 Environmental limits for *Limnoperna fortunei, Dreissena polymorpha,* and *Dreissena rostriformis bugensis.* Values given are those that allow for survival of adult individuals, but not necessarily reproduction or survival of larvae (unless otherwise noted)

Factors	*L. fortunei*	*D. polymorpha*	*D. r. bugensis*
Upper salinity limit (‰)	Continuous: 2 (Angonesi et al. 2008; Sylvester et al. 2013); discontinuous, punctuated by periods of fresh water: up to 23 (Sylvester et al. 2013)	6 (reviewed in Karatayev et al. 1998)	3.5 (reviewed in Lyakhnovich et al. 1994)
Lower temperature limit for adult survival (°C)	0 (Choi and Kim 1985; Choi and Shin 1985)	0 (Luferov 1965)[a]	0 (Orlova 1987)
Upper temperature limit (°C)	35 (Oliveira et al. 2011)	32–33 (reviewed in Karatayev et al. 1998, 2006, 2007b; Allen et al. 1999)	31 (reviewed in Karatayev et al. 1998, 2007b)
Lower temperature limit for reproduction (°C)	15–17 (Morton 1977; Cataldo and Boltovskoy 2000; Nakano et al. 2010a; Brugnoli et al. 2011)	12–15 (Sprung 1987; Borcherding 1991; Lvova et al. 1994a; Karatayev et al. 1998; Pollux et al. 2010; Garton et al. 2014)	5–7 (Roe and MacIsaac 1997; Nalepa 2010)
Lower pH limit	<6.0 (Oliveira et al. 2011)	7.3–7.5 (Sprung 1987; Ramcharan et al. 1992; Burlakova 1998; Hallstan et al. 2010)	No data
Lower calcium limit (mg/L)	1 (Oliveira et al. 2011)	23–28 (Ramcharan et al. 1992; Burlakova 1998)	No data
Lower oxygen limit at 20 °C (mg/L)	0.5 (Boltovskoy et al. 2006)[b]	1.8–2.4 (Spiridonov 1972; Shkorbatov et al. 1994)	1.5 (Shkorbatov et al. 1994)
Tolerance to pollution	High (Villar et al. 1999; Belaich et al. 2006; Boltovskoy et al. 2006; Bonel et al. 2013; Young et al. 2014)	Medium (bij de Vaate et al. 1992; Jantz and Neumann 1992; Burlakova 1998)	No data
Distribution within a waterbody	Littoral, sublittoral (Boltovskoy et al. 2009a; Karatayev et al. 2010a)[d]	Littoral, sublittoral (Karatayev et al. 1998, 2014c; Burlakova et al. 2006)	Littoral, sublittoral, and profundal (Patterson et al. 2005; Watkins et al. 2007; Nalepa et al. 2009a; Nalepa 2010; Karatayev et al. 2014b, 2015)

Table 2 (continued)

Factors	*L. fortunei*	*D. polymorpha*	*D. r. bugensis*
Substrate type	Natural or artificial hard substrata, aquatic macrophytes, avoids soft silt (Boltovskoy et al. 2006, 2009a; Rojas Molina 2010)[e]	Natural or artificial hard substrata, avoids soft silt (Karatayev et al. 1998, 2014c; Burlakova et al. 2006)	Natural or artificial hard substrata and soft silt (Nalepa 2010; Karatayev et al. 2015, 2014c)

[a] Based on records in the upper River Volga, which freezes in winter

[b] In experimental conditions, at 20 °C large (20 mm) mussels survive up to 29 days at 0.16 mg/L (Perepelizin and Boltovskoy 2011)

[c] In experimental conditions, at 17.5 °C mussels survive anoxia for up to 18 days (Matthews and McMahon 1994)

[d] Highest densities are normally recorded along the coastal fringe, where hard substrata are normally much more abundant than at depth. However, results with artificial substrata indicate higher densities of recruits in the deeper layers (5–18 m) than closer to the surface (Morton 1977, Nakano et al. 2010b, Brugnoli et al. 2011)

[e] See Chapter "*Limnoperna fortunei* Colonies: Structure, Distribution and Dynamics" in this volume

mussel's ability to maintain self-sustaining populations. Reproductive cycles (as evidenced by the presence of larvae in the water column) clearly show that temperature is the dominant factor for spawning (see Chapter "Reproductive Output and Seasonality of *Limnoperna fortunei*" in this volume). The shorter the periods of high temperature, the shorter is the spawning season. Thus, while in the Upper Paraná River, where water temperatures range around 18 to >30 °C, larvae are produced for 9–10 months each year, in Japan, at ~7–25 °C, larval output is restricted to 1–2 months, and in Korea, at 0–30 °C, reproduction is restricted to around 20 days (Choi and Shin 1985; Nakano et al. 2010a; Hamada 2011; Mata 2011; see Fig. 2, 3 and 10 in Chapter "Reproductive output and Seasonality of *Limnoperna fortunei*" in this volume). Interestingly, in all of these waterbodies, summer water temperatures are high. Even Paldang Reservoir, which freezes in the winter, reaches ~30 °C in the summer (Choi and Shin 1985). This suggests that the magnitude and duration of warm summer temperatures determine whether self-sustaining populations are possible, rather than minimum winter values. Data at hand indicate that the lowest temperatures at which *L. fortunei* spawns are around 15–18 °C (see Chapter "Reproductive Output and Seasonality of *Limnoperna fortunei*" in this volume), which suggests that waterbodies whose temperature is always below these values are unlikely to be colonized by this mussel. Therefore, Andean Patagonian lakes located south of ~38 °S, most of which do not freeze but never reach temperatures above 13–15 °C (Baigun and Marinone 1995; Díaz et al. 2000) are most probably not at risk of colonization by *L. fortunei*. In contrast, the North American Great Lakes, which may freeze in the winter, but usually have 3–4 month periods when water temperatures are above 16 °C (except Lake Superior; National Oceanic and Atmospheric Administration, NOAA 2014), are probably suitable for colonization by *L. fortunei*.

Salinity

In Europe and North America, *D. polymorpha* can form stable populations at salinities below 6 ‰, which is only slightly higher than the limit for quagga mussels (Table 2). For *L. fortunei,* constant salinities around 2 ‰ are the upper limit for extended survival (Huang et al. 1981; Angonesi et al. 2008; Barbosa and Melo 2009; Sylvester et al. 2013). However, at intermittent saltwater-freshwater conditions, such as those normally present in tidal estuaries, golden mussels can tolerate short periods (hours) of salinities up to 23 ‰ without significant mortality (Sylvester et al. 2013; see Chapter "Chemical Strategies for the Control of the Golden Mussel (*Limnoperna fortunei*) in Industrial Facilities" in this volume). This suggests that tests at constant salinity underestimate the tolerance of this species, and probably other freshwater molluscs, to saltwater exposure. Because estuarine ports represent ~70 % of nonmarine ports globally, they constitute major donor and recipient hotspots for the spread of nonnative species into continental aquatic ecosystems via shipping. It is probable that the tolerance of *L. fortunei* to estuarine conditions contributes to this species' success as an invader (Sylvester et al. 2013).

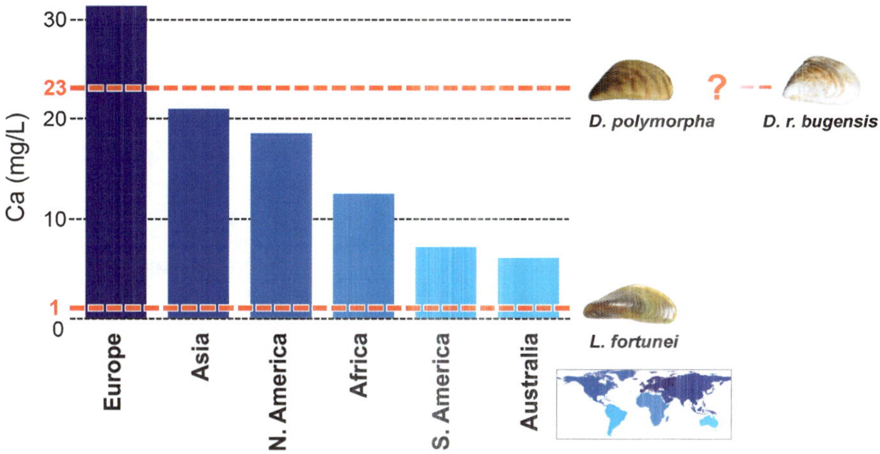

Fig. 1 Mean calcium concentrations in rivers on different continents (Wetzel 1975) and minimum calcium requirements for *Dreissena polymorpha* and *Limnoperna fortunei*

pH and Calcium

Zebra mussels are restricted to waters with neutral or alkaline pH (>7.3–7.5; Table 2). To our knowledge, there are no published data on pH limits for quagga mussels. Both in Europe and in North America, zebra mussels have colonized many more waterbodies than have quagga mussels. With few exceptions, almost all lakes had already been colonized by zebra mussels when quagga mussels invaded, suggesting that the pH limits for both species of *Dreissena* largely overlap. The threshold for calcium needed to support sustainable populations of zebra mussels is >23 mg/L (Fig. 1, Table 2), although values as low as 8–15 mg/L have been reported (Mellina and Rasmussen 1994; Jones and Ricciardi 2005). However, these lower values may reflect limits for the survival of adult mussels, rather than the establishment of locally sustainable populations (Sprung 1987).

The calcium limits for both species of *Dreissena* are substantially higher than those for *L. fortunei* (Fig. 1). Calcium is generally scarce in South American floodplain rivers colonized by the golden mussel (3-9 mg/L; Maglianesi 1973, Bonetto et al. 1998), and values as low as 1 mg/L of Ca and pH<6 have been reported from some areas successfully colonized by *L. fortunei*, such as the Upper Paraguay River (Oliveira et al. 2011).

Dissolved Oxygen

D. polymorpha is intolerant of even moderate hypoxia. Although it may colonize the deep oxygenated areas of some lakes, it usually is restricted to littoral and sublittoral zones (reviewed in Karatayev et al. 1998, 2015) (Table 2). In contrast, *D. r. bugensis* survives at lower oxygen concentrations than the zebra mussel (Shkorbatov et al.

1994), which may be related to its lower respiration rate (Stoeckmann 2003), and at least partially explains the ability of quagga mussels to colonize the profundal zone of deep lakes. However, both species of *Dreissena* are absent from hypoxic areas (e.g., central basin of Lake Erie; Karatayev et al. 2014c).

In contrast, *L. fortunei* survives in areas with very low oxygen concentrations, high organic loads, and industrial pollution (Villar et al. 1999; Belaich et al. 2006; Boltovskoy et al. 2006; Perepelizin and Boltovskoy 2011; Bonel et al. 2013; Young et al. 2014). In the delta of the Lower Paraná River, dense *L. fortunei* beds are present in the vicinity of urbanized and industrialized areas which discharge untreated domestic and industrial wastes. These waters and sediments contain pollutants at levels several times above those considered hazardous for aquatic life (e.g., Zn, Cr, Cu, Benzo[a] pyrene, polychlorinated biphenyls (PCBs), etc.), where other organisms (e.g., the Asian clam, *Corbicula fluminea*) do not survive (Cataldo et al. 2001a, 2001b).

Substrata

Within a waterbody, one of the main factors that affect the distribution of both *L. fortunei* and *Dreissena* spp. is the availability of suitable substrata. These mussels usually require hard substrate for attachment, and therefore their distribution is extremely patchy, with harder and coarser substrata yielding the highest densities and biomass of mussels. The most favorable substrata for these species are rocks, gravel, shells, and consolidated sediments (Karatayev et al. 1998, 2010a; Boltovskoy et al. 2006, 2009a; Burlakova et al. 2006).

In South American rivers, where hard substrata are scarce, plants may constitute important sites for attachment. Roots, rhizomes, and stolons of the water hyacinth (*Eichhornia crassipes, Eichhornia azurea*) seem to be particularly important substrata. Although densities of *L. fortunei* on these plants are comparatively low (see Chapter "*Limnoperna fortunei* Colonies: Structure, Distribution and Dynamics" in this volume), the abundance and widespread distribution of species of *Eichhornia* make them key elements of seeding sites (Callil et al. 2006; Marçal and Callil 2008, 2012; Rojas Molina 2010; Rojas Molina et al. 2010; Ohtaka et al. 2011; see Fig. 3e in Chapter "*Limnoperna fortunei* colonies: structure, distribution and dynamics" in this volume).

D. polymorpha and *L. fortunei* usually avoid pure mud, where they only occur on isolated hard objects, such as wood fragments, shells, stones, or artificial substrata (e.g., discarded debris). Mussels can use the hard fragments for initial attachment and subsequently attach to each other forming druses (Karatayev et al. 1998, 2010a, 2015).

Although the pattern of distribution of *L. fortunei* across substrate types is similar to *D. polymorpha,* the golden mussel appears to reach higher densities and especially higher biomass per unit area (Karatayev et al. 2010a). Both zebra and golden mussels are largely limited to the littoral zone and usually avoid soft sediments of the cold profundal zone (Karatayev et al. 1998, 2010a, 2015; Burlakova et al. 2006; Boltovskoy et al. 2009a). It is not clear, however, if golden mussels favor the shallow, coastal fringe because that is where hard substrata are most often found

(Boltovskoy et al. 2009a), or because they prefer shallower sites, regardless of substrate type. Colonization of artificial substrata suggests that recruits prefer settling at depth (6–18 m), rather than closer to the surface (Morton 1977; Nakano et al. 2010b; Brugnoli et al. 2011), but this pattern may also reflect differences in predation pressure. Comparison of samples scraped from the concrete wall of a penstock of the hydroelectric Yacyretá power plant (Upper Paraná River) yielded higher densities at 10 m (248,200 ind./m^2), than at the surface (170,400 ind./m^2), and at 40 m (54,400 ind./m^2) (Darrigran et al. 2007). While differences in densities at these three depths may reflect differences in hydrodynamics specific to this particular water intake structure, they still show that when offered adequate substrate, within these limits depth does not curtail the survival of *L. fortunei*.

In addition to the littoral zone, quagga mussels can colonize silty sediments, especially those found in the profundal zones of deep large lakes (Patterson et al. 2005; Watkins et al. 2007; Nalepa et al. 2009a; Nalepa 2010; Karatayev et al. 2015, 2014c). In these soft sediments, *D. r. bugensis* usually has a more even distribution across the bottom, and rarely forms large druses. Instead, single mussels or small aggregations almost float on the surface of the silty bottom (Nalepa 2010; Karatayev et al. 2014c). Therefore, in deep lakes with large profundal zones, quagga mussels may be found at higher overall numbers across the whole lake than either zebra or golden mussels.

Rate of Spread

Of the three species considered in this Chapter, *D. polymorpha* has by far the longest and the best-documented history of invasion. This species began to spread from its native range in Europe in the early 1800s (Karatayev et al. 2007b, 2011, 2015; Pollux et al. 2010; van der Velde et al. 2010; bij de Vaate et al. 2014).

At the global scale, three major phases in the spread of the zebra mussel can be recognized: (1) An initial exponential phase in the nineteenth century in Europe, where it spread at a rate of ~3.9 geographic regions (countries, or geographic provinces within large countries) per decade; (2) A period of extremely slow spread for almost a century during the industrial revolution and increased water pollution; and (3) A second period of exponential spread that started in the 1960s, and included expansion in both Europe and North America (where zebra mussels were introduced in the 1980s, Carlton 2008), when it spread at an average rate ~6.6 regions/decade (Karatayev et al. 2011).

Although there was extensive ship traffic between areas inhabited by *D. r. bugensis* (the Dnieper-Bug Liman and the lower reaches of the Southern Bug River in Ukraine) and other regions of eastern and western Europe through the middle of the twentieth century, quagga mussels remained restricted to their native range until the 1940s (Zhulidov et al. 2004; Karatayev et al. 2007b, 2011; Son 2007; van der Velde et al. 2010; Zhulidov et al. 2010). Starting in the mid-1980s, quagga mussels spread in Europe and North America (where this species was first discovered in 1989, Mills et al. 1993) at a rate of 7.4 regions/decade, which is significantly faster than the initial spread of zebra mussels in Europe, but similar to the current rate of spread of zebra mussels at a global scale (Karatayev et al. 2011, 2014a). The delay in the spread of *D. r. bugensis* was likely due to its inability to use mechanisms and vectors responsible for the spread as efficiently as *D. polymorpha*. Quagga mussels appear to be less resistant to dislodgment than zebra mussels (Mackie 1991; Dermott and Munawar 1993; Peyer et al. 2009, 2010). As a result, zebra mussels may be more likely to remain attached to boat hulls than quagga mussels, facilitating their transport to new habitats.

In Europe, most waterbodies had already been colonized by zebra mussels long before quagga mussels began to spread, making it difficult to compare their rates of spread. However, North America was colonized by both species at approximately the same time (1980s), in the same area (Lake Erie), making their rates of spread in North America directly comparable. By 2008, zebra mussels had colonized twice as many US states as quagga mussels, almost eight times more counties, and over 15 times more waterbodies (Karatayev et al. 2011). By 2010, 25 years after their introduction into North America, *D. polymorpha* had colonized 17 times more waterbodies than *D. r. bugensis* (Benson 2014). These differences clearly show that zebra mussels are far more efficient at colonizing new waterbodies than quagga mussels.

It has been shown that estimates of the rates of spread of exotic bivalves depend upon the spatial resolution of the scale of spread, and may be accelerated or slowed by various human activities (reviewed in Karatayev et al. 2007b; see Chapter "Distribution and Colonization of *Limnoperna fortunei:* Special Traits of an Odd Mussel" in this volume). In general, the rate of spread is slower at finer spatial scales. For example, aquatic exotic species may quickly spread along connected waterways within a recently invaded continent, and soon reach their maximum range across the continental scale. However, it takes much longer to colonize all regions within an invaded continent, and much longer again to spread to every isolated lake and river (waterbody scale) within a region. This difference in the rate of colonization across different spatial scales may be several orders of magnitude. For example, in the nineteenth century it took less than 40 years for *D. polymorpha* to spread across Europe, chiefly through canal systems, to present day Belarus, Poland, the Baltic states, Great Britain, the Netherlands, Germany, Belgium, and France (reviewed in Karatayev et al. 2007b). On the other hand, at the regional scale it took over 150 years for *D. polymorpha* to spread across geographical barriers to Alpine regions (Kinzelbach 1992), and almost 200 years to colonize Ireland (Minchin 2000) and Spain (bij de Vaate et al. 2002).

The spread of *L. fortunei* outside of its purported native range in China, south of the Yangtze River, into tropical Indochina (Cambodia, Laos, Thailand, Vietnam), likely occurred centuries ago (Morton and Dinesen 2010), but the first documented record of expansion was in 1965, when this species colonized Hong Kong (Morton 1975). In the late 1980s, it was recorded in Japan (Matsuda and Uenishi 1992). In the early 1990s, it spread to South America (Pastorino et al. 1993), and is presently found in Argentina, Uruguay, Paraguay, Bolivia, and Brazil (see Chapter "Colonization and Spread of *Limnoperna fortunei* in South America" in this volume). A rough comparison of the rates of spread shows that *D. polymorpha* spread ~2800 km

(Minneapolis to New Orleans in 7 years (1986–1993), whereas *L. fortunei* spread 3400 km (Río de la Plata estuary to the Pantanal wetland) in 8–9 years (1990–1998). Thus, the rates of expansion in these areas have been generally similar, but the major pathways used for expansion likely differed. Once *D. polymorpha* colonized the uppermost reaches of the Mississippi River system (in 1991; Benson 2014), it swiftly expanded southwards by means of its downstream drifting larvae (Stoeckel et al. 2004). In contrast, *L. fortunei* first invaded the outlet of the Río de la Plata watershed (the Río de la Plata estuary), and spread northwards and upstream. Upstream expansion was obviously facilitated by attachment of adult individuals to the hulls of commercial boats that operate along the Paraná-Paraguay waterway, thus fitting the "jump dispersal" mode (MacIsaac et al. 2001). For *L. fortunei,* the importance of boat traffic as a dispersal vector is reinforced by the fact that in the Uruguay River, much of which is not navigable, the upstream expansion has been much slower than in the Paraná-Paraguay system (Boltovskoy et al. 2006).

Rates of spread across river basins, on the other hand, have apparently been faster for *D. polymorpha,* especially in the USA, than for *L. fortunei*. In Japan, the golden mussel is still restricted to a rather limited part of the country (see Chapter "Colonization and Spread of *Limnoperna fortunei* in Japan" in this volume), whereas in South America in more than 20 years only one major basin has been colonized (the Río de la Plata basin), and a few minor ones (Mar Chiquita, Patos-Mirim, Guaíba, Tramandaí; see Chapter "Colonization and Spread of *Limnoperna fortunei* in South America" in this volume).

The main mechanisms for interbasin dispersal of freshwater mussels are man-made canals and aqueducts, and overland transport. Canals and aqueducts are partly responsible for the spread of *L. fortunei* in China (see Chapter "Distribution and Spread of *Limnoperna fortunei* in China" in this volume), and in Japan (see Chapter "Colonization and Spread of *Limnoperna fortunei* in Japan" in this volume), but, for different reasons, their impact has been limited. In China, many of the major hydraulic projects are very recent, suggesting that the effects of invasion are still underway. In Japan, there are 400,000 km of man-made canals, many of which connect watersheds (Ministry of Agriculture, Forestry and Fisheries 2003); however, because of the country's topography, watersheds are numerous and very small (Japan Commission on Large Dams 2009). Despite a millennium of efforts by man to reshape the drainage network (according to the International Commission on Large Dams, of the 20 oldest dams in the world, 15 are located in Japan), many are still isolated. This may explain why the overall spread of *L. fortunei* in Japan has been comparatively slow (see Chapter "Colonization and Spread of *Limnoperna fortunei* in Japan" in this volume). In contrast, there are many large, navigable rivers in the USA, and almost 20,000 man-made canals, including several major interbasin transfer aqueducts, some of which are known to have been instrumental for the rapid dispersal of dreissenids (Benson 2014). In comparison, natural basins in South America have suffered little modification (with the exception of dams, especially in the Río de la Plata watershed, see Chapter "Colonization and Spread of *Limnoperna fortunei* in South America" in this volume), and there are no man-made interbasin

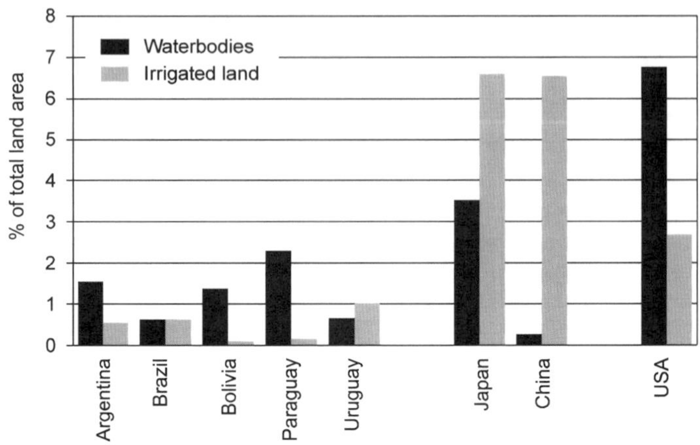

Fig. 2 Percentage of total land area occupied by waterbodies and artificially irrigated in countries invaded by *Limnoperna fortunei* or *Dreissena* spp.

connections (although plans to interconnect all major navigable waterways have been under consideration for years).

For freshwater byssate mussels, overland interbasin transfer is chiefly accomplished through fouling of recreational boats (Balcom 1994; Padilla et al. 1996; Buch and McMahon 2001; Johnson et al. 2001). Thus, invasion pressure on unconnected waterbodies is highly dependent on the number of boats (chiefly trailered), which in turn is associated with income and living standard levels. By all indices, the USA has higher economic development than China and the five South American countries where *L. fortunei* is invasive, and most probably has a significantly higher number of recreational, trailerable watercraft.

Another potentially important factor is the number and density of waterbodies. Areas where lakes and rivers are more numerous would be more susceptible to the dispersal of aquatic species than those where such features are scarcer. In the USA, the surface of lakes and rivers accounts for 6.8% of the total land area, which is 3–10 times higher than in any of the South American countries invaded by *L. fortunei,* and 2 and 24 times higher than in Japan and China, respectively (Fig. 2).

Potential for Future Spread

Dispersal of exotic species can be considered at different spatial scales, including the global, regional, local, and waterbody scales, each characterized by particular environmental constraints (Karatayev et al. 2007b). Because different exotic bivalves have different environmental limits (Table 2), their current and potential ranges are also different. Based on thermal tolerance alone, all three species have the potential to invade all continents except Antarctica, but none has fully reached

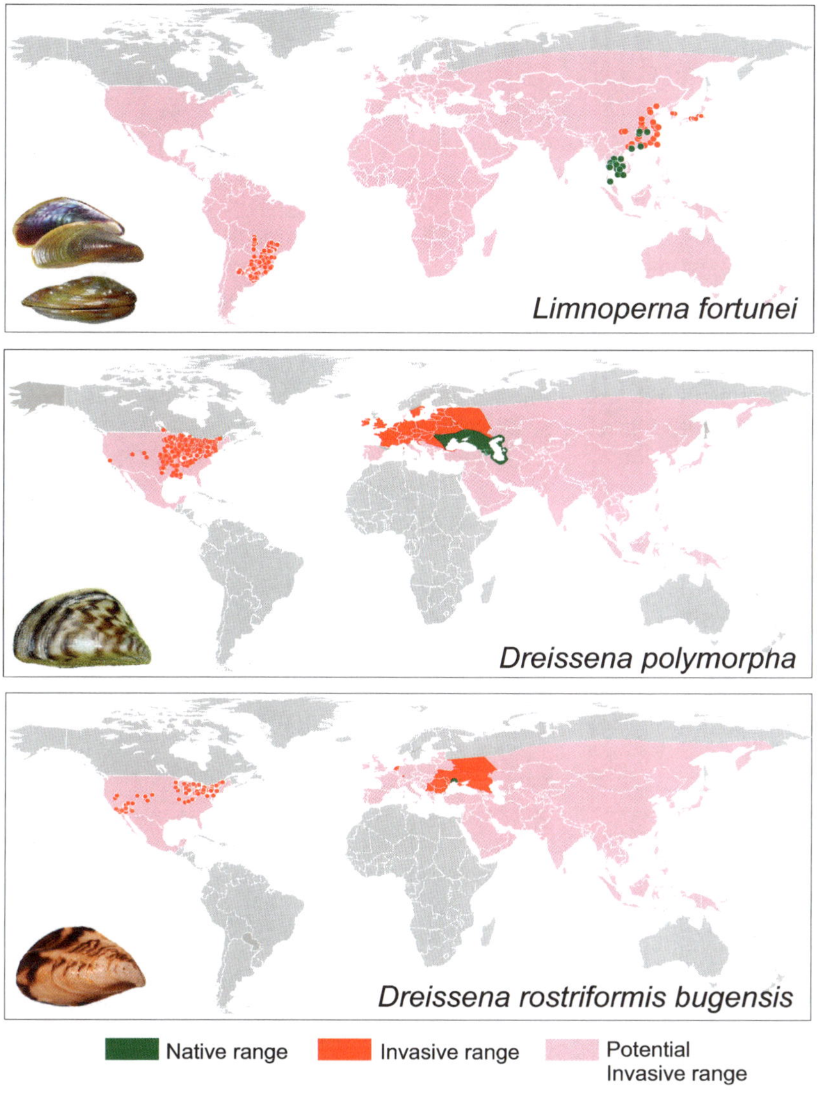

Fig. 3 Current and potential worldwide distribution of *Limnoperna fortunei, Dreissena polymorpha,* and *Dreissena rostriformis bugensis.* For *L. fortunei,* records in Indochina (denoted as native) are probably areas invaded before the twentieth century. (Distribution of *D. polymorpha* and *D. r. bugensis* in North America, courtesy of the United States Geological Service, Nonindigenous Aquatic Species)

this potential (Karatayev et al. 2007b). Both species of *Dreissena* are expanding their range in eastern and western Europe and have invaded North America. Neither has yet invaded Asia, Africa, South America, or Australia (Fig. 3). *L. fortunei* is spreading in Asia and has invaded South America (Fig. 3). Colonization of Asia and

Africa by *D. polymorpha* and *D. r. bugensis* has been anticipated (e.g., Starobogatov and Andreeva 1994), but high calcium requirements (Ramcharan et al. 1992; Burlakova 1998; Karatayev et al. 2007b) may curtail their spread into these continents. Most Australian and South American fresh waters have low concentrations of calcium, averaging 4 and 7 mg/L, respectively, while many North American and most European freshwaters have calcium concentrations that normally exceed 20–30 mg/L (Wetzel 1975; Payne 1986; Fig. 1). In South American Patagonia (Argentina and Chile), the temperature of many lakes located along the Andes cordillera south of ~38°S seems adequate for colonization by zebra and quagga mussels (2–3 to ~16 °C), but, again, the levels of dissolved calcium are most likely too low for these mussels. Of 21 lakes analyzed by Díaz et al. (2000), only two have Ca levels above 7 mg/L.

In contrast, it seems likely that *L. fortunei* may colonize many of the areas presently occupied by species of *Dreissena,* as well as those where *Dreissena* cannot live, including North America, Europe, Africa, and Australia (Karatayev et al. 2007b). Temperature may represent a deterrent in some regions, but the fact that the golden mussel thrives in Paldang Reservoir (Korea) suggests that as long as peak summer temperatures are high (> 18 °C) it might establish viable populations and survive winter temperatures as low as 0 °C. Colonization of other major South American watersheds, especially those that drain into the Atlantic Ocean (Tocantins, São Francisco, Amazonas, Orinoco, Magdalena), is probably inevitable (Boltovskoy et al. 2006; Oliveira et al. 2010), but so far no records of invasion by *L. fortunei* have been reported from these basins.

The spread of these three species is still far from complete. For example, in 2008, after more than 200 years of invasion in Belarus, only 33 % of all colonizable lakes were invaded by the zebra mussel (Karatayev et al. 2010a). Similarly, less than 10 years after the initial invasion of North America, zebra mussels had spread throughout most of the major connected river systems east of the continental divide; however, this spread has been much slower at the regional scale, and even slower at the waterbody scale (Padilla 2005). After more than 20 years of invasion, only 120 of more than 15,000 inland lakes in Wisconsin (< 1 %) were invaded by 2013 (reviewed in Karatayev et al. 2014a). To date, quagga mussels have not invaded any of the inland lakes in Belarus or in Wisconsin. Similarly, in Argentina, the golden mussel is present in a small fraction of the potentially colonizable waterbodies (Fig. 4). In Buenos Aires Province alone, < 10 % of the ~530 permanent lentic waterbodies (Toresani et al. 1994) are currently invaded by *L. fortunei.* This mussel has not yet expanded its range beyond the large Río de la Plata watershed, a few minor basins in Uruguay and southern Brazil, and a small endorheic basin (Mar Chiquita) located in central Argentina (see Chapter "Colonization and Spread of *Limnoperna fortunei* in South America" in this volume).

It should be noted that monitoring of South American waterbodies for the presence of golden mussels is nowhere as systematic and thorough as that for dreissenids in the northern hemisphere. None of the countries invaded has a comprehensive program aimed at the early detection of *L. fortunei,* and efforts at tracking its expansion are isolated and uncoordinated. Furthermore, while so far the golden mussel has

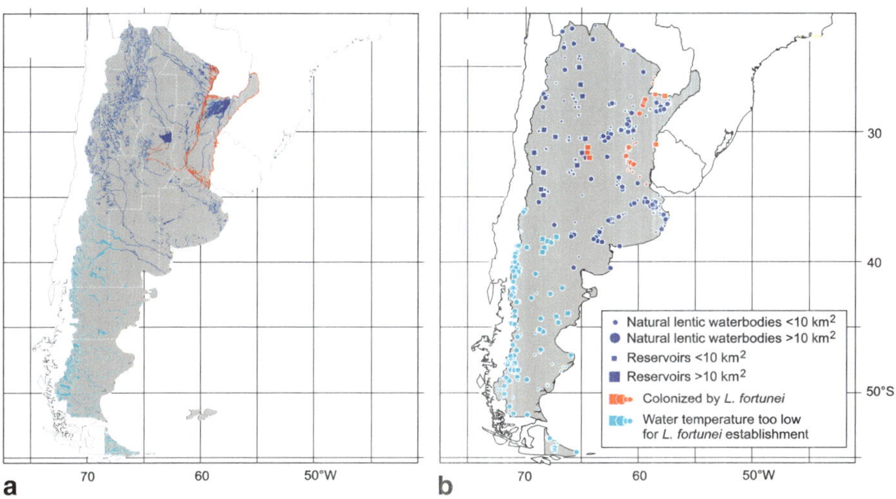

Fig. 4 Waterbodies (**a:** lakes and rivers, **b:** major lakes and reservoirs) colonized by the golden mussel in Argentina (*red*). *Light blue* denotes waterbodies where water temperatures are probably too low for the establishment of *Limnoperna fortunei*

been spreading in populated and industrialized areas of the Río de la Plata watershed, where the presence of this invader seldom went unnoticed, the next major watershed, the Amazon, is largely a very sparsely populated dense rainforest where most of the population lives in a few larger cities. Thus, the presence of the golden mussel is less likely to be noticed swiftly, and it is even less likely to be reported in the literature. On the other hand, because rivers are the main paths of transportation for people and produce, once colonization of the Amazon basin starts, the spread of golden mussels will likely be very fast.

Competition

The distributional ranges of zebra and quagga mussels overlap in Europe and in North America, and both species have the potential to overlap with *L. fortunei* in the future (Fig. 3). When co-occurring, species with similar habitat use will be expected to compete. Zebra and quagga mussels coexist in their native range in the Dnieper River delta and Dnieper-Bug Liman, Ukraine (reviewed in Zhulidov et al. 2010; Karatayev et al. 2011, 2014a). However, where the invasive ranges of both species overlap (Fig. 3), quagga mussels seem to outcompete zebra mussels over time (Nalepa 2010; Zhulidov et al. 2010; Karatayev et al. 2011, 2014a). This is especially typical for deep waterbodies and is likely due to the greater energetic efficiency of *D. r. bugensis* (Mills et al. 1999; Diggins 2001; Baldwin et al. 2002; Stoeckmann 2003; Karatayev et al. 2010c; Nalepa 2010). In addition, *D. r. bugensis* can colonize silty habitats, is more tolerant of low oxygen levels, has higher growth rates and lower mortality, and reproduces at lower temperatures than *D. polymorpha* (Mills et al. 1996; Claxton and Mackie 1998; Karatayev et al. 1998, 2010c, 2014c; Nalepa 2010). These demographic and physiological traits of quagga mussels allow them to colonize the large, cold profundal zone of deep lakes, which is unsuitable for zebra mussels. Thus, they can colonize the entire lake achieving much higher total population sizes, and outcompete zebra mussels by depleting food resources to levels that are too low for zebra mussels, but sufficient to support quagga mussels (Nalepa 2010; Karatayev et al. 2011, 2014a, 2015). In contrast, due to their tolerance of some abiotic factors (Table 2), greater rate of byssal thread production, and higher attachment strength (Peyer et al. 2009, 2010), zebra mussels are likely to be better adapted to the unstable, high-energy environment of the upper littoral zone, where fluctuations in temperature, currents, and wave action are prominent (reviewed in Karatayev et al. 2011, 2014a, 2015). Zebra mussels can still have an advantage in shallow lakes and rivers and coexist with quagga mussels (Zhulidov et al. 2004, 2010; Grigorovich et al. 2008; Peyer et al. 2009; Karatayev et al. 2011, 2015). Even in Lake Erie, where $\geq$95 % of the mussels in the central and eastern basins are quagga mussels, *D. polymorpha* still composes 30 % of the mussels in the shallow western basin after more than 20 years of coexistence (Karatayev et al. 2014c).

Because at present *L. fortunei* does not co-occur with either species of *Dreissena*, it is difficult to predict if they will compete, and if they do, under what conditions which species will prevail. Data at hand suggest that such co-occurrence is most likely to happen in Europe, Asia, and/or North America (Fig. 3). *D. polymorpha* has recently been found on barges imported to Argentina from the USA for grain transport along the Paraná-Paraguay rivers (Pablo Almada, personal observation), indicating that ballast water is not the only pathway for intercontinental transport of invaders. However, no live mussels were found in the few samples examined, and even if they had zebra mussels would probably not survive in the Río de la Plata watershed (see above). From their corresponding environmental tolerance ranges (Table 2), it seems likely that *L. fortunei* will outcompete the dreissenids in warmer, more polluted, less oxygenated and more acidic waters, as well as in waters with lower Ca concentrations. However, the outcome of their competitive interaction when conditions are suitable for all three species is unclear. The niche of *L. fortunei* within a waterbody appears to be more similar to that of zebra mussels than quagga mussels, suggesting that the competition may be stronger between the former two species (Table 2). However, similarities between zebra mussels and golden mussels also suggest that the impact of *D. r. bugensis* on *L. fortunei* may be similar to the one observed on *D. polymorpha*.

Impacts of Invasion

L. fortunei and both species of *Dreissena* are functionally similar, and as a consequence, many of their impacts on the systems they invade are also similar (Table 3). All three species are ecosystem engineers, sessile suspension feeders that attach to substrate with byssal threads. All three form druses, increasing habitat complexity for other benthic invertebrates, and affect planktonic communities, trophic relationships, and nutrient cycling via their feeding and filtering activities (Karatayev et al. 1997, 2002, 2007a, 2007b, 2015; Darrigran 2002; Beekey et al. 2004; Boltovskoy et al. 2006, 2009a; Burlakova et al. 2012; Boltovskoy and Correa 2015). However, the magnitude of these effects, and in some cases even their sign, depends on the invasive species, the other species present in the native community, and waterbody type.

Effects on Benthic Invertebrates

By creating reef-like three-dimensional structure, both species of *Dreissena* and *L. fortunei* change the physical habitat and provide refuge from predation and from physical stressors (waves, currents, desiccation) for benthic organisms that would otherwise be scarce or absent. In addition to increased habitat complexity, the impact of these byssate bivalves is compounded by their role as suspension feeders. All three species increase the rates of deposition of both inorganic and especially organic material on the bottom, providing an enhanced food subsidy for benthic deposit feeders. Many studies have shown that both *D. polymorpha* and *L. fortunei* have positive effects on most native invertebrates, which take advantage of both the structural complexity and food resources provided by zebra and golden mussels (Botts et al. 1996, Karatayev et al. 1997, 2002, 2007a, 2007b, 2010a; Darrigran et al. 1998; Stewart et al. 1998; Gutierrez et al. 2003; Beekey et al. 2004; Sardiña et al. 2008, 2011; Burlakova et al. 2012; Boltovskoy and Correa 2015; see Chapter "Relationships of *Limnoperna fortunei* with Benthic Animals" in this volume). At the same time, a few species of invertebrates have occasionally been found to be less abundant in mussel beds than in nearby bare sediments. Sardiña et al. (2011) reported that some snails, ostracods, nematodes, and chironomids may be less abundant in *L. fortunei* beds than in nearby bare sediments. However, the overall diversity, density, and biomass of native invertebrates is always higher in druses and mussel beds compared to nearby bare sediments (Table 3).

While in the littoral zone the effects of *D. r. bugensis* are probably similar to those of zebra and golden mussels (Bially and MacIsaac 2000; Yakovleva and Yakovlev 2011), in the cold profundal zone of deep lakes (where, unlike zebra and golden mussels, quagga mussels can be very abundant; Patterson et al. 2005; Watkins et al. 2007; Nalepa 2010; Karatayev et al. 2015, 2014c), their effects are quite different. Quagga mussels usually do not create large druses, but rather live individually or form small aggregates that float on the surface of soft silt (rather than sink), separated by the length of their siphons (Dermott and Kerec 1997, Karatayev and Burlakova, personal observations). Thus, they provide fewer refugia for benthic taxa and can compete with native invertebrates for space and food decreasing their overall diversity, density, and biomass (Dermott and Kerec 1997; Lozano et al. 2001;

Table 3 Effects of freshwater, byssate invasive mussels or mussel-related processes on invaded freshwater systems

Parameter	*L. fortunei*	*D. polymorpha*	*D. r. bugensis*
Water transparency	Increase (Boltovskoy et al. 2009a; Kawase 2011; Cataldo et al. 2012a; Boltovskoy and Correa 2015)[a]	Increase (Karatayev et al. 1997, 2002, 2007a, 2007b; Vanderploeg et al. 2002; Higgins and Vander Zanden 2010; Kelly et al. 2010)	Increase (Barbiero and Tuchman 2004; Nalepa 2010; Pothoven and Fahnenstiel 2014)
Seston concentration	Decrease (Boltovskoy et al. 2009a; Kawase 2011; Cataldo et al. 2012a, 2012b; Boltovskoy and Correa 2015)[a]	Decrease (Karatayev et al. 1997, 2007a, 2007b; Higgins and Vander Zanden 2010)	Decrease (Shevtsova 1989)
Nutrients	Alter nutrient cycling (Boltovskoy et al. 2009a; Kawase 2011; Cataldo et al. 2012a, 2012b; Boltovskoy and Correa 2015)[a]	Alter nutrient cycling (Karatayev et al. 2002, 2007a, 2007b; Vanderploeg et al. 2002; Higgins and Vander Zanden 2010; Kelly et al. 2010)	Alter nutrient cycling (Nalepa 2010)
Phytoplankton and chlorophyll	Decrease, changes in community composition (Boltovskoy et al. 2009a; Cataldo et al. 2012a; Di Fiori et al. 2012; Boltovskoy and Correa 2015)[a]	Decrease, changes in community composition (Karatayev et al. 2002 2007a, 2007b; Higgins and Vander Zanden 2010; Kelly et al. 2010)	Decrease, changes in community composition (Barbiero and Tuchman 2004; Fahnenstiel et al. 2010; Nalepa 2010; Pothoven and Fahnenstiel 2013)
Macrophytes, periphyton, benthic algae	Increase biomass and extent (Boltovskoy et al. 2009a; Cataldo et al. 2012b; Boltovskoy and Correa 2015)[a]	Increase biomass and extent (Karatayev et al. 1997; Nalepa et al. 1999; Vanderploeg et al. 2002; Hunter and Simons 2004, Karatayev et al. 2007a, Karatayev et al. 2007b; Higgins and Vander Zanden 2010)	Increase biomass and extent (Nalepa 2010)
Zooplankton	Decrease density, change community composition (Rojas Molina et al. 2010; Fachini 2011; Rojas Molina et al. 2011, 2012; Boltovskoy and Correa 2015)[b]	Decrease density, change community composition (Karatayev et al. 1997, 2007a, 2007b; Vanderploeg et al. 2002; Higgins and Vander Zanden 2010; Kelly et al. 2010)	Decrease? (reviewed in Nalepa 2010; Pothoven and Fahnenstiel 2014)

Table 3 (continued)

Littoral zoobenthos	Increase density and diversity; changes in community composition (Darrigran et al. 1998; Sylvester 2006; Sylvester et al. 2007a, 2007b; Sardiña et al. 2008, 2011, Karatayev et al. 2010a; Burlakova et al. 2012)[c]	Increase density; changes in community composition (reviewed in Karatayev et al. 1998, 2007b; Higgins and Vander Zanden 2010; Kelly et al. 2010)	Increase density; changes in community composition (Bially and MacIsaac 2000; Yakovleva and Yakovlev 2011)
Profundal zoobenthos	Normally *L. fortunei* absent or very scarce because of lack of adequate substrata[d]	*D. polymorpha* normally absent	Decrease, changes in community composition (Dermott and Kerec 1997; Nalepa et al. 1998, 2007, 2009a, 2009b; Lozano et al. 2001; Watkins et al. 2007; Nalepa 2010; Soster et al. 2011)
Unionids	Probably negative (Mansur et al. 2003; Scarabino 2004; Karatayev et al. 2010a)[e]	Negative (reviewed in Karatayev et al. 1997, 2007b; Burlakova et al. 2000; Lucy et al. 2014)	Negative (Schloesser and Masteller 1999; Zhulidov et al. 2010; Sherman et al. 2013; Lucy et al. 2014)
Adult and larval fishes	Probably positive from enhanced food resources (Montalto et al. 1999; Penchaszadeh et al. 2000; Boltovskoy et al. 2006; Paolucci et al. 2007; Paolucci et al. 2010a, 2010b; Boltovskoy and Correa 2015)[f]	Increase density of benthivorous fishes, changes in community composition (reviewed in Karatayev et al. 1997, 2002, 2007b; Molloy et al. 1997; Kelly et al. 2010)	Decrease density and changes in community composition in profundal zone (reviewed in Nalepa 2010; Karatayev et al. 2015)

[a] See Chapter "Nutrient Recycling, Phytoplankton Grazing, and Associated Impacts of *Limnoperna fortunei*" in this volume
[b] See Chapter "Impacts of *Limnoperna fortunei* on Zooplankton" in this volume
[c] See Chapter "Relationships of *Limnoperna fortunei* with Benthic Animals" in this volume
[d] See Chapter "*Limnoperna fortunei* Colonies: Structure, Distribution and Dynamics" in this volume
[e] *L. fortunei* have been observed to attach to unionid shells, but the potential impact has not been investigated
[f] See Chapters "Trophic relationships of *Limnoperna fortunei* with larval fishes" and "Trophic relationships of *Limnoperna fortunei* with adult fishes" in this volume

Nalepa et al. 2007, 2009a, 2009b; Watkins et al. 2007; Soster et al. 2011; Burlakova et al. 2014; Karatayev et al. 2015, 2014c).

Effects on the Water Column

The effects of *L. fortunei* and dreissenids on the water column are associated with their roles as suspension feeders, and effects can be system-wide, as opposed to effects on benthic invertebrates, which are mostly local. Suspension feeding not only affects nutrients and planktonic communities, it also transfers materials from the water column to the benthos, enhancing the coupling between planktonic and benthic components of the ecosystem, which can trigger a suite of changes that increase the relative importance of the benthic community—a process sometimes referred to as "benthification" (Mayer et al. 2014). The intensity and extent of these effects depend on many factors, including mussel population density and distribution in a waterbody, food resources available for the bivalves, water mixing rates, lake morphology, and plankton turnover rates (Karatayev et al. 1997, 2002; Kelly et al. 2010; Boltovskoy and Correa 2015). Because *D. polymorpha* is usually restricted to the littoral zone, its impacts may be significantly greater in small, shallow lakes than in large, deep ones (Karatayev et al. 2015). The impacts of *L. fortunei* may be similar, but this has not been confirmed by ad hoc studies. In contrast, quagga mussels are found throughout the entire waterbody, and, in deep lakes, they have larger total population sizes. As a consequence, they may have greater system-wide effects than golden or zebra mussels (reviewed in Karatayev et al. 2015).

Although there are more data on the system-wide impacts of zebra mussels than those of quagga and golden mussels, because of their functional similarity, their impacts on waterbodies are likely to be similar (Table 3), although the final outcome may differ depending on waterbody characteristics. The feeding activity of these invasive bivalves boosts nutrient concentrations and alters their proportions, in particular increasing the phosphorus to nitrogen (P:N) ratio (Conroy and Culver 2005; Cataldo et al. 2012b). Consumption of organic particles, including phyto- and zooplankton, and the rejection of organic and inorganic suspended matter as feces and pseudofeces decreases plankton densities and turbidity, which in turn favors light penetration and growth of macrophytes and periphyton. These effects have been described repeatedly in European, Asian, and North and South American waterbodies colonized by dreissenids or *L. fortunei* (see references in Table 3; see Chapter "Nutrient Recycling, Phytoplankton Grazing, and Associated Impacts of *Limnoperna fortunei*" in this volume). However, their net impacts on the systems investigated are not necessarily identical, especially when comparing cold-temperate North American lakes with tropical and subtropical South American freshwater habitats. The Paraguay-Paraná-Uruguay floodplain river system invaded by *L. fortunei* is quite different than the colder, clearer, and more oligotrophic North American waterbodies colonized by *Dreissena*. A particularly important contrast are the mean concentrations of particulate organic carbon (POC), which are much higher in South America (about 3.5 mg/L in the Paraná River, 20–40 % of it labile and available for biologic consumption; Depetris 1976; Depetris and Pasquini 2007) than in many of the waterbodies invaded by *Dreissena* (typically around 0.15–1 mg/L in the Great Lakes; Fanslow et al. 1995; Barbiero and Tuchman 2004; Johengen et al. 2008). Filtering organisms are generally scarce and probably not food limited in South America (Sylvester et al. 2005), which suggests that competitive impacts with suspension-feeding native animals, such as those described in North America (Bartsch et al. 2003; Thorp and Casper 2003; Raikow 2004), are less likely in South America. Furthermore, indigenous suspension-feeding organisms in the Río de la Plata watershed are scarce, and the main source of energy for animals is of detrital origin. Most of the suspended organic matter is flushed out into the ocean through the Río de la Plata estuary (~ 1,000,000–2,000,000 t of POC per year; Depetris and Kempe 1993; Guerrero et al. 1997). *L. fortunei,* the only abundant macrobenthic suspension-feeder, intercepts part of this organic matter and retains it in the system for use by a wide array of animals. The ecosystem-wide effects of this new energetic subsidy to the benthos have not been investigated, but are likely significant (Boltovskoy et al. 2006).

One of the most contentious questions is the impact of exotic bivalves on toxic cyanobacteria, in particular *Microcystis* spp. Several authors have suggested that *Dreissena* spp. promote toxic blooms via selective grazing and rejection of toxic strains of blue-green algae and excretion of soluble waste products at low nitrogen to phosphorus ratios (Conroy and Culver 2005; Bykova et al. 2006; Fishman et al. 2009). Other studies (in both North America and Europe) have found that zebra mussels may actively consume and reduce the density of *Microcystis* spp. (Baker et al. 1998; Strayer et al. 1999; Dionisio Pires et al. 2005, 2010). It was suggested that the positive effect of dreissenids on *Microcystis* spp. is restricted to lakes with low to moderate total phosphorus concentrations (<25 µg total P/L), whereas those with high nutrient loadings are not affected (Vanderploeg et al. 2001; Nicholls et al. 2002; Sarnelle et al. 2005; Knoll et al. 2008). In contrast, *L. fortunei* boosts *Microcystis* spp. growth at very high P concentrations (50–100 µg/L; Cataldo et al. 2012b).

Trophic Interactions with Fishes

In Europe and North and South America, dreissenids and *L. fortunei* provide an abundant food resource for fishes. At least 38 species of fish in Europe and in North America feed on *Dreissena* spp. (Molloy et al. 1997), and almost 50 species of fish consume *L. fortunei* in South America (see Chapter "Trophic Relationships of *Limnoperna fortunei* with Adult Fishes" in this volume). The importance of mussels in fish diets varies depending on the feeding mode and fish age, season of the year, and the morphology of the waterbody (Karatayev et al. 2002, 2007a; Strayer et al. 2004; Boltovskoy and Correa 2015).

In dreissenid-invaded areas, shortly after invasion there has been an increase in benthivorous fishes, especially in the littoral zone. This is true even for those that do not feed on dreissenids because of the increase in biomass of native benthic invertebrates that occurs with invasion (Karatayev et al. 2002, 2007a; Higgins and Vander Zanden 2010; Kelly et al. 2010; Burlakova et al. 2012). In Europe, a shift to dreissenid-based diets has resulted in increased growth, average and maximum sizes, and condition for some species of fish (Lyagina and Spanowskaya 1963; Poddubnyi 1966). In contrast, in the profundal zone of the Great Lakes, the introduction of zebra, and especially quagga mussels has been linked to the decline in the abundance, condition, and growth of several fish species. This effect has been associated with a decrease in their main food, the amphipod *Diporeia* spp., and to the lower energy content of the new food resource (mussels), which replaced the original forage base (Lozano et al. 2001; Hoyle et al. 2008; Nalepa et al. 2009a, 2009b; Rennie et al. 2009). Limited data suggest that dreissenids can have both negative and positive effects on planktivorous fishes. Suspension feeding by mussels can reduce planktonic food resources. Increased water transparency can result in increased predation on larval fish, but may also facilitate prey capture by visual fish predators (Francis et al. 1996; Mayer et al. 2001, 2014; Mills et al. 2003). In the long term, however, the effects of *Dreissena* spp. on fish were found to decrease with time (Strayer et al. 2014).

Following the introduction of the golden mussel in South America, several fish species shifted their diet from plants and detritus to the energetically more profitable *L. fortunei* (Boltovskoy et al. 2006; see Chapter "Trophic Relationships of *Limnoperna fortunei* with Adult Fishes" in this volume). *L. fortunei* is consumed not only by fishes that can detach mussels from a clump and grind their valves, but also by species that swallow whole individuals, and even others that nibble on extended siphons and mantle edges. Many of these midsized fishes are in turn consumed by larger, piscivorous species with high commercial value, suggesting that improved feeding conditions for their prey are likely to have a positive impact on these large species as well.

Consumption of *L. fortunei* veligers by fish larvae is probably even more significant than the consumption of adult mussels. Of 25 larval fish taxa surveyed in the Paraná, Paraguay, and Uruguay rivers, 18 feed on veligers, especially their earliest life stage (protolarvae) (Paolucci et al. 2007; Paolucci 2010; see Chapter "Trophic Relationships of *Limnoperna fortunei* with Larval Fishes" in this volume). Veligers are not only more abundant and easier to capture than crustacean zooplankton, but they also represent an energetically more profitable food resource yielding significantly higher growth rates than crustaceans (Paolucci et al. 2010b).

Concluding Remarks

Limnoperna fortunei was originally described in 1856 (as *Volsella fortunei;* Dunker 1856), and subsequently referred to under various different names including *Modiola lacustris, Limnoperna lacustris, Modiola siamensis, Limnoperna siamensis, Modiola cambodgensis, Modiola (Limnoperna) siamensis* (Morton and Dinesen 2010), in chiefly taxonomic and distributional studies. It was a species of little interest until it invaded Japan and South America around 1990. After that time, the number of publications dedicated to the golden mussel soared from <0.3/year as of

1992, to >20/year after 1993 (see "Preface" in this volume). The striking similarity between *L. fortunei* and species of *Dreissena* has been noticed since the very first detailed studies of the biology of the golden mussel, when it invaded Hong Kong ~1965 (Morton 1975). By then, *D. polymorpha* had been expanding across Europe for centuries, and there was abundant information on its biology, ecology, and impacts. Thus, using *Dreissena* as a model and a recurrent reference in subsequent literature on the golden mussel was an obvious outcome.

The growing body of information on *L. fortunei* clearly shows that, indeed, parallels with the dreissenids, in particular with *D. polymorpha,* are numerous and warranted. However, proven similarities also encouraged ascribing to *L. fortunei* processes, and particularly impacts, reported for zebra mussels in the northern hemisphere. Although many researchers were cautious in their conclusions, stating that such effects were merely a possibility, others were not. These assumptions had a snowball effect whereby subsequent publications indiscriminately extrapolated results on the impacts of zebra mussels in the northern hemisphere to those of *L. fortunei* in South America.

Boltovskoy and Correa (2015) noted that "Complications for interpreting the effects of *L. fortunei* on the ecosystem are even more critical when attempting to label the impacts as negative or positive. A basic precautionary principle and the long list of examples where introduced species have been shown to have devastating effects on the biota (Simberloff 2003) clearly support the need to make all efforts possible to keep biological invasions at bay, or to eradicate them if feasible. However, once a nonnative species has been introduced and its eradication is out of the question (as is the case of *L. fortunei*), analyses of its interactions with the local biota should be based on evidence, rather than on extrapolations from other invasives and geographic areas. Much of the literature on the golden mussel has been oriented at forcibly demonstrating the environmental harm caused by this invader, thus biasing if not the results, the interpretation of the evidence obtained (Bujes et al. 2007; Defeo et al. 2013)".

As shown in this review, impacts of these invasive mussels vary widely among geographic areas and waterbodies, and even in different sectors within the same waterbody. Furthermore, interactions with the local biota change as a function of mussel species, their densities, and with time after initial colonization. Using data on the much more thoroughly researched dreissenids furnished useful guidelines for defining potential interactions and fruitful research topics, but it has also tended to hinder assessment of differences between the golden mussel and *Dreissena* spp., many of which have been shown to be responsible for quite dissimilar environmental impacts (Boltovskoy et al. 2006; Boltovskoy and Correa 2015). We contend that in order to effectively widen our current knowledge, research on *L. fortunei* should center on identifying contrasts and dissimilarities with dreissenids, rather than on confirming parallels.

Future research should aim at shedding light on the many unknown aspects of the biology and ecology of the golden mussel, which are particularly critical for a comprehensive assessment of its interactions with the local biota. So far, only a few effects at local scales have been explored, whereas at the ecosystem scale our understanding of interactions of *L. fortunei* with the environment is still very limited. For example, although mussel densities are a key element for gauging the impacts of the invader on ecosystems, so far only one attempt has been made at assessing this parameter over an entire waterbody (Boltovskoy et al. 2009a). Several potential traits (e.g., fecundity, metabolism) and interactions of utmost importance (e.g., biomagnification and transfer of contaminants, thermal shifts due to changes in water transparency, the homogenization of faunal composition across environments, facilitation of other invasive species, changes in macrophyte growth, modifications in benthic oxygenation, overgrowth of other organisms, trophic relationships with waterfowl and aquatic vertebrates other than fishes, etc.) have practically not been addressed so far.

Acknowledgments This work was partially financed by grants from the University of Buenos Aires, Argentina (UBA X-020 and 20020100100035) and from the Argentine Agencia Nacional de Promoción Científica y Tecnológica, Argentina (PICT 2007 1968) to D. Boltovskoy.

References

Aldridge DW, Payne BS, Miller AC (1995) Oxygen consumption, nitrogenous excretion, and filtration rates of *Dreissena polymorpha* at acclimation temperature between 20 and 32 °C. Can J Fish Aquat Sci 52:1761–1767

Allen YC, Thompson BA, Ramcharan CW (1999) Growth and mortality rates of the zebra mussel, *Dreissena polymorpha*, in the Lower Mississippi River. Can J Fish Aquat Sci 56:748–759

Angonesi LG, da Rosa NG, Bemvenuti CE (2008) Tolerance to salinities shocks of the invasive mussel *Limnoperna fortunei* under experimental conditions. Iheringia (Série Zoologia) 98:66–69

Bacchetta R, Mantecca P, Vailati G (2010) Reproductive behaviour of zebra mussels living in shallow and deep water in the South Alps lakes. In: van der Velde G, Rajagopal S, Bij de Vaate A (eds) The Zebra Mussel in Europe. Backhuys Publishers, Leiden, pp 161–168

Baigun C, Marinone MC (1995) Cold-temperate lakes of South America: do they fit northern hemisphere models? Archiv Hydrobiologie-Beiheft Ergebnisse der Limnologie 135:23–51

Baker SM, Levinton JS, Kurdziel JP, Shumway SE (1998) Selective feeding and biodeposition by zebra mussels and their relation to changes in phytoplankton composition and seston load. J Shellfish Res 17:1207–1213

Balcom N (1994) Zebra mussel awareness and boat use patterns among boaters using three "high risk" Connecticut lakes. In: 4th International Zebra Mussel Conference, Madison (Wisconsin)

Baldwin BS, Mayer MS, Dayton J, Pau N, Mendilla J, Sullivan M, Moore A, Ma A, Mills EL (2002) Comparative growth and feeding in zebra and quagga mussels (*Dreissena polymorpha* and *Dreissena bugensis*): implications for North American lakes. Can J Fish Aquat Sci 59:680–694

Barbiero RP, Tuchman ML (2004) The deep chlorophyll maximum in Lake Superior. J Great Lakes Res 30(Supplement 1):256–268

Barbosa FG, Melo AS (2009) Modelo preditivo de sobrevivência do Mexilhão Dourado (*Limnoperna fortunei*) em relação a variações de salinidade na Laguna dos Patos, RS, Brasil. Biota Neotropica 9:407–412

Bartsch LA, Richardson WB, Sandheinrich MB (2003) Zebra mussels (*Dreissena polymorpha*) limit food for larval fish (*Pimephales promelas*) in turbulent systems: a bioenergetics analysis. Hydrobiologia 495:59–72

Beekey MA, McCabe DJ, Marsden JE (2004) Zebra mussels affect benthic predator foraging success and habitat choice on soft sediments. Oecologia 141:164–170

Belaich M, Oliver C, Pilloff M, Porta A (2006) Evaluation of a biomarker of Cd(II) exposure on *Limnoperna fortunei*. Environ Pollut 144:280–288

Benson AJ (2014) Chronological history of zebra and quagga mussels (Dreissenidae) in North America, 1988–2010. In: Nalepa TF, Schloesser DW (eds) Quagga and zebra mussels: biology, impacts, and control. 2nd edition CRC Press, Boca Raton, pp 9–32

Bially A, MacIsaac HJ (2000) Fouling mussels (*Dreissena* spp.) colonize soft sediments in Lake Erie and facilitate benthic invertebrates. Freshw Biol 43:85–97

bij de Vaate A, Greijdanus-Klaas M, Smit H (1992) Densities and biomass of zebra mussels in the Dutch part of the lower Rhine. In: Neumann D, Jenner HA (eds) The zebra mussel *Dreissena polymorpha*. Ecology, biology monitoring and first applications in the water quality management. Gustav Fisher, Stuttgart, pp 67–77

bij de Vaate A, Jazdzewski K, Ketelaars HAM, Gollasch S, van der Velde G (2002) Geographical patterns in range extension of Ponto-Caspian macroinvertebrate species in Europe. Can J Fish Aquat Sci 59:1159–1174

bij de Vaate A, van der Velde G, Leuven RSEW, Heiler KCM (2014) Spread of the quagga mussel (*Dreissena rostriformis bugensis*) in western Europe. In: Nalepa TF, Schloesser DW (eds) Quagga and zebra mussels: biology, impacts, and control. Second edition. CRC Press, Boca Raton, pp 83–92

Boltovskoy D, Cataldo D (1999) Population dynamics of *Limnoperna fortunei*, an invasive fouling mollusc, in the lower Paraná River (Argentina). Biofouling 14:255–263

Boltovskoy D, Correa N (2015) Ecosystem impacts of the invasive bivalve *Limnoperna fortunei* (golden mussel) in South America. Hydrobiologia. 746:81–95

Boltovskoy D, Correa N, Cataldo D, Sylvester F (2006) Dispersion and ecological impact of the invasive freshwater bivalve *Limnoperna fortunei* in the Río de la Plata watershed and beyond. Biol Invasions 8:947–963

Boltovskoy D, Karatayev A, Burlakova L, Cataldo D, Karatayev V, Sylvester F, Mariñelarena A (2009a) Significant ecosystem-wide effects of the swiftly spreading invasive freshwater bivalve *Limnoperna fortunei*. Hydrobiologia 636:271–284

Boltovskoy D, Sylvester F, Otaegui A, Leytes V, Cataldo D (2009b) Environmental modulation of the reproductive activity of the invasive mussel *Limnoperna fortunei* in South America. Austral Ecol 34:719–730

Bonel N, Solari LC, Lorda J (2013) Differences in density, shell allometry and growth between two populations of *Limnoperna fortunei* (Mytilidae) from the Río de la Plata basin, Argentina. Malacologia 56:43–58

Bonetto C, Villar C, De Cabo L, Vaithiyanathan P (1998) Hydrochemistry of a large floodplain river. Verhandlungen der Internationalen Vereinigung Theoretische und Angewandte Limnologie 26:899–902

Borcherding J (1991) The annual reproductive cycle of the freshwater mussel *Dreissena polymorpha* Pallas in lakes. Oecologia 87:208–218

Botts PS, Patterson BA, Schloesser DW (1996) Zebra Mussel effects on benthic invertebrates: physical or biotic? J N Am Benthol Soc 15:179–184

Brugnoli E, Dabezies MJ, Clemente JM, Muniz P (2011) *Limnoperna fortunei* (Dunker 1857) en el sistema de embalses del Rio Negro, Uruguay. Oecologia Australis 15:576–592

Buch KL, McMahon RF (2001) Assessment of potential for dispersal of aquatic nuisance species by recreational boaters into the western United States. The University of Texas at Arlington, Arlington, pp 1–70

Bujes CS, Ely I, Verrastro L (2007) *Trachemys dorbigni* (Brazilian Slider) diet. Herpetol Rev 38:335

Burlakova LE (1998) Ecology of *Dreissena polymorpha* (Pallas) and its role in the structure and function of aquatic ecosystems. PhD Thesis, Academy of Sciences of Belarus (Belarus), pp 1–167

Burlakova LE, Karatayev AY, Padilla DK (2000) The impact of *Dreissena polymorpha* (Pallas) invasion on unionid bivalves. Int Rev Hydrobiol 85:529–541
Burlakova LE, Karatayev AY, Karatayev VA (2012) Invasive mussels induce community changes by increasing habitat complexity. Hydrobiologia 685:121–134
Burlakova LE, Karatayev AY, Pennuto C, Mayer C (2014) Changes in Lake Erie benthos over the last 50 years: historical perspectives, current status, and main drivers. J Great Lakes Res. 40:560–573
Bykova O, Laursen A, Bostan V, Bautista J, Mc Carthy L (2006) Do zebra mussels (*Dreissena polymorpha*) alter lake water chemistry in a way that favours *Microcystis* growth? Sci Total Environ 371:362–372
Callil CT, Mansur MCD, Marcelo SM (2006) Bivalves invasores no Pantanal. In: XVII Encontro Brasileiro de Malacologia, Rio De Janeiro (Brazil)
Carlton JT (2008) The zebra mussel *Dreissena polymorpha* found in North America in 1986 and 1987. J Great Lakes Res 34:770–773
Cataldo D, Boltovskoy D (2000) Yearly reproductive activity of *Limnoperna fortunei* (Bivalvia) as inferred from the occurrence of its larvae in the plankton of the lower Paraná River and the Río de la Plata estuary (Argentina). Aquatic Ecol 34:307–317
Cataldo D, Boltovskoy D, Stripeikis J, Pose M (2001a) Condition index and growth rates of field caged *Corbicula fluminea* (Bivalvia) as biomarkers of pollution gradients in the Paraná River delta (Argentina). Aquatic Ecosyst Health Manag 4:187–201
Cataldo D, Colombo JC, Boltovskoy D, Bilos C, Landoni P (2001b) Environmental toxicity assessment in the Paraná River delta (Argentina): simultaneous evaluation of select pollutants and mortality rates of *Corbicula fluminea* (Bivalvia) early juveniles. Environ Pollut 112:379–389
Cataldo D, O'Farrell I, Paolucci E, Sylvester F, Boltovskoy D (2012a) Impact of the invasive golden mussel (*Limnoperna fortunei*) on phytoplankton and nutrient cycling. Aquat Invasions 7:91–100
Cataldo D, Vinocur A, O'Farrell I, Paolucci E, Leites V, Boltovskoy D (2012b) The introduced bivalve *Limnoperna fortunei* boosts *Microcystis* growth in Salto Grande Reservoir (Argentina): evidence from mesocosm experiments. Hydrobiologia 680:25–38
Choi SS, Kim JS (1985) Studies on the metamorphosis and the growth of larva in *Limnoperna fortunei*. Korean J Malacol 1:13–18 [In Korean]
Choi SS, Shin CN (1985) Study on the early development and larvae of *Limnoperna fortunei*. Korean J Malacol 1:5–12 [In Korean]
Claxton WT, Mackie GL (1998) Seasonal and depth variations in gametogenesis and spawning of *Dreissena polymorpha* and *Dreissena bugensis* in eastern Lake Erie. Can J Zool 76:2010–2019
Conroy J, Culver DA (2005) Do dreissenid mussels affect Lake Erie ecosystem stability processes? Am Midland Nat 153:20–32
Darrigran GA (2002) Potential impact of filter-feeding invaders on temperate inland freshwater environments. Biol Invasions 4:145–156
Darrigran GA, Martin SM, Gullo B, Armendariz L (1998) Macroinvertebrates associated with *Limnoperna fortunei* (Dunker, 1857) (Bivalvia, Mytilidae) in Río de la Plata, Argentina. Hydrobiologia 367:223–230
Darrigran GA, Penchaszadeh PE, Damborenea C (1999) The reproductive cycle of *Limnoperna fortunei* (Dunker, 1857) (Mytilidae) from a neotropical temperate locality. J Shellfish Res 18:361–365
Darrigran GA, Damborenea C, Greco N (2007) An evaluation pattern for antimacrofouling procedures: *Limnoperna fortunei* larvae study in a hydroelectric power plant in South America. Ambio 36:575–579
Defeo O, Castrejón M, Ortega L, Kuhn AM, Gutiérrez NL, Castilla JC (2013) Impacts of climate variability on Latin American small-scale fisheries. Ecol Soc 18:30
Depetris PJ (1976) Hydrochemistry of the Paraná River. Limnol Oceanogr 21:736–739
Depetris PJ, Kempe S (1993) Carbon dynamics and sources in the Paraná River. Limnol Oceanogr 38:382–395

Depetris PJ, Pasquini AI (2007) The geochemistry of the Paraná River: an overview. In: Iriondo MH, Paggi JC, Parma MJ (eds) The Middle Paraná River. Limnology of a subtropical wetland. Springer Verlag, Berlin, pp 143–174

Dermott R, Kerec D (1997) Changes to the deep water benthos of eastern Lake Erie since the invasion of *Dreissena*: 1979–1993. Can J Fish Aquat Sci 54:922–930

Dermott R, Munawar M (1993) Invasion of Lake Erie offshore sediments by *Dreissena*, and its ecological implications. Can J Fish Aquat Sci 50:2298–2304

Díaz M, Pedrozo F, Baccala N (2000) Summer classification of Southern Hemisphere temperate lakes (Patagonia, Argentina). Lakes & Reservoirs: Res Manag 5:213–229

Di Fiori E, Pizarro H, dos Santos Afonso M, Cataldo D (2012) Impact of the invasive mussel *Limnoperna fortunei* on glyphosate concentration in water. Ecotoxicol Environ Safety 81:106–113

Diggins TP (2001) A seasonal comparison of suspended sediment filtration by quagga (*Dreissena bugensis*) and zebra (*D. polymorpha*) mussels. J Great Lakes Res 27:457–466

Dionisio Pires LM, Bontes BM, Van Donk E, Ibelings BW (2005) Grazing on colonial and filamentous, toxic and non-toxic cyanobacteria by the zebra mussel *Dreissena polymorpha*. J Plankton Res 27:331–339

Dionisio Pires LM, Ibelings BW, van Donk E (2010) Zebra mussels as a potential tool in the restoration of eutrophic shallow lakes, dominated by toxic cyanobacteria. In: van der Velde G, Rajagopal S, Bij de Vaate A (eds) The zebra mussel in Europe. Backhuys Publishers, Leiden, pp 361–372

Dunker W (1856) Mytilacea nova collectionis Cumingianæ. Proc Zool Soc Lond 24:358–366

Dyga AK, Zolotareva VI (1976) Biology of *Dreissena bugensis* Andrusov of Zaporozhskoe Reservoir and their role in water purification. In: Andrushaitis GP et al. (eds) Abstracts of the Third Congress of the All-Union Hydrobiological Society. Zinatne Press, Riga, pp 237–239 [In Russian]

Fachini A (2011) Filtração do bivalve invasor *Limnoperna fortunei* (Dunker, 1857), o mexilhão dourado, sobre a comunidade planctônica natural e na presença de cianobactéria tóxica. MSc Thesis, Universidade Federal do Rio Grande do Sul (Brazil), pp 1–68

Fahnenstiel GL, Pothoven SA, Nalepa TF, Vanderploeg HA, Klarer DM, Scavia D (2010) Recent changes in primary production and phytoplankton in the offshore region of southeastern Lake Michigan. J Great Lakes Res 36:20–29

Fanslow DL, Nalepa TF, Lang GA (1995) Filtration rates of the zebra mussel (*Dreissena polymorpha*) on natural seston from Saginaw Bay, Lake Huron. J Great Lakes Res 21:489–500

Fishman DB, Adlerstein SA, Vanderploeg HA, Fahnenstiel GL, Scavia D (2009) Causes of phytoplankton changes in Saginaw Bay, Lake Huron, during the zebra mussel invasion. J Great Lakes Res 35:482–495

Francis JT, Robillard SR, Marsden JE (1996) Yellow perch management in Lake Michigan: a multi-jurisdictional challenge. Fisheries 21:18–20

Garton DW, McMahon R, Stoeckmann AM (2014) Limiting environmental factors and competitive interactions between zebra and quagga mussels in North America. In: Nalepa TF, Schloesser DW (eds) Quagga and zebra mussels: biology, impacts, and control, 2nd edition. CRC Press, Boca Raton, pp 383–402

Grigorovich IA, Angradi TR, Stepien CA (2008) Occurrence of the quagga mussel (*Dreissena bugensis*) and the zebra mussel (*Dreissena polymorpha*) in the upper Mississippi River system. J Freshw Ecol 23:429–435

Guerrero R, Acha EM, Framiñan MB, Lasta CA (1997) Physical oceanography of the Río de la Plata Estuary, Argentina. Cont Shelf Res 17:727–742

Gutierrez JL, Jones CG, Strayer DL, Iribarne OO (2003) Mollusks as ecosystem engineers: the role of shell production in aquatic habitats. Oikos 101:79–90

Hallstan S, Grandin U, Goedkoop W (2010) Current and potential distribution of the zebra mussel (*Dreissena polymorpha*) in Sweden. Biol Invasions 12:285–296

Hamada M (2011) Relationship between water temperature and period of occurrence and attachment in planktonic larvae of golden mussel *Limnoperna fortunei*, in the Yahagi River. Report of Yahagi River Institute 15:45–54 [In Japanese]

Higgins SN, Vander Zanden MJ (2010) What a difference a species makes: a meta–analysis of dreissenid mussel impacts on freshwater ecosystems. Ecol Monogr 80:179–196

Hoyle JA, Bowlby JN, Morrison BJ (2008) Lake whitefish and walleye population responses to dreissenid mussel invasion in eastern Lake Ontario. Aquat Ecosyst Health Manag 11:403–411

Huang Z, Li C, Zhang L, Li F, Zheng C (1981) The distribution of fouling organisms in Changjiang River estuary. Oceanologia et Limnologia Sinica 12:531–537

Hunter RG, Simons KA (2004) Dreissenids in Lake St. Clair in 2001: evidence for population regulation. J Great Lakes Res 30:528–537

Jantz B, Neumann D (1992) Shell growth and aspects of the population dynamics of *Dreissena polymorpha* in the River Rhine. In: Neumann D, Jenner HA (eds) The zebra mussel *Dreissena polymorpha*. Ecology, biology monitoring and first applications in the water quality management. Gustav Fisher, Stuttgart, pp 49–66

Japan Commission on Large Dams (2009) Dams in Japan. Past, present and future. CRC Press, Boca Raton, pp 1–232

Johengen TH, Biddanda BA, Cotner JB (2008) Stimulation of Lake Michigan plankton metabolism by sediment resuspension and river runoff. J Great Lakes Res 34:213–227

Johnson LE, Ricciardi A, Carlton JT (2001) Overland dispersal of aquatic invasive species: a risk assessment of transient recreational boating. Ecol Appl 11:1789–1799

Jones LA, Ricciardi A (2005) Influence of physicochemical factors on the distribution and biomass of invasive mussels (*Dreissena polymorpha* and *Dreissena bugensis*) in the St. Lawrence River. Can J Fish Aquat Sci 62:1953–1962

Karatayev AY, Burlakova LE, Padilla DK (1997) The effects of *Dreissena polymorpha* (Pallas) invasion on aquatic communities in Eastern Europe. J Shellfish Res 16:187–203

Karatayev AY, Burlakova LE, Padilla DK (1998) Physical factors that limit the distribution and abundance of *Dreissena polymorpha* (Pall.). J Shellfish Res 17:1219–1235

Karatayev AY, Burlakova LE, Padilla DK (2002) Impacts of zebra mussels on aquatic communities and their role as ecosystem engineers. In: Leppäkoski E, Gollasch S, Olenin S (eds) Invasive aquatic species of Europe: Distribution, impacts and management. Kluwer Academic Publishers, Dodrecht, pp 433–446

Karatayev AY, Burlakova LE, Padilla DK (2006) Growth rate and longevity of *Dreissena polymorpha* (Pallas): a review and recommendations for future study. J Shellfish Res 25:1–23

Karatayev AY, Boltovskoy D, Padilla DK, Burlakova LE (2007a) The invasive bivalves *Dreissena polymorpha* and *Limnoperna fortunei*: parallels, contrasts, potential spread and invasion impacts. J Shellfish Res 26:205–213

Karatayev AY, Padilla DK, Minchin D, Boltovskoy D, Burlakova LE (2007b) Changes in global economies and trade: the potential spread of exotic freshwater bivalves. Biol Invasions 9:161–180

Karatayev AY, Burlakova LE, Karatayev VA, Boltovskoy D (2010a) *Limnoperna fortunei* versus *Dreissena polymorpha*: Population densities and benthic community impacts of two invasive freshwater bivalves. J Shellfish Res 29:975–984

Karatayev AY, Burlakova LE, Padilla DK (2010b) *Dreissena polymorpha* in Belarus: history of spread, population biology, and ecosystem impacts. In: van der Velde G, Rajagopal S, bij de Vaate A (eds) The zebra mussel in Europe. Backhuys Publishers, Leiden, pp 45–58

Karatayev AY, Mastitsky SE, Padilla DK, Burlakova LE, Hajduk MH (2010c) Differences in growth and survivorship of zebra and quagga mussels: size matters. Hydrobiologia 668:183–194

Karatayev AY, Burlakova LE, Mastitsky SE, Padilla DK, Mills EL (2011) Contrasting rates of spread of two congeners, *Dreissena polymorpha* and *Dreissena rostriformis bugensis*, at different spatial scales. J Shellfish Res 30:923–931

Karatayev AY, Burlakova LE, Padilla DK (2014a) General overview of zebra and quagga mussels. What we do and do not know. In: Nalepa TF, Schloesser DW (eds) Quagga and zebra mussels: biology, impacts, and control, 2nd edn. CRC Press, Boca Raton, pp 695–703

Karatayev AY, Burlakova LE, Padilla DK (2015) Zebra versus quagga mussels: Spread, population dynamics and ecosystem impacts. Hydrobiologia 746:97–112

Karatayev AY, Burlakova LE, Pennuto C, Ciborowski J, Karatayev VA, Juette P, Clapsadl M (2014c) Twenty five years of changes in *Dreissena* spp. populations in Lake Erie. J Great Lakes Res 40:550–559

Kawase M (2011) Feeding and filtration in *Limnoperna lacustris*. Available via http://www.mizuho-c.ac.jp/images/library/kiyo_05/amckiyo-no05-05.pdf. Accessed 11 Feb 2014 [In Japanese]

Keller RP, Drake JM, Lodge DM (2007) Fecundity as a basis for risk assessment of nonindigenous freshwater molluscs. Conserv Biol 21:191–200

Kelly DW, Herborg L-M, MacIsaac HJ (2010) Ecosystem changes associated with *Dreissena* invasions: recent developments and emerging issues. In: van der Velde G, Rajagopal S, bij de Vaate A (eds) The zebra mussel in Europe. Backhuys Publishers, Leiden, pp 199–210

Kinzelbach R (1992) The main features of the phylogeny and dispersal of the zebra mussel *Dreissena polymorpha*. In: Neumann D, Jenner HA (eds) The zebra mussel *Dreissena polymorpha*. Ecology, biology monitoring and first applications in the water quality management. Gustav Fisher, Stuttgart, pp 5–17

Knoll LB, Sarnelle O, Hamilton SK, Kissman CEH, Wilson AE, Rose JB, Morgan MR (2008) Invasive zebra mussels (*Dreissena polymorpha*) increase cyanobacterial toxin concentrations in low-nutrient lakes. Can J Fish Aquat Sci 65:448–455

Lozano SJ, Scharold JV, Nalepa TF (2001) Recent declines in benthic macroinvertebrate densities in Lake Ontario. Can J Fish Aquat Sci 58:518–529

Lucy FE, Burlakova LE, Karatayev AY, Mastitsky SE, Zanatta DT (2014) Zebra mussel impacts on unionids. In: Nalepa TF, Schloesser DW (eds) Quagga and zebra mussels: biology, impacts, and control, 2nd edn. CRC Press, Boca Raton, pp 623–646

Luferov VP (1965) Organisms living anabiotically frozen in ice of the littoral zone of Rybinskoe Reservoir. Trudy Instituta Biologii Vnutrennikh Vod Akademii Nauk SSSR 8:151–154 [In Russian]

Lvova AA (1977) The ecology of *Dreissena polymorpha* (Pall.) in Uchinskoe Reservoir. PhD Thesis, Moscow State University (Russia), pp 1-116 [In Russian]

Lvova AA, Makarova GE (1994) Reproduction. In: Starobogatov JI (ed) Freshwater zebra mussel *Dreissena polymorpha* (Pall.) (Bivalvia, Dreissenidae). Systematics, ecology, practical meaning. Nauka Press, Moscow, pp 138–148 [In Russian]

Lvova AA, Karatayev AY, Makarova GE (1994a) Planktonic larva. In: Starobogatov JI (ed) Freshwater zebra mussel *Dreissena polymorpha* (Pall.) (Bivalvia, Dreissenidae). Systematics, ecology, practical meaning. Nauka Press, Moscow, pp 138–148 [In Russian]

Lvova AA, Makarova GE, Miroshnichenko MP (1994b) Growth of the *Dreissena* in different parts of the distribution area. In: Starobogatov JI (ed) Freshwater zebra mussel *Dreissena polymorpha* (Pall.) (Bivalvia, Dreissenidae). Systematics, ecology, practical meaning. Nauka Press, Moscow, pp 156–179 [In Russian]

Lyagina TN, Spanowskaya VD (1963) Morphological peculiarities of some fish in the Uchinskoe water storage. In: Uchinskoe and Mozhaiskoe Reservoirs: hydrobiological and ichthyological studies. Moscow State University, Moscow, pp 269–310 [In Russian]

Lyakhnovich VP, Karatayev AY, Andreev NI, Andreeva SI, Afanasiev SA, Dyga AK, Zakutskiy VP, Zolotareva VI, Lvova AA, Nekrasova MY, Osadchikh VF, Pligin YV, Protasov AA, Tischikov GM (1994) Living conditions. In: Starobogatov YI (ed) Freshwater zebra mussel *Dreissena polymorpha* (Pall.) (Bivalvia, Dreissenidae). Systematics, ecology, practical meaning. Nauka Press, Moscow, pp 109–119 [In Russian]

MacIsaac HJ, Grigorovich IA, Ricciardi A (2001) Reassessment of species invasions concepts: the Great Lakes basin as a model. Biol Invasions 3:405–416

Mackie GL (1991) Biology of the exotic zebra mussel, *Dreissena polymorpha*, in relation to native bivalves and its potential impact in Lake St. Clair. Hydrobiologia 219:251–268

Magara Y, Matsui Y, Goto Y, Yuasa A (2001) Invasion of the non-indigenous nuisance mussel, *Limnoperna fortunei*, into water supply facilities in Japan. J Water Supply: Res Technol—Aqua 50:113–124

Maglianesi RE (1973) Principales características químicas y físicas del las aguas del Alto Paraná y Paraguay Inferior. Physis 32:185–197

Mansur MCD, Santos CP, Darrigran G, Heydrich I, Callil CT, Cardoso FR (2003) Primeiros dados quali-quantitativos do mexilhão-dourado, *Limnoperna fortunei* (Dunker), no Delta do Jacuí, no Lago Guaíba e na Laguna dos Patos, Rio Grande do Sul, Brasil e alguns aspectos de sua invasão no novo ambiente. Revista Brasileira de Zoologia 20:75–84

Marçal SF, Callil CT (2008) Structure of invertebrates community associated with *Eichhornia crassipes* Mart. (Solms-Laubach) after the introduction of *Limnoperna fortunei* (Dunker, 1857) (Bivalvia, Mytilidae) in the Upper Paraguay River, MT, Brazil. Acta Limnologica Brasiliensia 20:359–371

Marçal SF, Callil CT (2012) *Limnoperna fortunei* associada a macrófitas aquáticas na bacia do Rio Paraguai, Mato Grosso. In: Mansur MCD, Santos CP, Pereira D, Padula Paz IC, Leite Zurita ML, Raya Rodriguez MT, Vilar Nehrke M, Aydos Bergonci PE (eds) Moluscos límnicos invasores no Brasil. Biologia, prevenção, controle. Redes Editora, Porto Alegre, pp 201–206

Mata FAR (2011) Abundância e distribuição temporal de *Limnoperna fortunei* Dunker, 1857 (Mollusca, Bivalvia) e os impactos da incrustação em usinas geradoras de energia elétrica. MSc Thesis, Universidade Federal de Ouro Preto (Brazil), pp 1–91

Matsuda M, Uenishi M (1992) A newly introduced mussel, *Limnoperna lacustris* (V. Martens) (Mollusca: Mytilidae) in Lake Biwa, Shiga Prefecture, Japan. Annual Rep Biwako Bunkakan 10:45 [In Japanese]

Matthews MA, McMahon RF (1994) The survival of zebra mussels (*Dreissena polymorpha*) and asian clams (*Corbicula fluminea*) under extreme hypoxia. In: 4th International Zebra Mussel Conference, Madison (USA)

Mayer CM, Rudstam LG, Mills EL, Cardiff SG, Bloom CA (2001) Zebra mussels (*Dreissena polymorpha*), habitat alteration and yellow perch (*Perca flavescens*) foraging: system-wide effects and behavioural mechanisms. Can J Fish Aquat Sci 58:2459–2467

Mayer CM, Burlakova LE, Eklöv P, Fitzgerald D, Karatayev AY, Ludsin SA, Millard S, Mills EL, Ostapenya AP, Rudstam LG, Zhu B, Zhukova TV (2014) Benthification of freshwater lakes. In: Nalepa TF, Schloesser DW (eds) Quagga and zebra mussels: biology, impacts, and control, 2nd edn. CRC Press, Boca Raton, pp 575–585

McMahon RF, Bogan AE (2001) Mollusca: Bivalvia. In: Thorn JH, Covich AP (eds) Ecology and classification of North American freshwater invertebrates. Academic Press, San Diego, pp 331–429

Mellina E, Rasmussen JB (1994) Patterns in the distribution and abundance of zebra mussel (*Dreissena polymorpha*) in rivers and lakes in relation to substrate and other physicochemical factors. Can J Fish Aquat Sci 51:1024–1036

Mills EL, Dermott RM, Mellina E, Conn DB, Spidle AP (1993) Colonization, ecology, and population structure of the "Quagga" mussel (Bivalvia: Dreissenidae). Can J Fish Aquat Sci 50:2305–2314

Mills EL, Rosenberg G, Spidle AP, Ludyanskiy M, Pligin Y (1996) A review of the biology of the quagga mussel (*Dreissena bugensis*), a second species of freshwater dreissenid introduced to North America. Am Zool 36:271–286

Mills EL, Chrisman JR, Baldwin B, Owens RW, O'Gorman R, Howell T, Roseman EF, Raths MK (1999) Changes in the dreissenid community in the Lower Great Lakes with emphasis on Southern Lake Ontario. J Great Lakes Res 25:187–197

Mills EL, Casselman JM, Dermott R, Fitzsimons JD, Gal G, Holeck KT, Hoyle JA, Johannsson OE, Lantry BF, Makarewicz JC, Millard ES, Munawar IF, Munawar M, O'Gorman R, Owens RW, Rudstam LG, Schaner T, Stewart TJ (2003) Lake Ontario: food web dynamics in a changing ecosystem (1970–2000). Can J Fish Aquat Sci 60:471–490

Minchin D (2000) Dispersal of zebra mussel in Ireland. Verhandlungen der Internationalen Vereinigung Theoretische und Angewandte Limnologie 27:1576–1579

Ministry of Agriculture, Forestry and Fisheries (2003) The global diversity of irrigation. Available via http://www.maff.go.jp/j/nousin/keityo/mizu_sigen/index.html. Accessed 5 April 2014

Molloy DP, Karatayev AY, Burlakova EB, Kurandina DP, Laruelle F (1997) Natural enemies of zebra mussels: predators, parasites, and ecological competitors. Rev Fish Sci 5:27–97

Montalto L, Oliveros OB, Ezcurra De Drago I, Demonte LD (1999) Peces del Río Parana medio predadores de una especie invasora: *Limnoperna fortunei* (Bivalvia, Mytilidae). Revista de la Facultad de Bioquímica y Ciencias Biológicas de la Universidad Nacional del Litoral 3:85–101

Morton B (1975) The colonization of Hong Kong's raw water supply system by *Limnoperna fortunei* (Dunker 1857) (Bivalvia: Mytilacea) from China. Malacol Rev 8:91–105

Morton B (1977) The population dynamics of *Limnoperna fortunei* (Dunker 1857) (Bivalvia: Mytilacea) in Plover Cove Reservoir, Hong Kong. Malacologia 16:165–182

Morton B, Dinesen G (2010) Colonization of Asian freshwaters by the Mytilidae (Bivalvia): A comparison of *Sinomytilus harmandi* from the Tonle-Sap River, Phnom Penh, Cambodia, with *Limnoperna fortunei*. Molluscan Res 30:57–72

Nakano D, Kobayashi T, Sakaguchi I (2010a) Differences in larval dynamics of golden mussel *Limnoperna fortunei* between dam reservoirs with and without an aeration system. Landsc Ecol Eng 6:53–60

Nakano D, Kobayashi T, Sakaguchi I (2010b) Predation and depth effects on abundance and size distribution of an invasive bivalve, the golden mussel *Limnoperna fortunei*, in a dam reservoir. Limnology 11:259–266

Nalepa TF (2010) An overview of the spread, distribution, and ecological impacts of the quagga mussel, *Dreissena rostriformis bugensis*, with possible implications to the Colorado River system. In: Colorado River basin science and resource management symposium. Coming together, coordination of science and restoration activities for the Colorado River ecosystem, Scottsdale (USA)

Nalepa TF, Hartson DJ, Fanslow DL, Lang GA, Lozano SJ (1998) Decline in benthic macroinvertebrate populations in southern Lake Michigan, 1980–1993. Can J Fish Aquat Sci 55:2402–2413

Nalepa TF, Fahnenstiel GL, Johengen TH (1999) Impacts of the zebra mussel (*Dreissena polymorpha*) on water quality: A case study in Saginaw Bay, Lake Huron. In: Claudi R, Leach JH (eds) Non-indigenous freshwater organisms in North America: their biology and impact. CRC Press, Boca Raton, pp 255–271

Nalepa TF, Fanslow DL, Pothoven SA, Foiley III AJ, Lang GA (2007) Long-term trends in benthic macroinvertebrate populations in Lake Huron over the past four decades. J Great Lakes Res 33:421–436

Nalepa TF, Fanslow DL, Lang GA (2009a) Transformation of the offshore benthic community in Lake Michigan: recent shift from the native amphipod *Diporeia* spp. to invasive mussel *Dreissena rostriformis bugensis*. Freshw Biol 54:466–479

Nalepa TF, Pothoven SA, Fanslow DL (2009b) Recent changes in benthic macroinvertebrate populations in Lake Huron and impact on the diet of lake whitefish (*Coregonus clupeaformus*). Aquat Ecosyst Health Manag 12:2–10

Nalepa TF, Fanslow DL, Pothoven SA (2010) Recent changes in density, biomass, recruitment, size structure, and nutritional state of *Dreissena* populations in southern Lake Michigan. J Great Lakes Res 36:5–19

Nicholls KH, Heintsch L, Carney E (2002) Univariate step-trend and multivariate assessments of the apparent effects of P loading reductions and zebra mussels on the phytoplankton of the Bay of Quinte, Lake Ontario. J Great Lakes Res 28:15–31

NOAA (2014) NOAA CoastWatch Great Lakes Node. Available via http://coastwatch.glerl.noaa.gov/. Accessed 6 June 2014

Ohtaka A, Tetsuya N, Takahiro K, Haruo K, Yuji A, Sokrithy IM, Rachna C, Shinji T (2011) Composition of aquatic invertebrates associated with macrophytes in Lake Tonle Sap, Cambodia. Limnology 12:137–144

Oliveira MD, Hamilton SK, Jacobi CM (2010) Forecasting the expansion of the invasive golden mussel *Limnoperna fortunei* in Brazilian and North American rivers based on its occurrence in the Paraguay River and Pantanal wetland of Brazil. Aquat Invasions 5:59–73

Oliveira MD, Calheiros DF, Jacobi CM, Hamilton SK (2011) Abiotic factors controlling the establishment and abundance of the invasive golden mussel *Limnoperna fortunei*. Biol Invasions 13:717–729

Orlova MI (1987) Differences in salinity tolerance of *Dreissena* in the lower Dnieper River and the Dnieper-Bug Liman. In: Likharev IM (ed) Abstracts of the 8th meeting on the investigation of molluscs. Nauka Press, Leningrad, pp 261–263 [in Russian]

Orlova MI, Muirhead J, Antonov PI, Shcherbina GK, Starobogatov I, Biochino GI, Therriault TW, MacIsaac HJ (2004) Range expansion of quagga mussels (*Dreissena rostriformis bugensis*) in the Volga River and Caspian Sea basin. Aquat Ecol 38:561–573

Padilla DK (2005) The potential for zebra mussels as a model for invasion biology. American Malacological Bulletin, 20:123–131

Padilla DK, Chotkowski MA, Buchan LAJ (1996) Predicting the spread of zebra mussels (*Dreissena polymorpha*) to inland waters using boater movement patterns. Glob Ecol Biogeogr Lett 5:353–359

Paolucci EM (2010) Impacto del molusco invasor *Limnoperna fortunei* sobre el ecosistema: interacción trófica y efectos sobre las larvas de peces nativos. PhD Thesis, Universidad de Buenos Aires (Argentina), pp 1–162

Paolucci EM, Cataldo DH, Fuentes CM, Boltovskoy D (2007) Larvae of the invasive species *Limnoperna fortunei* (Bivalvia) in the diet of fish larvae in the Paraná River, Argentina. Hydrobiologia 589:219–233

Paolucci EM, Cataldo DH, Boltovskoy D (2010a) Prey selection by larvae of *Prochilodus lineatus* (Pisces): indigenous zooplankton versus larvae of the introduced bivalve *Limnoperna fortunei*. Aquat Ecol 44:255–267

Paolucci EM, Thuesen EV, Cataldo DH, Boltovskoy D (2010b) Veligers of an introduced bivalve (*Limnoperna fortunei*) are a new food resource that enhances growth of larval fish in the Paraná River (South America). Freshw Biol 55:1831–1844

Park H-K, Cho K-H, Won DH, Lee J, Kong D-S, Jung D-I (2013) Ecosystem responses to climate change in a large on-river reservoir, Lake Paldang, Korea. Climatic Change 120:477–489

Pastorino G, Darrigran GA, Martín SM, Lunaschi L (1993) *Limnoperna fortunei* (Dunker, 1857) (Mytilidae), nuevo bivalvo invasor en aguas del Río de la Plata. Neotropica 39:101–102

Patterson MWR, Ciborowski JJH, Barton DR (2005) The distribution and abundance of *Dreissena* species (Dreissenidae) in Lake Erie, 2002. J Great Lakes Res 31(Supplement 2):223–237

Payne AI (1986) The ecology of tropical lakes and rivers. Wiley, New York, pp 1–310

Penchaszadeh PE, Darrigran GA, Angulo C, Averbuj A, Brögger M, Dogliotti A, Pírez N (2000) Predation of the invasive freshwater mussel *Limnoperna fortunei* (Dunker, 1857) (Mytilidae) by the fish *Leporinus obtusidens* Valenciennes, 1846 (Anostomidae) in the Rio de la Plata, Argentina. J Shellfish Res 19:229–231

Perepelizin PV, Boltovskoy D (2011) Resistance of the invasive pest mussel *Limnoperna fortunei* to anoxia. J Am Water Works Assoc 103:79–85

Peyer SM, McCarthy AJ, Lee CE (2009) Zebra mussels anchor byssal threads faster and tighter than quagga mussels in flow. J Exp Biol 212:2026–2035

Peyer SM, Hermanson JC, Lee CE (2010) Developmental plasticity of shell morphology of quagga mussels from shallow and deep-water habitats of the Great Lakes. J Exp Biol 2013:2602–2609

Poddubnyi AG (1966) Adaptive response of *Rutilus rutilus* to variable environmental conditions. Proc Institute of Biology of Inland Waters of the Academy of Science of the USSR 10:131–138 [In Russian]

Pollux BJ, van der Velde G, bij de Vaate A (2010) A perspective on global spread of *Dreissena polymorpha*: a review on possibilities and limitations. In: van der Velde G, Rajagopal S, bij de Vaate A (eds) The zebra mussel in Europe. Backhuys Publishers, Leiden, pp 45–58

Pothoven SA, Fahnenstiel GL (2013) Recent change in summer chlorophyll a dynamics of southeastern Lake Michigan. J Great Lakes Res 39:287–294

Pothoven SA, Fahnenstiel GL (2014) Lake Michigan after dreissenid mussel invasion. In: Nalepa TF, Schloesser DW (eds) Quagga and zebra mussels: biology, impacts, and control, 2nd edn. CRC Press, Boca Raton, pp 545–553

Raikow DF (2004) Food web interactions between larval bluegill (*Lepomis macrochirus*) and exotic zebra mussels (*Dreissena polymorpha*). Can J Fish Aquat Sci 61:497–504
Ramcharan CW, Padilla DK, Dodson SI (1992) Models to predict potential occurrence and density of the zebra mussel, *Dreissena polymorpha*. Can J Fish Aquat Sci 49:2611–2620
Rennie MD, Sprules WG, Johnson TB (2009) Resource switching in fish following a major food web disruption. Oecologia 159:789–802
Roe SL, MacIsaac HJ (1997) Deepwater population structure and reproductive state of quagga mussels (*Dreissena bugensis*) in Lake Erie. Can J Fish Aquat Sci 54:2428–2433
Rojas Molina F (2010) Efectos del molusco invasor *Limnoperna fortunei* (Dunker) sobre el zooplancton del paraná medio. PhD Thesis, Universidad de Buenos Aires (Argentina), pp 1–205
Rojas Molina F, Paggi JC, Devercelli M (2010) Zooplanktophagy in the natural diet and selectivity of the invasive mollusk *Limnoperna fortunei*. Biol Invasions 12:1647–1659
Rojas Molina F, Paggi JC, Devercelli MD (2011) Vulnerability of microcrustaceans to predation by the invasive filter-feeding mussel *Limnoperna fortunei* (Dunker). Marine and Freshwater Behaviour and Physiology 44:329–338
Rojas Molina F, Jose de Paggi SB, Frau D (2012) Impacts of the invading golden mussel *Limnoperna fortunei* on zooplankton: A mesocosm experiment. Zool Stud 51:733–744
Sardiña P, Cataldo D, Boltovskoy D (2008) The effects of the invasive mussel, *Limnoperna fortunei*, on associated fauna in South American freshwaters: importance of physical structure and food supply. Fund Appl Limnol/Archiv Hydrobiologie 173:135–144
Sardiña P, Chaves E, Marchese M (2011) Benthic community responses to invasion by the golden mussel, *Limnoperna fortunei* Dunker: biotic homogenization vs environmental driving forces. J N Am Benthol Soc 30:1009–1023
Sarnelle O, Wilson AE, Hamilton SK, Knoll LB, Raikow DF (2005) Complex interactions between the zebra mussel, *Dreissena polymorpha*, and the harmful phytoplankter, *Microcystis aeruginosa*. Limnol Oceanogr 50:896–904
Scarabino F (2004) Conservación de la malacofauna uruguaya. Comunicaciones de la Sociedad Malacológica del Uruguay 8:267–273
Schloesser DW, Masteller EC (1999) Mortality of unionid bivalves (Mollusca) associated with dreissenid mussels (*Dreissena polymorpha* and *D. bugensis*) in Presque Isle Bay, Lake Erie. Northeast Nat 6:341–352
Sherman JJ, Murray BA, Woolnough DA, Zanatta DT, Uzarski DG (2013) Assessment of remnant unionid assemblages in a selection of Great Lakes coastal wetlands. J Great Lakes Res 39:201–210
Shevtsova LV (1989) Importance of the Bug River *Dreissena* for decreasing organic matter and for the sedimentation of suspended substances. Hydrobiol J 25:47–51
Shkorbatov GL, Karpevich AF, Antonov PI (1994) Ecological physiology. In: Starobogatov JI (ed) Freshwater zebra mussel *Dreissena polymorpha* (Pall.) (Bivalvia, Dreissenidae). Systematics, ecology, practical meaning. Nauka Press, Moscow, pp 67–108 [In Russian]
Simberloff D (2003) Confronting introduced species: a form of xenophobia? Biol Invasions 5:179–192
Son MO (2007) Native range of the zebra mussel and quagga mussel and new data on their invasions within the Ponto-Caspian region. Aquat Invasions 2:169–179
Soster FM, McCall PL, Herrmann KA (2011) Decadal changes in the benthic invertebrate community in western Lake Erie between 1981 and 2004. J Great Lakes Res 37:226–237
Spiridonov YI (1972) Polarographic investigation of the respiration of *Dreissena*. In: Topics in physiological and population ecology, Saratov University Press, Saratov (Russia), pp 15–21 [In Russian]
Sprung M (1987) Ecological requirements of developing *Dreissena polymorpha* eggs. Archiv Hydrobiologie 79:69–86
Sprung M (1991) Costs of reproduction: A study on metabolic requirements of the gonads and fecundity of the bivalve *Dreissena polymorpha*. Malacologia 33:63–70

Starobogatov YI, Andreeva SI (1994) Range. In: Starobogatov YI (ed) Freshwater zebra mussel *Dreissena polymorpha* (Pall.) (Bivalvia, Dreissenidae). Systematics, ecology, practical meaning. Nauka Press, Moscow, pp 47–53 [In Russian]

Stewart TW, Miner JG, Lowe RL (1998) Quantifying mechanisms for zebra mussel effects on benthic macroinvertebrates: organic matter production and shell-generated habitat. J N Am Benthol Soc 17:81–94

Stoeckel JA, Rehmann CR, Schneider DW, Padilla DK (2004) Retention and supply of zebra mussel larvae in a large river system: importance of an upstream lake. Freshw Biol 49:919–930

Stoeckmann A (2003) Physiological energetics of Lake Erie dreissenid mussels: a basis for the displacement of *Dreissena polymorpha* by *Dreissena bugensis*. Can J Fish Aquat Sci 60:126–134

Strayer DL, Caraco NF, Cole JJ, Findlay S, Pace ML (1999) Transformation of freshwater ecosystems by bivalves. A case study of zebra mussels in the Hudson River. Bioscience 49:19–27

Strayer DL, Hattala KA, Kahnle AW (2004) Effects of an invasive bivalve (*Dreissena polymorpha*) on fish in the Hudson River estuary. Can J Fish Aquat Sci 61:924–941

Strayer DL, Hattala KA, Kahnle AW, Adams RD (2014) Has the Hudson River fish community recovered from the zebra mussel invasion along with its forage base? Can J Fish Aquat Sci. 71:1146–1157

Sylvester F (2006) Biología alimentaria y ecología del molusco invasor *Limnoperna fortunei* (Mytilidae) en el Paraná inferior y Río de la Plata. PhD Thesis, Universidad de Buenos Aires (Argentina), pp 1–160

Sylvester F, Dorado J, Boltovskoy D, Juárez A, Cataldo D (2005) Filtration rates of the invasive pest bivalve *Limnoperna fortunei* as a function of size and temperature. Hydrobiologia 534:71–80

Sylvester F, Boltovskoy D, Cataldo D (2007a) Fast response of freshwater consumers to a new trophic resource: Predation on the recently introduced Asian bivalve *Limnoperna fortunei* in the lower Parana River, South America. Austral Ecol 32:403–415

Sylvester F, Boltovskoy D, Cataldo D (2007b) The invasive bivalve *Limnoperna fortunei* enhances benthic invertebrate densities in South American floodplain rivers. Hydrobiologia 589:15–27

Sylvester F, Cataldo D, Notaro C, Boltovskoy D (2013) Fluctuating salinity improves survival of the invasive freshwater golden mussel at high salinity: implications for the introduction of aquatic species through estuarine ports. Biol Invasions 15:1355–1366

Thorp JH, Casper AF (2003) Importance of biotic interactions in large rivers: an experiment with planktivorous fish, dreissenid mussels and zooplankton in the St Lawrence River. River Res Appl 19:265–279

Toresani NI, López HL, Gómez SE (1994) Lagunas de la Provincia de Buenos Aires. Dirección de Intereses Marítimos, Dirección Provincial de Pesca e Intereses Maritimos, Subcretaría de Pesca, Intereses Maritimos y Producciones Intensivas, Ministerio de la Producción de la Provincia de Buenos Aires, La Plata, pp 1–109

Vanderploeg HA, Liebig JR, Carmichael WW, Agy MA, Johengen TH, Fahnenstiel GL, Nalepa TF (2001) Zebra mussel (*Dreissena polymorpha*) selective filtration promoted toxic *Microcystis* blooms in Saginaw Bay (Lake Huron) and Lake Erie. Can J Fish Aquat Sci 58:1208–1221

Vanderploeg HA, Nalepa TF, Jude DJ, Mills EL, Holeck KT, Liebig JR, Grigorovich IA, Ojaveer H (2002) Dispersal and emerging ecological impacts of Ponto-Caspian species in the Laurentian Great Lakes. Can J Fish Aquat Sci 59:1209–1228

van der Velde G, Rajagopal S, bij de Vaate A (2010) From zebra mussels to quagga musels: an introduction to the Dreissenidae. In: van der Velde G, Rajagopal S, bij de Vaate A (eds) The zebra mussel in Europe. Backhuys Publishers, Leiden, pp 1–10

Villar C, Stripeikis J, D'Huicque L, Tudino M, Troccoli O, Bonetto C (1999) Cd, Cu and Zn concentrations in sediments and the invasive bivalves *Limnoperna fortunei* and *Corbicula fluminea* at the Rio de la Plata basin, Argentina. Hydrobiologia 416:41–49

Watkins JM, Dermott R, Lozano SJ, Mills EL, Rudstam LG, Scharold JV (2007) Evidence for remote effects of dreissenids mussels on the amphipod *Diporea*: analysis of Lake Ontario benthic surveys, 1997–2003. J Great Lakes Res 33:642–657

Wetzel RG (1975) Limnology. Saunders, Philadelphia, pp 1–860

Wong WH, Gerstenberger S, Baldwin W, Moore B (2012) Settlement and growth of quagga mussels (*Dreissena rostriformis bugensis* Andrusov, 1897) in Lake Mead, Nevada-Arizona, USA. Aquat Invasions 7:7–19

Yakovleva AV, Yakovlev VA (2011) Impact of *Dreissena polymorpha* and *Dreissena bugensis* on zoobenthos structure in the upper reaches of the Kuybyshev Water Reservoir, Russia. Rus J Biol Invasions 2:312–319

Young SS, Yang HN, Huang DJ, Liu SM, Huang YH, Chiang CT, Liu JW (2014) Using benthic macroinvertebrate and fish communities as bioindicators of the Tanshui River Basin around the greater Taipei area—multivariate analysis of spatial variation related to levels of water pollution. Int J Environ Res Public Health 11:7116–7143

Zhulidov AV, Pavlov DF, Nalepa TF, Scherbina GH, Zhulidov DA, Gurtovaya TY (2004) Relative distributions of *Dreissena bugensis* and *Dreissena polymorpha* in the Lower Don River System, Russia. Int Rev Hydrobiol 89:326–333

Zhulidov AV, Kozhara AV, Scherbina GH, Nalepa TF, Protasov A, Afanasiev SA, Pryanichnikova EG, Zhulidov DA, Gurtovaya TY, Pavlov DF (2010) Invasion history, distribution, and relative abundances of *Dreissena bugensis* in the Old World: a synthesis of data. Biol Invasions 12:1923–1940

Part III
Distribution and Spread

Distribution and Colonization of *Limnoperna fortunei:* Special Traits of an Odd Mussel

Demetrio Boltovskoy

Abstract The invasional success of *Limnoperna fortunei* was largely determined by the combination of two key components. One of them is rare biological traits of the species, generally unfavorable in freshwater bivalves: the possession of planktonic larvae and a sessile, byssate adult. The other component is man's extensive modification of landscapes, in particular construction of reservoirs, large interbasin connections—canals and aqueducts and freshwater navigation. This combination was instrumental for the fast dispersal and success of a species that would otherwise have remained inconspicuous and restricted geographically. Environmental tolerance, while helpful, was probably of lesser importance. Possession of planktonic larvae results in a significant advantage for adults that manage to travel upstream, but is a major limitation for those that settle too close to the river outlet into the sea because their offspring are doomed due to expatriation into saline waters. Short rivers are therefore less vulnerable to colonization by self-sustaining populations, especially if there are no lakes or reservoirs along their path that can serve as refuge and seeding grounds for reproducing adults. In South America, interbasin spread has not occurred as fast as anticipated, but will most probably continue increasing.

Keywords *Limnoperna fortunei* · Golden mussel · Adaptation · Biological invasion · Geographic spread

D. Boltovskoy (✉)
Facultad de Ciencias Exactas y Naturales, Instituto de Ecología, Genética y Evolución de Buenos Aires (IEGEBA), Universidad de Buenos Aires-CONICET,
Ciudad Universitaria, 1428 Buenos Aires, Argentina
e-mail: demetrio@ege.fcen.uba.ar

Museo Argentino de Ciencias Naturales 'Bernardino Rivadavia'-CONICET,
Buenos Aires, Argentina

D. Boltovskoy (ed.), *Limnoperna fortunei,* Invading Nature - Springer Series in Invasion Ecology 10, DOI 10.1007/978-3-319-13494-9_16

Mussel Traits and Man's Alteration of Earth: A Key Combination for Expansion

The fast dispersal and widespread nuisance effects of freshwater byssate mussels in general, and of *Limnoperna fortunei* in particular, are largely the result of five converging factors. Two of these factors are biological traits of the bivalve: (1) free swimming larvae and (2) a sessile, byssate adult. The other three factors are driven by human activities: (3) increasing interbasin connectivity, (4) construction of reservoirs, and (5) growing navigation in continental waters (Fig. 1).

It is widely acknowledged that the lack of a free-swimming larval stage in the vast majority of freshwater invertebrates is an adaptation to avoid expatriation into hostile marine environments—the sea. In contrast to marine benthic animals, most of which have planktonic larvae (Thorson 1950), the majority of invertebrates that have adapted to freshwaters have lost a planktonic larval stage (Lopez 1988), indicating that in the trade-off between lowering expatriation rates versus maintaining the high dispersal potential of their marine ancestors, evolution favored the former.

Possession of a byssus is also a rarity among freshwater bivalves (it occurs in ~1 % of all North American species: McMahon and Bogan 2001; <1 % of all South American bivalves: Pereira et al. 2014), as most have an infaunal mode of life, burrowing into soft sediment. This, too, probably reflects an adaptation to the scarcity of hard surfaced bottoms in lakes and lowland rivers. When present, hard substrata are normally limited to the shoreline fringe and river bank. Thus, *L. fortunei* is an oddity that has adapted to and survived in freshwaters despite these unfavorable traits.

For millions of years, *L. fortunei* has been constrained to a limited area in continental Southeast Asia, most probably restricted to the south of the Yangtze River (Morton and Dinesen 2010; Ye et al. 2011; see Chapter "Distribution and Spread of *Limnoperna fortunei* in China" in this volume). It somehow succeeded in establishing permanent, seeding populations upstream far enough from the sea for its larvae to complete metamorphosis and settle before being flushed out into saline waters, but it did not expand beyond this region (colonization of Indochina, including Thailand, Laos, Cambodia, and Vietnam, is probably relatively recent and likely associated with the influence of human migrations: Morton and Dinesen 2010). From its present distribution area (Fig. 2a) and the characteristics of the waterbodies where it thrives, it is clear that the range of the golden mussel was limited not because of its narrow ecological requirements, but because it was unable to overcome the geographic barriers involved. Models of its potential worldwide distribution (Kluza and McNyset 2005; Campos 2014; Campos et al. 2014; Fig. 2b) reinforce this assumption. In other words, as shown for other invasives (Kraft et al. 2002; Karatayev et al. 2007b), interbasin transportation was the bottleneck, rather than survival in a different milieu. Indeed, *L. fortunei* is thought to be extremely tolerant of a wide spectrum of environmental conditions (see Chapter "Parallels and Contrasts Between *Limnoperna fortunei* and Species of *Dreissena*" in this volume; Karatayev et al. 2007a, 2010), which may represent a necessary counterbalance to its primitive, mytiloid,

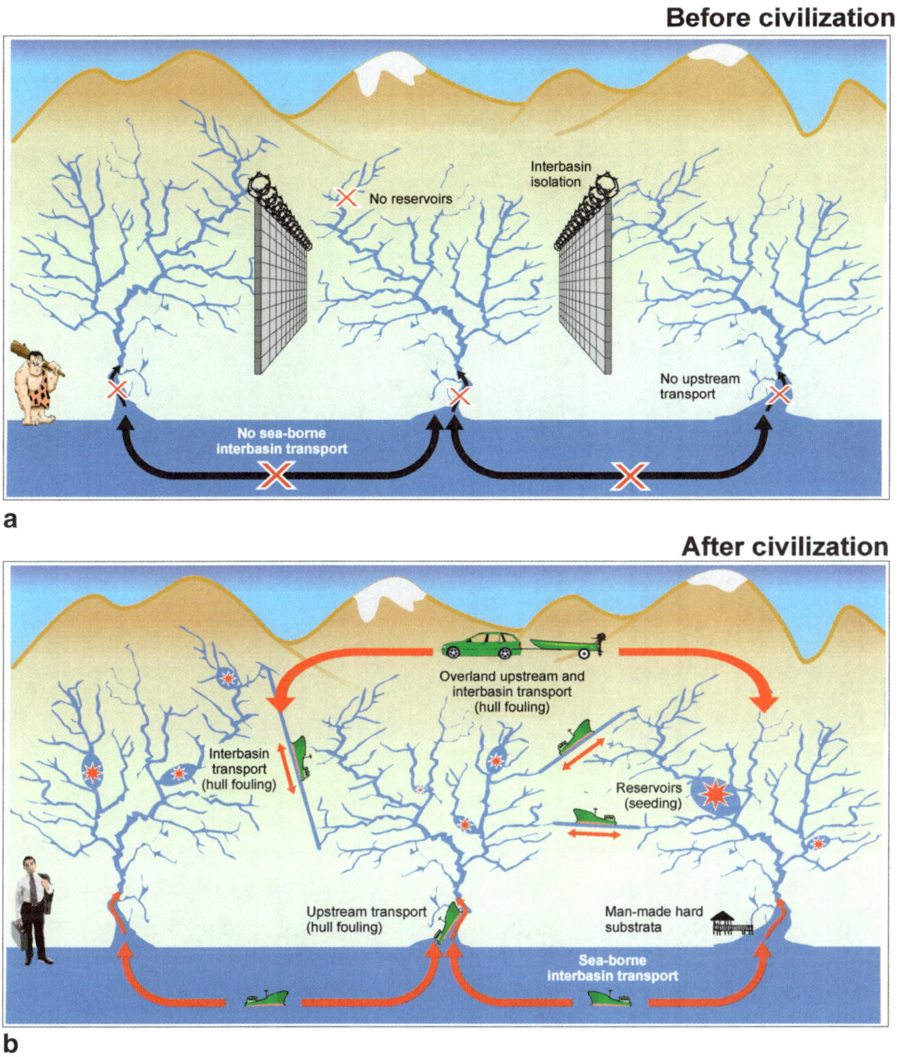

Fig. 1 Expansion vectors and mechanisms instrumental for the geographic spread of *Limnoperna fortunei* before (**a**) and after (**b**) large-scale modifications brought about by man's activities

freshwater-unfavorable traits (planktonic larvae and the requirement of hard substrata), allowing it to survive in a hostile habitat.

It should be stressed that invasional success does not necessarily imply broad environmental tolerance. Objective comparative studies of tolerance to environmental extremes between *L. fortunei* and native mussels have not been performed, but research in other geographic areas suggests that invaders may be less tolerant than native species. McMahon (2002), for example, concluded that *Dreissena polymorpha,* a highly successful invader, is not more resistant to ecological stress than

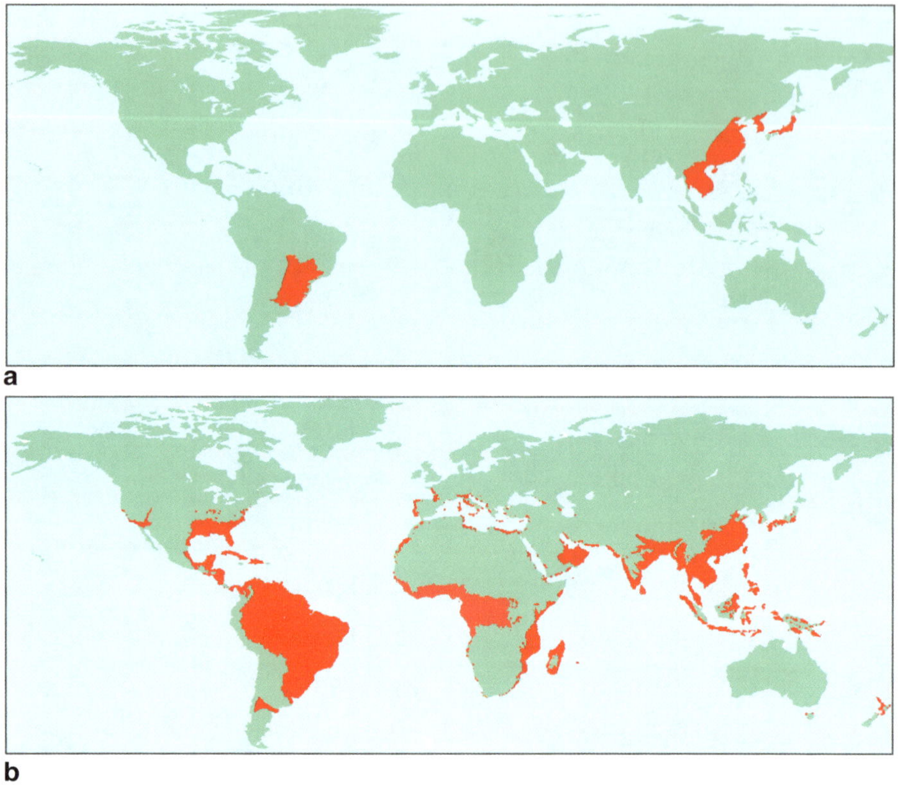

Fig. 2 Current worldwide distribution of *Limnoperna fortunei* (**a**) and potential distribution according to one of many possible modeling results (**b**). (b: modified from Kluza and McNyset 2005)

native unionids, its invasional success being largely explained by the fact that it is a typical r-strategist which recovers swiftly from catastrophic population reductions. *L. fortunei* is also an r-strategist, but unlike the zebra mussel it seems to be more resilient to stress than local mussels. Ancillary observations in various Chinese and Argentine waterbodies indicate that it is regularly present under extreme conditions, particularly in low oxygen and highly polluted waters, where many other organisms fail to survive (Boltovskoy et al. 2006; Contardo-Jara et al. 2009; Young et al. 2014).

The expansion of man is changing drastically the original rules of the game for much of the Earth's biota, but for some species, like *L. fortunei,* the changes are particularly profound and, unlike for many plants and animals, occasionally favorable. Increasing interbasin connectivity due to large hydraulic projects, such as irrigation and shipping canals, and aqueducts, is one significant shift that affects mussels (Minchin and Gollasch 2002; Nehring 2002; Karatayev et al. 2007b). However, large hydraulic projects are not particularly selective for *L. fortunei,* or for other byssate species, as they promote the spread more or less evenly of all freshwater organisms. On the other hand, shipping within continental waters, both natural and new man-made ones, became a key factor for organisms capable of traveling upstream by attaching to watercraft below the waterline (Fig. 1b). In the invasion of the Río de la Plata watershed by *L. fortunei,* the available evidence points at precisely this type of mechanism. After having been introduced in the freshwater estuarine area through the Río de la Plata estuary in ballast water, some adult specimens traveled upstream attached to the hulls of ships and barges that regularly sail on the Paraná-Paraguay rivers as far north as Cáceres (Brazil), over 2500 km upstream from the estuary (Boltovskoy et al. 2006). A similar dispersion vector has been found very important for other freshwater byssate mussels (Minchin et al. 2002; Karatayev et al. 2007b). Once established upstream, seeding the rest of the waterway with downstream drifting larvae was a natural and effortless outcome (see Chapter "Colonization and Spread of *Limnoperna fortunei* in South America" in this volume).

Admittedly, attachment to boat hulls (waterborne or carried upstream overland; Fig. 1b) is not the only vector possible. In Japan, where inland boat traffic is very limited, upstream transportation of *L. fortunei* is probably associated with fish culturing activities (Nomura et al. 2008; see Chapter "Colonization and Spread of *Limnoperna fortunei* in Japan" in this volume). Mussels can also be transported in mud caught in the limbs, wings, feathers, or fur of insects, amphibians, birds, or aquatic mammals (as shown for other freshwater invertebrates: Anastácio et al. 2014), or even in the stomachs of fishes that are known to egest intact animals (Belz et al. 2012), but these mechanisms are likely rare and ineffective for long-distance transport (Johnson and Carlton 1996).

An additional human activity which may have further facilitated geographic expansion of *L. fortunei,* particularly in areas where hard substrata are scarce, is modification of the coastline through the construction of docks, piers, spur dikes, groynes, pilings, breakwaters, revetments, rock armors, gabions, quay walls, etc., as well as deployment of buoys and other waterborne structures. In some areas, these objects are the only hard and stable substrate available for mussels, and although the populations they can harbor are comparatively small, they can be important as seeding areas or stepping stones for further expansion.

The Importance of River Length

Planktonic larvae are clearly advantageous for downstream dispersal. However, in order to benefit from their free-swimming veligers, reproducing adults must have reached safe areas far enough from the river mouth for their larvae to complete development before being flushed into the sea. For populations restricted to the middle or lower river reaches, drifting larvae represent a significant risk. Surface current speeds in major rivers of the Río de la Plata watershed can be around 1–3 m/s. At these speeds larvae, which spend around 10 to more than 20 days in the plankton (Cataldo et al. 2005), can drift downstream over 2000 km before they are ready to settle on a substratum. Thus, large, navigable rivers are more likely to succeed in maintaining reproducing populations of golden mussels than shorter, nonnavigable waterways. Models based on various environmental variables (calcium concentration, temperature, dissolved oxygen, pH, total suspended solids, slope, elevation, flow accumulation, flow direction, precipitation, etc.) simulating prospective spread of this mussel yielded interesting results (Kluza and McNyset 2005; Cohen 2009; Oliveira et al. 2010a, b; Campos 2014; Campos et al. 2014), but constraints associated with limitations imposed by planktonic larvae have not been explicitly addressed. For example, some models (Kluza and McNyset 2005; Campos 2014) indicate that *L. fortunei* could colonize areas along the western coast of South and Central America (Fig. 2b), most of which are characterized by short rivers with high current speeds. However, the above considerations suggest that this is unlikely.

Admittedly, the history of the geographic spread of *L. fortunei* shows that invasions are not restricted to large rivers. The mussel has been successfully colonizing Japan since 1990 (Kimura 1994), where rivers are short and mostly located in areas with steep topographic gradients (Yoshimura et al. 2005; Japan Commission on Large Dams 2009; see Chapter "Colonization and Spread of *Limnoperna fortunei* in Japan" in this volume). Research on expatriation of planktonic organisms by currents, dubbed as the "drift paradox," shows that in order to counterbalance downstream advection thus enhancing the chances of persistence, organisms rely on lateral diffusion, high dispersal rates, and the availability of refugia in hydrodynamically dead zones (Reynolds and Carling 1991; Reckendorfer et al. 1999; Speirs and Gurney 2001; Pachepsky et al. 2005). Local settings, such as embayments, inlets, or backwater areas where water movement is minimal, could significantly improve the ability of *L. fortunei* to survive in short, exorheic rivers, but it is probably the presence of dams that has had the greatest influence on the mussel's chances of establishing self-sustaining populations under these adverse conditions.

The presence of lentic waterbodies in the course of a river is particularly significant, as they serve as refugia for seeding populations (Havel et al. 2005), especially for large species with longer life cycles unable to compensate for downstream transit (Speirs and Gurney 2001; Pollux et al. 2004; Pachepsky et al. 2005; Allan and Castillo 2007). *L. fortunei* has managed to establish self-sustaining populations in these short, fast flowing, and turbulent Japanese rivers because most of them are punctuated by dams and reservoirs. Japan has more than 2600 dams higher than 15 m, and over 60,000 smaller irrigation ponds and dams (Japan Commission on Large Dams 2009). In terms of the number of dams, Japan occupies the fourth place in the world, and is third in number of reservoirs per unit surface (0.71/100 km^2; Yoshimura et al. 2005). Most Japanese rivers have been modified by man, chiefly for flood control and water use for agriculture and other purposes (Japan Commission on Large Dams 2009), thus creating thousands of refugia with significantly higher water residence times where planktonic organisms thrive best (Søballe and Kimmel 1987; Karatayev et al. 2007b).

In addition to dams, canals and pipelines connecting different rivers and watersheds have also greatly facilitated the spread and persistence of *L. fortunei* in Japan. The country is crossed by 400,000 km of man-made canals, chiefly for irrigation purposes, that connect most watersheds (Ministry of Agriculture, Forestry and Fisheries 2003; see Chapter "Colonization and Spread of *Limnoperna fortunei* in Japan" in this volume).

Why Did the Golden Mussel Take So Long to Start Expanding?

Given the fast and apparently effortless colonization of South America, an intriguing question is: "Why didn't it happen earlier?" While growing trade with Southeast Asia must play an important role (Darrigran and Pastorino 1995; Karatayev et al. 2007b), it is conceivable that the successful ~ 1990 colonization of the Río de la Plata watershed was but one of many previous attempts when mussels were seeded in the area, but whose offspring were entirely swept into the ocean (Boltovskoy et al. 2006). The notion that there might have been several unsuccessful introductions, both in South America and in Japan, before the mussel finally managed to establish local reproductive populations, is supported by biogeographic data (Boltovskoy et al. 2006), and by genetic evidence (Ito 2011; Zhan et al. 2012; Ghabooli et al. 2013) that suggest that there have been multiple successful introductions.

Ballast water-related invasions by overseas freshwater organisms are faced with the problem that most freshwater ports serving ocean-going ships are located in estuarine areas, in the vicinity of and upstream from brackish and saline waters. This may explain why freshwater invasions like those of the golden mussel do not occur more often. Interestingly, the unintentional introduction of *L. fortunei* in South America is exceptional in that practically all other freshwater nonindigenous species recorded in Argentina are the result of deliberate introductions associated with aquarium trade, angling, and aquaculture (Baigún and Quirós 1985; Vigliano and Darrigran 2002).

Future Spread

Since its first records in Hong Kong, and especially after having been found in Japan and South America in the early 1990s, many reports have forecast that *L. fortunei* will rapidly spread northwards to Central and North America, and elsewhere (Morton 1975; Ricciardi 1998; Boltovskoy et al. 2006; Karatayev et al. 2007b; Oliveira et al. 2010b). Comparison of its current geographic range with estimates of its potential worldwide distribution (Fig. 1) suggests that we are only witnessing the beginning of its expansion. However, almost 25 years after it invaded the Río de la Plata watershed, it has not yet been recorded in the next large South American basin—the Amazon, which is considered highly vulnerable and largely suitable for the mussel (Boltovskoy et al. 2006; Oliveira et al. 2010b). In Japan, the geographic expansion of the mussel's range is relentless but slow (see Chapter "Colonization and Spread of *Limnoperna fortunei* in Japan" in this volume). This may suggest that interbasin transfer is a more important bottleneck than originally anticipated (Karatayev et al. 2007b), but it is very unlikely that watershed limits will deter further colonization indefinitely. In South America, the Cuiabá River, a tributary of the Paraguay River, which has been colonized by *L. fortunei* at least since 2000 (Boltovskoy et al. 2006) is only 150 km from the Teles Pires River in the Tapajós River basin within the Amazon watershed (Calazans et al. 2013). Both this proximity and the fact that the Amazon is navigable to ocean liners of virtually any tonnage, including ships with ballast water from infested ports along the Paraná-Uruguay-Río de la Plata waterways and the Guaiba basin (where compliance with international water ballast regulations in rather loosely enforced; Boltovskoy et al. 2011), suggest that sooner or later *L. fortunei* will invade this basin and, eventually, other freshwater bodies worldwide.

Concluding Remarks

Bivalve traits that evolution allowed reluctantly to persist for millions of years tolerating the existence of such "outlyers" as *L. fortunei,* suddenly became a major asset for expanding a historically small geographic range when man created interbasin dispersion corridors and supplied upstream transportation vectors (Fig. 1).

These conclusions cast a shadow of doubt on the widely accepted assumption that the success of *L. fortunei* stems from the fact that it is particularly well adapted to freshwater habitats. While it obviously does possess many traits which became instrumental for its impressive geographic expansion (e.g., low requirements of calcium, very high fertility, comparatively short life span, early sexual maturation, extended reproductive period, etc.; Morton 1973, 1975), before man's intervention planktonic larvae and the requirement of hard substrata were most probably major limitations, rather than advantages.

Interestingly, this conclusion conflicts with the notion that the success of invasive species is due to their having the same general suite of traits exhibited by most successful organisms in general, irrespective of their alien or native status (Thompson and Davis 2011). While this has been shown for many invasives (Thompson et al. 1995), alteration of Earth by man is changing the adaptive value of traits acquired in the course of millions of years of evolution in pristine environments.

When contrasted with those of many other species, invasions by golden and zebra mussels indicate that invasive success may be associated with quite dissimilar intrinsic and environmental variables, and that unifying concepts in invasion ecology are far from foolproof. Lumping all invasive species in an attempt to synthesize unique settings that explain success and effects on the ecosystems is probably not only questionable, but may also be counterproductive. As noticed by Gurevitch and Padilla (2004, p. 474), "If we determine that domestic livestock are causing widespread plant extinctions, it is far more informative to focus on the impact of domestic livestock than to say, more generally, that aliens are causing these extinctions."

Acknowledgments This work was partially financed by grants from the University of Buenos Aires, Argentina (UBA X-020 and 20020100100035) and from the Argentine Agencia Nacional de Promoción Científica y Tecnológica, Argentina (PICT 2007 1968) to D. Boltovskoy.

References

Allan JD, Castillo MM (2007) Stream ecology. Structure and function of running waters. Springer, Dordrecht, pp 1–436

Anastácio PM, Ferreira MP, Banha F, Capinha C, Rabaça JE (2014) Waterbird-mediated passive dispersal is a viable process for crayfish (*Procambarus clarkii*). Aquat Ecol 48:1–10

Baigún CM, Quirós R (1985) Itroducción de peces exóticos en la República Argentina. INIDEP (Instituto Nacional de Investigación y Desarrollo Pesquero, Argentina), Departamento de Aguas Continentales, Informe Técnico. INIDEP (Instituto Nacional de Investigación y Desarrollo Pesquero, Argentina), Departamento de Aguas Continentales, Informe Técnico, pp 1–90

Belz CE, Darrigran GA, Netto OSM, Boeger WA, Ribeiro PJ (2012) Analysis of four dispersion vectors in inland waters: the case of the invading bivalves in South America. J Shellfish Res 31:777–784

Boltovskoy D, Correa N, Cataldo D, Sylvester F (2006) Dispersion and ecological impact of the invasive freshwater bivalve *Limnoperna fortunei* in the Río de la Plata watershed and beyond. Biol Invasions 8:947–963

Boltovskoy D, Almada P, Correa N (2011) Biological invasions: assessment of threat from ballast-water discharge in Patagonian (Argentina) ports. Environ Sci Policy 14:578–583

Calazans SHC, Americo JA, Fernandes FD, Aldridge DC, Rebelo MD (2013) Assessment of toxicity of dissolved and microencapsulated biocides for control of the golden mussel *Limnoperna fortunei*. Marine Environ Res 91:104–108

Campos MCS (2014) Modelagem matemática do processo de invasão de *Limnoperna fortunei* (Dunker, 1857). Fundação Centro Tecnológico de Minas Gerais (Brazil). Technical Report GT 343, Controle do Mexilhão Dourado: Bioengenharia e novos materiais para aplicações em ecossistemas e usinas hidrelétricas, pp 3–36

Campos MCS, Andrade AFA, Kunzmann B, Galvão DD, Silva FA, Cardoso AV, Carvalho MD, Mota HR (2014) Modelling of the potential distribution of *Limnoperna fortunei* (Dunker, 1857) on a global scale. Aquat Invasions 9:239–242

Cataldo D, Boltovskoy D, Hermosa JL, Canzi C (2005) Temperature-dependent larval development rates of *Limnoperna fortunei* (Bivalvia: Mytilidae). J Molluscan Stud 71:41–46

Cohen A (2009) Managing the exotic mussels *Dreissena polymorpha, Dreissena bugensis, Limnoperna fortunei* and *Mytilopsis leucophaeata* in SFPUC's Reservoirs. San Francisco Public Utilities Commission, Final Report. San Francisco (USA), pp 1–236

Contardo-Jara V, Galanti LN, Ame MV, Monferran MV, Wunderlin DA, Wiegand C (2009) Biotransformation and antioxidant enzymes of *Limnoperna fortunei* detect site impact in watercourses of Córdoba, Argentina. Ecotoxicol Environ Saf 72:1871–1880

Darrigran GA, Pastorino G (1995) The recent introduction of a freshwater Asiatic bivalve, *Limnoperna fortunei* (Mytilidae) into South America. The Veliger 38:171–175

Ghabooli S, Zhan A, Sardiña P, Paolucci E, Sylvester F, Perepelizin PV, Briski E, Cristescu ME, MacIsaac HJ (2013) Genetic diversity in introduced golden mussel populations corresponds to vector activity. PLoS ONE 8:e59328

Gurevitch J, Padilla DK (2004) Are invasive species a major cause of extinctions? Trends Ecol Evol 19:470–474

Havel JE, Lee CE, Vander Zanden MJ (2005) Do reservoirs facilitate invasion into landscapes? BioScience 55:518–525

Ito K (2011) Spatial distribution and status of the golden mussel, *Limnoperna fortunei* in Kanto region, Japan. Report of Yahagi River Institute 15:91–96 [In Japanese]

Japan Commission on Large Dams (2009) Dams in Japan. Past, present and future. CRC Press, Boca Raton, pp 1–232

Johnson LE, Carlton JT (1996) Post-establishment spread in large-scale invasions: dispersal mechanisms of the zebra mussel *Dreissena polymorpha*. Ecology 77:1686–1690

Karatayev AY, Boltovskoy D, Padilla DK, Burlakova LE (2007a) The invasive bivalves *Dreissena polymorpha* and *Limnoperna fortunei*: parallels, contrasts, potential spread and invasion impacts. J Shellfish Res 26:205–213

Karatayev AY, Padilla DK, Minchin D, Boltovskoy D, Burlakova LE (2007b) Changes in global economies and trade: the potential spread of exotic freshwater bivalves. Biol Invasions 9:161–180

Karatayev AY, Burlakova LE, Karatayev VA, Boltovskoy D (2010) *Limnoperna fortunei* versus *Dreissena polymorpha* : Population densities and benthic community impacts of two invasive freshwater bivalves. J Shellfish Res 29:975–984

Kimura T (1994) The earliest record of *Limnoperna fortunei* (Dunker) from Japan. Chiribotan (J Malacol Soc Jpn) 25:34–35 [In Japanese]

Kluza DA, McNyset KM (2005) Ecological niche modeling of aquatic invasive species. Aquat Invaders 16:1–7

Kraft CE, Sullivan PJ, Karatayev AY, Burlakova LE, Nekola JC, Johnson LE, Padilla DK (2002) Landscape patterns of an aquatic invader: assessing dispersal extent from spatial distributions. Ecol Appl 12:749–759

Lopez GR (1988) Comparative ecology of the macrofauna of freshwater and marine muds. Limnol Oceanogr 33:946–962

McMahon RF (2002) Evolutionary and physiological adaptations of aquatic invasive animals: r selection versus resistance. Can J Fish Aquat Sci 59:1235–1244

McMahon RF, Bogan AE (2001) Mollusca: Bivalvia. In: Thorn JH, Covich AP (eds) Ecology and classification of North American freshwater invertebrates, Academic Press, San Diego, pp 331–429

Minchin D, Gollasch S (2002) Vectors—How exotics get around. In: Leppakoski E, Gollasch S, Olenin S (eds) Invasive aquatic species of Europe: distribution, impacts and management, Kluwer Academic Publishers, Dordrecht, pp 183–194

Minchin D, Lucy F, Sullivan M (2002) Zebra mussel: Impacts and spread. In: Leppakoski E, Gollasch S, Olenin S (eds) Invasive aquatic species of Europe: distribution, impacts and management, Kluwer Academic Publishers, Dordrecht, pp 135–148

Ministry of Agriculture, Forestry and Fisheries (2003) The global diversity of irrigation. http://www.maff.go.jp/j/nousin/keityo/mizu_sigen/index.html. Accessed 5 April 2014

Morton B (1973) Some aspects of the biology and functional morphology of the organs of feeding and digestion of *Limnoperna fortunei* (Dunker) (Bivalvia: Mytilacea). Malacologia 12:265–281

Morton B (1975) The colonization of Hong Kong's raw water supply system by *Limnoperna fortunei* (Dunker 1857) (Bivalvia: Mytilacea) from China. Malacological Rev 8:91–105

Morton B, Dinesen G (2010) Colonization of Asian freshwaters by the Mytilidae (Bivalvia): a comparison of *Sinomytilus harmandi* from the Tonle-Sap River, Phnom Penh, Cambodia, with *Limnoperna fortunei*. Molluscan Res 30:57–72

Nehring S (2002) Biological invasions into german waters: an evaluation of the importance of different human-mediated vectors for nonindigenous macrozoobenthic species. In: Leppakoski E, Gollasch S, Olenin S (eds) Invasive aquatic species of Europe: distribution, impacts and management. Kluwer Academic Publishers, Dordrecht, pp 373–383

Nomura M, Kanai H, Takahashi K, Matsumoto I, Matsumoto H (2008) Golden mussel, *Limnoperna fortunei* from Tomioka to Fujioka area. Research Reports of the Gunma Museum of Natural History 4:40–47 [In Japanese]

Oliveira MD, Hamilton SK, Calheiros DF, Jacobi CM, Latini RO (2010a) Modeling the potential distribution of the invasive golden mussel *Limnoperna fortunei* in the Upper Paraguay River system using limnological variables. Braz J Biol 70:831–840

Oliveira MD, Hamilton SK, Jacobi CM (2010b) Forecasting the expansion of the invasive golden mussel *Limnoperna fortunei* in Brazilian and North American rivers based on its occurrence in the Paraguay River and Pantanal wetland of Brazil. Aquat Invasions 5:59–73

Pachepsky E, Lutscher F, Nisbet RM, Lewis MA (2005) Persistence, spread and the drift paradox. Theor Popul Biol 67:61–73

Pereira D, Mansur MCD, Duarte LDS, Oliveira AS, Pimpão DM, Callil CT, Ituarte C, Parada E, Peredo S, Darrigran G, Scarabino F, Clavijo C, Lara G, Miyahira IC, Rodriguez MTR, Lasso C (2014) Bivalve distribution in hydrographic regions in South America: historical overview and conservation. Hydrobiologia 735:15–44

Pollux BJA, Santamaria L, Ouborg NJ (2004) Differences in endozoochorous dispersal between aquatic plant species, with reference to plant population persistence in rivers. Freshw Biol 50:232–242

Reckendorfer W, Keckeis H, Winkler G, Schiemer F (1999) Zooplankton abundance in the River Danube, Austria: the significance of inshore retention. Freshw Biol 41:583–591

Reynolds CS, Carling PA (1991) Flow in river channels: new insights into hydraulic retention. Archiv Hydrobiologie-Beiheft Ergebnisse der Limnologie 121:171–179

Ricciardi A (1998) Global range expansion of the Asian mussel *Limnoperna fortunei* (Mytilidae): another fouling threat to freshwater systems. Biofouling 13:97–106

Søballe DM, Kimmel BL (1987) A large-scale comparison of factors influencing phytoplankton abundance in rivers, lakes, and impoundments. Ecology 68:1943–1954

Speirs DC, Gurney WSC (2001) Population persistence in rivers and estuaries. Ecology 82:1219–1237

Thompson K, Davis MA (2011) Why research on traits of invasive plants tells us very little. Trend Ecol Evol 26:155–156

Thompson K, Hodgson JG, Rich TCG (1995) Native and alien invasive plants: more of the same? Ecography 18:390–402

Thorson G (1950) Reproductive and larval ecology of marine bottom invertebrates. Biol Rev 25:1–45

Vigliano PH, Darrigran GA (2002) Argentina's freshwater systems: aliens in wonderland. In: 11th International Conference on Aquatic Invasive Species, Alexandria (USA)

Ye BM, Cao XW, Xu M, Wang ZY, Lin CC (2011) Study of *Limnoperna fortunei* invasion in long distance water transmission project. Water Wastewater Eng 37:99–102 [In Chinese]

Yoshimura C, Omura T, Furumai H, Tockner K (2005) Present state of rivers and streams in Japan. River Res Appl 21:93–112

Young SS, Yang HN, Huang DJ, Liu SM, Huang YH, Chiang CT, Liu JW (2014) Using benthic macroinvertebrate and fish communities as bioindicators of the Tanshui River Basin around the greater Taipei area—multivariate analysis of spatial variation related to levels of water pollution. Int J Environ Res Public Health 11:7116–7143

Zhan A, Perepelizin PV, Ghabooli S, Paolucci E, Sylvester F, Sardiña P, Cristescu ME, MacIsaac HJ (2012) Scale-dependent post-establishment spread and genetic diversity in an invading mollusc in South America. Divers Distrib 18:1042–1055

Distribution and Spread of *Limnoperna fortunei* in China

Mengzhen Xu

Abstract The native range of *Limnoperna fortunei* is the Pearl River basin in China. From there, the mussel was transferred by ships during the 1960s–1970s to the estuaries of the rivers in the Fujiang and Zhejiang regions and the Yangtze River. After 1980, its range expanded to the Huaihe, Yellow, and Haihe River basins. In 1980, *L. fortunei* was found in Tianjin, a city on the Bohai Sea in northern China, most probably introduced by coastal shipping activities. At present, golden mussels are present in the middle reaches of the Yellow River basin and even further north, around Beijing. Due to the highland topography in western China, dispersion of *L. fortunei* in this area will not take place without anthropogenic facilitation. Golden mussels might potentially colonize the Liao River basin and the Inland River basin in Northeast China if the water temperature increases due to climate change.

Keywords *Limnoperna fortunei* · Golden mussel · Biological invasion · Geographic spread · China

Although it is a native species of Southeast China (Dunker 1856), *Limnoperna fortunei* has been expanding its range throughout different regions of the country. The native range of *L. fortunei* appears to be the Pearl River basin, and from there it was transferred by ships from the estuary of the Pearl River to the estuaries of rivers in the Fujian and Zhejiang regions, and into the Yangtze, Huaihe, Yellow, and Haihe River basins (Fig. 1). It first spread downstream to Hong Kong around 1965 (Morton 1975) and to Taiwan probably in 1986 (Huang 2008), although the timing of this latter event is debated. Holikoshi (1935) mentioned having sampled *Volsella* (*Limnoperna*) *lacustris* in Taipei, and this record was later used by Kuroda (1941) who included the species in his inventory of Taiwanese mollusks. However, according to Huang (2008), Holikoshi (1935) record is dubious, and no original materials were preserved to confirm it.

According to Liu et al. (1976), *L. fortunei* colonized the middle to downstream areas of the Yangtze River and the southern parts of the central and eastern regions

---

M. Xu (✉)
State Key Laboratory of Hydroscience and Engineering, Tsinghua University, 100084 Beijing, China
e-mail: xumz07@gmail.com; mzxu@mail.tsinghua.edu.cn

D. Boltovskoy (ed.), *Limnoperna fortunei,* Invading Nature - Springer Series in Invasion Ecology 10, DOI 10.1007/978-3-319-13494-9_17

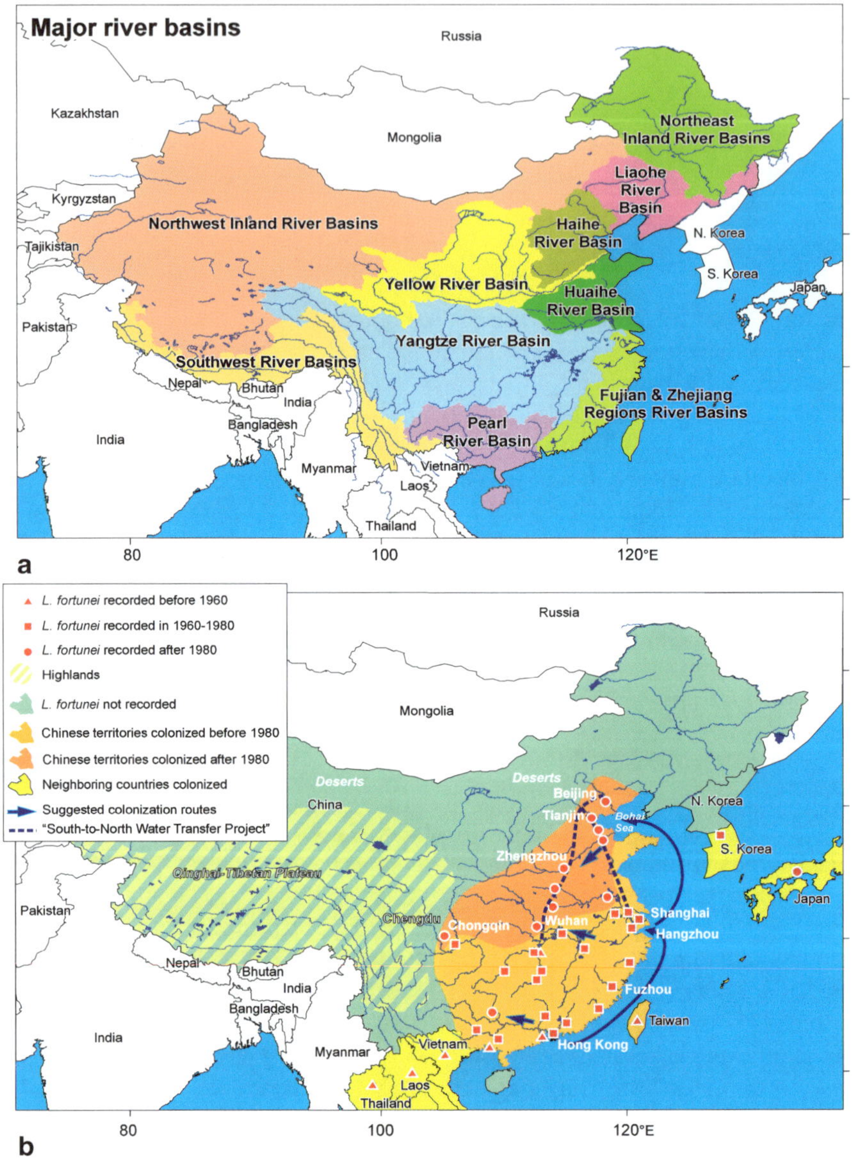

Fig. 1 **a** Major Chinese river basins or groups of basins. **b** Areas in China colonized by the golden mussel in different periods

of China with the Yangtze River as the northern boundary before 1980. In 1980, *L. fortunei* was found in Tianjin (Luo 2006), a coastal city of the Bohai Sea in northern China, presumably introduced by shipping activities along the coast. At present, *L. fortunei* is present in the middle reaches of the Yellow River basin in Zhengzhou (Yang et al. 2013), and even further north in the Shisanling reservoir in the northern suburbs of Beijing (Ye et al. 2011).

All of the available literature and field investigation reports describing the occurrence and colonization of *L. fortunei* in China are summarized in Table 1. There are ten major river basins in mainland China (Fig. 1a), and it is clear that the Pearl, Yangtze, and river basins in the Fujian and Zhejiang regions are the main habitat for golden mussels. Northwards, golden mussels are occasionally found in the Huaihe, Yellow, and Haihe River basins. The Liaohe, Northeast and Northwest Inland River basins, and the Southwest River basins (Fig. 1a) have not yet been invaded by *L. fortunei*.

Table 1 also lists the years when *L. fortunei* was first discovered in different river basins. *Limnoperna fortunei* was first described by Dunker (1856) from specimens collected in China (Huang 2008); however, the sampling location was not specified. The second earliest record was that of Miller and McClure (1931), according to whom golden mussels were sampled from the Pearl River in Guangdong Province in southern China in 1858 (Fig. 1b). After the 1950s, *L. fortunei* dispersed gradually to the first-order tributaries and then upstream to the second-order tributaries of the Pearl River. In the Yangtze River basin, the first report of *L. fortunei* was in the late 1950s, much later than in the Pearl River (Table 1). According to the discovery year, the invasion of *L. fortunei* proceeded through the connecting lakes and stem of the downstream and mid-downstream Yangtze River during the1950s to 1960s. Subsequently, the species expanded throughout the midstream and upstream tributaries from the 1970s through the 2000s (Fig. 1b).

The colonization of the river basins in the Fujian and Zhejiang regions occurred contemporaneously with the Yangtze River from 1960 through 1980, probably facilitated by the temperature rise after the 1950s and 1960s (Shen 2003). *Limnoperna fortunei* survived along the estuaries and expanded upstream, and consequently dispersed widely in the river basins in the Fujian and Zhejiang regions.

The colonization of golden mussels in the Huaihe, Yellow, and Haihe river basins occurred after 1980, 20 years later than the colonization in the Yangtze River and river basins of the Fujian and Zhejiang regions. The later date of these invasions might be related to the colder climate in northern China before 1980 (Liu et al. 2012). Nevertheless, the high suspended sediment concentration in the Yellow River mouth and instability of the Yellow River delta before 1980 may also have played important roles in preventing the invasion of *L. fortunei* into this area (Xu 2012).

In general, the dispersal pattern of *L. fortunei* in the different river basins is similar to that described by Boltovskoy et al. (2006) in South America. Golden mussels were possibly transferred by ships from the estuary of the Pearl River, and dispersed to estuaries of the rivers in the Fujian and Zhejiang regions and the Yangtze, Huaihe, Yellow, and Haihe River basins. Inside each river basin, *L. fortunei* expanded upstream (*short arrows* in Fig. 1b). These river basins are navigable by commercial traffic (Wu 2010); therefore, it is very likely that the expansion of *L. fortunei* is largely due to the upstream transport of adult individuals attached to the hulls of ships and associated structures, similar to the invasions in Argentina, Paraguay, and Brazil (see Chapter "Colonization and Spread of *Limnoperna fortunei* in South America" in this volume).

Table 1 Geographic distribution, dates, and references of recorded occurrences of *Limnoperna fortunei* in China

River basin	Area	Year recorded	Reference
?	China	1856	Dunker (1856)
Pearl	Pearl River (Guangdong Province)	1858	Miller and McClure (1931)
	Lower East River (1t of Pearl River; Guangdong Province)	1965	Morton (1975)
	Upper East River (1t of Pearl River; Guangdong Province)	1981–1982	Lai (1988)
	Lower Xizhijiang River (2t of Pearl River; Guangdong Province)	1981–1982	Lai (1988)
	Upper Nanshui River (2t of Pearl River; Guangdong Province)	1980s	Su et al. (1989), Luo (2006)
	Lower Xi River (1t of Pearl River; Fangcheng city, Guangxi Province)	1955	Huang (2008)
	Middle Xi River (1t of Pearl River; Wuzhou city, Guangxi Province)	2009	Pan et al. (2011)
	Lower Hongshuihe River (2t of Pearl River; Hechi and Liuzhou cities, Guangxi Province)	2007	Zhou et al. (2011)
Yangtze	Chaohu Lake (lower Yangtze River; Chaohu city, Anhui Province)	1963	Hu and Yao (1981)
	Poyang Lake (mid-lower Yangtze River; Nanchang, Duchang, Wucheng cities, and several others, Jiangxi Province)	1963	Tchang and Li (1965)
	Hwama Lake (mid-lower Yangtze River; Ezhou city, Hubei Province)	1959–1960	Chen (1979)
	Donghu Lake (mid-lower Yangtze River; Wuhan city, Hubei Province)	1960s	Pipeline Study Group (1973), Luo (2006)
	Dongting Lake (mid-lower Yangtze River; Yuanjiang, Xiangjiang, and other cities, Hunan Province)	1956–1963	Tchang et al. (1965)
	Lower Jialingjiang and Wujiang Rivers (1t of middle Yangtze River; Chongqing Province)	1978	Zeng et al. (1981)
	Hanjiang River (1t of middle Yangtze River; Xiangfan city, Hubei Province)	1994	Zhang et al. (2000)
	Fujiang River (1t of upper Yangtze River; vicinity of Chengdu city, Sichuan Province)	2007	Duan (2009)
	Wujiang River (1t of upper Yangtze River; from Nayong city in Guizhou Province to Fuling city in Chongqin Province)	2006	Chen et al. (2010)

Table 1 (continued)

River basin	Area	Year recorded	Reference
Fujian and Zhejiang regions	Nanxi River (Jinding, Longhai Prefecture, Fujian Province)	1961	Huang (2008)
	Tai Hu Lake and surrounding waterbodies (cities of Wuxi, Suzhou, Wujiang, Wuxian, Qingpu, Deqing, Jiaxing, and Tangqi in Zhejiang and Jiangsu provinces)	1973	Liu et al. (1980)
	Caoejiang River (cities of Shaoxing, Shangyu, Fengxian, Xinchang, and others in Zhejiang Province)	1981–1982	Huang et al. (1995)
	Yongjiang River (Yin County, Zhejiang Province)	1982–1986	Huang et al. (2002)
Huaihe	Middle and lower Huaije River, Nushan Lake (Anhui Province)	1991–1992	Zu et al. (1998)
	Upper Huaihe River tributary (Luohe, Henan Province)	1985	Zhao et al. (1986)
Yellow	Lakes and rivers connected with the lower Yellow River (Shangdong Province)	1979–1987	Ma and Sun (1997)
	Fuyang River (tributary of lower Yellow River; Handan city, Hebei Province)	1985	Xiang (1985)
	Middle Yellow River (Zhengzhou city, Henan Province)	2009	Yang et al. (2013)
Haihe	Xihe River (tributary of lower Haihe River, close to the river mouth in Tianjin City)	1980	Huang (2008)
	Tributary of the middle Haihe River (Shisanling, suburb of North Beijing)	2009	Ye et al. (2011)
Liaohe	Not recorded		Zhang (2013)
Northeast Inland River basins	Not recorded		Duan (2009), Pan et al. (2011)
Northwest Inland Rriver basins	Not recorded		Xue et al. (2011), Wang et al. (2012)
Southwest River basins	Not recorded		Duan (2009), Xu et al. (2012)

1t first-order tributary, *2t* second-order tributary

The sites colonized by golden mussels before 1960, in 1960–1980, and after 1980 are illustrated in Fig. 1b. All the locations are in lowland plains below 500 m above sea level and below a latitude roughly less than 40°N. Western and northwestern China are dominated by highlands and mountains associated with the Qinghai-Tibetan Plateau (Fig. 1b). Most of the major rivers in China originate from this plateau and flow eastward (Fig. 1b). Therefore, dispersal of *L. fortunei* has been restrained by high flow velocity and glacier meltwater in the rivers along the boundary of Qinghai-Tibetan Plateau and Chengdu, Central and North China plains. It is suggested that in order to cross this boundary and expand to the plateau, *L. fortunei* will have to be introduced by human activities. Moreover, many river basins in northwestern China are endorheic and have no connection with waterbodies colonized by golden mussels. Therefore, northwestern China will not be invaded by *L. fortunei* unless river basins are artificially connected with infected bodies of water. However, more and more interbasin water transfer projects are under construction or in the planning stages. The largest one is the "South-to-North Water Transfer Project" (Fig. 1b) that will transfer water from Hubei Province, which is heavily colonized by golden mussels, to northern China (Zhang 2009). Although the golden mussel is already present in the Beijing area where these aqueducts end (Fig. 1b), massive introduction of larvae and seeding adults will facilitate rapid dispersion of *L. fortunei* through the inland region of northern China.

Areas north of 40°N have very long and cold winters, and very short summers. The annual water temperature is generally below 3–5 °C (Duan 2009), which is lower than the threshold value of accumulated temperature for the reproduction of golden mussels (Xu et al. 2012). This is a primary factor explaining why *L. fortunei* has not yet expanded to northeast China. However, there is a possibility that the water temperature in northeast China might increase due to climate change. *L. fortunei* could colonize the Liaohe River basin and the Northeast Inland River basins if the water temperature rises sufficiently for it to survive and reproduce.

Acknowledgment This work was supported by the Natural Science Foundation of China (51409146).

References

Boltovskoy D, Correa N, Cataldo D, Sylvester F (2006) Dispersion and ecological impact of the invasive freshwater bivalve *Limnoperna fortunei* in the Río de la Plata watershed and beyond. Biol Invasions 8:947–963

Chen QY (1979) A report on mollusca in Lake Hwama, Hubei Province. Oceanologia et Limnologia Sinica 10:46–66 [In Chinese]

Chen X, Lin T, Qin FX (2010) Macroinvertebrate community structure and biological assessment to the water quality of Wujiang River. J Hydroecology 3:5–11 [In Chinese]

Duan DX (2009) Ecological impacts of water and sediment transportation on macroinvertebrate community in rivers. PhD Thesis, Tsinghua University (China), pp 1-194 [In Chinese]

Dunker W (1856) Mytilacea nova collectionis Cumingianæ. Proc Zool Soc Lond 24:358–366

GPS (Pipeline Study Group) (1973) Biology of golden mussel (*Limnoperna fortunei*). Animal's use and prevention. Institute of Aquatic Organisms of the Chinese Academy of Sciences, Beijing, pp 33–36 [In Chinese]
Holikoshi Y (1935) Ecology of freshwater Mollusca in Taiwan. Trans Nat Hist Soc Formosa 25:226–231 [In Japanese]
Hu JX, Yao WQ (1981) Zoobenthos of Chaohu Lake. J Anhui Univ (Natural Sciences) 2:159–173 [In Chinese]
Huang CW (2008) Population genetic structure of *Limnoperna fortunei* (Dunker, 1857) in Taiwan based on sequences of mitochondrial cytochrome C oxidase subunit I. MSc Thesis, National Changhua University of Education (Taiwan), pp 1–60
Huang WH, Li ZL, Liu YY, Zhang WZ, Wang YX (1995) Investigation report on freshwater molluscs in Caoejiang River. J Zhanjiang Fish Coll 15:17–24 [In Chinese]
Huang WH, Li ZL, Liu YY, Zhang WZ, Wang YX (2002) Freshwater bivalves in the Yong Jiang watershed. Fish Sci Technol Inf 29:51–55 [In Chinese]
Kuroda T (1941) A catalogue of molluscan shells from Taiwan (Formosa), with descriptions of new species. Memoirs of the Faculty of Science and Agriculture, Taihoku Imperial Univ 22:65–216
Lai ZX (1988) Investigations on benthic animals from the East River Basin in Canton. Freshw Fish 2:28–34 [In Chinese]
Liu YY, Wang YX (1976) Common economically important freshwater bivalves in China. Chin J Zool 2:45–48 [In Chinese]
Liu YY, Zhang WZ, Wang YX (1980) Bivalves (Mollusca) of the Tai Hu and its surrounding waters, Jiangsu Province, China. Acta Zool Sin 26:365–369 [In Chinese]
Liu Q, Yan CR, Zhang YQ, Yang JY, Zheng SH (2012) Variation in precipitation and temperature in the Yellow River Basin during the last 50 years. Chin J Agrometeorol 33:475–480 [In Chinese]
Luo FM (2006) Studies on biology and control techniques of *Limnoperna fortunei* (Duker) in water supply system of Shenzhen city. MSc Thesis, Nanchang University (China), pp 1-108 [In Chinese]
Ma J, Sun H (1997) Freshwater shellfishes of Shandong Province. Stud Mar Sin 39:27–31 [In Chinese]
Miller RC, McClure FA (1931) The fresh-water clam industry of the Pearl River. Lingnan Sci J 10:307–322
Morton B (1975) The colonization of Hong Kong's raw water supply system by *Limnoperna fortunei* (Dunker 1857) (Bivalvia: Mytilacea) from China. Malacol Rev 8:91–105
Pan BZ, Wang ZY, He XB (2011) Studies on assemblage characteristics of macrozoobenthos in the west river. Acta Hydrobiol Sin 35:851–856 [In Chinese]
Shen HY (2003) The decadal scale variation of rainfall-runoff in the Yangtze River Basin. J Lake Sci 15:90–96 [In Chinese]
Su BJ, Li WX, Lai ZX (1989) Investigations on benthic invertebrates in Pearl River in Canton. Chin J Zool 24:15–19 [In Chinese]
Tchang S, Li SC (1965) Bivalves (Mollusca) of Poyang Lake and surrounding waters, Kiangsi Province, China, with description of a new species. Acta Zool Sin 17:309–317 [In Chinese]
Tchang S, Li SC, Liu YY (1965) Bivalves (Mollusca) of Tung-Ting Lake and its surrounding waters, Hunan Province, China. Acta Zool Sin 17:212–213 [In Chinese]
Wang BQ, Xue JZ, Liu J, Zhu XY, Wu HX (2012) Community structure of macrobenthos in different salt ponds of Altay Prefecture, Xinjiang. Arid L and Geography 35:261–265 [In Chinese]
Wu P (2010) Opportunity and challenge of inland navigation development in China. Pt Waterw Eng 2:11–16 [In Chinese]
Xiang YL (1985) Study of the harm and prevention measures of mussel growth in pipelines in power generation plants. North China Electr Power 4:24–27 [In Chinese]
Xu M (2012) Experimental study of macroinvertebrate *Limnoperna fortunei* invasion and prevention in water transfer tunnels. PhD Thesis, Tsinghua University (China), pp 1-192 [In Chinese]
Xu M, Cao XW, Wang ZY, Wang XZ (2012) Attachment characteristics of golden mussels (*Limnoperna fortunei*) in water transport projects. J Tsinghua Univ [Qinghua Daxue Xuebao] 52:170–176 [In Chinese]

Xue JZ, Wang BQ, Wu HX, Chen WY, Huang SL (2011) A study of macrobenthic biodiversity in the upper Irtysh River. Shanghai Environ Sci 30:98–105 [In Chinese]
Yang W, Han HY, An W (2013) The importance of the water treatment plant laboratory in water quality control. Munic Eng Technol 2:157–160 [In Chinese]
Ye BM, Cao XW, Xu M, Wang ZY, Lin CC (2011) Study of *Limnoperna fortunei* invasion in long distance water transmission project. Water Wastewater Eng 37:99–102 [In Chinese]
Zeng HQ, Su LD, Xie SG (1981) Freshwater shellfish in Chongqing area. J Southwest Norm Univ, Nat Sci 2:78–82 [In Chinese]
Zhang Q (2009) The South-to-North Water Transfer Project of China: Environmental implications and monitoring strategy. J Am Water Resour Assoc 45:1238–1247
Zhang WZ (2013) The indices of biotic integrity and diversity of benthic macroinvertebrates in the Tanjiang, Dongjiang and Liaohe River. MSc Thesis, Hebei University (China), pp 1–57 [In Chinese]
Zhang JY, Luo L, Li CS, Ma RH, Wan DB, Tan YY, Kuang QJ (2000) Study of impact of the south-to-north water transfer project on the river ecological environment of the middle and lower reaches of Hanjiang. Environ Sci Technol 90:1–32 [In Chinese]
Zhao YH, Zhang PQ, Zhou YJ (1986) The occurrence of *Limnoperna lacustris* in Henan province. J Henan Norm Univ, Nat Sci 7:23 [In Chinese]
Zhou J, Han YQ, Lei JJ, He AY, Gan X, Qi YB, Huang HY, Wang XZ (2011) Aquaculture capacity planning study in Longtan Reservoir, Hongshui River. Fish Sci Technol Guangxi 1:2–31 [In Chinese]
Zu GZ, Gan YL, Hou GJ, Chen Y, Li HY (1998) Zoobenthos in the Nushan Lake. J Anhui Agric Univ 25:276–280 [In Chinese]

Colonization and Spread of *Limnoperna fortunei* in Japan

Kenji Ito

Abstract The golden mussel, *Limnoperna fortunei* (Dunker 1857), is an invasive freshwater bivalve native to continental China. In Japan, it was first found in Lake Biwa in 1992. In the 2000s, it began to be found in eastern Japan, including in the Tokai and Kanto regions. One possible route for the arrival of *L. fortunei* to Japan is believed to be among edible freshwater clams (the Asian clam, *Corbicula fluminea*) imported from China. DNA data suggest that the mussel's invasion of Japan has occurred on at least two separate occasions. As of 2013, the mussel has been found in 12 of Japan's 47 prefectures. Its routes of dispersal among water systems have not yet been fully identified. However, one of the most important known routes of its range expansion within and among water systems is via irrigation infrastructure such as canals and headrace channels. By using information on water current and data on the distribution of the mussel in irrigation facilities, we may be able to predict the areas that will likely be invaded.

Keywords *Limnoperna fortunei* · Golden mussel · Biological invasion · Geographic spread · Japan

Introduction

Many alien species have been introduced into Japanese freshwater systems. As of November 2013, 38 species of fish, 17 molluscs, and 7 crustaceans from overseas have established sustainable populations in Japanese freshwaters (National Institute for Environmental Studies 2013). These alien species may adversely affect not only Japan's fisheries, agriculture, and human health but also native ecosystems (Maezono and Miyashita 2003; Maezono et al. 2005; Usio et al. 2006).

The golden mussel, *Limnoperna fortunei* (Dunker 1857), is a small freshwater bivalve species that originally inhabited China (Miller and McClure 1931; Morton

---

K. Ito (✉)
National Institute for Agro-Environmental Sciences, Kannondai 3-1-3, Tsukuba, Ibaraki, 305-8604 Japan
e-mail: itoken@affrc.go.jp

D. Boltovskoy (ed.), *Limnoperna fortunei*, Invading Nature - Springer Series in Invasion Ecology 10, DOI 10.1007/978-3-319-13494-9_18

1973), and was probably introduced into several other southeast Asian countries (Cambodia, Laos, Thailand, Vietnam) as a result of human displacements (Morton and Dinesen 2010). This species has an epifaunal mode of life and exerts a harmful influence on the environment and man-made structures, such as water treatment systems and power stations (see "Impacts of *Limnoperna fortunei* on Man-made Structures and Control Strategies: General Overview" in this volume). It has extended its distribution to include many Asian (Hong Kong, Korea, Japan, Taiwan) and South American (Argentina, Bolivia, Brazil, Paraguay, and Uruguay) countries (Boltovskoy et al. 2006; Oliveira et al. 2006; Morton and Dinesen 2010). Here, I report the spatial distribution and present status of the mussel in Japanese freshwater habitats.

Introduction and Geographic Expansion of the Golden Mussel in Japan

In Japan, *L. fortunei* was first found in 1992 (Matsuda and Uenishi 1992). However, Kimura (1994) claimed to have recorded it in 1990 in the Ibi River; judging from the size of these mussels, the first infestation is likely to have occurred before 1989 (Kimura 1994).

The mussel is thought to have been transported from its place of origin by humans. One possible route for the arrival of *L. fortunei* in Japan is believed to have been among edible freshwater clams (Asian clam, *Corbicula fluminea*) imported from China and/or Korea. In 1987, *L. fortunei* was found among live clams imported into Japan from Lake Taihu in Jiangsu Province, China (Nishimura and Habe 1987). This clam is a pest species that was recently introduced to Japan and has spread throughout the country (Masuda and Habe 1988; Nemoto et al. 2003; Sonohara 2005). It is thought that dumping or deliberate release into rivers and lakes has caused its invasion and spread (Nemoto et al. 2003). *Limnoperna fortunei* is thought to have been introduced into various rivers and lakes in Japan by being inadvertently mixed in with these edible clams.

Before the early 2000s, *L. fortunei* was found in only two river systems in western Japan, namely the Kiso–Nagara–Ibi River system and the Lake Biwa–Yodo River system (Fig. 1a, 1b: Matsuda and Nakai 2002). In the 2000s, the mussel began to be found in eastern Japan. In 2004, it was recorded in the Yahagi River system in Aichi Prefecture (Fig. 1c; Uchida et al. 2007). After a year, huge infestations of mussels were reported in the Kabura-Gawa irrigation canal leading from the Ohshio Reservoir (Fig. 1e; Katayama et al. 2005), and in Lake Kasumigaura (Fig. 1f; Sunoh 2006, Ito 2007). In the Tone River, the mussel was found up to about 120 km from the river mouth in 2007 (Fig. 1h; Ito 2008). Kimura et al. (2011) have searched for the mussel on the Japan Sea coast and in the Kyushu region but have so far been unable to find it. Nevertheless, by 2013 the mussel was distributed in 12 of Japan's 47 prefectures.

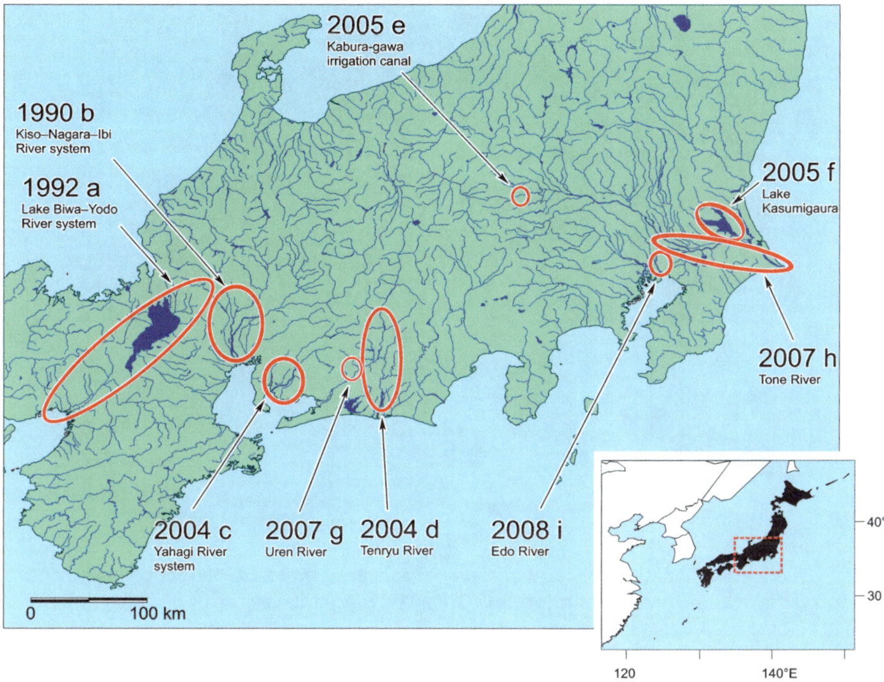

Fig. 1 Spatial and temporal distribution of *Limnoperna fortunei* in Japan

With the increase in its density and distribution, the mussel has been causing increasing damage to infrastructure used in human activities, including water purification plants (Nakanishi and Mukai 1997; Magara et al. 2001), hydraulic power plants (Magara et al. 2001), and irrigation facilities (Magara et al. 2001; Katayama et al. 2005; Yoshida 2006; Ito and Takimoto 2013).

Case Study: Expansion of the Distribution of the Golden Mussel in Lake Kasumigaura

Since it was first recorded in Japan in the 1990s, *L. fortunei* has been expanding relentlessly (Fig. 1), which suggests that there still are many uninvaded areas suitable for the mussel. Figure 2 shows the spatial distribution and abundance of the mussel in 2006 and 2012 along the shore of Lake Kasumigaura, Japan's second largest lake (Ito 2007; Ito and Takimoto 2013). From 2006 to 2012, the mussel expanded its distribution from 46 to 83 % of the shoreline of Lake Kasumigaura, and the density (as indicated by the number of mussels collected in 10 min by a single researcher) was on average 3.8 times higher in 2012 than in 2006 (Fig. 2). On the basis of these distribution maps, Ito and Takimoto (2013) predicted the expansion of the golden mussel by using a meta-population model (Koike 2006). The mussel

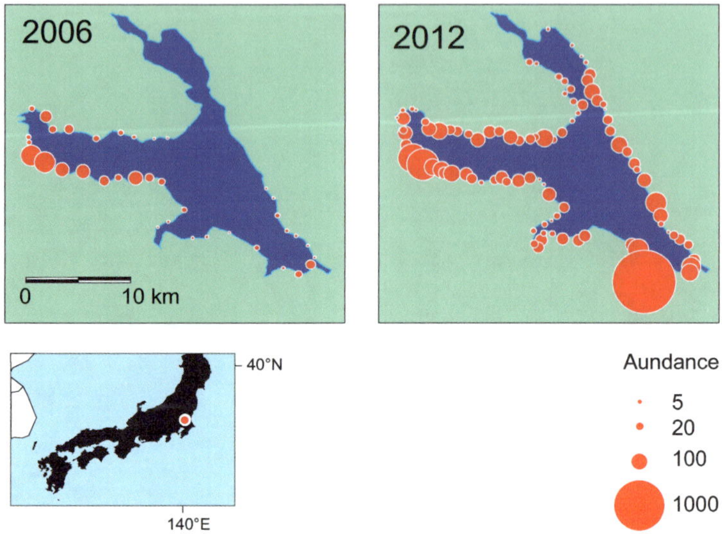

Fig. 2 Distribution and abundance of *Limnoperna fortunei* around Lake Kasumigaura. Abundance is defined as the number of mussels collected by one researcher in 10 min. (Modified from Ito and Takimoto 2013)

was observed in 2012 at all sampling sites where it had been found in 2006; therefore, local extinction was disregarded. The likelihood of colonization of a vacant site from the nearest inhabited site was primarily distance-dependent. In the 6 years elapsed, there was a 50 % probability that the golden mussels would colonize a vacant site as distant as ~ 11 km from the nearest inhabited site (Fig. 3). Given this colonization potential, golden mussels will spread over the entire shoreline of Lake Kasumigaura by 2018 at the latest (Ito and Takimoto 2013).

Fig. 3 Relationship between distance from a source site and colonization probability. Colonization probability decreased to only 50 % at a distance of 10.96 ± 0.62 km from a source site over 6 years. (Modified from Ito and Takimoto 2013)

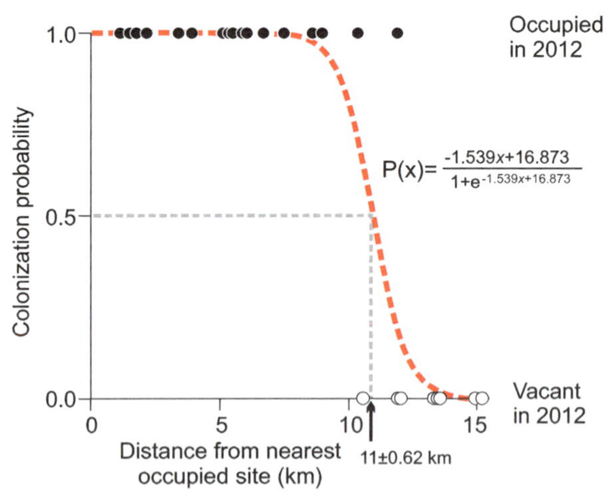

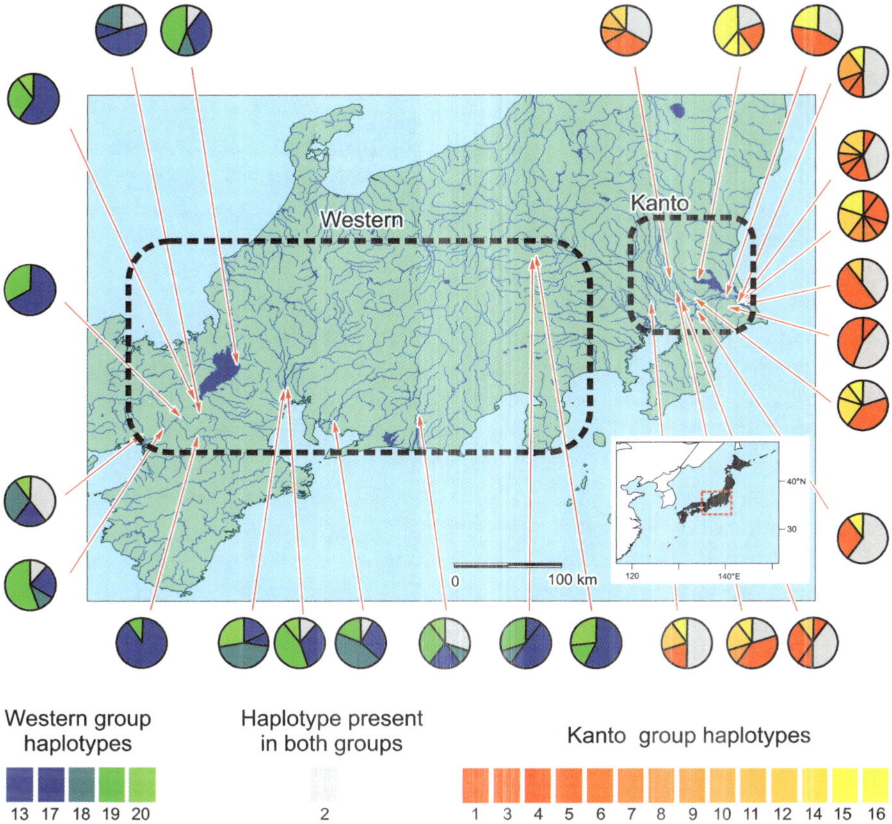

Fig. 4 Sampling localities and haplotype frequencies of *Limnoperna fortunei*. *Blue-green shadings:* haplotypes shared by the Western group and absent from the Kanto populations; *red-yellow shadings:* haplotypes endemic to the Kanto populations; *gray:* haplotype shared by both groups. (Based on data from Tominaga et al. 2009)

DNA-Based Study of the Expansion Process

Population genetic analyses are vital for tracing the processes by which alien species are spread. Tominaga et al. (2009) used mitochondrial DNA to survey the genetic structure of populations of golden mussels introduced in Japan. They detected a total of 20 unique haplotypes from 26 local populations. Fourteen of these were restricted to the lowlands of the Kanto District (Kanto group, Fig. 4). The other five haplotypes were not detected in the Kanto group but occurred only in the Western group (Fig. 4). Tominaga et al. (2009) also constructed a neighbor-joining tree among populations and deduced that populations from the Kanto group were genetically separated from the Western group. From these results, they concluded that the origin of the populations in the Kanto group differed from those of the other Japanese populations. In the Kanto group, the mussel was found later than in the other areas, namely in 2005 or later (Fig. 1). They concluded that the mussel has invaded Japan on at least two occasions (Tominaga et al. 2009). DNA data suggest that the origins of the Kanto group are Shanghai (China) and Korea (Tominaga and Kimura 2012).

As already mentioned, it is believed that the mussel was introduced as a contaminant of imported *C. fluminea* (Nishimura and Habe 1987; Magara et al. 2001). Because mussel quarantines have not yet been developed in Japan, the risk of further invasions is a major concern, which underscores the need to implement a system of quarantine for imported aquatic products.

Genetic analyses also revealed that there are many haplotypes shared among populations within the Kanto group and the Western group (Tominaga et al. 2009). These results imply that there is active gene flow among populations of these groups. These populations, especially those in the Western group, inhabit many water systems that are not directly connected to each other. Gene flow among populations from different water systems occurred as a result of long distance or "jump" dispersal (MacIsaac et al. 2001; Boltovskoy et al. 2006). Jump dispersal depends on the availability of long-distance dispersal agents, both natural and human mediated. The actual dispersal agent of the mussel among water systems, however, has not yet been identified in Japan. Nomura et al. (2008) suggested that accidental introduction of the mussel into fish nurseries might have caused its long-distance dispersal.

Routes of Expansion Within Water Systems: Rivers and Water Utility Systems

Natural rivers and waterways are among the most important routes for the geographic expansion of freshwater organisms (Leuven et al. 2009). Downstream dispersal of the mussel has been reported in Japanese rivers (Nakai 1996; Ito 2008). Upstream dispersal via attachment to the hulls of commercial ships and barges has been a key factor in the fast dispersal of *L. fortunei* throughout the Río de la Plata and Guaíba basins in South America (Boltovskoy et al. 2006; see "Colonization and Spread of *Limnoperna fortunei* in South America" in this volume). However, fast upstream dispersal has never been reported in Japanese freshwater systems. For example, the upstream limit of the mussel in the Kiso River system remained unchanged from 1993 to 1998 (Magara et al. 2001; Nakai 2001). One explanation for this difference may be that most Japanese rivers host little commercial boat traffic. In Japan, inland ship traffic was abandoned with the rapid expansion of railroad networks before World War II (Masuda 1993).

Recently, it became apparent that one of the most important routes of expansion of the mussel's geographic range is through irrigation facilities, such as irrigation canals and headrace channels. The Japanese landscape is characterized by high mountain ranges in the central section; the rivers are thus short (the longest river is the Shinano River, at 367 km), lie in very steep topographies, and are normally characterized by rather fast and turbulent flows (Yoshimura et al. 2005). Such topographies require the use of dams and canals for the control and efficient use of water resources for irrigation. In Japan, about 40,000 km of major irrigation canals (400,000 km including secondary canals) have been constructed to distribute irrigation water to farmlands (Ministry of Agriculture, Forestry and Fisheries 2003). Thus, most major rivers, lakes, and ponds have been connected by canals and pipelines. As shown below, there are several pieces of circumstantial evidence suggesting that golden mussels have expanded their distribution via these irrigation systems.

The Kokai River is a 118-km-long river that flows over the Kanto Plain. In this river, *L. fortunei* has been found only downstream of the Kokai River Water Division Facility, which is located on the Kasumigaura Canal (Fig. 5; Ito 2008). Most of the Kasumigaura Canal is represented by a headrace channel and reservoir, and it supplies water from Lake Kasumigaura to the southern part of Ibaraki Prefecture for irrigation, municipal, and industrial uses. *Limnoperna fortunei* was recorded after 2005 in Lake Kasumigaura and after 2006 in the Kasumigaura Canal (Sunoh 2006; Ito 2008). Larvae of the mussel were also found in the water division facility on Kasumigaura Canal. These results imply that the mussel was introduced from Lake Kasumigaura to the Kokai River via the Kasumigaura Canal (Fig. 5; Ito 2008, 2010). Similar distribution patterns of the mussel have been reported in the Edo River and the Kitachiba Water Conveyance Canal (Fig. 6). In the Edo River, *L. fortunei* has been found only downstream of the Matsudo Water Gate on the Kitachiba Water Conveyance Canal, which transfers water from the Tone River to the Edo River. Golden mussels have been present in the Kitachiba Water Conveyance Canal since 2007 (Ito 2010). DNA data also support the idea that the mussel's distribution expanded via these water utility systems (Tominaga et al. 2009).

There are still few reports of the golden mussel's presence in irrigation facilities. In Japan, 64 % of the water is used for agriculture (Yoshimura et al. 2005), and most irrigation facilities are managed by water users (i.e., farmers) associations, namely the Land Improvement District (Sato 2001). Therefore, in most of these facilities, scientific research has not yet been performed on the mussel. To control the mussel's spread, it may be effective to improve the management of the water supply facilities that are considered to be sources of the mussel and its geographic expansion routes. Water management strategies may affect the population dynamics of *L. fortunei* in irrigation systems (Nakano et al. 2010). By using information on water currents and data on the distribution of the mussel in irrigation facilities, we may be able to predict the areas it is likely to invade.

Conclusions

Because *L. fortunei* has both economic and environmental effects, Japan's Ministry of the Environment has officially designated it as an invasive alien species, and prohibits its culture and transportation in Japan without permission from the relevant authorities. The main purpose of such measures is to limit the intentional transport

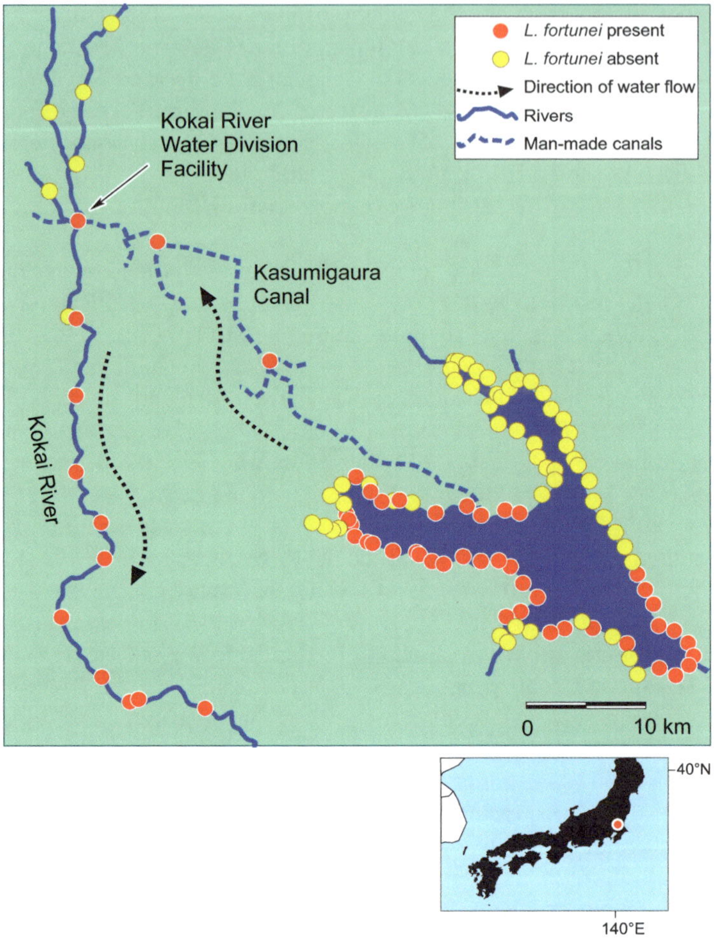

Fig. 5 Distribution of *Limnoperna fortunei* in the Kokai River and Kasumigaura Canal in 2006–2008. In the Kokai River, *L. fortunei* was found downstream of the Kokai River Water Division Facility on Kasumigaura Canal, which supplies water from Lake Kasumigaura to the southern part of Ibaraki Prefecture. (Modified from Ito 2010)

of invasive species. It is likely that introduction and expansion of the mussel occurred accidentally (e.g., through mixing with edible clam imports or by dispersal as larvae in rivers and canals). Therefore, the current regulations are not likely to be effective in controlling the mussel's expansion. To increase the effectiveness of this type of control, Japan needs regulations that limit non-intentional relocation.

Almost a quarter of a century has passed since the golden mussel was first recorded in Japan, and it seems to be spreading farther each year. However, there are still areas that have not been invaded. For example, the mussel has never been found in the Naka River system, where the search for it began in 2009 (Ito 2010; 2011, and unpublished data; Fig. 7), although this river system is adjacent to the Tone River

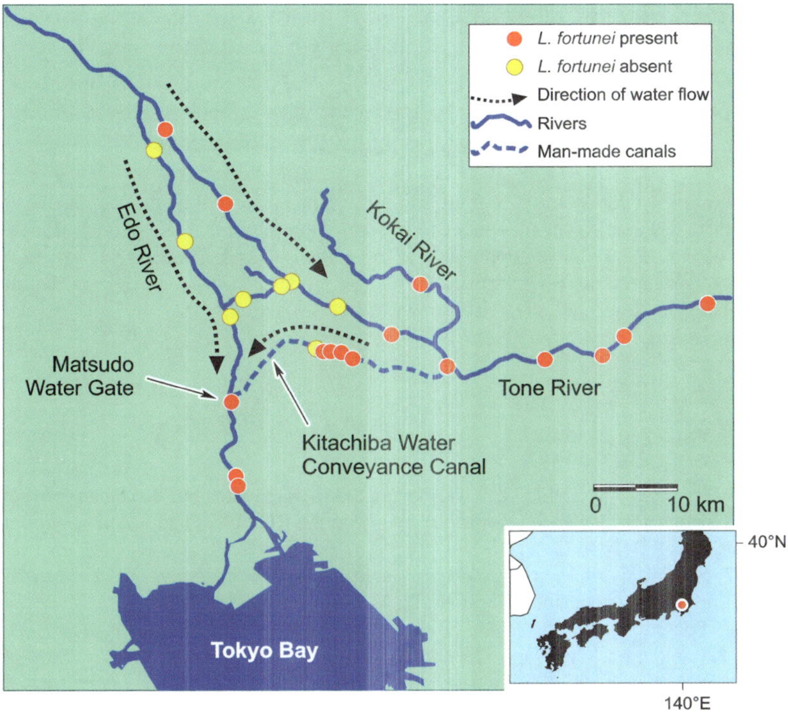

Fig. 6 Distribution of *L. fortunei* in the Edo River and Kitachiba Water Conveyance Canal in 2007–2008. (Modified from Ito 2010)

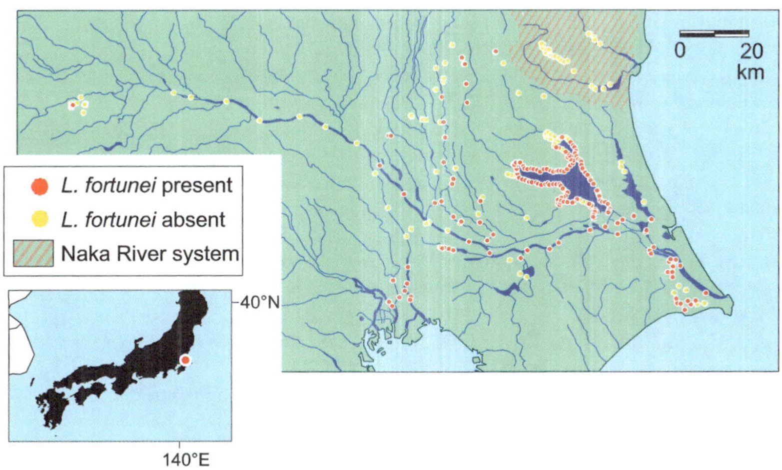

Fig. 7 Distribution of *L. fortunei* in the lowlands of the Kanto District in 2007–2013. Golden mussels have not been found in the Naka River system since searching began in 2009

basin, which *L. fortunei* inhabits since 2005 (Katayama et al. 2005; Sunoh 2006). Thus, it is likely that the golden mussel has not yet invaded this area. To reduce the mussel's impact, it is important to prevent its introduction and spread into these currently uninvaded areas. In order to achieve this, scientists will need to cooperate more closely with administrative bodies and water managers.

Acknowledgments I thank Drs. T. Kimura and A. Tominaga for their helpful comments and suggestions. Taylor & Francis is acknowledged for permission to use the data published in "Genetic structure of Japanese introduced populations of the golden mussel, *Limnoperna fortunei,* and the estimation of their range expansion process," authored by A. Tominaga, K. Goka, T. Kimura, and K. Ito, published in 2009 in Biodiversity, on which Fig. 4 is based.

References

Boltovskoy D, Correa N, Cataldo D, Sylvester F (2006) Dispersion and ecological impact of the invasive freshwater bivalve *Limnoperna fortunei* in the Río de la Plata watershed and beyond. Biol Invasions 8:947–963

Ito K (2007) Spatial distribution of golden mussel, *Limnoperna fortunei*, in Lake Kasumigaura, Ibaraki Prefecture, Japan. Jpn J Benthol 62:34–38 [In Japanese]

Ito K (2008) Spatial distribution of the golden mussel, *Limnoperna fortunei*, in the Tone River system, Kanto region, Japan. Jpn J Benthol 63:30–34 [In Japanese]

Ito K (2010) Spatial distribution and expansion process of the golden mussel, *Limnoperna fortunei* in Kanto region, Japan. Sess Org 27:17–23 [In Japanese]

Ito K (2011) Spatial distribution and status of the golden mussel, *Limnoperna fortunei* in Kanto region, Japan. Rep Yahagi River Inst 15:91–96 [In Japanese]

Ito K, Takimoto G (2013) Prediction of expansion of golden mussel, *Limnoperna fortunei*, in Lake Kasumigaura using a meta-population model. Jpn J Benthol 63:30–34 [In Japanese]

Katayama M, Shimizu R, Matsumoto H (2005) The first record of *Limnoperna fortunei* (Bivalvia, Mytilidae) in Gunma. Field Biol 14:35–40 [In Japanese]

Kimura T (1994) The earliest record of *Limnoperna fortunei* (Dunker) from Japan. Chiribotan (Journal of the Malacological Society of Japan) 25:34–35 [In Japanese]

Kimura T, Inoue A, Kimura S, Sato T (2011) Spatial distribution of an exotic mussel, *Limnoperna fortunei*, in Lake Biwa and connected waters in western Japan. Sess Org 28:9–18 [In Japanese]

Koike F (2006) Prediction of range expansion and optimum strategy for spatial control of feral raccoon using a metapopulation model. In: Koike F, Clout MN, Kawamichi M, De Poorter M, Iwatsuki K (eds) Assessment and control of biological invasion risks. Shoukadoh Book Sellers, Kyoto, pp 148–156

Leuven RSEW, van der Velde G, Baijens I, Snijders J, Van der Zwart C, Lenders HR, bij de Vaate A (2009) The river Rhine: a global highway for dispersal of aquatic invasive species. Biol Invasions 11:1989–2008

MacIsaac HJ, Grigorovich IA, Ricciardi A (2001) Reassessment of species invasions concepts: the Great Lakes basin as a model. Biol Invasions 3:405–416

Maezono Y, Miyashita T (2003) Community-level impacts induced by introduced largemouth bass and bluegill in farm ponds in Japan. Biol Conserv 109:111–121

Maezono Y, Kobayashi R, Kusahara M, Miyashita T (2005) Direct and indirect effects of exotic bass and bluegill on exotic and native organisms in farm ponds. Ecol Appl 15:638–650

Magara Y, Matsui Y, Goto Y, Yuasa A (2001) Invasion of the non-indigenous nuisance mussel, *Limnoperna fortunei*, into water supply facilities in Japan. J Water Supply Res Technol Aqua 50:113–124

Masuda H (1993) Coastal and river transport. In: Yamamoto H (ed) Technological innovation and the development of transportation in Japan. United Nations University Press, Tokyo, pp 186–195

Masuda O, Habe T (1988) *Corbicula insularis* Prime established in Kurashiki City, Okayama Pref., Honshu. Chiribotan (Journal of the Malacological Society of Japan) 19:30–40 [In Japanese]

Matsuda M, Nakai K (2002) Golden Mussel—The bivalve which adversely affects irrigation facilities. In: The Ecological Society of Japan Handbook of alien species in Japan, Chijin-Shokan, Tokyo (Japan), pp 173 [In Japanese]

Matsuda M, Uenishi M (1992) A newly introduced mussel, *Limnoperna lacustris* (V. Martens) (Mollusca: Mytilidae) in Lake Biwa, Shiga Prefecture, Japan. Annu Rep Biwako Bunkakan 10:45 [In Japanese]

Miller RC, McClure FA (1931) The fresh-water clam industry of the Pearl River. Lingnan Sci J 10:307–322

Ministry of Agriculture, Forestry and Fisheries (2003) The global diversity of irrigation. Available via http://www.maff.go.jp/j/nousin/keityo/mizu_sigen/index.html. Accessed 5 April 2014

Morton B (1973) Some aspects of the biology and functional morphology of the organs of feeding and digestion of *Limnoperna fortunei* (Dunker) (Bivalvia: Mytilacea). Malacologia 12:265–281

Morton B, Dinesen G (2010) Colonization of Asian freshwaters by the Mytilidae (Bivalvia): A comparison of *Sinomytilus harmandi* from the Tonle-Sap River, Phnom Penh, Cambodia, with *Limnoperna fortunei*. Molluscan Res 30:57–72

Nakai K (1996) Non-indigenous species in Lake Biwa: present status and problems, with particular reference to the mytilid mussel *Limnoperna fortunei* and the "Bass Problem". Kanasai Organization Nat Conserv 18:87–94 [In Japanese]

Nakai K (2001) Invasion of golden mussel into Japan. In: Kajihara T, Okutani T (eds) Kuroshozoku no shinnyusha: gairai fuchakusei nimaigai no saishingaku. Koseishakoseikaku, Tokyo (Japan), pp 71–85 [In Japanese]

Nakanishi M, Mukai S (1997) Growth and extermination of *Limnoperna fortunei* (Dunker 1857) at water purification facilities. Yousui to Haisui [Journal of Water and Waste] 39:1017–1020 [In Japanese]

Nakano D, Kobayashi T, Sakaguchi I (2010) Differences in larval dynamics of golden mussel *Limnoperna fortunei* between dam reservoirs with and without an aeration system. Landsc Ecol Eng 6:53–60

National Institute for Environmental Studies (2013) Invasive species of Japan. https://www.nies.go.jp/biodiversity/invasive/index_en.html. Accessed 22 November 2013

Nemoto T, Sugiura H, Komaru A (2003) Spatial distribution of introduced clams in Kitaura and Tone river. Bull Ibaraki Prefect Freshw Fish Exp Stn 38:32–41 [In Japanese]

Nishimura T, Habe T (1987) Chinese freshwater mussels mixed in the imported *Corbicula* clam. Chiribotan (Journal of the Malacological Society of Japan) 18:110–111

Nomura M, Kanai H, Takahashi K, Matsumoto I, Matsumoto H (2008) Golden mussel, *Limnoperna fortunei* from Tomioka to Fujioka area. Research Reports of the Gunma Museum of Nat Hist 4:40–47 [In Japanese]

Oliveira MD, Takeda AM, Fernandes de Barros L, Savio Barbosa D, Kawakami de Resende E (2006) Invasion by *Limnoperna fortunei* (Dunker, 1857) (Bivalvia, Mytilidae) of the Pantanal wetland, Brazil. Biol Invasion 8:97–104

Sato H (2001) The current state of paddy agriculture in Japan. Irrig Drain 50:91–99

Sonohara T (2005) Invasion of *Corbicula fluminea* into the Sagami river system and taxonomical change of Subfamily Corbicline. Actinia 16:11–19 [In Japanese]

Sunoh N (2006) Records of golden mussel *Limnoperna fortunei* in Lake Kasumigaura. Bull Ibaraki Prefect Freshw Fish Exp Stn 40:79 [In Japanese]

Tominaga A, Kimura T (2012) Estimation of range expansion process and invasion route of invasive alien mussel, *Limnoperna fortunei*. Chikyukankyo 17:143–150 [In Japanese]

Tominaga A, Goka K, Kimura T, Ito K (2009) Genetic structure of Japanese introduced populations of the golden mussel, *Limnoperna fortunei*, and the estimation of their range expansion process. Biodiversity 10:61–66

Uchida S, Shiragane A, Uchida A, Tanaka Y, Doi K, Matsuura Y (2007) A mass death after a high abundance of the invasive mussel, *Limnoperna fortunei*, in the Yahagi River, Honshu, Japan. Rep Yahagi River Inst 11:35–46 [In Japanese]

Usio N, Suzuki K, Konishi M, Nakano S (2006) Alien vs. endemic crayfish: roles of species identity in ecosystem functioning. Archiv Hydrobio–Beih Ergeb Limnol 166:1–21

Yoshida M (2006) Case study for removing the *Limnoperna fortunei* (Dunker, 1856) from the irrigation tunnel for canal in Kaburagawa LID. J Jpn Soc Irrig, Drain Reclam Eng 74:435–436 [In Japanese]

Yoshimura C, Omura T, Furumai H, Tockner K (2005) Present state of rivers and streams in Japan. River Res Appl 21:93–112

Colonization and Spread of *Limnoperna fortunei* in South America

Marcia D. Oliveira, Mônica C. S. Campos, Esteban M. Paolucci, Maria C. D. Mansur and Stephen K. Hamilton

Abstract The invasion of the exotic bivalve *Limnoperna fortunei* (Dunker 1857) in South America started with its introduction, presumably with ballast water from transoceanic ships trading with Southeast Asia, in the Río de la Plata estuary (Argentina) around 1990. From there, it spread swiftly to cover most of the Río de la Plata basin, as well as the basins of Guaíba and Tramandaí (Brazil), Patos–Mirim (Brazil–Uruguay), and Mar Chiquita (central Argentina). These smaller watersheds were most probably colonized as a result of secondary human-mediated introductions from waterbodies of the Río de la Plata basin. *L. fortunei* is now present in five South American countries: Argentina, Bolivia, Brazil, Paraguay, and Uruguay. Expansion was much faster along navigable waterways, especially the Paraná River and its tributaries (around 250 km/year), and slower elsewhere (Upper Paraguay

M. D. Oliveira (✉)
Embrapa Pantanal, Rua 21 de Setembro, 1880, Corumbá, MS, Brazil
e-mail: marcia.divina@embrapa.br

M. C. S. Campos
Instituto Senai de Tecnologia em Meio Ambiente, Centro de Inovação e Tecnologia SENAI FIEMG—Campus CETEC, Av. José Cândido da Silveira 2000, Belo Horizonte, MG 31035536, Brazil
e-mail: mosouza@fiemg.com.br

E. M. Paolucci
Facultad de Ciencias Exactas y Naturales, Instituto de Ecología, Genética y Evolución de Buenos Aires (IEGEBA), Universidad de Buenos Aires-CONICET, Ciudad Universitaria, 1428 Buenos Aires, Argentina

Museo Argentino de Ciencias Naturales 'Bernardino Rivadavia'-CONICET, Buenos Aires, Argentina
e-mail: estebanmpaolucci@gmail.com

M. C. D. Mansur
Universidade Federal do Rio Grande do Sul, Centro de Ecologia—CENECO, Av. Bento Gonçalves, 9500, setor 4, prédio 43411, Porto Alegre, RS 91501-970, Brazil
e-mail: mcrismansur@gmail.com

S. K. Hamilton
Kellogg Biological Station and Dept. of Zoology, Michigan State University, 3700 E. Gull Lake Dr., Hickory Corners, MI 49060, USA
e-mail: hamilton@kbs.msu.edu

D. Boltovskoy (ed.), *Limnoperna fortunei,* Invading Nature - Springer Series in Invasion Ecology 10, DOI 10.1007/978-3-319-13494-9_19

and Uruguay rivers and their tributaries). Over 20 years after its introduction in South America, new waterbodies are still being colonized by *L. fortunei* (e.g., the first records of the mussel in the Peixoto and Itapeva lakes, Tramandaí River system, date from 2013). Mussels have not been recorded in a few rivers of the Río de la Plata basin where their survival seems to be limited by excessively high suspended solid loads or salinities, or by the fact that they periodically dry out (e.g., the rivers Bermejo, Pilcomayo, and Salado del Norte, in north-central Argentina). South American populations display a relatively high genetic differentiation, confirming that geographic spread is strongly dependent on human activities: vessel and barge traffic is the main vector that helps to disperse the mussel locally through upstream "jumps" of adults attached to ship hulls. Genetic studies also suggest that there have been multiple introductions. By early 2014, *L. fortunei* had not yet been reported from any of the other major South American watersheds (Amazon, São Francisco, Orinoco), but colonization of these basins is probably inevitable. Modeling of potential distribution based on habitat fitness indicates that extensive regions could support *L. fortunei,* including much of lowland South America, southern Mexico, the southeastern USA, Europe, and Africa.

Keywords *Limnoperna fortunei* · Golden mussel · Biological invasion · Geographic spread · South America · Brazil · Argentina · Río de la Plata

Introduction

The golden mussel *Limnoperna fortunei* is an exotic bivalve that was established in southern South America around 1990. It presumably reached South America in the ballast water of transoceanic ships from Asia, and spread into the inland waters following the navigation routes, most probably attached to ship and barge hulls. The infestation of the golden mussel in South American waters has caused many economic impacts, albeit not well accounted for, affecting the operation of power generation (hydropower, nuclear, thermal), various other industrial facilities, fish farming, tourism, etc. (see "Impacts of *Limnoperna fortunei* on Man-made Structures and Control Strategies: General Overview" in this volume). The spread of *L. fortunei* has also had significant impacts on the native biota and ecosystems (Boltovskoy and Correa 2015). This chapter reviews the introduction, dispersal history, current distribution, and potential expansion of the golden mussel in South America.

Introduction and Dispersion of *L. fortunei* in South American Basins

The major hydrographic basins of South America are shown in Fig. 1. Northern South America contains the Amazon and the Orinoco basins, which carry the world's largest and third largest water discharges, respectively. The Río de la Plata

Fig. 1 Major drainage basins and rivers in South America

basin, which drains the central–eastern part of the subcontinent, is the fifth largest basin in the world and the second largest drainage basin in South America. Other important basins in northern and eastern Brazil are Tocantins and São Francisco, as well as many minor watersheds along the southern Atlantic coast that include the Guaíba, Patos-Mirim, and Tramandaí River basins. In Argentina, the Mar Chiquita endorheic basin includes several relatively small rivers draining into the saline Mar Chiquita Lake. Farther south, the climate is dryer and rivers have a west-to-east orientation, each comprising a discrete basin (Colorado, Negro, Chubut, Santa Cruz, etc.; Depetris et al. 2005).

Introduction and Dispersal in the Río de la Plata Basin

The Río de la Plata Basin: Rivers, Waterways, and Dams

The Río de la Plata basin drains 3.1 million km^2 extending over five countries: Brazil (44% of the basin), Argentina (32%), Paraguay (13%), Bolivia (6%), and Uruguay (5%; Fig. 1). The main drainage channel is the Paraná River, formed by the junction of the Grande and Paranaíba rivers, in Brazil. At about 4900 km, the Paraná River is the second longest river in South America. Its main tributaries are the Carcarañá, Salado del Norte, Paraguay, Grande, Tietê, Paranapanema, Ivaí, and Iguaçu (=Iguazú) rivers. The Paraná River flows roughly north to south; at about 31°S, it starts splitting into several arms creating a network of islands and wetlands. This delta, with an overall area of about 14,000 km^2, stretches from the cities of Santa Fe and Rosario (Argentina) to its outlet into the Río de la Plata estuary (Fig. 2). This basin hosts several very large cities, including Buenos Aires, Asunción, and São Paulo, among others. Its rivers comprise a large network of navigable waterways spanning 1930 km on the Paraná River, 2260 km on the Paraguay River, and 500 km on the Uruguay River. The annual volume of cargo transported along these waterways is around 10 million tons (ANTAQ (Agência Nacional de Transportes Aquaviários) 2011).

The Paraná River is navigable from Buenos Aires to Itaipu Dam, and from Itaipu to São Simão Dam (Paraná–Tietê waterway). It is interrupted at the Itaipu Dam because there are no navigation locks to transfer boats across Itaipu (Fig. 2). Between Itaipu and São Simão dams, the Paraná–Tieté waterway includes 2400 km of navigable routes, and many of the dams on the Tietê, uppermost Paraná, Paranaíba, Grande, and Paranapanema rivers are provided with locks allowing passage to waterborne traffic (Fig. 2 and 3). About 85% of the navigable waters are in impounded reaches associated with 13 dams, built in sequence with ten locks to compensate for a difference of level of around 230 m between Barra Bonita and Itaipu dams (Fig. 3; ANTAQ 2011). The Tietê–Paraná waterway connects the most important Brazilian industrial and urban areas in São Paulo State. The city of São Paulo (Fig. 2) is the most populous city in South America, and economically the most developed in Brazil; most of its population is concentrated in the vicinity of the Tietê and Grande rivers.

The Río de la Plata basin has a very high hydropower potential (Palomino Cuya et al. 2013). The energy produced by hydroelectric plants in the basin represents about 67% of all installed power generation capacity in the countries of the Río de la Plata watershed. This high percentage makes the five countries involved highly

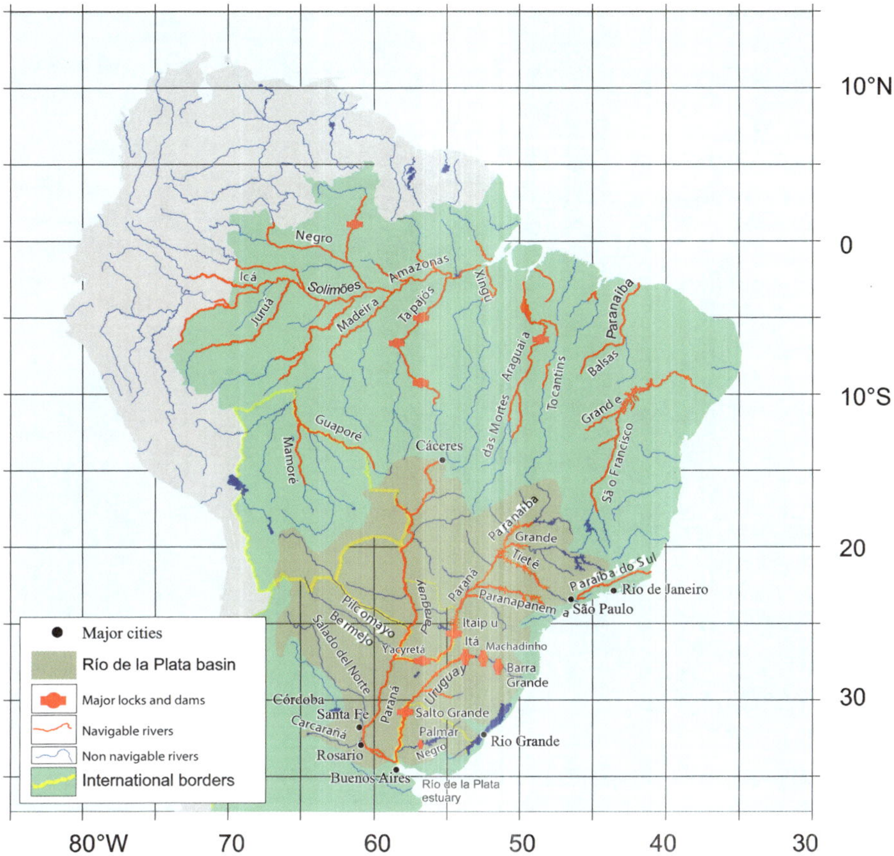

Fig. 2 Navigable South American rivers and major dams

dependent on water resources to satisfy their electricity demands. Paraguay obtains 100 % of its electricity from hydropower; Brazil 80 %; and Argentina, Bolivia, and Uruguay around 40–60 %. Seventy-six percent of Brazilian hydropower (whose total installed capacity is by far the largest in the region) is produced in the Río de la Plata basin. Large dams have been built on the Paraná (Itaipu, Yacyretá) and Uruguay (Salto Grande) rivers. In the Upper Paraná and its tributaries, there are around 57 hydropower dams (Fig. 3), and the highest energy demands are also concentrated in this region. Macrofouling by *L. fortunei* has been observed in all hydroelectric plants located in watercourses where this species is present, requiring costly control measures.

While hydropower plants are probably impacted the most by mussel fouling, they are not the only facilities affected. Most industries that use raw water from colonized waterbodies have problems, including water processing facilities, nuclear and thermal power plants, refineries, steel mills, food processing plants, fish culture

Fig. 3 Major reservoirs (*red dots*) in the areas colonized by the golden mussel

facilities, etc. (see "Impacts of *Limnoperna fortunei* on Man-made Structures and Control Strategies: General Overview" in this volume).

Distribution of *L. fortunei* in the Río de la Plata Basin

The current distribution of *L. fortunei* in South America is shown in Fig. 4. After its first record at Bagliardi Beach, ca. 20 km south of the city of La Plata, in the Río de la Plata estuary (Buenos Aires Province, Argentina) in 1991 (Fig. 4a; Pastorino et al. 1993), *L. fortunei* spread upstream and colonized the Argentine side of the estuary, reaching Martín García Island (34°10'S, 58°15'W) and the lower delta of the Paraná River by 1995 (Darrigran and Pastorino 2004; Boltovskoy et al. 2006). In 1994, the first macrofouling episode caused by the golden mussel was reported from a water treatment plant in La Plata city, Argentina (Darrigran 1995; Fig. 4b). The first record of *L. fortunei* on the Uruguayan side of Río de la Plata estuary was in 1994 (Scarabino and Verde 1994). In 1996, *L. fortunei* was recorded in the Santa Lucía River (Uruguay; Brugnoli et al. 2005). In 2001, the mussel was found in the port of Nueva Palmira (Uruguay), and in 2002 farther east along the coast of the Río

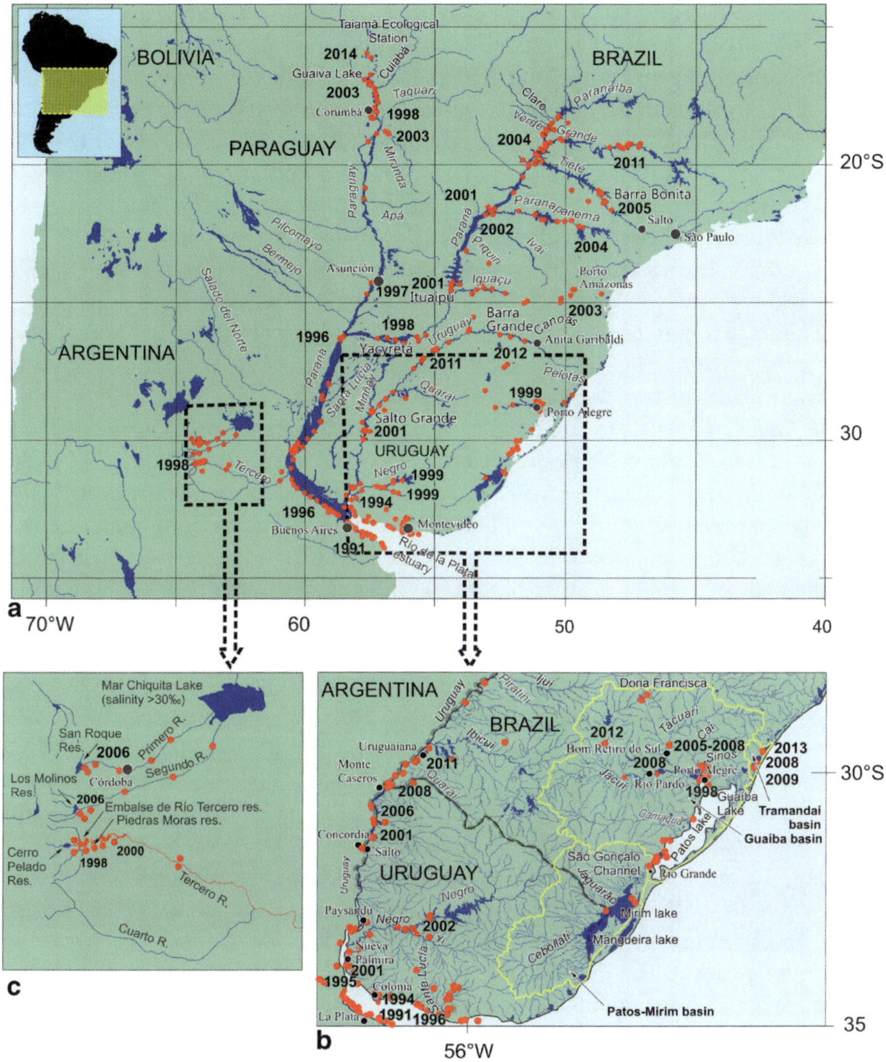

Fig. 4 Reported distribution of *Limnoperna fortunei* in South America with indications of some first sightings

de la Plata estuary (Punta Espinillo and the city of Montevideo), waning seawards as salinities increase (Brugnoli et al. 2005; Fig. 4b).

From 1991 to 1998, *L. fortunei* colonized an extensive area from the Río de la Plata estuary, upstream to the cities of Rosario (1995) and Corrientes (1996), reaching Yacyretá Dam, where it was first recorded in 1998 (Fig. 4a; Darrigran and Pastorino 2004; Boltovskoy et al. 2006). In 2001, *L. fortunei* was recorded at the Itaipu Dam in Brazil (Zanella and Marenda 2002), likely transported overland by a fouled fishing or leisure boat from an infested downstream location. Upstream from Itaipu, the Paraná

River has three large dams (Porto Primavera, Jupiá, and Ilha Solteira, Fig. 3), all of which were colonized by the mussel between 2002 and 2004 (Rückert et al. 2004; Fig. 4a). Spreading upstream along the Paraná–Tietê waterway, *L. fortunei* expanded to the north reaching the Paranaíba River in 2004 (currently its most upstream record), where it is restricted to the navigable section just below São Simão Dam (Fig. 4a; Campos et al. 2012). Along the Paranaíba River, the occurrence of *L. fortunei* is associated with portions of the river channel that have lower topographic gradients. In the narrower, faster, and more turbulent reaches, larval densities are significantly lower or absent, despite intense traffic by barges encrusted with adults (Campos 2013). Water levels and flow rates in this area are controlled by hydroelectric plants, resulting in fluctuations that may hinder the permanence of self-sustaining mussel populations. As of 2008, *L. fortunei* was not present in the tributaries of the Paranaíba River (e.g., Claro and Verde rivers; Campos and Silva 2008).

The Grande River, 1360 km long, is impounded by nine dams; all of them downstream from the Furnas Dam were recently colonized by the mussels (Fig. 3). In this river, the first record of *L. fortunei* was in 2011 in the Volta Grande Dam (CEMIG, personal communication; Fig. 4a). Upstream colonization has been comparatively slow when compared with other areas (e.g., around 250 km/year in the Paraná–Paraguay rivers; Boltovskoy et al. 2006), expanding around 500 km in about 7 years (probably from Ilha Solteira Dam, Paraná River, where it was first recorded in 2004). The Grande River is not navigable for commercial vessels, but other vectors of dispersion are present, including small leisure and fishing boats, fish farming, and use of sand retrieved in areas infested with mussels, such as the Tietê River and other sites in the state of São Paulo.

The occurrence of *L. fortunei* along the left bank tributaries of the Paraná River is shown in Fig. 4a. On the Tietê River there are seven major dams, all of them colonized by *L. fortunei* from 2002 to 2006 (Avelar et al. 2004; Pareschi et al. 2008). As of 2014, the most upstream record is Barra Bonita Dam, where it has been present since 2005. In the Paranapanema River, *L. fortunei* was recorded in 2002 at Rosana Dam, and subsequently at Xavantes, Capivara, and Jurumirim dams (observed on fish farming cages). The Piquiri River was also colonized by *L. fortunei,* but the Ivaí River has apparently not yet been invaded (Pestana et al. 2010). *L. fortunei* has been present in the Iguaçu River since 2003 and reached its headwaters (Porto Amazonas city, Brazil), around 850 km upstream (Takeda et al. 2003; Pestana et al. 2010).

A major tributary of the Paraná is the Paraguay River (Fig. 1), which drains the Pantanal wetland, one of the largest wetlands in the world and considered an area of major ecological importance. The Paraguay–Paraná waterway is navigable for commercial vessels for 3442 km from Cáceres (Brazil) to Buenos Aires (Argentina). In a period of about 7 years, *L. fortunei* extended its range 1200 km upstream in the Paraguay River, most probably as adult mussels encrusted on vessels operating in and out of the many ports along the Paraná and Río de la Plata estuary (Oliveira et al. 2011). In 1997, *L. fortunei* was first recorded in the Paraguay River in Asunción (Paraguay), and in 1998 in Corumbá (Brazil), in the southern Pantanal. As of 2014, its northernmost record is at the Taiamã Ecological Station, in the northern Pantanal, Brazil (C.T. Callil, personal communication; Fig. 4a). Along the main channel of the Paraguay River, the distribution of the golden mussel is heterogeneous, being more frequent in areas with rocky outcrops. The species also inhabits extensive lakes connected to the Paraguay River and the middle and lower reaches of some tributaries, including the Apa (since 2000), the Miranda (since 2003), and the Cuiabá (since 2008) rivers (Fig. 4a). In the Miranda River, it was detected in 2003 close to the mouth, moving about 100 km upstream in 7 years (Oliveira et al. 2011; Fig. 4a).

In addition to the Paraguay River itself, which defines the border between Brazil and Bolivia, some other Bolivian waterbodies have also been invaded, although the distribution of the mussel in Bolivia is practically unknown. Cáceres Lake, which is connected to the Paraguay River by the Tamengo Channel, has been colonized since 1998 (Fig. 4a). Farther upstream in the Paraguay River, the mussel is present on both the Bolivian and the Brazilian sides of Guaiva Lake (Fig. 4a).

Although *L. fortunei* has been widely distributed throughout the Lower Paraguay River for at least 15 years, as of 2005 it apparently was not present in two large tributaries originating in the Andes mountains—the Pilcomayo and Bermejo rivers, or in the Salado del Norte River, which flows into the Middle Paraná (Fig. 4a; Darrigran et al. 2011). High concentrations of suspended sediments (100–4500 mg/L: Bermejo), intermittent flow (Pilcomayo), and high salinities (Salado del Norte) have been tentatively identified as factors that likely restrict the colonization of these rivers by the mussel (Drago et al. 2008; Darrigran et al. 2011).

The Uruguay River originates at the confluence of the Pelotas and Canoas rivers in Brazil, and flows for 1770 km before ending in the Río de la Plata estuary (Fig. 1 and 4a). It is navigable from the Río de la Plata estuary to the cities of Concordia (Argentina) and Salto (Uruguay), at the head of the Salto Grande Dam (Fig. 2). Introduction of the golden mussel in the Uruguay River and its tributaries has been summarized by Brugnoli et al. (2005); Boltovskoy et al. (2006); Darrigran and Mansur (2006); and Darrigran et al. (2012). *L. fortunei* was detected at Salto Grande Dam in 2001, which is ca. 350 km upstream from the Río de la Plata estuary (Fig. 2 and 4a). In 2006, it was recorded at the tail end of the 120-km-long Salto Grande Reservoir, and around 2008 it was reported in the Quaraí River (a tributary of the Uruguay River; Lima et al. 2008). At Uruguayana city (Brazil), it was found in 2011 (E. M. Paolucci, personal observation; Fig. 4b). The farthest upstream records in the Uruguay River are at the Itá Dam (M.C.D. Mansur, personal observation), and between Machadinho city and Barra Grande Dam (Pelotas River), about 1600 km from the river mouth into the Río de la Plata (Agudo-Padrón et al. 2012; Fig. 4a). These records suggest that the entire Uruguay River is already colonized by the mussel.

The main tributaries of the Uruguay River are the Negro River (and its tributary the Yí River), in Uruguay; and the Ijuí, Piratini, Ibicuí, and Quaraí rivers in Brazil (Fig. 4b). Between 1999 and 2002, *L. fortunei* was recorded in the Negro River at the Palmar Dam, and in the Yí River (Clemente and Brugnoli 2002; Conde et al. 2002). In 2002, mussels were also collected at the Baygorria Dam on the Negro River, and at a meat plant in Paysandú city, on the Uruguay River (Brugnoli et al. 2005; Fig. 4b). Until 2006, *L. fortunei* was not present upstream in the Argentine tributaries of the Uruguay River (e.g., the Mocoretá River, which drains the Iberá Wetlands, Fig. 1; Darrigran et al. 2012), but it was present at the mouth of the

Miriñay River (Fig. 4b). Environmental conditions for the establishment of *L. fortunei* seem adequate in these waterbodies, which suggests that their colonization has either already occurred or will occur soon (Darrigran et al. 2012).

The Tercero–Carcarañá rivers drain an area in central Argentina (Córdoba Province), including several reservoirs, into the Middle Paraná (Fig. 4c). *L. fortunei* was first recorded in one of these reservoirs (Embalse de Río Tercero) in 1998 (Boltovskoy et al. 2006, 2009). Neither of these rivers is navigable, which indicates that *L. fortunei* was introduced by a land route, most probably attached to the hull of a recreational boat trailered overland from some location on the Paraná or Uruguay rivers (Boltovskoy et al. 2006). Hydropower (Embalse de Río Tercero, Piedras Moras) and nuclear (Central Embalse) facilities associated with these dams have been experiencing fouling problems due to the mussel since around 2000.

Introduction and Dispersion in the Guaíba and Patos–Mirim Basins

Along the southern Brazilian and Uruguayan coasts there is a complex system of lakes including the Guaíba, Patos, Mirim, and Mangueira lakes (Fig. 4b). The major tributaries in this system include the Jacuí and Sinos rivers, which discharge into the Guaíba–Patos Lake system, and the Camaquã River that flows into the Patos Lake, all in Brazil, as well as the Jaguarão, Taquari, and Cebollati rivers that flow into Mirim Lake in Uruguay (Fig. 4b). The rivers and lakes in this complex are interconnected, draining to the Atlantic Ocean through a single outlet located in the vicinity of the city of Rio Grande, Brazil. In addition to the lakes themselves, there are several navigable waterways, including the main channels of Guaíba and Patos lakes, lower parts of the Jacuí, Taquari, Gravataí, Sinos, and Caí rivers, and the São Gonçalo Channel (which connects Mirim and Patos lakes). Four locks exist in the Jacuí and Taquarí rivers to facilitate navigation (ANTAQ 2011).

L. fortunei was first recorded in Guaíba Lake in 1998, attached to the water hyacinth *Eichhornia azurea,* in the port of Porto Alegre city (Mansur et al. 1999; Burns et al. 2006; Fig. 4b). Presently, the mussel extends over most of the Jacuí River, Guaíba and Patos lakes in Brazil, and the northeast section of Mirim Lake in Uruguay. The introduction of *L. fortunei* in the Arroio Pelotas River and the São Gonçalo Channel, around 2002, resulted in the colonization of Mirim Lake, where the mussels were first recorded in 2005 (Langone 2005; Burns et al. 2006). Until 2008, the furthest upstream record of *L. fortunei* in the Jacuí River was near the city of Rio Pardo (Fig. 4b). In 2009, mussels were found at the Capinguí 1 Dam (Fig. 3), in the upper part of the Jacuí River (Santos et al. 2012); and in 2013 a little farther downstream at the Dona Francisca Dam (Fig. 4b). In the Taquari River, the most upstream record is at Bom Retiro do Sul city (Fig. 4b). In the other tributaries of Guaíba Lake, including the Caí, Sinos, and Gravataí rivers, *L. fortunei* has so far only been observed close to the mouth (Darrigran and Mansur 2009; Colling et al. 2012; Fig. 4b).

The introduction of *L. fortunei* in this basin is probably the result of secondary seeding by ballast water from ships operating with infested freshwater Argentine and/or Uruguayan ports. Much of the incoming traffic into the port of Porto Alegre originates from sites along the lower Paraná River. These ships sail along the Uruguayan and Brazilian coast, enter Patos Lake through the Rio Grande Channel, proceed north to Porto Alegre (on Guaíba Lake), and from there to the grain port of Estrela, through the Jacuí and Taquarí rivers (Fig. 4b). Infested freshwater ballast discharged anywhere along this route has most probably been the source of *L. fortunei* in this watershed (Mansur et al. 2004; Darrigran and Mansur 2009).

Introduction and Dispersion in the Tramandaí River Basin

The Tramandaí River basin (Fig. 1) comprises two main rivers and several associated coastal lakes (Quadros, Peixoto, Itapeva), which drain into the Atlantic through the Tramandaí River estuary and the Armazém and Tramandaí lakes, both of which are brackish (Silva 2001; Fig. 4b). In 2008, *L. fortunei* was observed in Quadros Lake, and in 2009 in the Tramandaí River (Fagondes de Freitas et al. 2009). Sampling in January 2013 revealed the presence of the mussel in parts of Peixoto, Quadros, and Itapeva lakes, and in the channel between Quadros and Itapeva lakes (Mansur et al. 2014).

Introduction and Spread in the Mar Chiquita Basin

Mar Chiquita is an endorheic basin located in central Argentina, where a few small rivers (Dulce, Primero, Segundo) drain into the large saline Mar Chiquita Lake (Fig. 1). Some of these rivers are impounded and have hydroelectric power plants. *L. fortunei* was recorded at the San Roque (Primero River) and Los Molinos (Segundo River) dams in 2006 (Darrigran et al. 2009), most probably transferred overland attached to a fishing or leisure boat from the nearby river Río Tercero or the reservoir Embalse de Río Tercero, which have been colonized by the mussel since 1998 (Fig. 4c). This basin includes the metropolitan area of Córdoba city, with a population of ca. 1.5 million, whose water supply comes mainly from the San Roque Dam. High microcystin levels in this reservoir have historically been a major concern, often to the point of precluding its use for drinking (Antenucci and Alexander 2001; Ruibal Conti et al. 2005). It is likely that colonization by the mussel, which is known to boost growth of *Microcystis* spp. (Cataldo et al. 2012), has worsened microcystin-related problems.

Origin, Genetic, and Morphological Variation of the Introduced Populations in South America

While there is little doubt *L. fortunei* was introduced into South America from Southeast Asia (Darrigran and Pastorino 2004; Boltovskoy et al. 2006), the original source population (or populations) and the dynamics of the introduction process

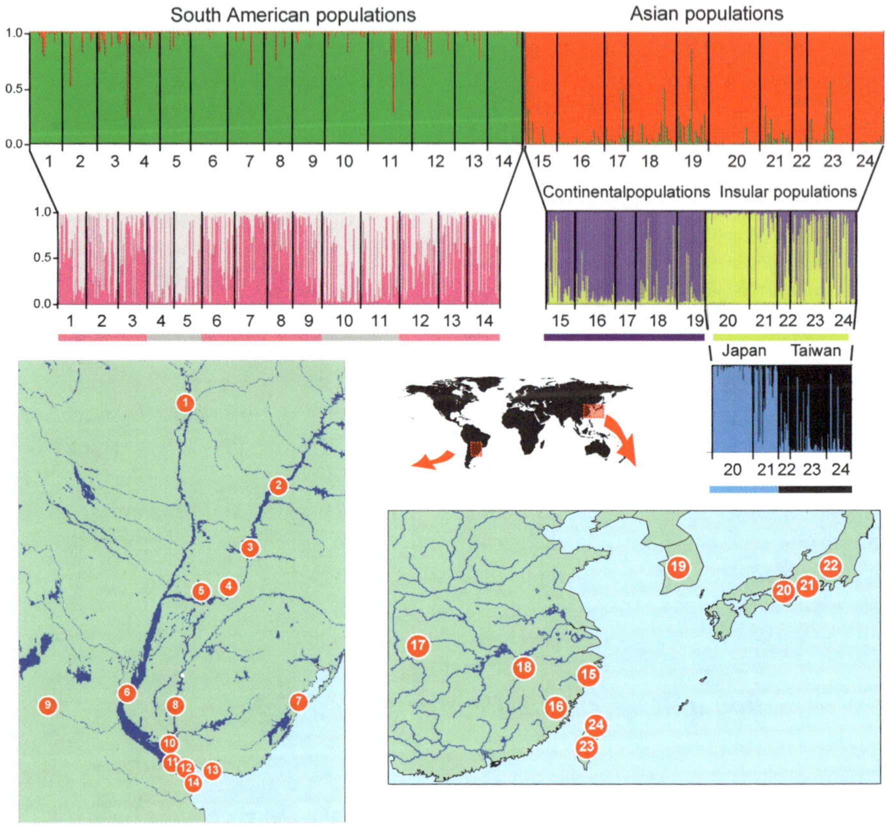

Fig. 5 Genetic similarity (Bayesian clustering based on eight polymorphic microsatellites) between 24 populations of *Limnoperna fortunei* from South America (1–14), continental Asia (15–19), and insular Asia (20–24). Each genotype is represented by a *thin vertical line,* with proportional membership in different clusters indicated by different colors. *Black vertical lines* separate collection sites, with site identifications indicated below the plot and in the maps. (Modified from Ghabooli et al. 2013)

are not known. By the time when *L. fortunei* was first discovered (1991), native or introduced populations of this mussel had already become established in mainland China, Thailand, Laos, Cambodia, Vietnam, Hong Kong, Taiwan, Korea, and Japan (see "Distribution and Spread of *Limnoperna fortunei* in China" in this volume), which implies that any or several of these countries may have been the source of South American immigrants.

Genetic analyses performed on multiple South American populations (Pereira da Silva 2012; Zhan et al. 2012; Ghabooli et al. 2013) suggest that several introduction events from different geographic sources have occurred. Bayesian clustering and three-dimensional factorial correspondence analyses of both native and introduced populations of *L. fortunei* based on microsatellite markers clearly separate South American from Asian populations (Fig. 5; Ghabooli et al. 2013). These analyses

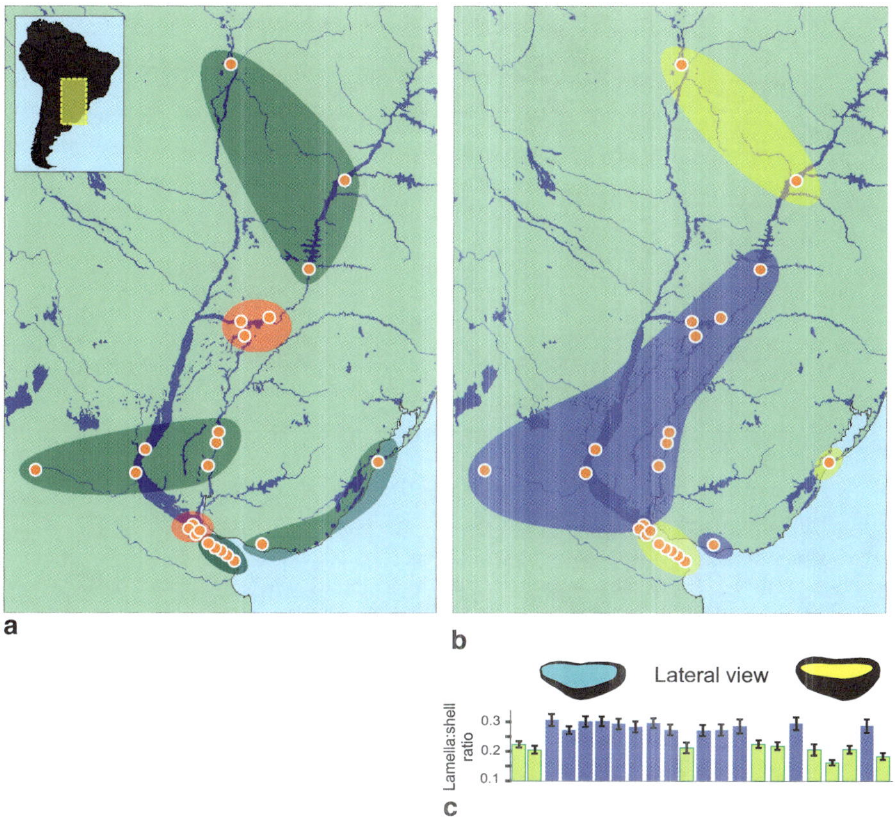

Fig. 6 **a** Groups of South American *Limnoperna fortunei* populations plotted according to the three-dimensional factorial correspondence and Bayesian clustering analyses. *Green and red populations* belong to the two different main clusters shown in Fig. 5 (Modified from Zhan et al. 2012). **b** Geographic distribution of populations with low (*yellow*) and high (*blue*) lamella area to shell area ratios, as detailed in (**c**). (Modified from Paolucci et al. 2014, by permission from the Association for the Sciences of Limnology and Oceanography, Inc.)

also distinguish two clear-cut clusters in South American populations (Zhan et al. 2012). These clusters (Fig. 6a) show a discontinuous geographical distribution, whereby several geographically close populations are split into genetically distinct clusters (Zhan et al. 2012; Ghabooli et al. 2013). This supports the hypothesis of multiple introductions and human-mediated upstream "jumps," as also suggested for Japan (Tominaga et al. 2009; see "Colonization and Spread of *Limnoperna fortunei* in Japan" in this volume).

Genetic analyses also furnish hints on the type and importance of the vectors involved in the mussel's dispersion around the world. The lower genetic diversity of South American populations, as compared to the introduced populations in Asia (Japan and Taiwan), indicates fewer propagule transfers in the South American introduction, which agrees with the much lower shipping activity in Argentina compared to Japan and Taiwan (Ghabooli et al. 2013). Furthermore, the only likely vector for

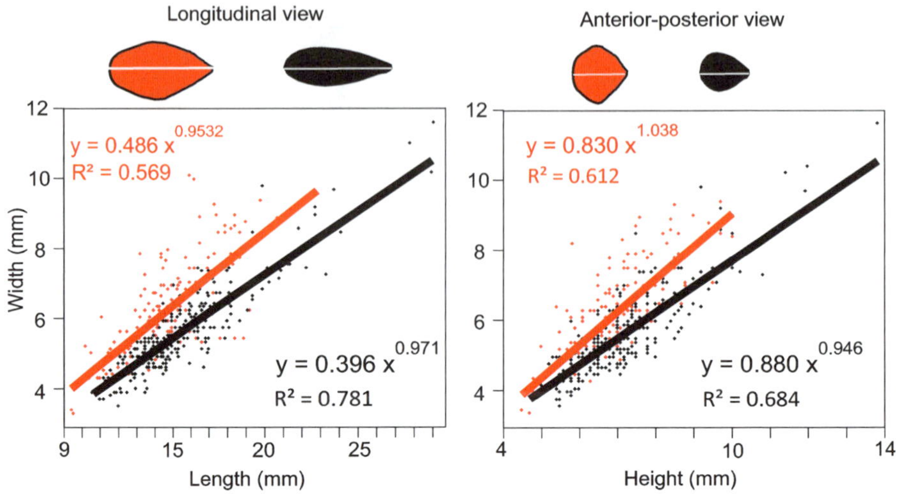

Fig. 7 Width to length and width to height ratios of *Limnoperna fortunei* populations studied across South America. *Red and black points* correspond to individuals from populations with low and high relative gill area ratio, respectively, as shown in Fig. 6. (Based on data from Paolucci et al. 2014, by permission from the Association for the Sciences of Limnology and Oceanography, Inc.)

South American introduction is ballast water, whereas in Japan and Taiwan aquaculture and even hull fouling may play important roles (Tominaga et al. 2009; see "Colonization and Spread of *Limnoperna fortunei* in Japan" in this volume).

South American populations of *L. fortunei* display a relatively high genetic differentiation that cannot be explained by natural downstream dispersion only (Pereira da Silva 2012; Zhan et al. 2012). Vessel and barge traffic on the Paraná–Paraguay and Paraná–Tietê waterways are clearly the main vector that helped to spread the mussel locally (Boltovskoy et al. 2006). Larvae produced by adults transported upstream can drift a long distance downriver since they spend up to over 20 days in the plankton before settling (Cataldo et al. 2005). Population genetic analyses (Fig. 6a) confirm this assumption (Pereira da Silva 2012; Zhan et al. 2012; Ghabooli et al. 2013). Upstream jumps through hull fouling and subsequent downstream larval drift is also supported by the massive heterozygote deficiency relative to the Hardy–Weinberg equilibrium (i.e., the constancy of both genotype and allele frequencies in a population through successive generations in the absence of disturbing factors) observed in 76 % of the analyzed cases. This result is actually a violation of the Hardy–Weinberg equilibrium assumptions, and may be explained by the Wahlund effect and inbreeding. The Wahlund effect (reduction of heterozygosity caused by subpopulation structure), in particular, stresses the importance of "jump dispersal," likely by commercial vessels.

These genetic studies were recently complemented with analyses of morphological variation in 24 South American populations (Paolucci et al. 2014). Significant differences in the relative gill area, in the shell width to length ratio, and in the mean density of gill cilia in these populations allowed differentiation of two discrete groups (Paolucci et al. 2014; Fig. 6b-c and 7). Interestingly, higher relative gill areas and lower gill cilia densities were significantly associated with sites with lower total suspended solids. Genetic structure, ascertained using both microsatellites and mitochondrial evidence, was not associated with these differences in gill and shell morphology (Paolucci et al. 2014), and neither were these morphologically different groups associated with the genetically distinct groups defined by Zhan et al. (2012; Fig. 6). These results suggest that consistent morphologic dissimilarities are the result of phenotypic plasticity, but further studies are needed to assess the importance of the genetic component of shell and gill morphology in this species. Morphological variation, especially in gill dimensions, of South American populations of *L. fortunei* appears to result from developmental plasticity, mostly in relation to total suspended sediments (Payne et al. 1995; Sousa et al. 2007; Dutertre 2009), and perhaps in relation to dissolved oxygen and contaminant levels as well. This phenotypic plasticity may play an important role in the successful spread and establishment of this species in a wide range of habitats.

Potential Expansion of *L. fortunei* Throughout South America

Potential habitat distribution (ecological niche) models have been applied to predict the worldwide spread of *L. fortunei* (Kluza and McNyset 2005; Campos 2014; Campos et al. 2014), and at local to regional scales (Oliveira et al. 2010b). These models are usually developed by examining habitat characteristics in order to define the species environmental tolerances, and comparing them with the characteristics of potential new areas, taking into account routes and opportunities of introduction.

Kluza and McNyset (2005) used climate and topographical data layers and the genetic algorithm for rule set production (GARP) model to predict the potential distribution of *L. fortunei* throughout the world. They considered occurrence data from Southeast Asia, and validated their predictions with data on its invasive range in South America. Their model predicted potential colonization of small areas in Europe and Africa, Central America, southeast North America, and an extended area in South America. Campos (2014) and Campos et al. (2014), also working at the global scale, used climatic layers related to temperature and rainfall and four algorithms (Mahalanobis distance, Domain, GARP, and MAXENT) to develop predictions based on occurrence records from Asia and South America. Models based on all algorithms predicted extensive areas around the world with high potential for colonization by *L. fortunei* (Fig. 8), including areas in Europe, southern Africa, Central and South America, and southeastern USA. Parts of South America concluded to be at high risk of colonization by *L. fortunei* include the Río de la Plata basin (practically all of Uruguay and large areas in Argentina, Bolivia, and Perú), as well as the Orinoco and Magdalena watersheds.

In Argentina, only a small fraction of the waterbodies fit for colonization have so far been invaded by *L. fortunei* (see "Parallels and Contrasts Between *Limnoperna fortunei* and Species of *Dreissena*" in this volume; Fig. 15.4). As mentioned above,

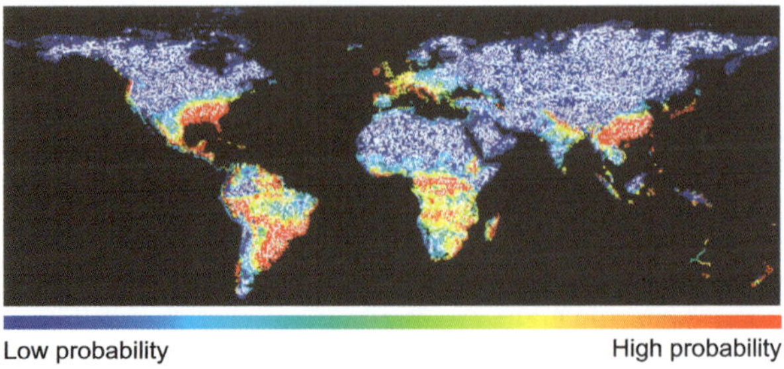

Fig. 8 Potential distribution of *Limnoperna fortunei* as suggested by modeling with the algorithms MAXENT, GARP, Mahalanobis distance, and Domain based on its known distribution in Asia and South America (results shown are the average of all four algorithms). (From Campos 2014)

some major rivers flowing into the Paraná and Paraguay are probably unfit for survival of the mussel (e.g., Salado del Norte, Bermejo, Pilcomayo; Darrigran et al. 2011), but there still are many waterbodies potentially adequate for its establishment which have not yet been invaded. In Buenos Aires Province alone, $<10\%$ of the ~ 530 permanent lentic waterbodies (Toresani et al. 1994) are currently invaded by *L. fortunei*. South of approximately 36°S all large rivers originate in the Andes mountains and flow eastwards toward the Atlantic Ocean (Colorado, Negro, Chubut, Deseado, Chico, Santa Cruz, Coyle, Gallegos, Fig. 1; Depetris et al. 2005). Calcium levels in most of these waterbodies are well above those needed for *L. fortunei* (Pedrozo et al. 1993; Rogora et al. 2008), but none of them are navigable, which hinders the possibilities of upstream dispersal through hull fouling. Nevertheless, some of these rivers are dammed or have lakes along their course, which could serve as seeding areas if the mussel manages to invade sites upstream from these lentic waterbodies. Winter water temperatures are well below those in the Río de la Plata basin, especially in the southernmost streams, but *L. fortunei* is abundant in lakes where water temperature drops to ca. 5 °C (Magara et al. 2001; Nakano et al. 2011), and has been reported from reservoirs that even freeze during the winter (Choi and Kim 1985; Choi and Shin 1985). As opposed to rivers farther north, hard substrata are widespread in these waterbodies, which could facilitate establishment of the mussel. These considerations suggest that invasion of some Northern Patagonian freshwaters whose temperatures rise above 15–18 °C (the threshold values for reproduction in *L. fortunei,* see "Reproductive Output and Seasonality of *Limnoperna fortunei*" in this volume) by *L. fortunei* is probable.

Uruguay seems to be even more vulnerable, as most areas free of the mussel seem potentially fit for colonization (Darrigran et al. 2012), with the exception of the short rivers draining directly into the Río de la Plata estuary and the Atlantic Ocean.

Working on different geographic scales, Oliveira et al. (2010b) and Campos (2014) investigated the potential spread of *L. fortunei* in Brazil and worldwide, respectively. Oliveira et al. (2010b) used ecological niche modeling at regional scales assessing environmental tolerance on the basis of limnological variables (water temperature, dissolved oxygen, conductivity, pH, calcium concentration, total suspended solids, and the calcite saturation index). Their results showed that most Brazilian rivers are suitable for the establishment of *L. fortunei*. Mussel environmental tolerance ranges were derived from Paraguay River data, where *L. fortunei* is found in waters with conductance >50 μS/cm, pH>6, calcium concentrations >1 mg/L, and calcite saturation indices above −4 (Fig. 9). These limits may underestimate the mussel's potential spread because most Brazilian rivers have better conditions for calcification and shell growth than the Paraguay River. Minimum calcium and pH for the species survival have been defined at >1.8 mg/L and 5.7–9.0, respectively (Campos 2013). Calcium concentrations below 0.1 mg/L produce 100% mortality in 72 h, but extremely high values are deleterious as well: at >100 mg/L mussels die after 336 h (Campos 2013). Available data for Brazilian waterbodies indicate that calcium concentrations in hydrographic regions will not represent a barrier to the spread of *L. fortunei* (Campos 2013).

The results of Campos (2014; Fig. 8) generally agree with those of Oliveira et al. (2010b), in that the Amazon and Tocantins rivers (Fig. 9) are highly vulnerable to invasion by the golden mussel. Smaller sections of the Tapajós and Xingú rivers (major Amazon River tributaries), and the Negro and Araguaia rivers (Fig. 9) seem less suitable for colonization by *L. fortunei* because their waters are more acidic and have lower calcium concentrations. The Amazon River is navigable for ocean liners for two thirds of its course (Fig. 2), making it very vulnerable to the introduction of *L. fortunei* with ballast water. Transoceanic ships regularly reach the city of Manaus (Brazil), nearly 1600 km upstream, and smaller ships can reach Iquitos (Peru), 3700 km from the river mouth (the farthest point from sea of any port serving ocean traffic; Boltovskoy et al. 2006; Fig. 1). If *L. fortunei* reaches Iquitos or Manaus, it will spread swiftly over a huge region because the Amazon River network, with ca. 17,000 km of navigable waterways, is the main means of transportation for both people and goods in this remote area.

The São Francisco watershed may be less vulnerable than the Amazon to colonization by the mussel because commercial navigation is restricted in the São Francisco River (Boltovskoy et al. 2006; Fig. 2). However, as many other watersheds, the São Francisco basin can be colonized through many other vectors, including sports and commercial fishing and fish farming in cages, in particular from the nearby Grande River (Fig. 2), where *L. fortunei* is present. The waters of the São Francisco River are fit for the mussel (Oliveira et al. 2010b).

Another important area in Brazil which is at risk of invasion by *L. fortunei* is the Paraíba do Sul River (Fig. 2; Oliveira et al. 2010b). The Paraíba do Sul has a catchment area of about 55,500 km², and drains one of the most developed and populated (>5 million inhabitants) regions in Brazil. It is the main source of water for the metropolitan region of Rio de Janeiro, serving a population of over 8 million inhabitants. Approximately two thirds of the river flow (~160 m^3/s) are captured and pumped into the Guandu River, where the water treatment plant is located. Introduction of *L. fortunei* in this system is likely to cause severe fouling problems in the water intake system.

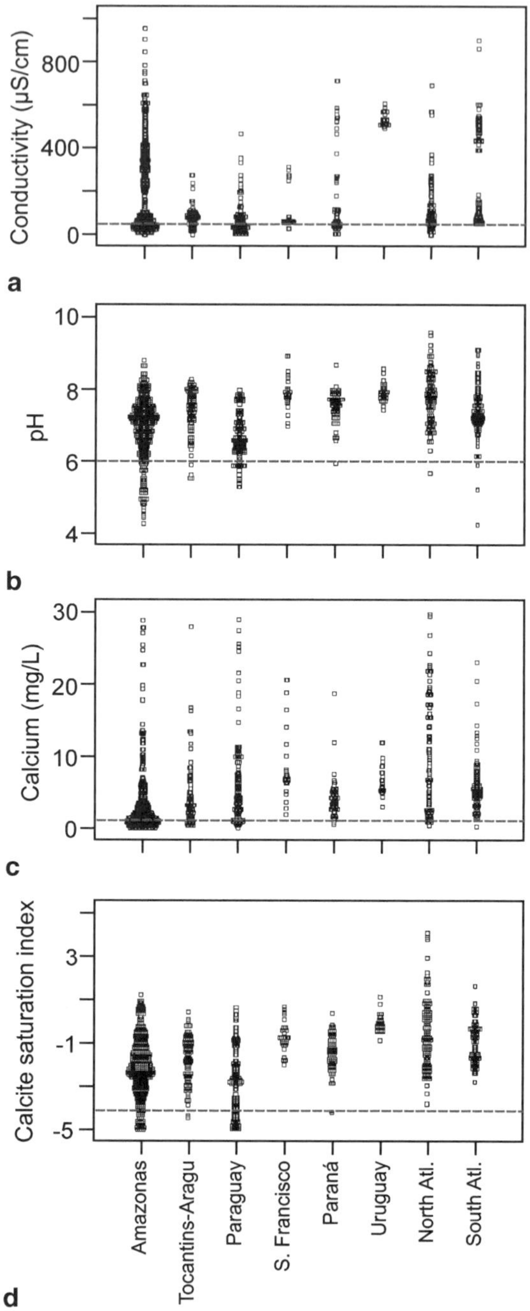

Fig. 9 Ranges of conductivity (**a**), pH (**b**), calcium concentration (**c**), and calcite saturation index (**d**) in Brazilian river systems. *Dashed red lines* depict approximate lower thresholds for *Limnoperna fortunei* establishment. (From Oliveira et al. 2010b)

In addition to water quality, hydrological variables and substrate availability may be important limiting factors at local scales, but they are usually not incorporated in the models. The absence of *L. fortunei* in some sectors of the Paranaíba River has been ascribed to hydrological traits, including changes in flow downstream of the São Simão Dam acting as barriers for the establishment of the species, hindering larval settlement, and decreasing survival and recruitment (Campos et al. 2012; Campos 2013). Water velocity was concluded to be the main factor precluding larval settlement and absence of larvae upstream, which is supported by evidence indicating that high turbulence and flow speeds above ~2 m/s are effective in controlling mussel establishment (Matsui et al. 2002; Xu et al. 2012). Oliveira et al. (2011) suggested that a combination of water velocity and high concentrations of suspended solids affected larval settlement in the Miranda River. As discussed elsewhere (see "*Limnoperna fortunei* Colonies: Structure, Distribution And Dynamics" in this volume), hard substrata facilitate mussel establishment and growth, but in their absence mussels can settle on many other surfaces, including stabilized silt, various aquatic plants, several hard-shelled organisms, including mollusks, crustaceans, etc. Thus, the absence of rock outcrops does not represent an insurmountable obstacle for the mussel's spread.

South American countries along the Pacific coast are much less likely to host permanent *L. fortunei* populations because their short, nonnavigable, fast-flowing rivers offer few chances of upstream dispersal for a species with planktonic larvae (see "Distribution and Colonization of *Limnoperna fortunei*: Special Traits of an Odd Mussel" in this volume).

Final Remarks

For several decades, politicians, administrators, and engineers have contemplated the possibility of linking the river systems of South America to open up navigation across the entire continent. There are several projects involving various rivers and watersheds, but the ultimate goal is to connect all major hydrographic basins in order to foster trade and socioeconomic integration across the subcontinent (e.g., ANTAQ 2011; Fig. 1). While unarguably positive for trade and regional integration, such a project would have significant and unpredictable consequences for the ecology of these rivers, lakes, and reservoirs, in particular boosting the dispersal rates of both native and introduced species. In fact, much of the dispersal of many significant freshwater invaders worldwide, including golden and zebra mussels, has been made possible by the elimination of natural barriers between watersheds (Karatayev et al. 2007). Interbasin connections are not the only features that facilitate dispersal of species like *L. fortunei*; dams and their associated reservoirs also seem to be instrumental in enhancing the dispersal of the mussel, most probably because they represent stepping stones and seeding areas for further range extensions and the lentic conditions in reservoirs can maintain high mussel populations in fluvial systems (see "Distribution and Colonization of *Limnoperna fortunei*: Special Traits of an

Odd Mussel" in this volume). The much higher mussel densities in the Paraná than in the Paraguay River are partly due to the scarcity of hard substrata and seasonal oxygen depletion events in the latter (Oliveira et al. 2010b; Marçal and Callil 2012), but also probably to the fact that, unlike the Paraguay, the Paraná is punctuated by many dams (Fig. 3; Johnson et al. 2008; Oliveira et al. 2010a). New dams that are projected and under construction on several Amazon River tributaries (e.g., Tapajós and Madeira rivers, Fig. 2), as well as elsewhere in South America, will increase the potential spread of *L. fortunei* in the Amazon basin and elsewhere.

References

Agudo-Padrón AI, Porto Filho É, Saalfeld K (2012) First confirmed record of *Limnoperna fortunei* (Dunker, 1857) in Santa Catarina State, Southern Brazil. In: XI International Congress on Medical and Applied Malacology, Rio de Janeiro (Brazil)

ANTAQ (Agência Nacional de Transportes Aquaviários) (2011) Panorama aquaviário, vol 6. Assessoria de Comunicação Social ANTAQ, Brasilia, pp 1–151

Antenucci J, Alexander R (2001) San Roque Reservoir. Contract Research Group, Centre For Water Research, Argentina, p 2

Avelar WE, Martim SL, Vianna MP (2004) A new occurrence of *Limnoperna fortunei* (Dunker 1856) (Bivalvia, Mytilidae) in the State of São Paulo, Brazil. Braz J Biol 64:739–742

Boltovskoy D, Correa N (2015) Ecosystem impacts of the invasive bivalve *Limnoperna fortunei* (golden mussel) in South America. Hydrobiologia 746:81–95

Boltovskoy D, Correa N, Cataldo D, Sylvester F (2006) Dispersion and ecological impact of the invasive freshwater bivalve *Limnoperna fortunei* in the Río de la Plata watershed and beyond. Biol Invasions 8:947–963

Boltovskoy D, Karatayev A, Burlakova L, Cataldo D, Karatayev V, Sylvester F, Mariñelarena A (2009) Significant ecosystem-wide effects of the swiftly spreading invasive freshwater bivalve *Limnoperna fortunei*. Hydrobiologia 636:271–284

Brugnoli E, Clemente J, Boccardi L, Borthagaray A, Scarabino F (2005) Golden mussel *Limnoperna fortunei* (Bivalvia: Mytilidae) distribution in the main hydrographical basins of Uruguay: update and predictions. Anais da Academia Brasileira de Ciências 77:235–244

Burns MDdM, Geraldi RM, Garcia AM, Bemvenuti CE, Capitoli RR, Vieira JP (2006) Primeiro registro de ocorrência do mexilhão dourado *Limnoperna fortunei* na bacia de drenagem da Lagoa Mirim, RS, Brasil. Biociências 14:83–84

Campos MCS (2013) Fatores ambientais determinantes da ocorrência de *Limnoperna fortunei* (Dunker, 1857) no baixo rio Paranaíba—bases para uma avaliação de risco. PhD Thesis, Universidade Federal de Ouro Preto (Brazil), pp 1–193

Campos MCS (2014) Modelagem matemática do processo de invasão de *Limnoperna fortunei* (Dunker, 1857). Fundação Centro Tecnológico de Minas Gerais (Brazil). Technical Report GT 343, Controle do Mexilhão Dourado: Bioengenharia e novos materiais para aplicações em ecossistemas e usinas hidrelétricas, pp 3–36

Campos MCS, Silva FA (2008) Monitoramento e controle do mexilhão dourado. Agência Goiana de Meio Ambiente (Brazil). Ecogeraes—Programa Básico Ambiental Monitoramento e Controle do Mexilhão Dourado—MB08, pp 1–33

Campos MCS, Lanzer R, Castro PT (2012) Hydrological stress as a limiting factor of the invasion of *Limnoperna fortunei* (Dunker, 1857) in the Upper Paraná River (Brazil). Acta Limnologica Brasiliensia 24:64–82

Campos MCS, Andrade AFA, Kunzmann B, Galvão DD, Silva FA, Cardoso AV, Carvalho MD, Mota HR (2014) Modelling of the potential distribution of *Limnoperna fortunei* (Dunker, 1857) on a global scale. Aquat Invasions 9:253–265

Cataldo D, Boltovskoy D, Hermosa JL, Canzi C (2005) Temperature-dependent larval development rates of *Limnoperna fortunei* (Bivalvia: Mytilidae). J Molluscan Stud 71:41–46

Cataldo D, Vinocur A, O'Farrell I, Paolucci E, Leites V, Boltovskoy D (2012) The introduced bivalve *Limnoperna fortunei* boosts *Microcystis* growth in Salto Grande Reservoir (Argentina): evidence from mesocosm experiments. Hydrobiologia 680:25–38

Choi SS, Kim JS (1985) Studies on the metamorphosis and the growth of larva in *Limnoperna fortunei*. Korean J Malacol 1:13–18 [In Korean]

Choi SS, Shin CN (1985) Study on the early development and larvae of *Limnoperna fortunei*. Korean J Malacol 1:5–12 [In Korean]

Clemente JM, Brugnoli E (2002) First record of *Limnoperna fortunei* (Dunker, 1857) (Bivalvia: Mytilidae) in continental waters of Uruguay (Río Negro and Río Yi). Boletin de la Sociedad Zoológica del Uruguay 13:29–33

Colling LA, Pinotti RM, Bemvenuti CE (2012) *Limnoperna fortunei* na Bacia da Lagoa dos Patos e Lagoa Mirim. In: Mansur MCD, Santos CP, Pereira D, Padula Paz IC, Leite Zurita ML, Raya Rodriguez MT, Vilar Nehrke M, Aydos Bergonci PE (eds) Moluscos límnicos invasores no Brasil. Biologia, prevenção, controle. Redes Editora, Porto Alegre, pp 187–191

Conde D, Paradiso M, Gorga J, Brugnoli E, De Leon L, Mandiá M (2002) Problemática de la calidad de água em el sistema de grandes embalses del Río Negro (Uruguay). Revista CIER 39:51–68

Darrigran GA (1995) *Limnoperna fortunei*: ¿un problema para los sistemas naturales de agua dulce del Mercosur? Museo (Revista de la Fundación Museo de La Plata "Francisco Pascasio Moreno") 5:85–87

Darrigran GA, Mansur MCD (2006) Distribución, abundancia y dispersión. In: Darrigran GA, Damborenea C (eds) Bioinvasión del mejillón dorado en el continente americano. Editorial Universitaria de La Plata, La Plata, pp 93–110

Darrigran G, Mansur MCD (2009) Introdução e dispersão do *Limnoperna fortunei*. In: Darrigran G, Damborenea C (eds) Introdução a biologia das invasões. O mexilhão dourado na América do Sul: biologia, dispersão, impacto, prevenção e controle. Cubo Editora, São Carlos, pp 89–109

Darrigran GA, Pastorino G (2004) Distribution of the golden mussel *Limnoperna fortunei* (Dunker, 1857), (Bivalvia: Mytilidae) after 10 years invading America. J Conchol (Special Publication) 3:95–101

Darrigran GA, Boeger W, Damborenea C, Maroñas M (2009) Evaluation of sampling and analysis techniques for early detection of *Limnoperna fortunei* (Mytilidae) in limit areas of its distribution. Braz J Biol 69:979–980

Darrigran GA, Damborenea C, Drago EC, Ezcurra de Drago I, Paira A (2011) Environmental factors restrict the invasion process of *Limnoperna fortunei* (Mytilidae) in the Neotropical region: A case study from the Andean tributaries. Annales de Limnologie–Int J Limnol 47:221–229

Darrigran GA, Damborenea C, Drago EC, Ezcurra de Drago I, Paira A, Archuby F (2012) Invasion process of *Limnoperna fortunei* (Bivalvia: Mytilidae): The case of Uruguay River and emissaries of the Esteros del Iberá Wetland, Argentina. Zoologia 29:531–539

Depetris PJ, Gaiero DM, Probst JL, Hartmann J, Kempe S (2005) Biogeochemical output and typology of rivers draining Patagonia's Atlantic seaboard. J Coastal Res 214:835–844

Drago EC, Wantzen KM, Paira AR (2008) The Lower Paraguay River-floodplain habitats in the context of the fluvial hydrosystem approach. Ecohydrology and Hydrobiology 8:125–142

Dutertre M (2009) Variations in the pallial organ sizes of the invasive oyster, *Crassostrea gigas*, along an extreme turbidity gradient. Estuarine, Coastal and Shelf Science 85:431–436

Fagondes de Freitas SM, Kapusta SC, Würdig NL (2009) Invertebrados bentônicos das lagoas da planície costeira do Rio Grande do Sul. In: Würdig NL, Freitas SMF (eds) Ecossistemas e biodiversidade do litoral norte do RS. Editora Nova Prova, Porto Alegre, pp 120–141

Ghabooli S, Zhan A, Sardiña P, Paolucci E, Sylvester F, Perepelizin PV, Briski E, Cristescu ME, MacIsaac HJ (2013) Genetic diversity in introduced golden mussel populations corresponds to vector activity. PLoS One 8:e59328

Johnson PTJ, Olden JD, Vander Zanden MJ (2008) Dam invaders: impoundments facilitate biological invasions into freshwaters. Front Ecol Environ 6:357–363

Karatayev AY, Padilla DK, Minchin D, Boltovskoy D, Burlakova LE (2007) Changes in global economies and trade: the potential spread of exotic freshwater bivalves. Biol Invasions 9:161–180

Kluza DA, McNyset KM (2005) Ecological niche modeling of aquatic invasive species. Aquat Invaders 16:1–7

Langone JA (2005) Notas sobre el mejillón dorado *Limnoperna fortunei* (Dunker, 1857) (Bivalvia, Mytilidae) en Uruguay. Publicacion Extra del Museo Nacional De Historia Natural y Antropología (Montevideo, Uruguay) 1:1–18

Lima LL, Querol E, Oliveira EV (2008) Ocorrência de *Limnoperna fortunei* (Dunker,1857) (Mollusca, Mytilidae) no Rio Quaraí, município de Barra do Quaraí, Rio Grande do Sul, Brasil. Biodiversidade Pampeana 6:50–52

Magara Y, Matsui Y, Goto Y, Yuasa A (2001) Invasion of the non-indigenous nuisance mussel, *Limnoperna fortunei*, into water supply facilities in Japan. J Water Supply: Res Technol—Aqua, 50:113–124

Mansur MCD, Zani Richinitti LM, Santos CP (1999) *Limnoperna fortunei* (Dunker, 1857), spread of a bivalve mollusc to Guaiba Bay, Rio Grande do Sul, Brazil. Biociências 7:147–149

Mansur MCD, Barboza Quevedo C, Santos CP, Callil CT (2004) Prováveis vias de introdução de *Limnoperna fortunei* (Dunker,1857) (Mollusca, Bivalvia, Mytilidae) na bacia da Laguna dos Patos, Rio Grande do Sul e novos registros de invasão no Brasil, pelas bacias do Alto Paraná e Paraguai. In: Silva JVC, Souza RCCL (eds) Água de Lastro e Invasão. Editora Interciência, Rio de Janeiro, pp 33–38

Mansur MCD, Miyahira IC, Clavijo C (2014) Avanços do mexilhão dourado, *Limnoperna fortunei*, nas bacias do Grande do Sul, Brasil. In: XXX Congresso Brasileiro de Zoologia, Porto Alegre (Brazil)

Marçal SF, Callil CT (2012) *Limnoperna fortunei* associada a macrófitas aquáticas na bacia do Rio Paraguai, Mato Grosso. In: Mansur MCD, Santos CP, Pereira D, Padula Paz IC, Leite Zurita ML, Raya Rodriguez MT, Vilar Nehrke M, Aydos Bergonci PE (eds) Moluscos límnicos invasores no Brasil. Biologia, prevenção, controle. Redes Editora, Porto Alegre, pp 201–206

Matsui Y, Nagaya K, Funahashi G, Goto Y, Yuasa A, Yamamoto H, Ohkawa K, Magara Y (2002) Effectiveness of antifouling coatings and water flow in controlling attachment of the nuisance mussel *Limnoperna fortunei*. Biofouling 18:137–148

Nakano D, Kobayashi T, Endo N, Sakaguchi I (2011) Growth rate and settlement of *Limnoperna fortunei* in a temperate reservoir. J Molluscan Stud 77:142–148

Oliveira MD, Hamilton SK, Calheiros DF, Jacobi CM (2010a) Oxygen depletion events control the invasive golden mussel (*Limnoperna fortunei*) in a tropical floodplain. Wetlands 30:705–716

Oliveira MD, Hamilton SK, Jacobi CM (2010b) Forecasting the expansion of the invasive golden mussel *Limnoperna fortunei* in Brazilian and North American rivers based on its occurrence in the Paraguay River and Pantanal wetland of Brazil. Aquat Invasions 5:59–73

Oliveira MD, Calheiros DF, Jacobi CM, Hamilton SK (2011) Abiotic factors controlling the establishment and abundance of the invasive golden mussel *Limnoperna fortunei*. Biol Invasions 13:717–729

Palomino Cuya DG, Brandimarte L, Popescu I, Alterach J, Peviani M (2013) A GIS-based assessment of maximum potential hydropower production in La Plata basin under global changes. Renew Energy 50:103–114

Paolucci EM, Sardiña P, Sylvester F, Perepelizin PV, Zhan A, Ghabooli S, Cristescu ME, Oliveira MD, MacIsaac HJ (2014) Morphological and genetic variability in an alien invasive mussel across an environmental gradient in South America. Limnol Oceanogr 59:400–412

Pareschi DC, Matsumura-Tundisi T, Medeiros GR, Luzia AP, Tundisi JG (2008) First occurrence of *Limnoperna fortunei* (Dunker, 1857) in the Rio Tietê watershed (São Paulo State, Brazil). Braz J Biol 68:1107–1114

Pastorino G, Darrigran GA, Martín SM, Lunaschi L (1993) *Limnoperna fortunei* (Dunker, 1857) (Mytilidae), nuevo bivalvo invasor en aguas del Río de la Plata. Neotropica 39:101–102

Payne BS, Lei J, Miller AC, Hubertz ED (1995) Adaptive variation in palp and gill size of the zebra mussel (*Dreissena polymorpha*) and Asiatic clam (*Corbicula fluminea*). Can J Fish Aquat Sci 52:1130–1134

Pedrozo F, Chillrud S, Temporetti P, Diaz M (1993) Chemical composition and nutrient limitation in rivers and lakes of Northern Patagonian Andes (39.5°–42°S; 71°W) (Rep. Argentina). Verhandlungen der Internationale Vereinigung Theoretische und Angewandte Limnologie 25:207–214

Pereira da Silva E (2012) Genética da bioinvasão do mexilhão-dourado. In: Mansur MCD, Santos CP, Pereira D, Padula Paz IC, Leite Zurita ML, Raya Rodriguez MT, Vilar Nehrke M, Aydos Bergonci PE (eds) Moluscos límnicos invasores no Brasil. Biologia, prevenção, controle. Redes Editora, Porto Alegre, pp 51–57

Pestana D, Ostrensky A, Tschá MK, Boeger WA (2010) Prospecção do molusco invasor *Limnoperna fortunei* (Dunker, 1857) nos principais corpos hídricos do estado do Paraná, Brasil. Papeis Avulsos de Zoologia 50:553–559

Rogora M, Massaferro J, Marchetto A, Tartari G, Mosello R (2008) The water chemistry of some shallow lakes in Northern Patagonia and their nitrogen status in comparison with remote lakes in different regions of the globe. J Limnol 67:75–86

Rückert G, Campos MCS, Rolla ME (2004) Alimentação de *Limnoperna fortunei* (Dunker 1857): taxas de filtração com ênfase ao uso de Cyanobacteria. Acta Scientiarum 26:421–429

Ruibal Conti AL, Guerrero JM, Regueira JM (2005) Levels of microcystins in two Argentinean reservoirs used for water supply and recreation: differences in the implementation of safe levels. Environ Toxicol 20:263–269

Santos SB, Thiengo SC, Fernandez MA, Miyahira IC, Gonçalves ICB, Ximenes RF, Mansur MCD, Pereira D (2012) Espécies de moluscos límnicos invasores no Brasil. In: Mansur MCD, Santos CP, Pereira D, Padula Paz IC, Leite Zurita ML, Raya Rodriguez MT, Vilar Nehrke M, Aydos Bergonci PE (eds) Moluscos límnicos invasores no Brasil. Biologia, prevenção, controle. Redes Editora, Porto Alegre, pp 25–49

Scarabino F, Verde M (1994) *Limnoperna fortunei* (Dunker, 1857) en la costa Uruguaya del Rio de la Plata (Bivalvia, Mytilidae). Comunicaciones de la Sociedad Malacológica del Uruguay 7:374–375

Silva AJM (2001) Subsídios teóricos para gestão integrada de bacia hidrográfica e zona costeira: estudo de caso da bacia hidrográfica do rio Tramandaí/RS. MSc Thesis, Universidade Federal de Santa Catarina (Brazil), pp 1–211

Sousa R, Freire R, Rufino M, Mendez J, Gaspar M, Antunes C, Guilhermino L (2007) Genetic and shell morphological variability of the invasive bivalve *Corbicula fluminea* (Müller, 1774) in two Portuguese estuaries. Estuarine, Coastal and Shelf Sci 74:166–174

Takeda AM, Mansur MCD, Fujita DS, Bibian JPM (2003) Occurrence of an invasive species, the golden mussel *Limnoperna fortunei* (Dunker, 1857), in two small reservoirs near Curitiba, PR. Acta Biologica Leopoldensia 25:251–254

Tominaga A, Goka K, Kimura T, Ito K (2009) Genetic structure of Japanese introduced populations of the golden mussel, *Limnoperna fortunei*, and the estimation of their range expansion process. Biodiversity 10:61–66

Toresani NI, López HL, Gómez SE (1994) Lagunas de la Provincia de Buenos Aires. Dirección de Intereses Marítimos, Dirección Provincial de Pesca e Intereses Marítimos, Subsecretaría de Pesca, Intereses Marítimos y Producciones Intensivas, Ministerio de la Producción de la Provincia de Buenos Aires. La Plata (Argentina), pp 1–109

Xu M, Cao XW, Wang ZY, Wang XZ (2012) Attachment characteristics of golden mussels (*Limnoperna fortunei*) in water transport projects. J Tsinghua Univ (Qinghua Daxue Xuebao) 52:170–176 [In Chinese]

Zanella O, Marenda LD (2002) Ocorrência de *Limnoperna fortunei* na Central Hidrelétrica de Itaipu. In: V Congresso Latinoamericano de Malacologia, São Paulo (Brazil)

Zhan A, Perepelizin PV, Ghabooli S, Paolucci E, Sylvester F, Sardiña P, Cristescu ME, MacIsaac HJ (2012) Scale-dependent post-establishment spread and genetic diversity in an invading mollusc in South America. Divers Distrib 18:1042–1055

Strategies and Measures to Prevent Spread of Invasive Species

Gustavo Darrigran and Cristina Damborenea

Abstract On the basis of the fouling caused by the golden mussel, *Limnoperna fortunei,* in industrial facilities and power plants (human environment), and its impacts on the ecosystem (natural environment), several strategies and measures for the mitigation of problems and prevention of further spread are discussed. At the local level, monitoring and early detection of the golden mussel can be accomplished through different methods, including those aimed at juveniles and adults, and also those aimed at their planktonic larvae. Priorities for designing bioinvasion management strategies should focus on generation of scientific knowledge, management, and actions at the sociopolitical level. Mitigation measures are closely related to the invasion phase, whereby the cost increases and the probability of eradication decreases over time.

Keywords *Limnoperna fortunei* · Golden mussel · Biological invasion · Prevention · Monitoring

Introduction

The golden mussel, *Limnoperna fortunei* (Dunker 1857), is one of the most aggressive freshwater invaders to have spread in Asia (Korea, Japan, Taiwan, Indochina) (Morton and Dinesen 2010) and in South America (Argentina, Bolivia, Brazil, Paraguay, and Uruguay) (Darrigran 2010). This invasive species has negative impacts on man-made structures (the "human environment" hereinafter), including power plants and industrial facilities. Plants are faced with problems derived from the blockage of pipelines, decreased water velocity, accumulation of shells, contamination of water by dead mussels, and clogged filters (Darrigran and Damborenea

G. Darrigran (✉) · C. Damborenea
División Zoología Invertebrados, Museo de La Plata, Universidad Nacional de La Plata, CONICET, Paseo del Bosque s/n, 1900 La Plata, Argentina
e-mail: invasion@fcnym.unlp.edu.ar

C. Damborenea
e-mail: cdambor@fcnym.unlp.edu.ar

D. Boltovskoy (ed.), *Limnoperna fortunei,* Invading Nature - Springer Series in Invasion Ecology 10, DOI 10.1007/978-3-319-13494-9_20

2005; see "Impacts of *Limnoperna fortunei* on Man-made Structures And Control Strategies: General Overview" in this volume).

The golden mussel also has an ecological impact on the "natural environment," where it affects the water column and changes the structure of available substrata. As a result, it impacts other members of the biota and their interactions, and it consequently modifies overall ecosystem processes, thus fulfilling the role of an "ecosystem engineer" (Jones and Lawton 1994; Darrigran and Damborenea 2011; Boltovskoy and Correa 2015).

Figure 1 summarizes the mechanisms by which an alien species can be transported elsewhere with the aid of man. For *L. fortunei,* expansion is associated with both "transportation-related" pathways and "commerce in living organisms." In the first case, there are two primary vectors. One is ballast water. Introduction of the mussel from Southeast Asia to South America is believed to have happened via bal-

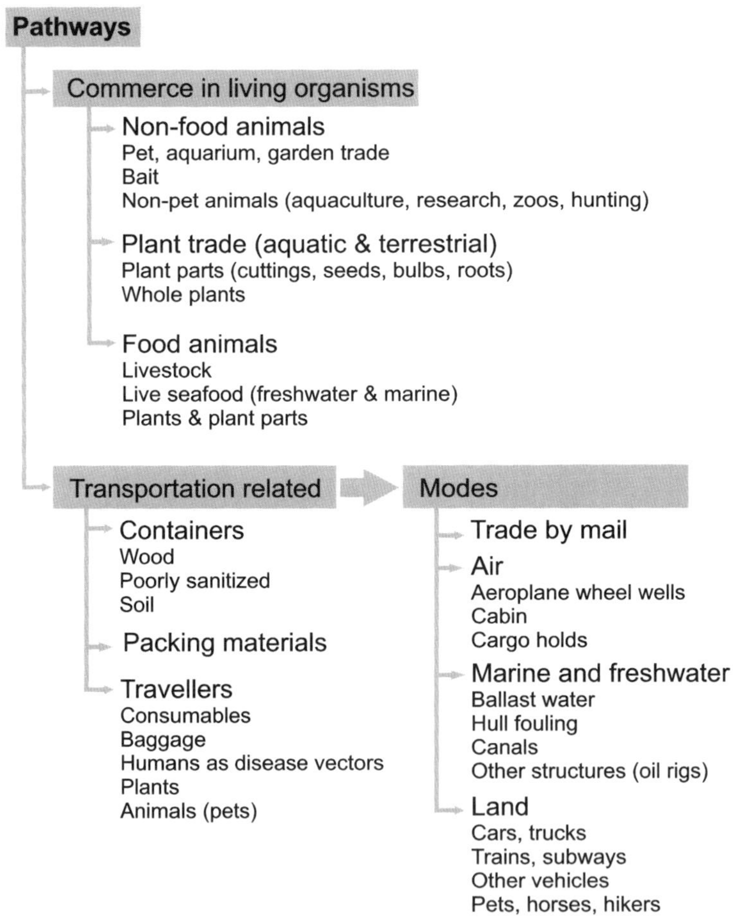

Fig. 1 Pathways responsible for the introduction of alien species

last water (Darrigran and Pastorino 1995), and its secondary spread to the Guaíba basin in Brazil likely occurred via ballast water originating in Argentina (Darrigran and Mansur 2009). The other primary vector is hull fouling. Hull fouling is a key mechanism for intra- and interbasin expansion (see "Distribution and Colonization of *Limnoperna fortunei*: Special Traits of an Odd Mussel" in this volume; Boltovskoy et al. 2006; Belz et al. 2012). Commerce in living organisms (Fig. 1), in turn, is thought to have been instrumental in the introduction of the golden mussel to Japan, presumably from China and/or Korea with edible freshwater Asian clams (*Corbicula fluminea*) (Nishimura and Habe 1987; see "Colonization and Spread of *Limnoperna fortunei* in Japan" in this volume). Secondary spread is also likely when larvae are inadvertently transported between basins with live fish bait of sport fishermen, or in the guts of fishes that swallow but do not digest mussels (Belz et al. 2012).

In South America, physiographic, hydrosedimentological, and chemical conditions of the large floodplain rivers and associated water bodies of the Río de la Plata watershed are particularly favorable for the spread of *L. fortunei* (Darrigran et al. 2012). "Jump dispersal," which combines upstream "jumps" of adults byssally attached to ship hulls followed by downstream passive drift of planktonic larvae (MacIsaac et al. 2001), has likely been instrumental in the swift dispersal of *L. fortunei* throughout this watershed (Boltovskoy et al. 2006). This mechanism has allowed the golden mussel to disperse upstream at a rate of approximately 240 km per year along the Paraná–Paraguay rivers since its introduction (Darrigran and Damborenea 2011).

It is now clear that *L. fortunei* has come to stay in both Asia and South America, and we must learn to live with it. This involves developing appropriate strategies to handle this bioinvasion. An adequate management program should take into consideration that *L. fortunei* has invaded two types of environments: the human environment and the natural environment. In both cases, the ultimate goal should be prevention of new invasions and controlling the spread of existing ones. Actions should be implemented at various levels, from the highest global level to the local level (DePoorter 2009).

Actions in the Human Environment

Invasive species can be harmful to environmental services, with negative effects on food production, biodiversity, the health of plants, animals, and man, public infrastructures, and therefore, human welfare (Pimentel 2002). The golden mussel is no exception. It interferes with fish culturing activities (e.g., in China, Uruguay, and Brazil), and water potabilization and irrigation systems, and it affects the operation of cooling systems in power plants and industries (Darrigran 2010; see "Impacts of *Limnoperna fortunei* on Man-made Structures And Control Strategies: General Overview" in this volume). Thus, every installation using raw water from water bodies in areas affected by the invasion must take measures to prevent and control biofouling of *L. fortunei*. Planning and implementation of these measures require knowledge of the mussel's biological traits, in particular its life cycle, and the climate of the region, as well as a thorough understanding of structural and functional details of the plant in question (Mackie and Claudi 2010, Darrigran and Pereyra 2011).

Early detection of *L. fortunei* is critical in establishing prevention and control measures. Due to the biological characteristics of this species (planktonic larval stages and sessile benthic adults), monitoring for the presence of the golden mussel can be accomplished through various methods: (1) detection of adults, either by direct observation of suitable existing substrata where the mussels develop, or through deployment of artificial substrata that provide surface for colonization and (2) detection of the larval planktonic stage with the aid of plankton samples subsequently analyzed under the microscope or processed using molecular methods.

The presence of adults indicates that the invasion is already underway, also furnishing information on the population dynamics of the mussel. Monitoring of the larvae, on the other hand, may allow detection of the species before it has become firmly established and achieved extensive spread.

Detection of Juveniles and Adults

Invasion of the golden mussel created a new scenario for many ecosystem compartments, and particularly for the benthos (Darrigran 2002; Boltovskoy and Correa 2015; see "Relationships of *Limnoperna fortunei* with Benthic Animals" in this volume). Monitoring of golden mussel populations in order to quantify densities, establish their structure and reproductive status, and describe their growth can be carried out in the vicinity of water intakes that feed the cooling systems of industrial facilities, or even within the cooling systems themselves (see Pereira et al. 2012 for a selection of applicable methods).

Artificial substrates provide suitable surfaces for colonization by benthic organisms and are widely used in the study of freshwater sessile macroinvertebrates (see see Fig. 2 in Chapter "Population dynamics and growth of *Limnoperna fortunei*" in this volume). Experimental frames of various designs and materials (polyvinyl chloride (PVC), concrete, wood, ceramics, various plastics, nettings, fabrics, etc.) were used in several studies, both in South America (Boltovskoy and Cataldo 1999; Fontes et al. 2002; Darrigran et al. 2007; Queiroz et al. 2007; Sylvester et al. 2007; Santos et al. 2008; Sardiña et al. 2008; Bergonci et al. 2009; Mansur et al. 2009; Belz et al. 2010; Pereira et al. 2010; Volkmer Ribeiro et al. 2010; Bonel 2011) and in Asia (Morton 1977; Ohkawa et al. 1999; Matsui et al. 2001, 2002; Nagaya et al. 2001; Nakano et al. 2010, 2011; Xu et al. 2013). The use of artificial substrates has both advantages and disadvantages. Among the former, ease of sample retrieval and density estimates, as well as the ability to ascertain the initial time of colonization, are important. On the other hand, they can be vandalized, lost, broken, exposed to air thereby killing adhering organisms, or undergo excessive siltation thus precluding mussel settlement or even causing death of already established colonizers (Sylvester 2006). In order to preclude predation of the mussels settling on the exposed artificial substrata (predation can eliminate up to over 90 % of *L. fortunei* biomass: Sylvester et al. 2007; Nakano et al. 2010), they can be protected with mesh or plastic netting. These enclosures can either be deployed barren of mussels, allowing the protected substrata to be colonized by drifting veligers and develop mussel beds, or stocked with mussels whose number and sizes have been pre-established, prior to deployment.

Detection of Larvae

Light Microscopy

Larvae of *L. fortunei* in the water column can be sought and quantified using plankton samples, obtained either by towing a net or by filtering a known volume of water with the aid of a suction pump (recommended when quantitative data are needed; see Boltovskoy 1981; Harris et al. 2000; Alder and Morales 2007; Suthers and Rissik 2009; Santos et al. 2012 for details on zooplankton sampling methods). This approach has been used extensively, usually to investigate seasonal changes in the reproductive activity of the mussel (see "Reproductive Output and Seasonality of *Limnoperna fortunei*" in this volume). However, precise abundance estimates are time consuming, and may require that large volumes of water be filtered, especially when larvae are scarce. If other molluscan species with planktonic larvae are present in the area, identification of different species will be also necessary (Ezcurra de Drago et al. 2006; Mansur et al. 2012a).

Molecular Methods

Pie et al. (2006) and Boeger et al. (2007) described a polymerase chain reaction (PCR)-based molecular method for the detection of *L. fortunei* larvae (see Tscha et al. 2012 for a detailed description). This method is very sensitive, detecting as little as 0.041 ng of *L. fortunei* DNA (a single larva yields ~28 ng of DNA), or the equivalent of one larva in 200 L of water (Pie et al. 2006; Tscha et al. 2012). Typical larval densities range around 1 to >20 larvae/L (see "Reproductive Output and Seasonality of *Limnoperna fortunei*" in this volume). Thus, this technique allows confirmation of the presence of mussels at a very early stage, before they have been detected by direct observation (Darrigran et al. 2009).

Although the molecular PCR-based method allows assessing the presence of *L. fortunei* larvae with high accuracy, larval densities cannot be quantified. In the past decade, the use of real-time quantitative PCR (qPCR) became widespread for detecting and quantifying specific DNA in a DNA complex solution. This technique can be used for the quantification of planktonic organisms, including larvae of *L. fortunei* (Endo et al. 2009). Field-collected plankton is subjected to DNA extraction and qPCR analysis using specific primers for the golden mussel (developed from CO1). Highly specific amplification, formed by 138 bases, indicates the presence of larvae in a sample and the number of larvae can be assessed based on the amplification factor (Endo and Nogata 2012).

These molecular techniques are operationally faster than traditional methods, more accurate, and they do not require taxonomic knowledge for the identification of larvae. On the other hand, the economic costs of these analyses, especially when considering the hardware required, are significantly higher.

Actions in the Natural Environment

The mechanisms involved in biological invasions comprise two complementary components: one is associated with the ability of a given species to invade a new range ("invasiveness"), and the other is the susceptibility of a given environment to be invaded ("invasibility") (Hicks 2004). Both must be taken into account when designing suitable strategies for curtailing bioinvasions.

Criteria for Defining an Invasive Species

Scoring of harmful invasive species as a function of their nuisance may be necessary when priorities in resource allocation are unavoidable. Thus, management efforts focus on the most problematic invasive species. However, deciding which species should be fought first is often not a straightforward issue. Increasing awareness of bioinvasions and the problems they cause have resulted in a growing volume of scientific publications on this topic (Kolar and Lodge 2001), as well as an accumulation of technical terms, often synonymous (Colautti and MacIsaac 2004; Lockwood et al. 2007). This has hindered both investigation and decision-making processes associated with the management of bioinvasions. The Convention on Biological Diversity defines invasive species as those that thrive unaided by humans and threaten natural or seminatural habitats outside of their normal area of distribution. Some native species can become invasive when they are transported to other areas within the same country (Simkanin et al. 2009), or even to nearby regions within the same ecosystem, especially if conditions for their survival, growth, and expansion have been modified (see "Distribution and Colonization of *Limnoperna fortunei*: Special Traits of an Odd Mussel" in this volume) (Xu et al. 2014).

Valéry et al. (2008) pointed out that the most widespread criteria to define invasive species have been the biogeographic criteria and the criteria related with their impact. The biogeographic approach is fairly practical, since it is based on an objective assessment of geographic distribution. An introduced species is one that manages to overcome a geographic barrier. This criterion thus allows for fast implementation of management options at an early stage of transfer from the original range. The impact criterion, on the other hand, requires that, in order to be considered invasive, a species must have a larger impact in the invaded ecosystem than the one it has in its native range. Assessment of the magnitude of impact, however, is unclear and questionable.

A more comprehensive definition of invasive species should be based on a series of requirements that the invasive species meets. As pointed out above, a biological invasion occurs when a species, aided by human activities, overcomes a number of barriers (Fig. 3), colonizes a new area where it acquires a competitive advantage, and grows rapidly in density and distribution. This may result in functional dominance of the system, which in turn can threaten native biodiversity and cause economic damage

Designing a Plan for Sustainable Management of Invasive Species

Cowie (2004) proposed three basic components for the conservation of biodiversity, which can also be useful for strategies of management of bioinvasions. The components are: (1) generation of scientific knowledge, (2) management, and (3) sociopolitical environment.

Scientific knowledge must reach the level of local and regional environmental organizations in order to help shape an adequate management plan. However, even when the requirements of scientific knowledge and management are met, without social awareness and support from environmental policies, conservation of biodiversity and management of bioinvasions can only succeed in small areas and for short periods. Social awareness is a sine qua non condition for active social involvement in the problem. Without it, success will be regionally restricted and temporally short. The social component is obviously the most complex of the three.

Generation of Scientific Knowledge

The results of scientific research aimed at assessing the risk of invasion should be explored and used by managers and decision makers. After the pioneering work of B. Morton in Hong Kong (Morton 1973, 1975, 1977, 1982; Morton et al. 1976), most subsequent information on *L. fortunei* has originated from South America (Argentina, Brazil, including several review books: Penchaszadeh 2005; Darrigran and Damborenea 2006, 2009; Mansur et al. 2012b) and Japan. In this context, investigations oriented at evaluating invasion risk and pathways, as well as possibilities of secondary spread (e.g., Darrigran and Pereyra 2011; Belz et al. 2012; Campos et al. 2012; Darrigran et al. 2012; Sylvester et al. 2013) are of particular significance. As noted by Hicks (2004), work on predicting future invasions is strongly lagging behind when compared to efforts dedicated to assessing impact of invasions that have already occurred. Investigations aimed at prediction, pathway analysis, and ecological scenario building that can be used in decision support systems are urgently needed (Hicks 2004). Some of this type of work has been done with *L. fortunei* (e.g., Kluza and McNyset 2005; Belz 2006, 2009; Oliveira et al. 2010; Campos 2014), but much more is needed.

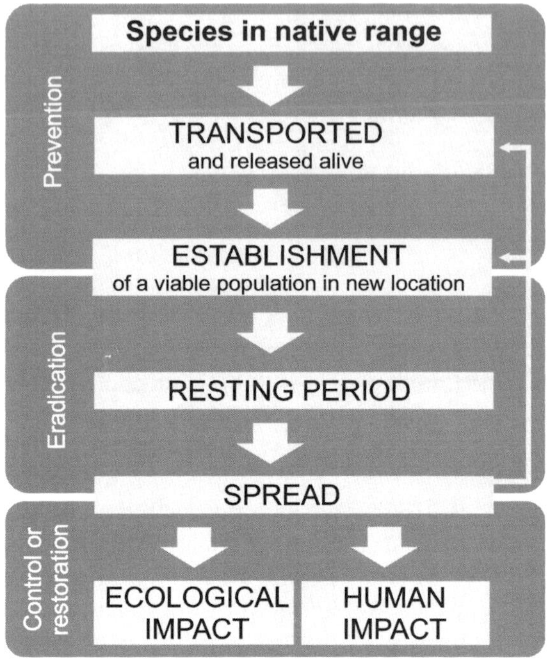

Fig. 2 Stages in the invasion process of a nonnative species, pointing out activities that need to be carried out at each stage

Management Plan

Human actions in response to impacts derived from biological invasions are often too late and too weak to significantly mitigate harm (Lodge et al. 2006). This is especially true for the golden mussel, where more than two decades after the introduction and extensive damage to many industrial installations (see "Impacts of *Limnoperna fortunei* on Man-made Structures And Control Strategies: General Overview" in this volume), none of the South American countries colonized by the mussel has developed a centralized, coordinated management strategy, and investment into research and management of this mussel remains very low. In Japan, no mussel quarantines have yet been implemented (see "Colonization and Spread of *Limnoperna fortunei* in Japan" in this volume).

Reversal of this trend will require coordinated management at various levels (regional, global) with federal or national leadership and the close cooperation of state and local government bodies (DePoorter 2009). Effectiveness in the early detection and prevention of biological invasions must increase significantly, thus allowing quick responses to new, potentially harmful introductions, and reducing the spread of existing invasions (Lodge et al. 2006). Actions and investments should be centralized and coordinated by a national center to maximize cost-effectiveness and sustainability of the control efforts.

Analysis of the components of an invasion (invasiveness and invasibility, Hicks 2004), and the stages of the invasion process (Figs. 2 and 3), indicate that each stage is characterized by very dissimilar attributes as to the feasibility and cost of

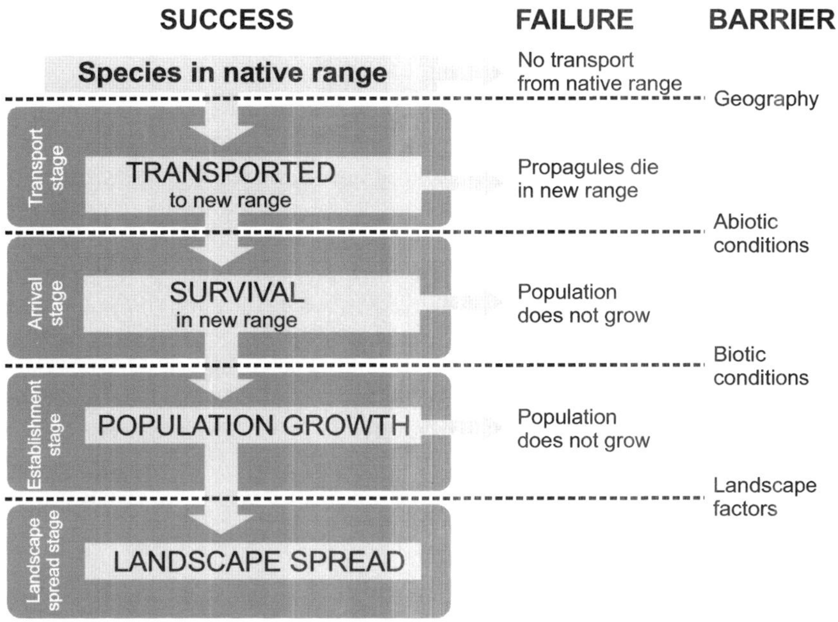

Fig. 3 A conceptual model of the invasion process of a species, highlighting the transition steps between the four stages of invasion

eradicating the invader, which depend, among others, on the ecosystem involved. These stages obviously also determine the best management option available. Management criteria and options eventually adopted, however, are often flawed and ineffective, partly because they do not take these variables into consideration, and partly because prior knowledge based on ad hoc studies is ignored or overlooked. Failure to consider conditions typical of each stage of the invasion can finally result in a useless management strategy. As suggested by Hicks (2004), the longer it takes to start taking action after a new species is introduced, the lower are the probabilities of its eradication and the higher the costs of coping with it. Thus, the best cost-to-benefit combination and the highest chances of eradication are restricted to the early stages of this process, which underscores the importance of early detection. By contrast, once the species has become conspicuous and widespread, the period of high cost-to-benefit ratio has already expired, and the probability of successful eradication is close to zero. By this time, the only remaining option is management of the invasion. Management efforts may be significant when the invasion affects human activities. On the other hand, when biodiversity or other ecosystem attributes are negatively affected but perceived by managers as of minor practical importance, the resulting management efforts are often minor.

Actions associated with reduction of the risk of entry, establishment, and dispersion of invasive species, including *L. fortunei,* fall into three main categories: (1) pre-border actions (adopted by the potential donor country/area aimed at precluding export of the species), (2) border actions, and (3) post-border (emergency) actions (Maynard and Nowell 2009). Each category, in turn, has two main components: physical (infrastructure, materials, finance) and human (legislation, procedures, capabilities). The first action arising from category (2) is quarantine. In order for quarantines to be effective, there must be a permanent compromise of physical and human resources guaranteeing infrastructure capacity, technical experience, communication capacity, and personnel training. Research on strategies aimed at mitigating bioinvasion-related problems often centers the attention on the receptor country or area, thus minimizing the responsibility of the donor side (Maynard and Nowell 2009). No safeguards are established enforcing the donor country to ensure that products and vectors (e.g., ships and trucks) leaving its ports comply with minimum requirements of safety as far as bioinvasions are concerned.

Lach et al. (2003) performed a survey exploring scientists' and the general public's expectations toward the role of scientists in communication, management, and policy. The five potential roles that research scientists might play were concluded to be: (1) reporting scientific results that others use in making decisions on natural resource management issues, (2) reporting and then interpreting scientific results for others who are involved in natural resource management decisions, (3) working closely with managers and others in integrating scientific results into management decisions, (4) actively advocating for specific and preferred natural resource management decisions, and (5) making decisions about natural resource management and policy. While scientists slightly preferred the interpretive role for themselves, other groups tended to prefer an integrative role for scientists. Most respondents were in favor of scientists getting involved in interpreting and helping to integrate the results of their science into policy decisions. This clearly is the best alternative for the sound management of many environmental issues, including bioinvasions. However, the administrative and political scenarios in many of the countries invaded by *L. fortunei* strongly hinder integration between administration and science.

Developing Social Awareness

Implementation of an integrated and sustainable management plan of biological invasions requires consideration of several components, including scientific knowledge and a blueprint for sound actions (Darrigran et al. 2008). Action must be taken at two levels: (1) in society in general, through formal and informal education with publicity strategies, documentaries, etc., seeking to generate demand at the following level and (2) among public officials, advocating for the development of norms and management programs to cope with bioinvasions in a timely manner and soundly.

An educational program aimed at the general public should include the following components: (1) it should inform society timely and thoroughly, making it aware of the ecological, economic, and social impacts that invasive species generate, including specifics as to the status and vulnerability of the different sectors endangered, (2) it should foster interinstitutional cooperation for the formation of specific work groups, timely participation of the communication media, and, above all, the cooperation of the different sectors involved, including civil organizations, and (3) it should implement programs in environmental education aimed at generating social awareness that will help prevent new introductions and facilitate early detection of nonnative species.

The economic burden of these actions is often lighter than that of control and eradication programs.

Acknowledgments This work was financially supported by the Consejo Nacional de Investigaciones Científicas y Técnicas (CONICET), Argentina, and by Projects N670 (Facultad de Ciencias Naturales y Museo, Universidad Nacional de La Plata, Argentina), and H648 (Facultad de Humanidades y Ciencias de la Educación, Universidad Nacional de La Plata, Argentina).

References

Alder VA, Morales CE (eds) (2007) Manual de métodos para el estudio de sistemas planctónicos marinos. Eudeba, Buenos Aires, pp 1–266

Belz CE (2006) Análise de risco de bioinvasão por *Limnoperna fortunei* (Dunker, 1857): um modelo para a bacia do rio Iguaçu, Paraná. PhD Thesis, Universidade Federal do Paraná (Brazil), pp 1–102

Belz C (2009) Análise de risco aplicada às bioinvasões. In: Darrigran GA, Damborenea C (eds) Introdução a biologia das invasões. O mexilhão dourado na América do Sul: biologia, dispersão, impacto, prevenção e controle. Cubo Editora, São Carlos, pp 229–245

Belz CE, Darrigran GA, Bonel N, Mader Netto OS (2010) Density, recruitment, and shell growth of *Limnoperna fortunei* (Mytilidae), an invasive mussel in tropical South America. J Freshw Ecol 25:227–233

Belz CE, Darrigran GA, Netto OSM, Boeger WA, Ribeiro PJ (2012) Analysis of four dispersion vectors in inland waters: the case of the invading bivalves in South America. J Shellfish Res 31:777–784

Bergonci PEA, Mansur MCD, Pereira D, Santos CP (2009) Population sampling of the golden mussel, *Limnoperna fortunei* (Dunker, 1857), based on artificial ceramic substrate. Biotemas 22:85–94

Boeger WA, Pie MR, Falleiros RM, Ostrensky A, Darrigran GA, Mansur MCD, Belz CE (2007) Testing a molecular protocol to monitor the presence of golden mussel larvae (*Limnoperna fortunei*) in plankton samples. J Plankton Res 29:1015–1019

Boltovskoy D (1981) Redes. In: Boltovskoy D (ed) Atlas del Zooplancton del Atlántico Sudoccidental y métodos de trabajo con el zooplancton marino. Instituto Nacional de Investigación y Desarrollo Pesquero, Mar del Plata, pp 15–35

Boltovskoy D, Cataldo D (1999) Population dynamics of *Limnoperna fortunei*, an invasive fouling mollusc, in the lower Paraná River (Argentina). Biofouling 14:255–263

Boltovskoy D, Correa N (2015) Ecosystem impacts of the invasive bivalve *Limnoperna fortunei* (golden mussel) in South America. Hydrobiologia 746:81–95

Boltovskoy D, Correa N, Cataldo D, Sylvester F (2006) Dispersion and ecological impact of the invasive freshwater bivalve *Limnoperna fortunei* in the Río de la Plata watershed and beyond. Biol Invasions 8:947–963

Bonel N (2011) Demografía experimental de un bivalvo invasor, *Limnoperna fortunei* (Dunker, 1857), en la cuenca del Plata. PhD Thesis, Universidad Nacional de La Plata (Argentina), pp 1–180

Campos MCS (2014) Modelagem matemática do processo de invasão de *Limnoperna fortunei* (Dunker, 1857). Fundação Centro Tecnológico de Minas Gerais (Brazil). Technical Report

GT 343, Controle do Mexilhão Dourado: Bioengenharia e novos materiais para aplicações em ecossistemas e usinas hidrelétricas, pp 3–36

Campos MCS, Lanzer R, Castro PT (2012) Hydrological stress as a limiting factor of the invasion of *Limnoperna fortunei* (Dunker, 1857) in the Upper Paraná River (Brazil). Acta Limnologica Brasiliensia,24:64–82

Colautti RI, MacIsaac HJ (2004) A neutral terminology to define 'invasive' species. Divers Distrib 10:135–141

Cowie RH (2004) Disappearing snails and alien invasions: the biodiversity/conservation interface in the Pacific. J Conchol (Special Publication) 3:23–37

Darrigran GA (2002) Potential impact of filter-feeding invaders on temperate inland freshwater environments. Biol Invasions 4:145–156

Darrigran GA (2010) Summary of the distribution and impact of the golden mussel in Argentina and neighboring countries. In: Mackie GL, Claudi R (eds) Monitoring and control of macrofouling mollusks in fresh water systems. CRC Press, Boca Raton, pp 389–396

Darrigran GA, Damborenea C (2005) A South American bioinvasion case history: *Limnoperna fortunei* (Dunker, 1857), the golden mussel. Am Malacol Bull 20:105–112

Darrigran GA, Damborenea C (eds) (2006) Bioinvasión del mejillón dorado en el continente americano. Editorial Universitaria de La Plata, La Plata, pp 1–221

Darrigran GA, Damborenea C (eds) (2009) Introdução a biologia das invasões. O Mexilhão Dourado na América do Sul: biologia, dispersão, impacto, prevenção e controle. Cubo Editora, Sao Carlos, pp 1–245

Darrigran GA, Damborenea C (2011) Ecosystem engineering impact of *Limnoperna fortunei* in South America. Zool Sci 28:1–7

Darrigran G, Mansur MCD (2009) Introdução e dispersão do *Limnoperna fortunei*. In: Darrigran G, Damborenea C (eds) Introdução a biologia das invasões. O mexilhão dourado na América do Sul: biologia, dispersão, impacto, prevenção e controle. Cubo Editora, São Carlos, pp 89–109

Darrigran GA, Pastorino G (1995) The recent introduction of a freshwater Asiatic bivalve, *Limnoperna fortunei* (Mytilidae) into South America. The Veliger 38:171–175

Darrigran GA, Pereyra P (2011) Estandarización de métodos de control para *Limnoperna fortunei*. In: Tópicos em malacologia, Ecos do XIX Encontro Brasileiro de Malacologia, Rio de Janeiro (Brazil)

Darrigran GA, Damborenea C, Greco N (2007) An evaluation pattern for antimacrofouling procedures: *Limnoperna fortunei* larvae study in a hydroelectric power plant in South America. Ambio 36:575–579

Darrigran GA, Vilches A, Legarralde T (2008) Desinterés del pasado, decisiones del futuro: educación para prevenir las invasiones biológicas. Revista Educación en Biología 11:39–52

Darrigran GA, Boeger W, Damborenea C, Maroñas M (2009) Evaluation of sampling and analysis techniques for early detection of *Limnoperna fortunei* (Mytilidae) in limit areas of its distribution. Braz J Biol 69:979–980

Darrigran GA, Damborenea C, Drago EC, Ezcurra de Drago I, Paira A, Archuby F (2012) Invasion process of *Limnoperna fortunei* (Bivalvia: Mytilidae): The case of Uruguay River and emissaries of the Esteros del Iberá Wetland, Argentina. Zoologia 29:531–539

DePoorter M (2009) International legal instruments and frameworks for invasive species. In: Clout MN, Williams PA (eds) Invasive species management: a handbook of principles and techniques. Oxford University Press, New York, pp 108–140

Endo N, Nogata Y (2012) Método de detecção e quantificação de larvas do mexilhão-dourado *Limnoperna fortunei*, usando PCR quantitativo em tempo real. In: Mansur MCD, Santos CP, Pereira D, Padula Paz IC, Leite Zurita ML, Raya Rodriguez MT, Vilar Nehrke M, Aydos Bergonci PE (eds) Moluscos límnicos invasores no Brasil. Biologia, prevenção, controle. Redes Editora, Porto Alegre, pp 149–153

Endo N, Sato K, Nogata Y (2009) Molecular based method for the detection and quantification of larvae of the golden mussel *Limnoperna fortunei* using real-time PCR. Plankton Benthos Res 4:125–128

Ezcurra de Drago I, Montalto L, Oliveros OB (2006) Desarrollo y ecología larval de *Limnoperna fortunei*. In: Darrigran G, Damborenea C (eds) Bio-invasión del mejillón dorado en el continente americano. Editorial de la Universidad de La Plata, La Plata, pp 85–93
Fontes HM, Takeda AM, Fugita DS (2002) Experience in Itaipu reservoir (Brazil) with artificial substrate (in situ): is it best way of monitoring an invading species like *Limnoperna fortunei* (Bivalvia, Mytilidae)? In: International Symposium on Reservoir Management in Tropical and Sub-Tropical Regions, Foz do Iguaçu (Brazil)
Harris RP, Wiebe PH, Lenz J, Skjoldal HR, Huntley M (eds) (2000) ICES Zooplankton methodology manual. Academic Press, San Diego, pp 1–684
Hicks G (2004) Turning the tide: Is aquatic bioinvaders research heading in the right direction? Aquat Invaders 15:9–20
Jones CG, Lawton JH (1994) Organisms as ecosystem engineers. Oikos 69:373–386
Kluza DA, McNyset KM (2005) Ecological niche modeling of aquatic invasive species. Aquat Invaders 16:1–7
Kolar CS, Lodge DM (2001) Progress in invasion biology: predicting invaders. Trends Ecol Evol 16:199–204
Lach D, List P, Steel B, Shindler B (2003) Advocacy and credibility of ecological scientists in resource decisionmaking: a regional study. BioScience 53:170–178
Lockwood J, Hoopes M, Marchetti M (2007) Invasion ecology. Blackwell, Malden, pp 1–304
Lodge DM, Williams S, MacIsaac HJ, Hayes KR, Leung B, Reichard S, Mack RN, Moyle PB, Smith M, Andow DA, Carlton JT, McMichael A (2006) Biological invasions: recommendations for U.S. policy and management. Ecol Appl 16:2035–2054
MacIsaac HJ, Grigorovich IA, Ricciardi A (2001) Reassessment of species invasions concepts: the Great Lakes basin as a model. Biol Invasions 3:405–416
Mackie GL, Claudi R (2010) Monitoring and control of macrofouling mollusks in fresh water systems. CRC Press, Boca Raton, pp 1–508
Mansur MCD, Pereira D, Santos CP, Bergonci PEA, Thormann BM, Takeda AM (2009) Colonização de substrato artificial pelo mexilhão dourado, *Limnoperna fortunei* (Dunker, 1857) (Bivalvia, Mytiloida, Mytilidae), no Delta do Rio Jacuí (RS, Brasil). Biotemas 22:75–80
Mansur MCD, Mansur Pimpao D, Bergonci PEA, Santos CP, Santos de Figueiredo GC (2012a) Morfologia e ciclo larval comparados de bivalves límnicos invasores e nativos. In: Mansur MCD, Santos CP, Pereira D, Padula Paz IC, Leite Zurita ML, Raya Rodriguez MT, Vilar Nehrke M, Aydos Bergonci PE (eds) Moluscos límnicos invasores no Brasil. Biologia, prevenção, controle. Redes Editora, Porto Alegre, pp 95–110
Mansur MCD, Santos CP, Pereira D, Padula Paz IC, Leite Zurita ML, Raya Rodriguez MT, Vilar Nehrke M, Aydos Bergonci PE (eds) (2012b) Moluscos límnicos invasores no Brasil. Biologia, prevenção, controle. Redes Editora, Porto Alegre, pp 1–411
Matsui Y, Nagaya K, Yuasa A, Naruto H, Yamamoto H, Ohkawa K, Magara Y (2001) Attachment strength of *Limnoperna fortunei* on substrates, and their surface properties. Biofouling 17:29–39
Matsui Y, Nagaya K, Funahashi G, Goto Y, Yuasa A, Yamamoto H, Ohkawa K, Magara Y (2002) Effectiveness of antifouling coatings and water flow in controlling attachment of the nuisance mussel *Limnoperna fortunei*. Biofouling 18:137–148
Maynard G, Nowell D (2009) Biosecurity and quarantine for preventing invasive species. In: Clout MN, Williams PA (eds) Invasive species management: a handbook of principles and techniques. Oxford University Press, Oxford, pp 1–18
Morton B (1973) Some aspects of the biology and functional morphology of the organs of feeding and digestion of *Limnoperna fortunei* (Dunker) (Bivalvia: Mytilacea). Malacologia 12:265–281
Morton B (1975) The colonization of Hong Kong's raw water supply system by *Limnoperna fortunei* (Dunker 1857) (Bivalvia: Mytilacea) from China. Malacol Rev 8:91–105
Morton B (1977) The population dynamics of *Limnoperna fortunei* (Dunker 1857) (Bivalvia: Mytilacea) in Plover Cove Reservoir, Hong Kong. Malacologia 16:165–182

Morton B (1982) The reproductive cycle in *Limnoperna fortunei* (Dunker, 1857) (Bivalvia: Mytilidae) fouling Hong Kong's raw water sypply system. Oceanologia et Limnologia Sinica 13:312–324

Morton B, Dinesen G (2010) Colonization of Asian freshwaters by the Mytilidae (Bivalvia): A comparison of Sinomytilus harmandi from the Tonle-Sap River, Phnom Penh, Cambodia, with *Limnoperna fortunei*. Molluscan Res 30:57–72

Morton B, Au CS, Lam WW (1976) Control of *Limnoperna fortunei*. J Inst Water Eng Sci 30:147–156

Nagaya K, Matsui Y, Ohira H, Yuasa A, Yamamoto H, Ohkawa K, Magara Y (2001) Attachment strength of an adhesive nuisance mussel, *Limnoperna fortunei*, against water flow. Biofouling 17:263–274

Nakano D, Kobayashi T, Sakaguchi I (2010) Predation and depth effects on abundance and size distribution of an invasive bivalve, the golden mussel *Limnoperna fortunei*, in a dam reservoir. Limnology 11:259–266

Nakano D, Kobayashi T, Endo N, Sakaguchi I (2011) Growth rate and settlement of *Limnoperna fortunei* in a temperate reservoir. J Molluscan Stud 77:142–148

Nishimura T, Habe T (1987) Chinese freshwater mussels mixed in the imported *Corbicula* clam. Chiribotan (J Malacol Soc Jpn) 18:110–111

Ohkawa K, Nishida A, Honma R, Matsui Y, Nagaya K, Yuasa A, Yamamoto H (1999) Studies on fouling by the freshwater mussel *Limnoperna fortunei* and the antifouling effects of low energy surfaces. Biofouling 13:337–350

Oliveira MD, Hamilton SK, Jacobi CM (2010) Forecasting the expansion of the invasive golden mussel *Limnoperna fortunei* in Brazilian and North American rivers based on its occurrence in the Paraguay River and Pantanal wetland of Brazil. Aquat Invasions 5:59–73

Penchaszadeh PE (ed) (2005) Invasores. Invertebrados bentónicos introducidos en el Río de la Plata y región costera marina aledaña. Eudeba, Buenos Aires, pp 1–384

Pereira D, Mansur MCD, Volkmer-Ribeiro C, Oliveira MD, Santos CP, Bergonci PEA (2010) Colonização de substrato artificial por macroinvertebrados límnicos, no delta do Rio Jacuí (RS, Brasil). Biotemas 23:101–110

Pereira D, Arruda JA, Bergonci PEA, Oliveira AS, Postiglione R, Mansur MCD (2012) Como monitorar moluscos límnicos invasores bentônicos e macroinvertebrados associados? In: Mansur MCD, Santos CP, Pereira D, Padula Paz IC, Leite Zurita ML, Raya Rodriguez MT, Vilar Nehrke M, Aydos Bergonci PE (eds) Moluscos límnicos invasores no Brasil. Biologia, prevenção, controle. Redes Editora, Porto Alegre, pp 155–184

Pie MR, Boeger WA, Patella L, Falleiros RM (2006) A fast and accurate molecular method for the detection of larvae of the golden mussel *Limnoperna fortunei* (Mollusca: Mytilidae) in plankton samples. J Molluscan Stud 72:218–219

Pimentel D (ed) (2002) Biological invasions. Economic and environmental costs of alien plant, animal, and microbe species. CRC Press, Boca Raton, pp 1–369

Queiroz JF, Silveira MP, Sitton M, Marigo ALS, Zambon GV, Silva JR, Carvalho MP, Ribacinko DB (2007) Coletor de macroinvertebrados bentônicos com substrato artificial para monitoramento da qualidade de água em viveiros de produção de tilápia. EMBRAPA(Brazil). Circular Técnica 16, pp 1–5

Santos CP, Mansur MCD, Wurdig NL (2008) Variaçoes no comprimento dos individuos de uma populacão do mexilhão dourado, *Limnoperna fortunei* (Mollusca: Bivalvia: Mytilidae), ao longo do ano, na Praia do Veludo, Lago Guaiba, Rio Grande do Sul, Brasil. Braz J Biol 25:389–396

Santos CP, Vilar Nehrke M, Mansur MCD, Gazulha V (2012) Como monitorar bivalves invasores no plâncton? Método da microscopia óptica. In: Mansur MCD, Santos CP, Pereira D, Padula Paz IC, Leite Zurita ML, Raya Rodriguez MT, Vilar Nehrke M, Aydos Bergonci PE (eds) Moluscos límnicos invasores no Brasil. Biologia, prevenção, controle. Redes Editora, Porto Alegre, pp 139–142

Sardiña P, Cataldo D, Boltovskoy D (2008) The effects of the invasive mussel, *Limnoperna fortunei*, on associated fauna in South American freshwaters: importance of physical structure and food supply. Fund Appl Limnol/Archiv Hydrobiologie 173:135–144

Simkanin C, Davidson I, Falkner M, Sytsma M, Ruiz G (2009) Intra-coastal ballast water flux and the potential for secondary spread of non-native species on the US West Coast. Marine Pollut Bull 58:366–374

Suthers IM, Rissik D (eds) (2009) Plankton. A guide to their ecology and monitoring for water quality. CSIRO, Collingwood, pp 1–256

Sylvester F (2006) Biología alimentaria y ecología del molusco invasor *Limnoperna fortunei* (Mytilidae) en el Paraná inferior y Río de la Plata. PhD Thesis, Universidad de Buenos Aires (Argentina), pp 1–160

Sylvester F, Boltovskoy D, Cataldo D (2007) Fast response of freshwater consumers to a new trophic resource: Predation on the recently introduced Asian bivalve *Limnoperna fortunei* in the lower Parana River, South America. Austral Ecol 32:403–415

Sylvester F, Cataldo D, Notaro C, Boltovskoy D (2013) Fluctuating salinity improves survival of the invasive freshwater golden mussel at high salinity: implications for the introduction of aquatic species through estuarine ports. Biol Invasions 15:1355–1366

Tscha MK, Patella R, Ostrensky A, Boeger WA (2012) O método molecular de prospecção do mexilhão-dourado. In: Mansur MCD, Santos CP, Pereira D, Padula Paz IC, Leite Zurita ML, Raya Rodriguez MT, Vilar Nehrke M, Aydos Bergonci PE (eds) Moluscos límnicos invasores no Brasil. Biologia, prevenção, controle. Redes Editora, Porto Alegre, pp 143–148

Valéry L, Fritz H, Lefeuvre J, Simberloff D (2008) In search of a real definition of the biological invasion phenomenon itself. Biol Invasions 10:1345–1351

Volkmer Ribeiro C, Parolin M, Fürstenau-Oliveira K, De Menezes ER (2010) Colonization of hydroelectric reservoirs in Brazil by freshwater sponges, with special attention on Itaipu. Interciencia 35:340–347

Xu M, Wang ZY, Lin CC, Pan BZ, Zhao N (2013) Experimental study of invasion and biofouling of freshwater mussel *Limnoperna fortunei*. Int J Geosci 4:1–7

Xu M, Darrigran G, Wang Z, Zhao N, Lin CC (2014) Experimental study on control of *Limnoperna fortunei* biofouling in water transfer tunnels. J Hydro-environment Res (in press)

Part IV
Impacts on Man-Made Structures and Biofouling Control

Impacts of *Limnoperna fortunei* on Man-Made Structures and Control Strategies: General Overview

Demetrio Boltovskoy, Mengzhen Xu and Daisuke Nakano

Abstract In China and South America, severe fouling problems caused by *Limnoperna fortunei* have been reported for a number of industrial facilities, including water and wastewater processing plants, power plants (nuclear, hydroelectric, thermal), refineries, steel mills, fish culture facilities, water transfer canals and aqueducts, and watercraft. In Japan, biofouling chiefly affects agricultural irrigation systems, balancing reservoirs and balancing tanks. However, most available reports furnish little detail on the components affected and on the measures taken to cope with the nuisance. Objective estimates of the economic losses involved are extremely rare. Although fouling by the golden mussel has occasionally derived in operation at below-standard regimes and even temporary plant shutoffs, as maintenance personnel acquired experience in curtailing mussel growth in sensitive areas, serious incidents have become less common. Fouling by *L. fortunei* has not caused a single definitive plant shutdown. Control methods assessed (either in laboratory settings or in actual plant operating conditions) include antifouling materials and coatings, chemical treatments, manual/mechanical cleaning, filtration, thermal shock, anoxia and hypoxia, desiccation, ozonation, ultraviolet treatment, electric currents, ultrasound, manipulations of flow speed, biological control, and various miscellaneous methods.

D. Boltovskoy (✉)
Facultad de Ciencias Exactas y Naturales, Instituto de Ecología, Genética y Evolución de Buenos Aires (IEGEBA), Universidad de Buenos Aires-CONICET, Ciudad Universitaria, 1428 Buenos Aires, Argentina
e-mail: demetrio@ege.fcen.uba.ar

Museo Argentino de Ciencias Naturales 'Bernardino Rivadavia'-CONICET, Buenos Aires, Argentina

M. Xu
State Key Laboratory of Hydroscience and Engineering, Tsinghua University, 100084 Beijing, China
e-mail: mzxu@mail.tsinghua.edu.cn

D. Nakano
Central Research Institute of Electric Power Industry (CRIEPI), 1646 Abiko, Chiba 270-1194, Japan
e-mail: d-nakano@criepi.denken.or.jp

D. Boltovskoy (ed.), *Limnoperna fortunei,* Invading Nature - Springer Series in Invasion Ecology 10, DOI 10.1007/978-3-319-13494-9_21

Keywords *Limnoperna fortunei* · Golden mussel · Biological invasion · Biofouling · Impact · Industrial plants · Power plants

Introduction

In contrast to its known effects on ecosystems, which often differ from those of the zebra mussel (Boltovskoy and Correa 2015; see Chapter "Ecology and Environmental Impact of *Limnoperna fortunei*: Introduction" in this volume), the impacts of *Limnoperna fortunei* on man-made structures and facilities are almost identical to those described for dreissenid species. Hundreds of papers and several excellent reviews have been published concerning these impacts during the long history of the invasion of Western Europe and North America by the zebra and quagga mussels that began in the nineteenth century (Claudi and Mackie 1994; McMahon et al. 1994; Sprecher and Getsinger 2000; Rajagopal et al. 2000; Nalepa and Schloesser 2014). The comprehensive manual by Mackie and Claudi (2010) provides a detailed review of the problems caused by freshwater fouling mussels around the world, for which reason we will not repeat this information here.

History, Spread, and General Appraisal of Macrofouling Problems Caused by the Golden Mussel

Although in its native range (China) the golden mussel has likely been a nuisance for centuries, information in the older Chinese scientific literature is oddly absent regarding *L. fortunei* and reports on macrofouling-related problems are restricted to marine mussels (e.g., Lou and Liu 1958). The presence of *L. fortunei* had been mentioned (Tchang et al. 1965), but information on its impacts on man-made structures is only found in isolated internal reports (GPS (Pipeline Study Group) 1973). This is partly due to the fact that until the first half of the twentieth century, the geographic range of the golden mussel was restricted to southern China, Thailand, Laos, Cambodia, and Vietnam. In the latter four countries, it was probably introduced through human actions (Morton and Dinesen 2010), and the range of *L. fortunei* started expanding greatly in the 1980s (Fig. 1). According to Xu (2013), *L. fortunei* was originally only found in southern China with the Yangtze River as its northern boundary. In the 1980s, it appeared in the Yellow River basin, and recently it has been found in the waters in and around Beijing (Ye et al. 2011) (see Chapter "Distribution and Spread of *Limnoperna fortunei* in China" in this volume). Thus, the first publications to point out the potentially harmful nature of the golden mussel (Morton 1973, 1975), as well as the first investigation on alternatives to control its fouling of raw water pipelines (Morton et al. 1976) are associated with the colonization of Hong Kong's freshwater supply system through water diversion works from the East River, a tributary of the Pearl River, which forms part of the native range of the mussel. At approximately the same time, fouling by *L. fortunei* was reported in

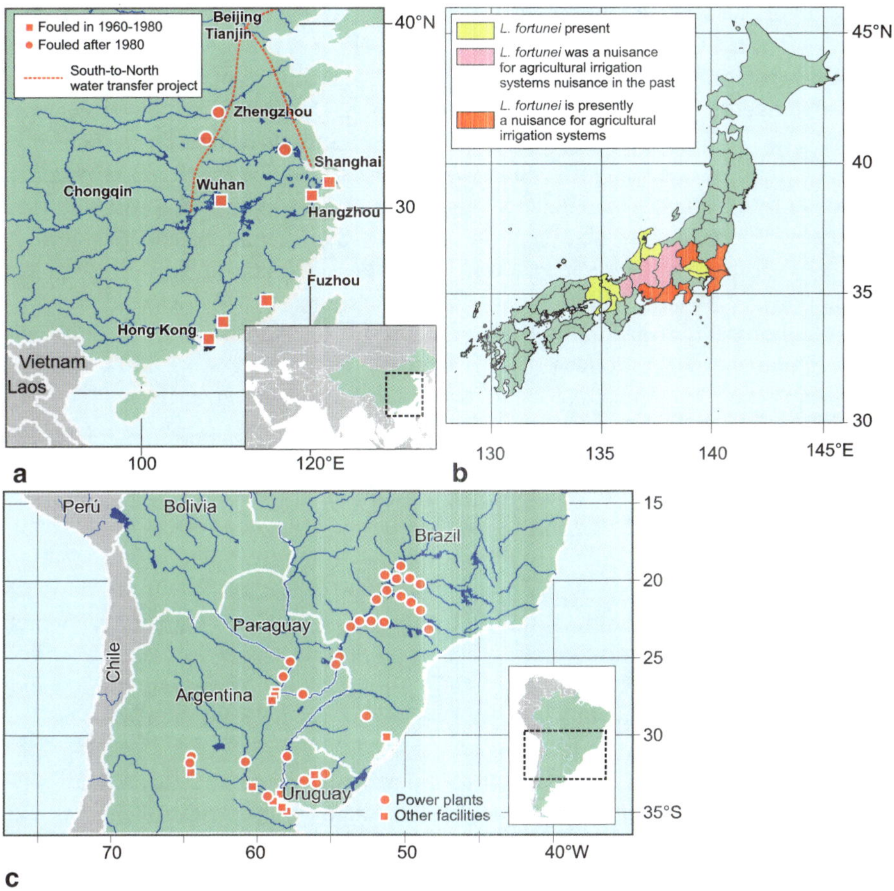

Fig. 1 Location of some industrial facilities in **a** China and **c** South America that have experienced mussel-fouling problems. Power plants include hydroelectric, thermal and nuclear facilities. "Other facilities" include refineries, steel factories, water and food processing plants, etc. **b** Japanese prefectures where *L. fortunei* is present but does not cause major harm (*yellow*), where it was a nuisance in the past (*pink*), and where it presently represents a problem (*red*)

cooling water pipes of one of the largest steel mills in China—the Wuhan Iron and Steel Corporation (GPS (Pipeline Study Group) 1973), and in the 1980s problems became widespread affecting many industrial and water transfer facilities, like those of Handan in Hebei Province (Xiang 1985), Xiamen in Fujian Province, Wuxi and Suzhou in Jiangsu Province. Some, like the water treatment plant in Suzhou, had temporary shutoffs due to pipe clogging by *L. fortunei* (Luo 2006).

In order to facilitate the fast industrialization and modernization process of the last decades, numerous inter-basin water diversion projects have been undertaken in China, thus improving the distribution of water across the country. While beneficial for the country's development, these aqueducts became a major vector for the mussel's geographic spread (Xu et al. 2009a). Vivid examples of these problems are the water diversion works that draw water from the East River, a tributary of the Pearl River (Morton 1975; Xu 2013). One of these pipelines supplies water to Hong Kong and the western part of Shenzhen (Morton 1975), while another, "the East River Water Source Project," transfers water to Huizhou, Dongguan, and the eastern part of Shenzhen (Xu 2013). All the reservoirs, lakes, and water transfer systems connected with these two water diversion works have already been colonized by golden mussels, thus requiring maintenance and cleaning tasks which, according to some estimates, amount to over US$ 1 million per year. Significantly, there are at least 30 such long-distance water diversion works in the Pearl river basin, built at a total estimated cost of US$ 12.4 billion (Guangdong Government 2010), almost all of which are potentially at risk of fouling by the golden mussel. Additional large aqueducts, in particular the huge "South-to-North Water Transfer Project" (Fig. 1), aimed at connecting the water-rich areas of the middle and lower Yangtze River with water-poor Beijing (see Fig. 1 in Chapter "Distribution and spread of *Limnoperna fortunei* in China" in this volume), will further worsen biofouling problems by *L. fortunei*.

In recent years, *L. fortunei*-related fouling became increasingly common in hydroelectric power stations. One of us (M. Xu) has had first hand contact with four of these power plants: Shisanling (in Beijing), Langyashan (in Anhui Province), Tianhuangping (in Zhejiang Province), and Guangxu (in Canton Province). This last one requested assistance for controlling the mussel. Fouling is common on concrete underwater structures, valves, trash racks, gates, etc. Dense mussel beds ca. 10 cm in thickness significantly increase resistance to water flow, enhance corrosion, clog pipes, jam mobile components, and pose serious safety risks for the plant's personnel (Li and Su 2007; Yao and Xu 2013).

This issue has drawn increasing attention since 2000. As of 2013, at least eight major scientific and technological projects associated with freshwater mussel fouling are underway. Several control strategies have been attempted, either on experimental scales, and/or under plant operating conditions, with variable success (Table 1).

Outside of mainland China, reports of problems associated with fouling by *L. fortunei* in industrial and water-treatment facilities started appearing a few years after the mussel's invasion. In Taiwan, Tan et al. (1987) reported heavy clogging of water intake grates at a water treatment facility, but they also mentioned that *L. fortunei* was first described from the island in 1941 (Kuroda 1941, as *Volsella (Limnoperna) lacustris*; although this record was subsequently questioned by Huang (2008). Interestingly, it was not included in the catalogue of Taiwanese terrestrial and freshwater mussels produced by Pilsbry and Hirase in 1905, which suggests the possibility that it was introduced between the turn of the century and 1940. In Korea, where it was introduced in the late 1970s to early 1980s, Kojima (1982) was the first to review its biology, fouling-related problems, and control alternatives. Some years later, it was reported from South America (Pastorino et al. 1993), and from Japan (Kimura 1994; Nakai et al. 1994).

In Japan, *L. fortunei*-related fouling problems are similar to those in other areas invaded. Affected facilities are chiefly water intake structures of drinking water treatment plants, hydroelectric power plants, and agricultural irrigation facilities.

Table 1 Control alternatives used in China for curtailing *L. fortunei* fouling

Control strategy	Type of facility and place	Setting	Assessment of effectiveness	Reference
Filtration	WT pipeline (Huizhou, Guangdong Prov.)	Plant	Filters failed in retaining veligers and became overgrown by mussels after a few months	Zhuang (2006)
Flow speed	Power plant (Handan, Hebei Prov.), WT pipeline (Huizhou, Guangdong Prov.)	Exp.	Only feasible where flow speed can be adjusted to > 2 m/s	Xiang (1985); Ye et al. (2011)
Antifouling coatings	Raw water pipelines (Huizhou and Shenzhen, Guangdong Prov.)	Exp.	Effective over limited time periods. Coatings release antifouling toxic substances, limiting applications in drinking water systems	Luo et al. (2006); Zhuang (2006)
Anoxia	Raw water pipelines (Shenzhen, Guangdong Prov.)	Exp.	Suitable for pipe sections that can be sealed off temporarily. Unfeasible for long-distance aqueducts	Liu et al. (2006)
Desiccation	WT pipeline (Shenzhen, Guangdong Prov.)	Exp.	Requires pipeline shut down for 10–15 days to kill the mussels, which is often unfeasible due to operational constraints	Luo et al. (2006)
Ozonation	WT pipeline (Huizhou, Guangdong prov.)	Exp.	Effective and environmentally friendly in small pipe sections. Continuous ozonization is cost-ineffective for long-distance pipelines	Xu et al. (2009b)
Hydrogen peroxide	WT pipeline (Huizhou, Guangdong Prov.)	Exp.	Concentrations of ~ 1.2–1.8 g/l are effective and economically feasible in restricted sealed off pipe sections. In long pipelines, lethal concentrations are difficult to maintain	Xu et al. (2009b)
Chlorination	WT pipeline (Shenzhen & Huizhou, Guangdong Prov.)	Exp.	Generally effective, but not widely applied in China because it deactivates biological nitrification along pipelines and affects water quality	Liu et al. (2006); Zhuang (2006)

Table 1 (continued)

Control strategy	Type of facility and place	Setting	Assessment of effectiveness	Reference
Manual and mechanical cleaning	Most of the infected man-made facilities in Guangdong Province	Plant	Effective in the short-term, but fouling recovers rapidly, requiring frequent cleaning operations, which are expensive and affect substrata	Ye et al. (2011)
Biological control (cultured fishes)	WT pipeline (Shenzhen, Yao Autonomous County of Ruyuan, Guangdong Prov.)	Exp.& plant	Environmentally friendly and sustainable when applied at the intake end of water transfer systems, but its efficacy is limited	Luo et al. (2006)
Attraction-attachment pools	WT pipeline (Huizhou, Guangdong Prov.)	Exp.	Environmentally friendly and sustainable, effective at the experimental scales tested. Not yet applied in actual operating conditions	Xu (2013)

Exp. experimental, *WT* water transfer

Several water treatment plants have been affected by *L. fortunei* biofouling in the Yodo River system in Japan (Nakanishi and Mukai 1997). Problems included mass attachment of mussels on raw water screening structures, obstruction in strainers and pipes for water quality monitoring, accumulation of dead mussels in settling and flocculation chambers, and blockage of cooling system pipes for intake pumps. Control and mitigation strategies included several approaches. Where possible, mussels obstructing pipes and other components were removed by manual or mechanical cleaning with subsequent disposal of the mussels. Further clogging was deterred by (1) treating the water with chlorine, (2) adding filters and strainers at the intake in order to prevent larvae from entering the system, and (3) duplicating some components in order to allow for the decommissioning and cleaning of one while maintaining the second in operation.

As almost all thermal and nuclear power plants in Japan are located on the coast and use salt water for cooling purposes, biofouling by *L. fortunei* is restricted to hydroelectric power plants. Here, the main problems are mussel growth on intake screens and headrace channels. Manual and mechanical cleaning has traditionally been used to cope with the problem, although disposal of waste materials is still a major cause of concern. Obstruction of bulwark pipes, which protect water-level monitoring instruments, has resulted in gauge malfunctions and failures. These problems were solved by replacing regular bulwark pipes with new ones made of copper alloy. Blockage of electric generator cooling water pipes has occurred in some plants. Mechanical cleaning during routine inspection and maintenance operations has mitigated the problems, but some plants plan on installing alternative control methods such as self-cleaning microstrainers and thermal shock treatments, depending on plant characteristics.

Biofouling by *L. fortunei* has been extensive in agricultural irrigation systems of the Kanto (Katayama et al. 2005) and Tokai (Akehoshi 2011) regions. Mass attachment of *L. fortunei* has occurred on intake screens, irrigation channels, balancing reservoirs, and balancing tanks. Obstruction by *L. fortunei* is also common in strainers and pipeworks (especially small diameter, terminal pipes) of irrigation systems (Ministry of Agriculture Forestry and Fisheries of Japan 2012). Fouling of gauging instruments for water-level monitoring has resulted in their malfunction. Manual and mechanical removal of the mussels is the main control method so far, although alternative strategies, such as desiccation and antifouling coatings, are being considered (Ministry of Agriculture Forestry and Fisheries of Japan 2012).

In South America, the earliest cases of fouling by *L. fortunei* date from around 1994 (Darrigran 1995), ca. 3–4 years after its introduction in Argentina (see Chapter "Colonization and Spread of *Limnoperna fortunei* in South America" in this volume). Subsequently, literature mentioning problems in various industrial facilities grew exponentially, but many of these reports furnish little detail on the components affected and on the measures taken to cope with the nuisance. Many of these publications are based on circumstantial and ancillary data or on previous reports, and are basically restricted to general comments on the perils and harm brought about by the golden mussel. Given the fact that most of the available information on *L. fortunei* comes from biological journals, and that in-depth analyses of its impacts are more technological than biological in nature, this outcome is hardly surprising. Indeed, biological journals shun strictly technological issues, whereas technologically oriented publications normally cover topics of wider interest, rather than in-depth accounts of problems of a single industrial plant. Thus, detailed information on fouling by *L. fortunei* in man-made installations is largely restricted to internal reports of limited distribution, most of which are not accessible to the scientific community.

Furthermore, plant engineers, the chief actors in possession of first hand information, are not encouraged (and often not allowed to), trained for, or interested in publishing descriptions of the problems they encounter or their solutions. Availability of these internal reports is further restricted by the fact that control measures are in some cases potentially harmful to the environment (especially in the case of chemical control methods), in which case disclosure of these operations is avoided. There are a few notable exceptions, where researchers or technical personnel of infested facilities described their experience at regional meetings (Cepero 2003; Bendati et al. 2004; Oviedo Antunes and de Madrinag 2005; Figueiredo de Resende and Martinez 2008; Glaser 2011) or in various other publications (GPS (Pipeline Study Group) 1973; Magara et al. 2001; Portella Kleber et al. 2009; Mata 2011; Netto 2011). However, most literature where fouling-related problems in Argentina (Darrigran et al. 2002; Darrigran et al. 2007b); Brazil (Colares et al. 2002; Simeão et al. 2006; Rolla and Mota 2010; Borges et al. 2013); China (Morton 1975; Liu et al. 2011a; Liu et al. 2011b; Ye et al. 2011; Xu et al. 2012; see Table 1); Japan (Magara et al. 1999; Magara et al. 2001; Goto 2002; Yoshida 2006; Hamada 2008; 2010; Sawada and Nakamura 2010; Akehoshi 2011); Korea (Kojima 1982) and Uruguay (Brugnoli et al. 2012) are addressed (see also publications on control methods listed below) furnish limited information on affected plant components. Objective estimates of economic losses involved are extremely rare.

In South America, the large majority of the plants impacted by mussel fouling are located along the upper Río de la Plata estuary, Paraná, Paraguay, and Uruguay rivers, and their tributaries. There are also infested facilities (including a nuclear power station) in central Argentina (Córdoba province), chiefly along reservoirs connected with the Paraná waterway by the Tercero-Carcarañá rivers (Río de la Plata basin), as well as San Roque Reservoir (31.37°S, 64.47°W), which is fed by a small endorheic basin (see Chapter "Colonization and Spread of *Limnoperna fortunei* in South America" in this volume). Although practically all facilities using river, lake, or reservoir waters colonized by the mussel experience some difficulties, most acute problems affect those that use raw, untreated water, usually for cooling purposes. Water used for potabilization or closed-system cooling circuits (like the Argentine thermal power plant Termoeléctrica General Belgrano, in Campana, Buenos Aires Province) is subject to treatment (addition of chemicals, filtration) immediately after intake, thus eliminating or significantly reducing densities of mussel larvae before they reach other plant components. In this case, clogging and overgrowth problems are normally restricted to the intake piping systems, pumps, trash racks, grates, and screens. Typically, power plants (hydroelectric, thermal, nuclear) use untreated water to eliminate excess heat, and are therefore affected the most by mussel fouling (O'Neill 1997).

While problems have been described throughout the area colonized by the mussel, Brazil has been impacted more severely because of the large number of hydroelectric plants involved. As of 2013, the mussel was present in at least 33 Brazilian hydroelectric power plants along the Paraná River and its tributaries (Borges et al. 2013; Fig. 1c). Netto (2011) estimated that shutoff of a single 40 MW turbine for servicing may cost as much as US$ 17,000 per day in lost power generation alone.

It is noteworthy that not all installations using raw water from mussel-infested areas are equally prone to fouling. The mode in which water is used may largely determine whether a system will get fouled or not. For example, large gated communities in the vicinity of major cities, like Buenos Aires, often process their own water when a nearby source is available. These settlements usually have separate lines for potable water and irrigation water; the latter is only filtered, but not made potable. Thus, irrigation pipes are in principle vulnerable to mussel fouling. However, normally they are not affected because, as opposed to potable water, which is used permanently, irrigation water is used intermittently, often with long gaps during rainy weather. During these stagnant periods, water in the pipes becomes anoxic rapidly, thus killing organisms trapped inside, including *L. fortunei* larvae and settled individuals (Boltovskoy and Correa 2006).

Facilities and Components Affected

The problems caused by *L. fortunei* are practically identical to those reported for *Dreissena* species (but not the efficacy of many of the control methods; see following chapters), which allows extrapolating from the extensive literature for the dreissenids (Mackie and Claudi 2010; Prescott et al. 2014). Any facility drawing raw water from a surficial source colonized by the mussel (rivers, lakes, reservoirs) can be affected. In Asia and South America, some of the installations that reported problems associated with mussel fouling include the following (Fig. 1):

Water and wastewater processing plants. China: Suzhou; Japan: Hanshin Water Supply Authority, Lake Biwa-Yodo River system, Osaka Prefectural Water Works Department; Argentina: AySA La Plata, AySA Palermo, AySA Bernal, Aguas Santafesinas, Aguas Cordobesas; Taiwan: Jyr-Tan pumping station

Nuclear power plants. Argentina: Central Nuclear Embalse, Central Nuclear Atucha I

Hydroelectric plants. China: Shisanling (Beijing), Langyashan (Anhui Province), Tianhuangping (Zhejiang Province), Guangxu (Shenzhen); Japan: Yahagi River; South America: Itaipu (Brazil/Paraguay), Yacyretá (Argentina/Paraguay), Salto Grande (Argentina/Uruguay), Fitz Simon, Cassafousth, Reolín, Piedras Moras, San Roque, La Calera (Argentina), Constitución (Uruguay), over 30 plants on the upper Paraná River and its tributaries (Paranaíba, Aporé, Claro, etc., Brazil)

Thermal power plants. Argentina: Central Puerto

Refineries. Argentina: Shell CAPSA (Dock Sud), ESSO (Campana)

Steel mills. China: Wuhan Iron and Steel Corporation; Argentina: Acindar

Food processing plants. Argentina: Tres Cruces

Fish culture facilities. China: Longtan Reservoir in Guangxi Province; South America: Itaipu Reservoir (clogging of net cages for pacu—*Piaractus mesopotamicus*), Esturiones del Río Negro (clogging of fish farming components for sturgeon—*Acipenser baerii baerii* in Río Negro, Uruguay)
Irrigation canals. Widespread in China and Japan
Water transfer canals, pipelines, drainage systems, and aqueducts. China: Shenzhen Dongjiang, East River to Plover Cove, Xizhijiang River, and many others
Navigation dams. Brazil
Stream level gauging components. Widespread in Japan
Watercraft (commercial and leisure boats, ships). Widespread in Argentina, Brazil, Paraguay, Uruguay
Fish diversion components. Widespread in Japan

The most common raw water components affected by mussel fouling include the following (see McMahon et al. 1994; Prescott et al. 2014):

Heat exchangers and condensers (Fig. 2a, b, j, l);
Pipes (Fig. 2f, i);
Strainers, filters, trash racks, grates, screens (Fig. 2c, d, g, n);
Penstocks;
Holding ponds, storage tanks, pump suction chambers, pump wells;
Water intake tunnels (Fig. 2f);
Sand filtration systems;
Pumps, nozzles, and sprinklers;
Vent lines, and air release valves;
Fire protection equipment;
Grit chambers, flocculators (Fig. 2e);
Submerged monitoring instrumentation, level gauges;
Pump and turbine shafts, seals, and wear rings;
Boat engines (cooling water ducts, filters, pumps) and submerged rudder and propulsion components (Fig. 2h).

The problems most commonly reported are associated with the following:

Clogging (by colonies of living *L. fortunei* and/or by dead, dislodged shell clusters), pressure loss, overheating;
Corrosion, erosion, and abrasion;
Deterioration of metal, concrete and other materials from fouling by organisms associated with mussel beds (bacteria, fungi);
Wear (pump/turbine shaft seals, pumps and turbine wear rings, slurry pump seals);
Jamming of moving components, poor sealing (stop logs, valves, boat underwater rudder and propulsion components);
Sediment accumulation;
Accumulation of dead specimens (e.g., in grit chambers, flocculators);
Nuisance to bathers (in recreational areas from colonization of submerged rocks);
Promotion of *Microcystis* growth, hindering use of the waterbodies for recreation, causing fish mortality, hampering potabilization, etc.;
Pollution, water quality deterioration from decomposition of dead mussels and mussel waste products.

While the economic losses involved are probably significant, with very few exceptions detailed information on the extra costs of dealing with the golden mussel have not been reported. Fouling problems invariably involve an increase in the number of man-hours devoted to cleaning and other maintenance procedures. For

Fig. 2 Various industrial plant components fouled by *L. fortunei*. **a** and **b** Heat exchange elements clean (**a**) and fouled (**b**) (Embalse de Río Tercero nuclear power plant, Argentina). **c** and **d** Steel grate protecting the raw water intake at the Atucha I nuclear power plant (Argentina) clean (**c**) and heavily fouled (**d**). **e** Grit chamber at Yonghu Pump Station in Huizhou, Guangdong Province, China (notice accumulation of dead *L. fortunei* shells on the bottom). **f** Internal wall of water transfer pipe of the East River Water Source Project. **g** Screen at the water intake of Dongjiang

example, at the Salto Grande (Argentina/Uruguay) hydroelectric power plant each turbine is subject to a cleaning and maintenance routine every 7 years. Before the introduction of *L. fortunei* cleaning the penstock walls took 2 days of work. With the establishment of the mussel in the reservoir, the same operation now requires 10 days (Glaser 2011). Additional maintenance procedures not only involve higher costs from increased man-hour expenditures, but they also affect the lifetimes of the components that decrease from additional handling. In some cases, partial blockage of cooling systems involves operation at below-standard regimes, which may significantly affect power production and, consequently, revenues (e.g., the Central Puerto thermal power plant in Buenos Aires, Argentina). Chemical control strategies involve the costs of initial design and installation of the dosage components, and the costs of the chemical products. In many cases, detoxification of the water may be required before returning it to the lake or river, which further increases costs. Environmentally friendlier treatments, like UV, are limited by the turbidity of the water and, when viable, may use large amounts of electric power (Perepelizin and Boltovskoy 2014).

Through consulting work and personal contacts, we know that some plants have had serious fouling-related problems leading to temporary shutoffs (e.g., a water treatment plant in Suzhou, China; the nuclear power plants Atucha I and Embalse in Argentina; the hydroelectric plant Yacyretá in Argentina/Paraguay, a water treatment plant in the Yodo River area, Osaka, Japan, etc.). However, as maintenance personnel became familiar with the problem and acquired experience in curtailing mussel growth in sensitive areas, serious incidents have become less common. To our knowledge, fouling by *L. fortunei* has not caused a single definitive plant shut down. All plants have developed alternatives for curtailing the impacts of fouling and remain operational (see Chapter "Control of *Limnoperna fortunei* Fouling: Antifouling Materials and Coatings" in this volume).

Control Strategies Assayed

As of 2013, there are around 100 publications dedicated specifically to the investigation of various methods aimed at eliminating *L. fortunei* fouling in industrial installations, over half of them centered on chemical treatments and antifouling materials and coatings. Most are based on laboratory studies, and a smaller num-

Pump Station in Huizhou (Guangdong Province, China). **h** Propeller shaft and supporting structure fouled by *L. fortunei* (leisure boat, Embalse de Río Tercero, Argentina). **i** Raw water pipe (drinking water processing plant in Villa del Dique, Argentina; from Anonymous 2006). **j** Fouling of hydroelectric plant components in Itaipú (Brazil/Paraguay; from http://sosriosdobrasil.blogspot.com.ar/2011/04/praga-do-mexilhao-dourado-deixa-em.html). **k** Fouled gate slots at Xizhijiang Pump Station in Huizhou (Guangdong Province, China). **l** Fouled filters of a transformer cooling unit, Salto Grande hydroelectric plant (Argentina/Uruguay). **m** Clogged hole in butterfly valve at Xizhijiang Pump Station in Huizhou (Guangdong Province, China). **n** Steel grates protecting the raw water intake at the Embalse de Río Tercero nuclear power plant (Argentina)

ber present studies in actual operating plants. The following listing summarizes the most relevant works, grouped thematically (see also Table 1):

Antifouling materials and coatings: Stupak et al. (1996); Gemini (1999); Matsui et al. (1999); Ohkawa et al. (1999); Garcia Sola et al. (2000); Matsui et al. (2001); Nagaya et al. (2001); Ohkawa et al. (2001); Caprari and Lecot (2002); Matsui et al. (2002); Caprari (2006); Faria et al. (2006); Luo (2006); Zhuang (2006); Perez Bergmann et al. (2010a); Perez Bergmann et al. (2010b)

Chemical treatments: Morton et al. (1976); Darrigran et al. (2001); Cataldo et al. (2003); Luo et al. (2006); Zhuang (2006); Darrigran et al. (2007a); Xu et al. (2009b); Asolkar et al. (2010); Kim et al. (2011); Netto (2011); Pereyra et al. (2011); Calazans and Fernandes (2012); Calazans et al. (2012); Godoy Fernandes et al. (2012); Liu et al. (2012); Pereira and Soares (2012); Pereyra et al. (2012); Calazans et al. (2013); Netto (2013); Mata et al. (2013); Montresor et al. (2013)

Manual/mechanical cleaning: Glaser (2011); Ye et al. (2011); Ministry of Agriculture Forestry and Fisheries of Japan (2012)

Filtration: Zhuang (2006)

Thermal shock: Montalto and Marchese (2003); Perepelizin (2011); Perepelizin and Boltovskoy (2011a); Perepelizin and Boltovskoy (2011c)

Anoxia and hypoxia: Liu et al. (2006); Perepelizin (2011); Perepelizin and Boltovskoy (2011b); Ye et al. (2011)

Desiccation: Iwasaki (1997); Montalto and Ezcurra de Drago (2003); Darrigran et al. (2004)

Ozonation: Xu et al. (2009b)

Ultraviolet treatment: Santos (2011); Santos et al. (2012a); Perepelizin and Boltovskoy (2014)

Electric currents: Maeda et al. (2003); Katsuyama et al. (2005)

Ultrasound: Santos et al. (2012b)

Manipulations of flow speed: Xiang (1985); Nagaya et al. (2001); Matsui et al. (2002); Oviedo Antunes and de Madrinag (2005); Ye et al. (2011); Xu et al. (2012)

Biological control: Luo et al. (2006); Xu (2013)

Miscellaneous methods: Ratkiewicz (2006); Padula Paz et al. (2012); Rackl et al. (2012); Xu (2013); Xu et al. (2013); Dengo and Carraro (2013)

These methods are treated in detail in the sections that follow.

Acknowledgements This work was partially financed by grants from the University of Buenos Aires, Argentina (UBA X-020 and 20020100100035) and from the Argentine Agencia Nacional de Promoción Científica y Tecnológica, Argentina (PICT 2007 1968) to DB.

References

Akehoshi Y (2011) Study of ecological features and prevention of damage of *Limnoperna fortunei* in the irrigation institution. Rep Yahagi Riv Inst 15:55–63 [In Japanese]

Anonymous (2006) El mejillón dorado (*Limnoperna fortunei*). Los serios problemas que origina en la planta potabilizadora. NotiCoop (Boletín Informativo de la Cooperativa de Obras y Servicios Públicos Limitada de Villa del Dique, Provincia de Córdoba, Argentina) 42:22–23

Asolkar R, Dow H, Huang M, Koivunen D, Marrone P, Shu S (2010) Chemical and biological agents for the control of molluscs. The United States Patent and Trademark Office (Pub. No.: 2010/0266717A1)

Bendati MMA, Suminski M, Colares ERC (2004) Avaliação da operação dos sistemas de abastecimento de água do DMAE (Porto Alegre, Brasil) em decorrência da presença do mexilhão-

dourado (*Limnoperna fortunei*). In: I Congresso Interamericano de Saúde Ambiental, Porto Alegre (Brazil)
Boltovskoy D, Correa G (2006) Informe sobre la vulnerabilidad potencial de los sistemas de agua de riego de Nordelta a las incrustaciones del bivalvo *Limnoperna fortunei*. Nordelta, pp 1–77
Boltovskoy D, Correa N (2015) Ecosystem impacts of the invasive bivalve *Limnoperna fortunei* (golden mussel) in South America. Hydrobiologia. 746:81–95
Borges P, Mäder O, Ludwig S, Duarte J, Bastos L, Lima F, Cardoso C, Nakayama D, Borba R (2013) Atual distribuição do molusco invasor *Limnoperna fortunei* (Dunker, 1857) no Brasil: registros em usinas hidrelétricas. In: 1 Congreso Argentino de Malacología, La Plata (Argentina)
Brugnoli E, Russo R, Mandiá M, Muñiz P (2012) Estudios para la mitigación del macrofouling ocasionado por especies invasoras acuáticas en áreas industriales y de servicio en Uruguay. In: Aber A, Ferrari G, Porcile JF, Rodríquez E, Zerbino S (eds) Identificación de prioridades para la gestión nacional de las especies exóticas invasoras, Comité Nacional de Especies Exóticas Invasoras, Montevideo, pp 62–83
Calazans SHC, Fernandes FC (2012) Microencapsulados. In: Mansur MCD, Santos CP, Pereira D, Padula Paz IC, Leite Zurita ML, Raya Rodriguez MT, Vilar Nehrke M, Aydos Bergonci PE (eds) Moluscos límnicos invasores no Brasil. Biologia, prevenção, controle, Redes Editora, Porto Alegre, pp 299–302
Calazans SHC, Godoy Fernandes LV, Fernandes FC (2012) Outros compostos. In: Mansur MCD, Santos CP, Pereira D, Padula Paz IC, Leite Zurita ML, Raya Rodriguez MT, Vilar Nehrke M, Aydos Bergonci PE (eds) Moluscos límnicos invasores no Brasil. Biologia, prevenção, controle, Redes Editora, Porto Alegre, pp 311–315
Calazans SHC, Americo JA, Fernandes FD, Aldridge DC, Rebelo MD (2013) Assessment of toxicity of dissolved and microencapsulated biocides for control of the golden mussel *Limnoperna fortunei*. Mar Environ Res 91:104–108
Caprari JJ (2006) Pinturas antiincrustantes. In: Darrigran G, Damborenea C (eds) Bio-invasión del mejillón dorado en el continente americano. Editorial de la Universidad de La Plata, La Plata, pp 183–213
Caprari JJ, Lecot CJ (2002) El control de bivalvos invasores *Limnoperna fortunei* (Dunker, 1857) en la central hidroeléctrica Yaciretá mediante pinturas. In: Seminario Internacional—Gestión Ambiental e Hidroelectricidad, Salto Grande
Cataldo D, Boltovskoy D, Pose M (2003) Toxicity of chlorine and three non-oxidizing molluscicides to the invasive pest mussel *Limnoperna fortunei*. J Am Water Work Assoc 95:66–78
Cepero E (2003) Control de *Limnoperna fortunei*. Infestación de plantas potabilizadoras de agua. In: 13 Congreso Argentino de Saneamiento y Medio Ambiente, Buenos Aires (Argentina)
Claudi R, Mackie GL (1994) Controlling zebra mussels at water treatment plant intakes. CRS Press, Boca Raton, pp 1–227
Colares ERC, Suminski M, Bendati MMA (2002) Diagnóstico e controle do mexilhão-dourado, *Limnoperna fortunei*, em sistemas de tratamento de água em Porto Alegre (RS, Brasil). In: VI Simpósio Ítalo-Brasileiro de Engenharia Sanitária e Ambiental, Vitória (Brazil)
Darrigran GA (1995) *Limnoperna fortunei*: ¿un problema para los sistemas naturales de agua dulce del Mercosur? Museo (Revista de la Fundación Museo de La Plata "Francisco Pascasio Moreno") 5:85–87
Darrigran GA, Maroñas ME, Colautti DC (2001) Primeras estimaciones de concentraciones letales de un biocida para el molusco invasor *Limnoperna fortunei* (Mytilidae). In: Seminario Internacional—Gestión Ambiental e Hidroelectricidad, Salto Grande
Darrigran GA, Penchaszadeh PE, Damborenea C, Greco N (2002) Abundance and distribution of golden mussel (*Limnoperna fortunei*) larvae in a hydroelectric power plant in South America. In: 11th International Conference on Aquatic Invasives Species, Alexandria
Darrigran GA, Maroñas ME, Colautti DC (2004) Air exposure as a control mechanism for the golden mussel, *Limnopema fortunei*, (Bivalvia: Mytilidae). J Freshw Ecol 19:461–464
Darrigran GA, Colautti DC, Maroñas ME (2007a) Potential biocide for control of the golden mussel, *Limnoperna fortunei*. J Freshw Ecol 22:359–360

Darrigran GA, Damborenea C, Greco N (2007b) An evaluation pattern for antimacrofouling procedures: *Limnoperna fortunei* larvae study in a hydroelectric power plant in South America. Ambio 36:575–579

Dengo RR, Carraro I (2013) Controle preventivo do mexilhão dourado em sistemas de abastecimento de água da CORSAN. Trabalho prévio de seleção de tecnologias de controle. CORSAN Report. Porto Alegre (Brazil), pp 1–38

Faria EA, Tavares Branco JR, Campos MCS, Oliveira MD, Rolla ME (2006) Estudo das características antiincrustantes de materiais. Relatorio Escola de Minas, Ouro Preto 59:233–238

Figueiredo de Resende M, Martinez CB (2008) Impacto da infestação de condutos forçados de PCH's pelo *Limnoperna fortunei*. In: VI Simpósio Brasileiro sobre Pequenas e Médias Centrais Hidrelétricas, Belo Horizonte (Brazil)

Garcia Sola E, Leiva Azuaga A, Niveyro G (2000) Evolución de los moluscos invasores "*Limnoperna fortunei*" en la región. Universidad Nacional del Nordeste. Comunicaciones Científicas y Tecnológicas de la Universidad Nacional del Nordeste, Universidad Nacional del Nordeste, pp 1–4

Gemini H (1999) Ensayo con pintura antifouling en calle UE2. Nuceoeléctrica Argentina Sociedad Anónima, Central Nuclear Atucha I. Lima (Argentina), pp 1–12

Glaser R (2011) Descripción de la invasión del mejillón dorado en instalaciones mecánicas de la represa de Salto Grande. In: Jornada de Difusión, Programas Ambientales de Salto Grande, Salto Grande (Uruguay)

Godoy Fernandes LV, Calazans SHC, Fernandes FC (2012) Formas de Cloro. In: Mansur MCD, Santos CP, Pereira D, Padula Paz IC, Leite Zurita ML, Raya Rodriguez MT, Vilar Nehrke M, Aydos Bergonci PE (eds) Moluscos límnicos invasores no Brasil. Biologia, prevenção, controle, Redes Editora, Porto Alegre, pp 303–306

Goto Y (2002) Behavior of nuisance mussel, *Limnoperna fortunei*, in water supply facilities. Water Sci Technol 46:45–50

GPS (Pipeline Study Group) (1973) Biology of golden mussel (*Limnoperna fortunei*). Animal's use and prevention. Institute of Aquatic Organisms of the Chinese Academy of Sciences, Beijing, pp 33–36 [In Chinese]

Guangdong Government (2010) The pearl river delta infrastructure integration planning (2009–2020). http://www.gd.gov.cn/govpub/rdzt/ghgy2/wjzl/201008/t20100816_127497.htm. Accessed: 20 Jan 2014 [In Chinese]

Hamada M (2008) The ecology of the golden mussel that has Invaded the Yahagi River System. Technol Dev News 130:21–22 [In Japanese]

Hamada M (2010) History of invasion and present status of the golden mussel *Limnoperna fortunei* in Central Japan. Sessile Org 27:7–16 [In Japanese]

Huang CW (2008) Population genetic structure of *Limnoperna fortunei* (Dunker, 1857) in Taiwan based on sequences of mitochondrial cytochrome C oxidase subunit I. MSc Thesis, National Changhua University of Education (Taiwan), pp 1–60

Iwasaki K (1997) Climbing behaviour and tolerance to aerial exposure of a freshwater mussel, *Limnoperna fortunei*. Venus (Jpn J Malacol) 56:15–25

Katayama M, Shimizu R, Matsumoto H (2005) The first record of *Limnoperna fortunei* (Bivalvia, Mytilidae) in Gunma. Field Biol 14:35–40 [In Japanese]

Katsuyama I, Satuito CG, Maeda T, Oonishi M, Kumagai T (2005) The effect of DC-pulse electric stimulus on the swimming behavior of larvae of the freshwater mussel *Limnoperna fortunei* in flowing water within a pipe. Sessile Org 2:1–5 [In Japanese]

Kim D, Suok H, Eunjae J, Changsoo S, Youngbeom Y, Seungkwan H (2011) Prechlorination at water intake for the quality improvement of raw water. J Korean Soc Water Qual 27:110–114 [In Korean]

Kimura T (1994) The earliest record of *Limnoperna fortunei* (Dunker) from Japan. Chiribotan (J Malacol Soc Jpn) 25:34–35 [In Japanese]

Kojima S (1982) The trouble by fresh water shell (*Limnoperna fortunei*) and its control. Jpn J Water Treat Biol 18:29–33 [In Japanese]

Kuroda T (1941) A catalogue of molluscan shells from Taiwan (Formosa), with descriptions of new species. Mem Fac Sci Agric, Taihoku Imp Univ 22:65–216

Li MJ, Su XM (2007) Analysis and evaluation of countermeasures to mussel fouling in the long distance water transport canal, Pearl River. Pearl River 3:29–30 [In Chinese]

Liu L, You Z, Luo F, Zhang J, Ouyang S (2006) Killing and removing *Limnoperna fortunei* from raw water pipe by isolating oxygen. Water Wastewater Eng 22:40–42, 47

Liu D, Hong J, Wang R, Cui F (2011a) Current solution to *Limnoperna fortunei* problem in water and pipelines. In: Second international conference on artificial intelligence, management science and electronic commerce (Aimsec 2011), Deng Feng

Liu D, Wang R, Cui F, Zhao Y, Hong J, Chen W, Wang C (2011b) The harm and control technology of *Limnoperna fortunei*. In: International conference on remote sensing, environment and transportation engineering, Nanjing [In Chinese]

Liu D-M, Wang R, Hong J, Cui F-Y, Chen W-X, Wang C-H, Chen C-X (2012) Experimental studies of inactivation effect on *Limnoperna fortunei* with potassium permanganate. J Harbin Inst Technol 44:46–49

Lou KH, Liu J (1958) Study on prevention and extermination of mussel growths in pipelines. Oceanologia et Limnologia Sinica 1:316–324 [In Chinese]

Luo FM (2006) Studies on biology and control techniques of Limnoperna fortunei (Duker) in water supply system of Shenzhen city. MSc Thesis, Nanchang University (China), pp 1–108 [In Chinese]

Luo F, Liu L, You Z, Huang W, Deng H, Ouyang S (2006) Study on solvability of *Limnoperna fortunei* byssus cohering in raw water pipe. Water Wastewater Eng 3:29–32 [In Chinese]

Mackie GL, Claudi R (2010) Monitoring and control of macrofouling mollusks in fresh water systems. CRC Press, Boca Raton, pp 1–508

Maeda T, Matsui N, Onishi M (2003) Development of prevention system against pollution by bivalve, *Limnoperna lacustris*. R & D News Kansai 417:20–21

Magara Y, Matsui Y, Nagaya K, Yuasa A (1999) Japanese experience with the freshwater adhesive mussel, *Limnoperna fortunei*. In: 9th international zebra Mussel and aquatic nuisance species conference, Duluth

Magara Y, Matsui Y, Goto Y, Yuasa A (2001) Invasion of the non-indigenous nuisance mussel, *Limnoperna fortunei*, into water supply facilities in Japan. J Water Supply: Res Technol—Aqua 50:113–124

Mata FAR (2011) Abundância e distribuição temporal de *Limnoperna fortunei* Dunker, 1857 (Mollusca, Bivalvia) e os impactos da incrustação em usinas geradoras de energia elétrica. MSc Thesis, Universidade Federal de Ouro Preto (Brazil), pp 1–91

Mata FAR, Dias C, Mader O (2013) Utilização do remediador MXD-100 para controle de incrustação em sistemas de resfriamento de usinas hidrelétricas, ocasionada por *Limnoperna fortunei* (Dunker, 1857). In: 1 Congreso Argentino de Malacología, La Plata (Argentina)

Matsui Y, Nagaya K, Yuasa A, Yamamoto H, Ohkawa K, Magara Y (1999) Control efficacy of the Asian freshwater mussel *Limnoperna fortunei* for coating and materials. In: 9th international zebra mussel and aquatic nuisance species conference, Duluth

Matsui Y, Nagaya K, Yuasa A, Naruto H, Yamamoto H, Ohkawa K, Magara Y (2001) Attachment strength of *Limnoperna fortunei* on substrates, and their surface properties. Biofouling 17:29–39

Matsui Y, Nagaya K, Funahashi G, Goto Y, Yuasa A, Yamamoto H, Ohkawa K, Magara Y (2002) Effectiveness of antifouling coatings and water flow in controlling attachment of the nuisance mussel *Limnoperna fortunei*. Biofouling 18:137–148

McMahon RF, Ussery TA, Clarke M (1994) Review of zebra mussel control methods (Technical Note ZMR-2-14). US Army Corps of Engineers, Waterways Experiment Station (Vicksburg, MS), Report ZMR-2-14. Vicksburg, pp 1–12

Ministry of Agriculture Forestry and Fisheries of Japan (2012) Practical manual for *Limnoperna fortunei*. http://www.maff.go.jp/j/nousin/kankyo/kankyo_hozen/k_hozen/pdf/kawahibarimanual.pdf. Accessed: 9 May 2014 [In Japanese]

Montalto L, Ezcurra de Drago I (2003) Tolerance to desiccation of an invasive mussel, *Limnoperna fortunei* (Dunker, 1857) (Bivalvia, Mytilidae), under experimental conditions. Hydrobiologia 498:161–167

Montalto L, Marchese M (2003) *Limnoperna fortunei* (Dunker, 1857) (Bivalvia, Mytilidae) tolerance to temperature and pH in experimental conditions. Neotropica 49:26–34

Montresor LC, Miranda-Filho KC, Paglia A, Luz DM, Araujo JM, Silva MJS, Gerhard L, Martinez CB, Vidigal TH (2013) Short-term toxicity of ammonia, sodium hydroxide and a commercial biocide to golden mussel *Limnoperna fortunei* (Dunker, 1857). Ecotoxicol Environ Saf 92:150–154

Morton B (1973) Some aspects of the biology and functional morphology of the organs of feeding and digestion of *Limnoperna fortunei* (Dunker) (Bivalvia: Mytilacea). Malacol 12:265–281

Morton B (1975) The colonization of Hong Kong's raw water supply system by *Limnoperna fortunei* (Dunker 1857) (Bivalvia: Mytilacea) from China. Malacol Rev 8:91–105

Morton B, Dinesen G (2010) Colonization of Asian freshwaters by the Mytilidae (Bivalvia): a comparison of *Sinomytilus harmandi* from the Tonle-Sap River, Phnom Penh, Cambodia, with *Limnoperna fortunei*. Molluscan Res 30:57–72

Morton B, Au CS, Lam WW (1976) Control of *Limnoperna fortunei*. J Inst Water Eng Sci 30:147–156

Nagaya K, Matsui Y, Ohira H, Yuasa A, Yamamoto H, Ohkawa K, Magara Y (2001) Attachment strength of an adhesive nuisance mussel, *Limnoperna fortunei*, against water flow. Biofouling 17:263–274

Nakai K, Niimura Y, Yamada J (1994) Distribution of the freshwater mytilid mussel, *Limnoperna fortunei*, in the rivers, Nagara-gawa and Ibi-gawa, central Japan. Venus (Jpn J Malacol) 53:139–140 [In Japanese]

Nakanishi M, Mukai S (1997) Growth and extermination of *Limnoperna fortunei* (Dunker 1857) at water purification facilities. Yousui to Haisui [J Water Waste] 39:1017–1020 [In Japanese]

Nalepa TF, Schloesser DW (eds) (2014) Quagga and zebra mussels: biology, impacts, and control, 2nd edn. CRC Press, Boca Raton, pp 1–815

Netto OSM (2011) Controle da incrustação de organismos invasores em materiais de sistemas de resfriamento de usinas hidrelétricas. MSc Thesis, Universidade Federal de Curitiba (Brazil), pp 1–113

Netto OSM (2013) Impactos econômicos e métodos de controle de moluscos invasores em plantas industriais. In: 1 Congreso Argentino de Malacología, La Plata (Argentina)

O'Neill JR (1997) Economic impact of zebra mussels: results of the 1995 national zebra mussel information clearinghouse study. Gt Lakes Res Rev 3:35–42

Ohkawa K, Nishida A, Honma R, Matsui Y, Nagaya K, Yuasa A, Yamamoto H (1999) Studies on fouling by the freshwater mussel *Limnoperna fortunei* and the antifouling effects of low energy surfaces. Biofouling 13:337–350

Ohkawa K, Ichimiya K, Nishida A, Yamamoto H (2001) Synthesis and surface chemical properties of adhesive protein of the Asian freshwater mussel, *Limnoperna fortunei*. Macromol Biosci 1:376–386

Oviedo Antunes H, de Madrinag TG (2005) La presencia del bivalvo *Limnoperna fortunei* (mejillón dorado) en la Central Hidroeléctrica de Itaipu. Itaipú Binacional. Informe Técnico. Itaipu (Brazil-Praguay), pp 1–7

Padula Paz IC, Pereira D, Souza AMS (2012) Controle Microbiano. In: Mansur MCD, Santos CP, Pereira D, Padula Paz IC, Leite Zurita ML, Raya Rodriguez MT, Vilar Nehrke M, Aydos Bergonci PE (eds) Moluscos límnicos invasores no Brasil. Biologia, prevenção, controle, Redes Editora, Porto Alegre, pp 351–356

Pastorino G, Darrigran GA, Martín SM, Lunaschi L (1993) *Limnoperna fortunei* (Dunker, 1857) (Mytilidae), nuevo bivalvo invasor en aguas del Río de la Plata. Neotropica 39:101–102

Pereira D, Soares MF (2012) Sulfato de Cobre. In: Mansur MCD, Santos CP, Pereira D, Padula Paz IC, Leite Zurita ML, Raya Rodriguez MT, Vilar Nehrke M, Aydos Bergonci PE (eds) Moluscos límnicos invasores no Brasil. Biologia, prevenção, controle, Redes Editora, Porto Alegre, pp 307–309

Perepelizin PV (2011) Alternativas ambientalmente inocuas (no químicas) para el control del mejillón incrustante *Limnoperna fortunei* en instalaciones industriales. PhD Thesis, Universidad de Buenos Aires (Argentina), pp 1–126
Perepelizin PV, Boltovskoy D (2011a) Hot water treatment (chronic upper lethal temperature) mitigates biofouling by the invasive Asian mussel *Limnoperna fortunei* in industrial installations. Environ Sci Technol 45:7868–7873
Perepelizin PV, Boltovskoy D (2011b) Resistance of the invasive pest mussel *Limnoperna fortunei* to anoxia. J Am Water Works Assoc 103:79–85
Perepelizin PV, Boltovskoy D (2011c) Thermal tolerance of *Limnoperna fortunei* to gradual temperature increase and its applications for biofouling control. Biofouling 27:667–674
Perepelizin PV, Boltovskoy D (2014) Effects of 254 nm UV irradiation on the mobility and survival of larvae of the fouling invasive mussel *Limnoperna fortunei*. Biofouling 30:197–202
Pereyra PJ, Rossini GB, Darrigran GA (2011) Toxicity of three commercial tannins to the nuisance invasive species *Limnoperna fortunei* (Dunker, 1857): implications for control. Fresenius Environ Bull 20:1432–1437
Pereyra PJ, Rossini GB, Darrigran GA (2012) Toxicity of Neem's oil, a potential biocide against the invasive mussel *Limnoperna fortunei* (Dunker 1857). Anais da Academia Brasileira de Ciências 84:1065–1071
Perez Bergmann C, Mansur MCD, Bergonci PEA, Pereira D, Santos CP, Basegio T, Vicenzi J, Santos SCA (2010a) Seleção de materiais e revestimentos para o controle de incrustações do mexilhão dourado na Usina Hidrelétrica de Ibitinga (SP, Brasil). Revista Matéria 15:21–30
Perez Bergmann C, Mansur MCD, Pereira P, Bergonci PEA, Santos CP, Basegio T, Vicenzi J, Santos SCA (2010b) Avaliação de materiais e revestimentos para o controle de incrustações do mexilhão dourado *Limnoperna fortunei* (Dunker, 1857). Biotemas 23:87–100
Pilsbry HA, Hirase Y (1905) Catalogue of the land and fresh-water Mollusca of Taiwan (Formosa), with descriptions of new species. Proc Acad Nat Sci Phila 57:720–752
Portella Kleber F, Joukoski A, Sheffer da Silva A, Machuca Brassac N, Belz CE (2009) Biofouling e biodeterioração química de argamassa de cimento portland em reservatório de usina hidroelétrica. Quimica Nova 32:1047–1051
Prescott TH, Claudi R, Prescott KL (2014) Impact of dreissenid mussels on the infrastructure of dams and hydroelectric power plants. In: Nalepa TF, Schloesser DW (eds) Quagga and zebra mussels: biology, impacts, and control, 2nd edn. CRC Press, Boca Raton, pp 244–257
Rackl S, Koivunen M, Marrone P, Huang H, Asolkar R (2012) Agents for the control of *Limnoperna* sp. United States Patent and Trademark Office (Pub. No. 20120121745A1)
Rajagopal S, Jenner HA, Venugopalan VP (eds) (2012) Operational and environmental consequences of large industrial cooling water systems. Springer, New York, pp 1–477
Ratkiewicz MLA (2006) Relatório da experiência piloto dos equipamentos da LGM-DANSUL para controle da infestação por mexilhões dourados na EBAB e incrustações na adutora de água bruta da US de Barra do Ribeiro. DEPEDET/SUMOP/CORSAN. Porto Alegre, pp 1–18
Rolla ME, Mota HR (2010) Response of a major Brazilian utility to the golden mussel invasion. In: Mackie GL, Claudi R (eds) Monitoring and control of macrofouling mollusks in fresh water systems. CRC Press, Boca Raton, pp 396–403
Santos CP (2011) Desenvolvimento de metodologia para controle das larvas de *Limnoperna fortunei* com o uso de radiação ultravioleta e seus impactos sobre *Microscystis aeruginosa* potencialmente presentes na água superficial. PhD Thesis, Universidade Federal do Rio Grande do Sul (Brazil), pp 1–83
Santos CP, Raya Rodriguez MT, Mansur MCD, Vilar Nehrke M, Leite Zurita ML (2012a) Controle de *Limnoperna fortunei* com aplicação de radiação ultravioleta. In: Mansur MCD, Santos CP, Pereira D, Padula Paz IC, Leite Zurita ML, Raya Rodriguez MT, Vilar Nehrke M, Aydos Bergonci PE (eds) Moluscos límnicos invasores no Brasil. Biologia, prevenção, controle, Redes Editora, Porto Alegre, pp 335–338
Santos CP, Vicenzi J, Berutti FA, Mansur MCD, Perez Bergmann C, Raya Rodriguez MT, Vilar Nehrke M, Leite Zurita ML (2012b) Controle de bivalves com a utilização do ultrassom. In: Mansur MCD, Santos CP, Pereira D, Padula Paz IC, Leite Zurita ML, Raya Rodriguez MT,

Vilar Nehrke M, Aydos Bergonci PE (eds) Moluscos límnicos invasores no Brasil. Biologia, prevenção, controle, Redes Editora, Porto Alegre, pp 339–341
Sawada T, Nakamura Y (2010) Breeding damages and adhesive characteristics of *Limnoperna fortunei* in an irrigation canal. In: Rural engineering society national conference, Tokyo [In Japanese]
Simeão CMG, Barreira Martinez C, Formagio PS (2006) *Limnoperna fortunei* : situação atual e perspectivas futuras. In: V Simpósio Brasileiro Sobre Pequenas e Médias Centrais Hidrelétricas, Florianópolis (Brazil)
Sprecher SL, Getsinger KD (2000) Zebra mussel chemical control guide. Erdc/El Tr-00-1, U.S. Army Engineer Research and Development Center, Vicksburg (USA). Erdc/El Tr-00-1, U.S. Army Engineer Research and Development Center, Vicksburg (USA), pp 1–89
Stupak M, Pérez MC, García MT, Solá EG, Azuaga AL, Mendivil A, Niveyro G (1996) Preliminary study of the biofouling of the Parana River (Argentina). Corros Rev 14:145–155
Tan T-H, Pai J-Y, Hsha KC (1987) The recovery of fouling clam, *Limnoperna fortunei* from Taiwan. Bull Malacol, Repub China 13:97–100 [In Chinese]
Tchang S, Li SC, Liu YY (1965) Bivalves (Mollusca) of Tung-Ting Lake and its surrounding waters, Hunan Province, China. Acta Zoologica Sinica 17:212–213 [In Chinese]
Xiang YL (1985) Study of the harm and prevention measures of mussel growth in pipelines in power generation plants. N China Electr Power 4:24–27 [In Chinese]
Xu M (2013) Experimental study of bio-fouling control of *Limnoperna fortunei* in water transfer tunnels. In: 35 international association of hydraulic engineering & research congress, Chengdu
Xu M, Wang ZY, Duan XH, Zhuang MQ, de Souza FT (2009a) Ecological measures of controlling invasion of golden mussel (*Limnoperna fortunei*) in water transfer systems. In: 33 international association of hydraulic engineering & research congress: water engineering for a sustainable environment, Vancouver
Xu M, Wang ZY, Duan XZ (2009b) Prevention and treatment of the freshwater mussel in water-transmission line. Water Wastewater Eng 35:205–208
Xu M, Cao XW, Wang ZY, Wang XZ (2012) Attachment characteristics of golden mussels (*Limnoperna fortunei*) in water transport projects. J Tsinghua Univ [Qinghua Daxue Xuebao] 52:170–176 [In Chinese]
Xu M, Wang ZY, Lin CC, Pan BZ, Zhao N (2013) Experimental study of invasion and biofouling of freshwater mussel *Limnoperna fortunei*. Int J Geosci 4:1–7
Yao GY, Xu M (2013) *Limnoperna* fouling and water quality impact in Guangxu water-storage hydropower station in Shenzhen. Tsinghua University, Research Project Proposal Tsinghua, China, pp 1–18 [In Chinese]
Ye BM, Cao XW, Xu M, Wang ZY, Lin CC (2011) Study of *Limnoperna fortunei* invasion in long distance water transmission project. Water Wastewater Eng 37:99–102 [In Chinese]
Yoshida M (2006) Case study for removing the *Limnoperna fortunei* (Dunker, 1856) from the irrigation tunnel for canal in Kaburagawa LID. J Jpn Soc Irrig Drain Reclam Eng 74:435–436 [In Japanese]
Zhuang MQ (2006) Study of growth rhythm and controlling of mollusks in long distance water transfer pipelines. Shenzhen (China). East river water source project management division of Shenzhen. Report, pp 1–142 [In Chinese]

Control of *Limnoperna fortunei* Fouling: Antifouling Materials and Coatings

Kousaku Ohkawa and Takaomi Nomura

Abstract Biochemical and molecular biological studies on *Limnoperna fortunei* adhesive proteins indicate that marine and freshwater mussels share several essential molecular traits involving oxidation of tyrosine residues to generate both adhesive force (surface coupling) and cohesive force (protein cross-linking in adhesive plaque matrices). The amino acid compositions of freshwater mussels show higher levels of electrolytic residues, such as aspartic and glutamic acids. With the purpose of developing antifouling strategies based on the knowledge of the adhesive mechanism of *L. fortunei*, several laboratory experiments have been conducted that minimize polar or hydrogen-bonding surface components. Chemical modifications and substrate coatings were investigated to search for low-energy surfaces and examine biocidal effects of metal ions. Field experiments indicate that water flow velocity is correlated with the attachment force of young, newly recruited mussels, and provide useful information on sites most vulnerable to mussel fouling in water treatment facilities.

Keywords *Limnoperna fortunei* · Golden mussel · Biological invasion · Biofouling control · Impact · Industrial plants · Power plants · Irrigation facilities · Antifouling coatings · Antifouling materials

Introduction

The Asian freshwater mussel, *Limnoperna fortunei*, is a well-known nuisance organism which attaches at tremendous numbers to man-made structures in freshwater systems (Boltovskoy et al. 2006; Oliveira et al. 2006). The extracellular organ

K. Ohkawa (✉)
Division of Biological and Medical Fibers (DBMF), Institute for Fiber Engineering (IFES), Interdisciplinary Cluster for Cutting Edge Research (ICCER),
Shinshu University, Tokida 3-15-1, Ueda, NAGANO, Japan
e-mail: kohkawa@shinshu-u.ac.jp

T. Nomura
Division of Applied Biology, Faculty of Textile Science and Technology,
Shinshu University, Tokida 3-15-1, Ueda, NAGANO, Japan
e-mail: nomurat@shinshu-u.ac.jp

D. Boltovskoy (ed.), *Limnoperna fortunei,* Invading Nature - Springer Series in Invasion Ecology 10, DOI 10.1007/978-3-319-13494-9_22

of the mussel used for underwater attachment, the byssus, is composed of proteinaceous threads. Each thread comprises an adhesive plaque, a distal thread, and a proximal thread, each of which is considered to have its own specific function in order to achieve higher resistance to detachment from the underwater substrata (Waite et al. 1998). The adhesive plaque is also referred to as the adhesive pad, holding the distal end of each thread to the substrate surface (Waite and Qin 2001). Inside the adhesive plaque matrix, many fibrils near the distal end of the byssal thread radiate apart from each other, and then are fixed by glue proteins to make a pad-like structure (Ohkawa et al. 2009). Chemical modification of the glue protein molecules results in underwater adhesion (Waite and Andersen 1980). Glue proteins are stored in a special organ called the phenol gland near the tip of the mussel's foot (Waite 1990).

Biochemical characterization of the glue proteins is important to devise antifouling strategies aimed at curtailing *L. fortunei* fouling, because the glue protein molecules directly contact the substrate surface (Ohkawa et al. 2001). Investigations of molecular-level interactions between glue proteins and substrate coatings are profitable research areas that can improve the antifouling abilities of the coatings or chemically modified surfaces. Investigations on glue proteins are proceeding for marine mussels, especially for species of Mytilidae, which include *Mytilus edulis*, *M. galloprovincialis*, *M. californianus*, etc. (Waite et al. 1985; Rzepecki et al. 1991). Oxidation processes in the byssus precursor proteins have been suggested by several histochemical methods beginning in the 1950s (Brown 1950), and the first molecular-level understanding was brought about in the early 1980s (Waite 1983), using *M. edulis* foot samples.

The *M. edulis* glue protein, Mefp-1 (*M. edulis* foot protein), has been shown to have some interesting specific characteristics. Mefp-1 consists of periodic amino acid sequences composed of 6–10 residues. These are not completely identical, but rather they are homologous and usually referred to as "motifs." Oxidized prolines (L-*trans*-4-hydroxyproline, 4-Hyp and another unidentified Hyp) and Tyr (L-β-3,4-dihydroxyphenyl-α-alanine, DOPA) are involved in all the periodic amino acid sequences investigated so far. In 1994, another Hyp was identified as L-trans-2,3-*cis*-3,4-dihydroxyproline (Taylor et al. 1994). The IUPAC nomenclature of DOPA indicates that DOPA has one catechol (diphenol) side chain, which can reduce oxidative chemicals. The first discovery of Mefp-1 and a series of related studies evoked a systematic understanding of the chemical structures of the glue proteins from mussels. DOPA undergoes further oxidation to bring about bimolecular coupling, resulting in intermolecular cross-linkages *via* covalent bonds between the glue proteins inside the adhesive plaque matrix (Stewart et al. 2011). DOPA can interact with metal oxide components of the substrate surface to produce relatively strong adhesive forces between the adhesive plaque and the substrate. This process is called "surface coupling" (Waite et al. 1992). Amino acid side chains other than DOPA also contribute to both cohesion and adhesion through hydrophobic interactions, electrostatic associations, and hydrogen-bond formation (Kamino 2008).

Limnoperna fortunei Foot Protein-1

A candidate glue protein in the foot of the Asian freshwater mussel, *L. fortunei*, was first purified in 1999 (Ohkawa et al. 1999b). The DOPA-containing protein from the *L. fortunei* foot was named Lffp-1. The purification procedure usually requires a screening method to detect DOPA, which is a characteristic amino acid of glue proteins. Detection of DOPA has been done by silver-staining, whereby DOPA-containing protein reduces ammoniacal silver nitrate to yield a brownish band using sodium dodecyl sulfate-polyacrylamide gel electrophoresis (SDS-PAGE) (Ohkawa et al. 1999b). Nitroblue tetrazolium staining for flavin-containing proteins was used to detect DOPA (Paz et al. 1991), and this method has become popular to assay for DOPA-containing proteins in other animals (Stewart et al. 2004; Zhao et al. 2005).

Lffp-1 was first extracted from foot tissue using an acetic acid solution containing high concentration of urea, and then condensed by salting out with 26% ammonium sulfate. Subsequent gel filtration and reversed-phase high performance liquid chromatography yielded an Lffp-1 preparation with an electrophoretic homogeneity. The estimated molecular mass of Lffp-1 was 96 kDa determined by electrophoretic mobility.

DOPA-containing proteins have been isolated from the foot of the zebra mussel, *Dreissena polymorpha* (Rzepecki and Waite 1993b, a), and the foot protein, Dpfp-1, has amino acid sequence patterns different from those found in marine mussels (Anderson and Waite 2001). Table 1 indicates the consensus amino acid sequences (motifs) found in the foot proteins of six marine and two freshwater mussels. Shaded residues represent the conserved identities on eight motifs. The fp-1-type proteins are composed of the repetitive motifs as "–KP*–Y*—Y*K", where *P* and *Y* denote Hyp/Pro and DOPA/Tyr, respectively. Other amino acids are specific for each species; for example, in the green mussel *Perna viridis*, (2)-hexosylated tryptophan and (2)-hexosylated hydroxytryptophan (W*) have been found recently (Zhao et al. 2009). A remarkable difference in amino acid adaptations other than the conserved ones is that non-charged S, T, G, A, and P are found in marine mussels, while charged D and E are frequently found in freshwater mussels (*L. fortunei* and *D. polymorpha*).

Limnoperna fortunei Foot Protein-2

The fp-2-type protein family is another group of DOPA-containing proteins in mussels (Rzepecki et al. 1992). In 1993, a protein named Dpfp-2 was discovered in the *D. polymorpha* foot (Rzepecki and Waite 1993b), and later, a cDNA microarray was constructed (Xu and Faisal 2009). One of the EST-sequences was assigned as the Dpfp-2 gene, which means the entire structure of Dpfp-2 was revealed. The repetitive motifs in the deduced amino acid sequence were, however, rather different from the fp-2-type proteins of marine mussels, which are homologous to the EGF factors (Inoue et al. 1995). Therefore, the numbering of the fp-2 in *D. polymorpha* is

Table 1 Repetitive sequences of fp-1-type glue proteins from various mussels. Amino acids are denoted with one-letter codes. X, hydrophobic residues, Z, oligopeptides. *Boldened letters* represent electrolytic residues, E and D, which are glutamic and aspartic acids, respectively. *Shadowed* are common parts of the sequences. "–" are gaps to match the sequences. *Y* and *P* are DOPA/Tyr and Pro/Hyp, respectively. For W*, see text. (Sources, *Perna viridis*: Ohkawa et al. (2004), Zhao et al. (2009); *Mytilus edulis* and *M. californianus*: Waite et al. (1985); *Brachidontes exustus* and *Septifer bifurcates*: Rzepecki et al. (1991); *Geukensia demissa*: Waite et al. (1989); *Dreissena polymorpha*: Rzepecki and Waite (1993b); *Limnoperna fortunei*: Ohkawa et al. (1999b))

	Sequences																
Species	1	2	3	4	5	6	7	8	9	10	11	12	13	14	15	16	17
Perna viridis	A	*P*	*P*	K	*P*	–	–	–	–	–	W*	T	A	–	W*	–	K
Mytilus edulis	A	–	–	K	*P*	S	–	*Y*	P	P	–	T	–	–	*Y*	–	K
M. californianus	P	–	–	K	X	T	–	*Y*	P	P	–	T	–	–	*Y*	–	K
Brachidontes exustus	G	–	–	K	P	S	P	*Y*	D	P	–	–	G	–	*Y*	–	K
Septifer bifurcatus	Z	–	–	K	*P*	S	S	*Y*	G	–	–	T	G	–	*Y*	–	K
Geukensia demissa						T	G	*Y*	S	A	–	–	G	–	*Y*	–	K
Dreissena polymorpha				K	P	G	P	*Y*	**D**	Y	–	**D**	G	P	Y	**D**	K
Limnoperna fortunei				K	*P*	T	Q	*Y*	S	–	–	**E**	**E**	–	Y	–	K

```
                                                                   Y  P  +PC
Lffp2 (Clone1-14)   MESVMIIAMLVCYAYA -----------------------MDYDPKDPCRT
Mgfp2 (Q25464)      MLFSFFLLLTCTQLCLGTNRPDYNDDEEDDYKPPVYKPSPSKYRPVNPCLK
Mefp2 (Q1XBT8)      MLFSFFLLLTCTQLCLGTSPRPY-DDDEDDYTPPVYKPSPSRYRPVNPCLK     repetitive regions

                      PCKY G C P      YKC CK GY+G NC  K+ C  CK+    C          + C C+   +G+ CE  N C P
Lffp2 (Clone1-14)   NPCKYGGYCSPTEYG-YKCKCKDGYHGENCE-KDRCHS  QPCKHGGTCTVTDDG-YSCFCKVDYWGKHCEYDNQCYP
Mgfp2 (Q25464)      KPCKYNGVCKPRGGS-YKCFCKGGYYGYNCNLKNACKP  NQCKNKSRCVPVGKT-FKCVCRNGNFGRLCE-KNVCSP
Mefp2 (Q1XBT8)      KPCKYNGVCKPSGGS-YKCVCKGGYYGYNCNLKNACKP  NQCKNKSRCIPVGKT-FKCECRNGNFGRLCE-RNVCSP

                      NPCKN  KC  +G   YKCTC  GY G  C + + C  NPCKN  +C   GK  YKC C  GY G  C ++
Lffp2 (Clone1-14)   NPCKNGAKCERVGINDYKCTCEIGYYGRHCDDDNKCYI  NPCKNGAKCERVGKDDYKCTCEKGYYGRHCDDDRRYTP*
Mgfp2 (Q25464)      NPCKNNGKCSPLGKTGYKCTCSGGYTGPRC-EVHACKP  NPCKNKGRCFPDGKTGYKCRCVDGYSGPTCQE-NACKP
Mefp2 (Q1XBT8)      NPCKNKGKCAPLGKTGYKCTCSGGYTGPRC-EVHACKP  NPCKNNGRCYPDGKTGYKCKCVGGYSGPTCQE-NACKP
```

Fig. 1 Comparison of repetitive regions of Lffp-2, Mgfp-2, and Mgfp-2

not homologous with the counterpart fp-2 genes in the other species. The fp-2-type proteins in the two marine mussels are considered to function as matrix components in the adhesive plaque. There are two similar motif patterns with 38–42 residues in the EGF-like repetitive motifs, and one has the motif "–PCY-G-C-P– – -YKC–CK–GY-G-NC–K–C–." This type of motif has not yet been found in foot proteins of freshwater mussels.

We have constructed a foot-derived cDNA library using *L. fortunei* specimens from Japan, which were provided by the Environmental Science Research Laboratory, Central Research Institute of the Electric Power Industry. An fp-2 homologue has been screened from 192 randomly chosen clones. The entire nucleic acid sequence was obtained with a start codon ATG and a stop codon TGA. This fp-2 homologue was registered in the DNA Data Bank of Japan with accession code, AB910939 as Lffp-2. The alignments of the deduced amino acid sequence from the EST of the fp-2 homologue and those from two marine mussels (Mefp-2 and Mgfp-2) are shown in Fig. 1.

Lffp-2 has a repetitive motif similar to Mefp-2 and Mgfp-2, as seen in the homology measures; E-values $= 2.0 \times 10^{-38}$ with respect to Mgfp-2 (identity, 44 %) and $= 6.0 \times 10^{-38}$ with respect to Mefp-2 (44 %). The deduced amino acid composition of Lffp-2 without the signal region has a specific characteristic whereby Lffp-2 contains a higher level of the acidic residue Asp compared to Mgfp-2 and Mefp-2. The calculated isoelectric point (pI) of Lffp-2 is 6.92, while those of Mgfp-2 and Mefp-2 are 9.21 and 9.16, respectively (Fig. 2). This indicates that the elevated level of Asp in Lffp-2 contributes to a decrease in the pI value approaching neutral.

The key component in underwater adhesion is the presence of DOPA, and this might be conserved both in marine and freshwater mussels. In addition, glue proteins have probably evolved with the appropriate amino acids to express more stable adhesive strength on a variety of substrates, including man-made structures in nature.

Underwater Adhesion of *Limnoperna fortunei*

It should be noted that the complete mechanism of adhesion by *L. fortunei* is not fully revealed, and the descriptions below comprise a possible working hypothesis. At the time of this writing, Lffp-1 is the only molecule present in the foot of *L. fortunei*

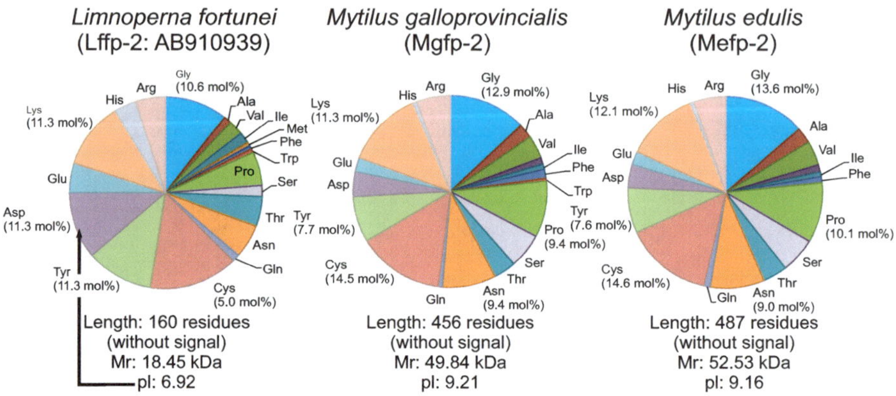

Fig. 2 Comparisons of amino acid compositions between mussel species

to have been experimentally proven to contain a DOPA-containing protein. Lffp-2 is another possible adhesion molecule, but this suggestion is based on knowledge concerning *D. polymorpha*. The Tyr residues in Lffp-2 have a high potential to be oxidized to DOPA, which is supported by experimental data that indicate the role of polyphenol oxidase in *L. fortunei* adhesion.

Figure 3 shows the current working hypothesis describing the byssus components of *L. fortunei* based on our knowledge of marine mussels. The potential roles of Lffp-1 are: (1) distal thread coating, (2) adhesive plaque cuticle, and (3) adhesion of limbic plaque. Probably, Lffp-2 is a matrix protein, which generates cohesive force in the adhesive plaque by fixing collagen-like protein fibrils in the distal thread (Waite et al. 1998). In marine mussels, adhesive primer proteins have been investigated, suggesting that unknown proteins are involved in the adhesion of *L. fortunei*.

Antifouling Assays

Laboratory Experiments

Experimental evidence on *L. fortunei* adhesion can inspire design of effective antifouling techniques for this species; however, laboratory experimental results are not always reproduced in field experiments. Despite this issue, laboratory experiments have been performed for the following reasons: (1) biologically clean conditions for screening responses of *L. fortunei* on the test substrates, (2) ability to measure adhesive strengths of attached mussels, and (3) providing for a variety of surface chemical modifications. Since February 2006, when *L. fortunei* was included in the Japanese Invasive Alien Species Act (Act No. 78 of 2004), laboratory experiments in Japan have been strictly controlled.

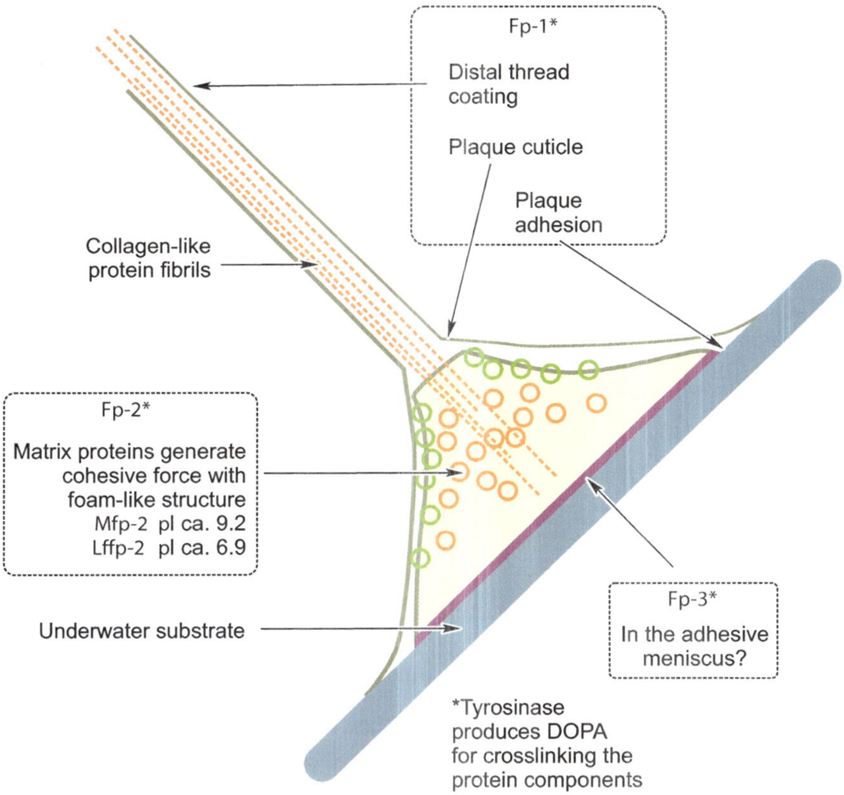

Fig. 3 A working hypothesis of *L. fortunei* byssus components

Ohkawa et al. (1999a) carried out laboratory experiments investigating *L. fortunei* settlement on several kinds of chemically modified glass surfaces from 1997 to 2000. At that time, all of the *L. fortunei* specimens and tank water used in the experiments were disposed of in a landfill, and then all experimental tools were sterilized by immersion in sodium hypochlorite solution for several days or autoclaved to ensure security, as seen in similar publications on *D. polymorpha*.

Chemical modifications of glass surfaces were performed using silane-coupling agents having a variety of functional groups, including halogens, cationic, mercapto, cyano, and glycidoxyl groups. Ten specimens of *L. fortunei* were placed on each test plate, and then the numbers of secreted byssal threads and attached mussels were counted after 1 week. Table 2 summarizes part of the results. Among ten kinds of test plates, fluoro-silane-coupled glass exhibited the lowest attachment (8 %, 3 out of 44 specimens). The total number of byssal threads secreted was 57, indicating that the mean number of byssal threads was 1.4 out of 44 specimens or 7.5 of the 3 attached specimens.

Other surfaces, including slate, rubber, polyamide, poly(tetrafluoroethylene) (PTFE), and silicone rubber, were also examined for *L. fortunei* settlement preference. As observed for fluoro-silane-coupled glass, the PTFE plate exhibited a

Table 2 Attachment of *L. fortunei* to various substrata, including untreated glass and glass surfaces treated with nine different silane-coupling agents. (Modified from Ohkawa et al. 1999b)

				Surface free energy (mN/m)	
Substrate	N	Attached	*P*	Dispersion (γ^d)	Polar (γ^p)
Slate	40	22 [55 %]	<0.05	n.d.	n.d.
Glass	138	28 [20 %]		21.5	42.9
Rubber	60	9 [15 %]	>0.05	26.1	0.5
Nylon	50	6 [12 %]	<0.05	34.1	9.1
PTFE	50	4 [8 %]	<0.05	18.6	0.5
Silicone rubber	50	1 [2 %]	<0.05	21.7	1.1
Aminoethylamino-silane	44	14 [32 %]	<0.05	22.1	34.3
Amino-silane	44	12 [27 %]	>0.05	26.7	9.2
Mercapto-silane	44	12 [27 %]	>0.05	29.3	4.9
Chloro-silane	44	9 [20 %]	>0.05	29.5	6.6
Cyano-silane	44	8 [18 %]	>0.05	27.2	9.3
Glycydoxy-silane	44	6 [14 %]	<0.05	28.9	14.9
Bromo-silane	40	5 [13 %]	<0.05	31.6	2.9
Iodo-silane	44	5 [11 %]	<0.05	34.4	3.1
Fluoro-silane	40	3 [8 %]	<0.05	23.5	3.8

P indicates significant difference in proportion of attached individuals with respect to plain glass (Mann-Whitney's U-test), *n.d.* no data

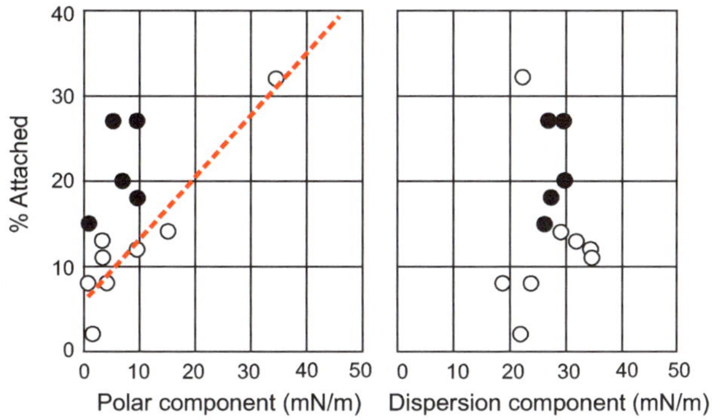

Fig. 4 Correlation between the polar component of the surface and the percentage of *L. fortunei* attachment. *Open and closed* circles represent the significant and nonsignificant (respectively) surfaces in Table 2

similar low attachment (8 %). Among six kinds of plates tested, attachment on the silicone rubber plate was the lowest at 2 %, with a mean number of secreted byssal threads at 0.8 out of a total of 50 specimens with only 2.0 threads in the one attached specimen. The surface free energies (SFE) of test plate surfaces were analyzed by means of the extended Fowkes' and Young-Dupré's equations (Yamamoto et al. 1996) in order to estimate the dispersion (γ^d) and polar (γ^p) components, where the SFE (γ) is represented as $\gamma=\gamma^d+\gamma^p$. The γ values were 53.6, 22.8, 19.1, and 27.3 mN/m for glass, silicone rubber, PTFE, and fluoro-silane-coupled glass, respectively. This indicates that *L. fortunei* has a tendency to avoid attaching to plates having lower SFE values. A weak correlation was observed between attachment percentage and γ^p value ($R=0.686$, $p<0.01$) but not with γ^d value.

Furthermore, examination of the relationship between attachment success and the components of surface free energy only found significant correlation between attachment to significant surfaces (Table 2) and γ^p ($R^2=0.8750$, $p<0.05$; Fig. 4). No correlations were found between nonsignificant surfaces and γ^p or for γ^d (Fig. 4).

Matsui et al. (2001) investigated the attachment strength of *L. fortunei* as a function of substrate surface properties, including SFE and physical roughness. Experimental animals were tied loosely to test coupons with a stainless steel thread and then kept in a tank for 7 days allowing them to byssally attach to the coupons. Detachment tests were conducted with the aid of a tensile loading apparatus pulling the mussels upwards while recording tensile force and the energy required for detachment, as well as the distance displaced before total detachment (Fig. 5). They employed a more detailed equation: $\gamma=\gamma^d+\gamma^p+\gamma^h$, where γ^h is the hydrogen-bonding component. A novel parameter was introduced, the "adhesive pad failure percentage," which represents the relative number of detachments between adhesive plaque and the substrate surface at the meniscus level (Fig. 3) as a function of the total number of byssal threads (Fig. 5). Thus, when all the adhesive plaques remain glued to the substrate and detachment is the result of byssus thread breakage alone, the pad adhesive failure percentage is 0 %; however, if forced detachment results in the separation of all adhesive plaques from the surface, the adhesive pad failure percentage is 100 % (Fig. 5).

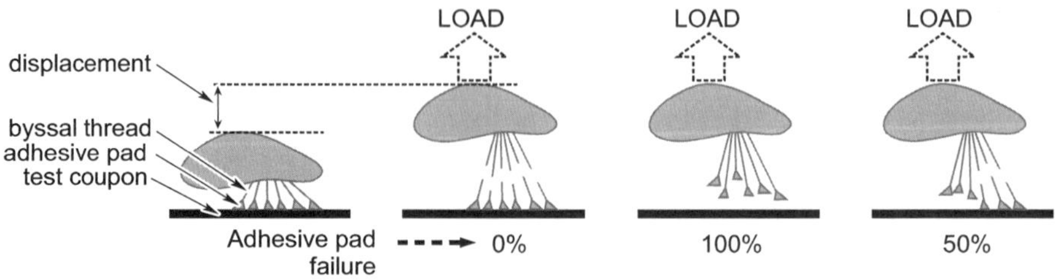

Fig. 5 Diagram of experiments on mussel detachment

Figures 6, 7 and 8 show the results reported by Matsui et al. (2001). The test coupons were classified into three categories: (1) commonly used materials (16 different materials without chemical modifications; Fig. 6), (2) coatings (13 coatings in 5 categories; Fig. 7), and (3) silane-coupled glass (10 different silane-coupling agents used for chemical modification of the glass surface; Fig. 8). Matsui et al. (2001) measured mussel displacement (in mm) upon detachment, estimated the number of broken byssal threads and adhesive pad failures, and the force (in N) and energy (in mJ, representing an integrated value of the force-displacement curve) needed to detach the mussels from the substrate. In general, the mechanical load was homogenously distributed among byssal threads and the adhesive pad. If the tensile strength of the byssal thread was higher than the adhesive force at the meniscus between the pad and the substrate surface, adhesive pad failure occurred, whereas if the byssal thread tensile strength was lower than the adhesive force of the pad, byssal threads would break. Hence, adhesive pad failure percentage yields an indication of the strength of the attachment allowed by different substrata.

The average displacement values for test coupons employing commonly used materials (Fig. 6) did not differ significantly, ranging at around 10–15 mm. Thus, detachment force and detachment energy were well correlated. Glass, Pyrex glass, stainless steel, and aluminum exhibited higher numbers of broken threads, which means that the adhesive forces between the pads and these materials were comparatively higher. The numbers of detached adhesive pads were higher in poly(methyl metacrylate) (PMMA), PTFE, and 6-nylon, indicating that adhesive force of the pads is weaker on these surfaces.

Detachment force is also affected by the number of byssal threads secreted, which differed among materials. The number of secreted threads was higher on stainless steel than on Pyrex glass. Stainless steel and Pyrex glass exhibited similar values of detached adhesive pads, but the number of broken threads was higher on stainless steel than on Pyrex glass, suggesting that attachment between adhesive pad and substrate is stronger on the former.

Results with 13 coatings in 5 material-based categories (Fig. 7) indicate that adhesive pad failures were more frequent on the three silicone resin-based coatings.

Tests with silane-coupled glass coupons (Fig. 8) showed that on fluoro-silane coupled glass the number of adhesive pad failures was considerably higher than on the other coupons. Although the number of secreted byssal threads on fluoro-silane

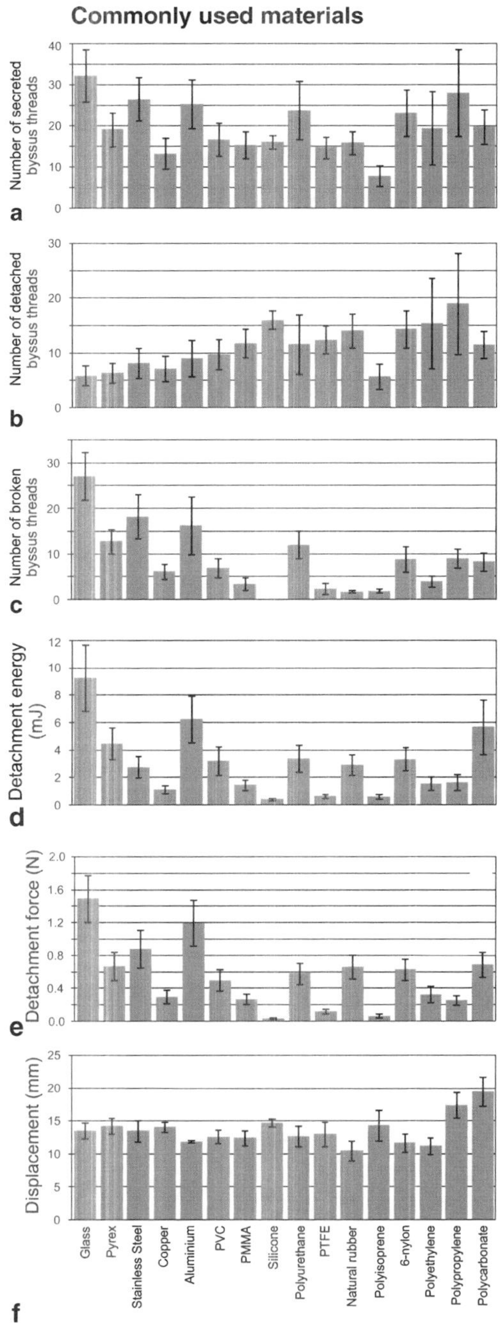

Fig. 6 Results of tests of forced detachment of *L. fortunei* from test coupons made of different commonly used materials. For each tested material the following information is illustrated (means and standard errors): **a** numbers of byssal threads per mussel produced by the experimental individuals, **b** numbers of byssal threads per mussel detached at the adhesive pad–substrate interface upon forced detachment, **c** numbers of byssal threads per mussel broken upon forced detachment, **d** and **e** mean energy and force (respectively) needed to detach the mussel from the test coupon, and **f** displacement upon detachment (see Fig. 5 for experimental scheme). (Based on data from Matsui et al. 2001)

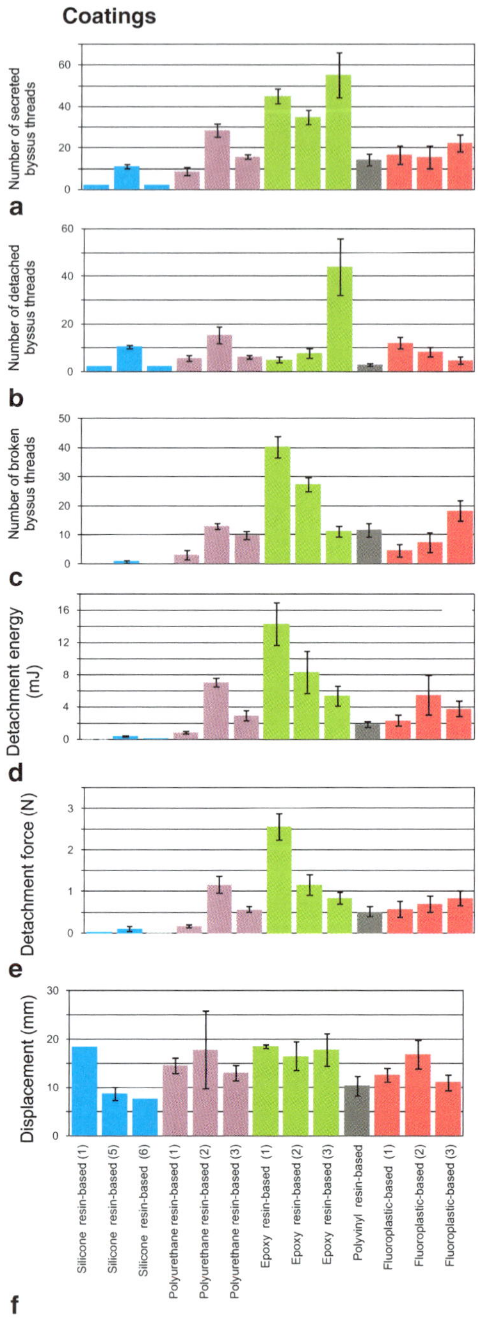

Fig. 7 Results of tests of forced detachment of *L. fortunei* from test coupons covered with different coatings (see Fig. 6 for detailed information and Fig. 5 for experimental scheme). Silicone resin-based coatings: (1) AF Biosuper SI (Usagida Chemical Industries Ltd, Kobe, Japan), (5) S-1 (a trial product) (Nippon Paint Company Ltd, Osaka, Japan), (6) Nippe Sleek (Nippon Paint Company Ltd, Osaka, Japan). Polyurethane resin-based coatings: (1) Ureall RIM PD-300 (Kawakami Paint MFG Company Ltd, Amagasaki, Japan), (2) ML3000 (S.F.C) (Dai-Nippon Toryo Company Ltd, Osaka, Japan), (3) Vtop Kai Free (a trial product) (Dai-Nippon Toryo Company Ltd, Osaka, Japan). Epoxy resin-based coatings: (1) ETON2300NB (Kawakami Paint MFG Company Ltd, Amagasaki, Japan), (2) Chosui Coat (Sigma Shinto Coatings Company Ltd, Amagasaki, Japan), (3) Kubota Powder TK 1413 (Kubota Corporation, Osaka, Japan). Polyvinyl resin-based coating: Suiyou 4000 (Kawakami Paint MFG Company Ltd, Amagasaki, Japan). Fluoroplastic-based coatings: (1) Duflon K300, (2) Duflon 100 Fresh, (3) Duflon Fresh Top Clear (Nippon Paint Company Ltd, Osaka, Japan). (Based on data from Matsui et al. 2001)

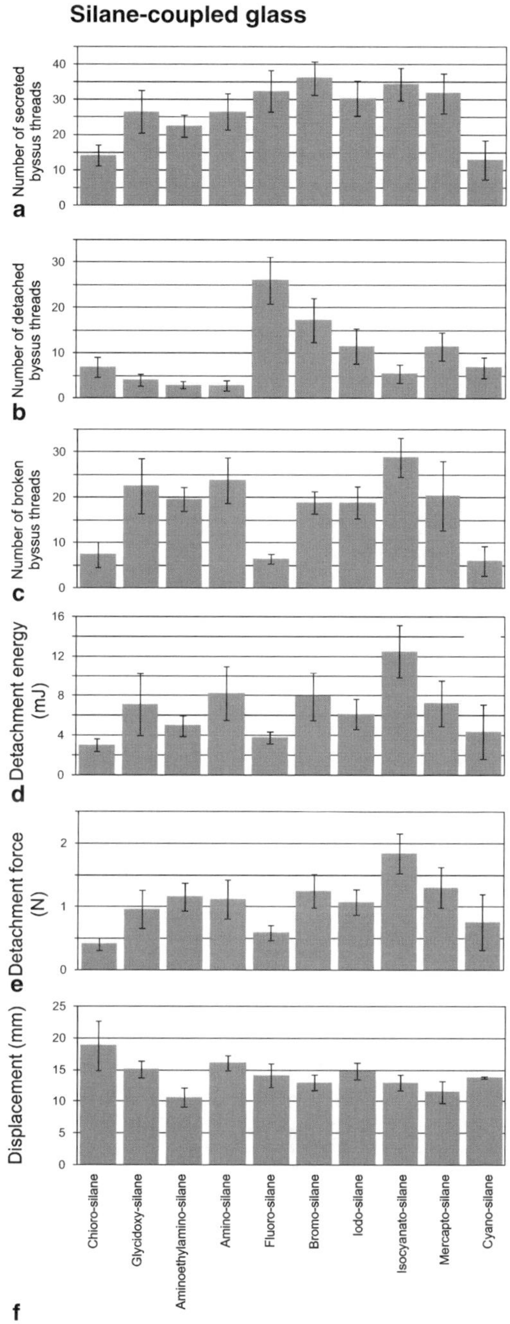

Fig. 8 Results of tests of forced detachment of *L. fortunei* from glass test coupons modified with silane-coupling agents. See Fig. 6 for detailed information and Fig. 5 for experimental scheme. (Based on data from Matsui et al. 2001)

coupled glass was relatively high compared to the other substrata, both the detachment force and detachment energy were significantly lower.

These results clearly show the importance of different materials and surface treatments for both the production of byssal threads by the mussels, and the strength of the resulting attachment of the animal to the substrate.

The most significant findings of the work of Matsui et al. (2001) are: (1) Detachment force correlates with detachment energy, and both are also associated with the number of secreted byssal threads; (2) adhesive pad failure tends to be higher for low γ^h surfaces, which suggest that the adhesive force between high γ^h surfaces and adhesive plaque is higher leading to thread breakage; (3) detachment energy increased chiefly as a function of γ^h values lower than ca. 7 mN/m, indicating that surfaces with higher γ^h values yield stronger adhesive plaque-substrate binding, as shown by lower adhesive pad failure (Nagaya et al. 2001); and (4) no clear correlation was observed between surface roughness and adhesive pad failure.

These laboratory studies strongly suggest that materials with lower SFE values, especially those having lower γ^p or γ^h, yield better antifouling surfaces. The reasons that lower SFE surfaces exhibit better antifouling effects are still unclear. The attachment of *L. fortunei* involves multiple steps. First, the mussels evaluate the surface using their foot in order to decide whether or not it is appropriate for attachment; if the result is negative, they crawl away in search of a better place. Second, physiological processes take place whereby mussels secrete the byssus precursor proteins. The amount of precursor proteins secreted to form the byssal thread varies depending on the physiological condition of the mussels and other factors. Third, a biochemical process occurs as the secreted proteins solidify with the aid of phenol oxidase catalytic action, turning into an insoluble byssus thread and adhesive plaque. Fourth, a concurrent chemical process takes place at the substrate surface in which the oxidized glue proteins coagulate on the surface and then cross-link to make an adhesion meniscus with higher resistance against detachment.

Once all these processes are completed, the mussel opens the central groove again to release a new thread toward the substrate. The fact that during these processes the mussel's foot is permanently in contact with the substrate suggests that the outcome of the steps detailed above is somehow affected by the surface chemical properties of the substrate. The physiological sensory mechanisms involved, however, are still unknown, and their elucidation will probably require much additional multidisciplinary work. As mentioned earlier, involvement of DOPA in the adhesion mechanism, especially for its surface coupling role, suggests that the interaction between the catechol of DOPA and the hydrogen-acceptor molecule on a given surface will enhance the adhesive force between the adhesive plaque and the surface. Low SFE values are associated with hydrophobic (water-repellent) and oil-repellent surfaces. Thus, adhesion of the plaque to low SFE surfaces is affected due to the absence of covalent, coordinate, or charge (permanent dipole moment) interactions, which means that the induced dipole is a unique force. As a result, low SFE surfaces enhance the frequency of adhesive pad failures and are therefore less vulnerable to fouling.

Field Experiments

Matsui et al. (2002) performed two field experiments on biofouling by *L. fortunei*. Both were carried out in a fishway of the Nagara River estuary barrage. Experimental coupons with different coatings were immersed and exposed to *L. fortunei* colonization for periods of 424 d (July 1997 to September 1999) and 513 d (May 1999 to October 2000). At the end of each experiment, mussels on each coupon were counted and measured.

In the July 1997–September 1999 experiment, mussel densities upon recovery differed greatly between coatings, with some silicone resin-based materials and copper-containing resins yielding the best antifouling results (Fig. 9). Mussel densities on these coatings were <5000 ind./m^2, in contrast to most others that yielded densities to over 13,000 ind./m^2 (Fig. 9). Furthermore, mussel densities were positively associated with the final mussel size ($R=0.51, p<0.05$), suggesting that coatings with highest antifouling properties were also more detrimental to mussel growth. Metal-containing resins were thus concluded to be toxic to the mussels, with copper-containing resins exhibiting the highest antifouling activities due to their biocidal effects (Fig. 9). Other coatings with reduced numbers of mussels were concluded to be nontoxic, but their surface chemical properties were concluded to discourage the secretion of byssal threads. Silicone-based coatings 2, 4, and 6 were among the latter, and their surface chemical properties, with particular attention to the hydrogen-bonding component (γ^h), are discussed below.

Colonization of test coupons deployed between May 1999 and October 2000 was much lower (max.: 89 ind./m^2) than during the previous experiment, but results were nonetheless significantly correlated with a similar experiment carried out in the laboratory ($R=0.93, p<0.05$). The average number of byssal threads secreted per mussel on each coating (which was estimated in the laboratory experiment), was also correlated with colonization densities on the corresponding coatings in both the first ($R=0.82, p<0.03$) and second field experiment ($R=0.86, p<0.02$).

One of the SFE components (γ^h) was strongly correlated with attachment density in the laboratory ($R=0.70, p<0.001$), but they were not correlated in field experiments ($R=0.27$ and 0.36, $p>0.1$). It seems most likely that the biofilm that developed on the surface of the field-deployed experimental coupons suppressed the antifouling effect of this SFE component. For silicone resin-based surfaces, test coupons with a surface roughness below 20–25 μm had better antifouling properties than those with a surface roughness above ca. 40 μm.

In order to evaluate the effects of fluid velocity on *L. fortunei* fouling, Matsui et al. (2002) performed another experiment at the Daido Water Intake Pumping Station, where test coupons were immersed in a grit chamber for a total of 888 days (June 1997 to December 1999). At the end of exposure, colonization density was highly correlated with γ^h ($R=0.71, p<0.005$). Two silicone resin-based coatings had no mussels on them and, in agreement with previously described results, increasing surface roughness was again associated with higher colonization densities.

At the same facility, densities of *L. fortunei* inside a water transfer pipe were surveyed along 35 m once in 1999 and once again in 2000. Spatial distribution

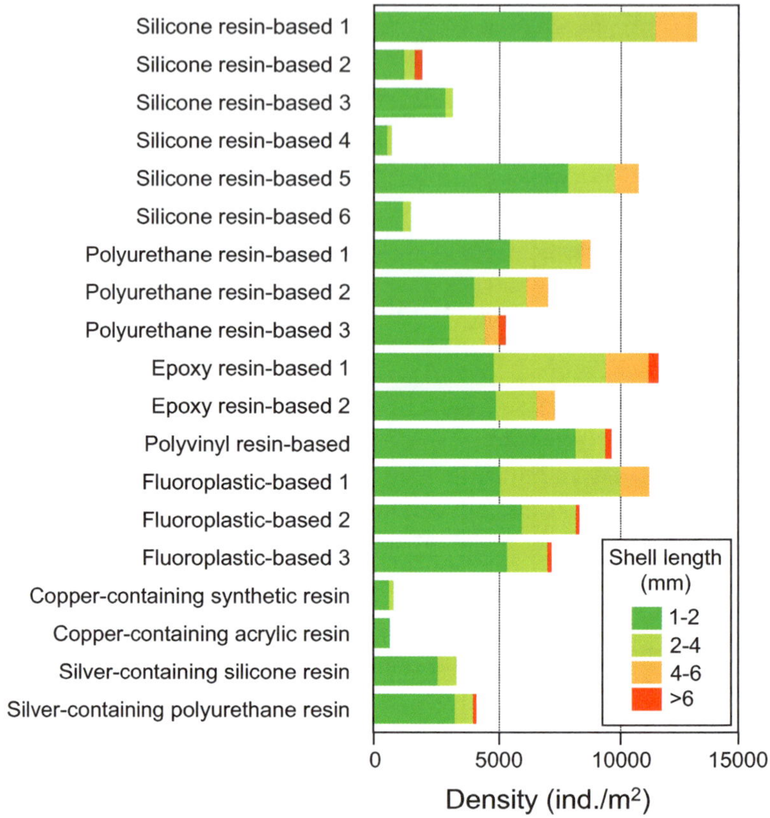

Fig. 9 Numbers and size of *L. fortunei* recorded on test coupons covered with different coatings after exposure to colonization in a fishway in the Nagara River estuary barrage (Japan) for 424 d (July 1997 to September 1999). Silicone resin-based coatings: (1) AF Biosuper SI (Usagida Chemical Industries Ltd, Kobe, Japan), (2) Sigma LSE Finish (Sigma Shinto Coatings Company Ltd, Amagasaki, Japan), (3) Kai Clean (Dai-Nippon Toryo Company Ltd, Osaka, Japan), (4) Bioclean (SPG) (Chugoku Marine Paints, Ltd, Tokyo, Japan), (5) S-1 (a trial product) (Nippon Paint Company Ltd, Osaka, Japan), 6, Nippe Sleek (Nippon Paint Company Ltd, Osaka, Japan). Polyurethane resin-based coatings: (1) Ureall RIM PD-300 (Kawakami Paint MFG Company Ltd, Amagasaki, Japan), (2) ML3000 (S.F.C) (Dai-Nippon Toryo Company Ltd, Osaka, Japan), (3) Vtop Kai Free (a trial product) (Dai-Nippon Toryo Company Ltd, Osaka, Japan). Epoxy resin-based coatings: (1) ETON2300NB (Kawakami Paint MFG Company Ltd, Amagasaki, Japan), (2) Chosui Coat (Sigma Shinto Coatings Company Ltd, Amagasaki, Japan). Polyvinyl resin-based coatings: Suiyou 4000 (Kawakami Paint MFG Company Ltd, Amagasaki, Japan). Fluoroplastic-based coatings: (1) Duflon K300, (2) Duflon 100 Fresh, (3) Duflon Fresh Top Clear (Nippon Paint Company Ltd, Osaka, Japan). Copper-pigmented synthetic resin: Sigmaplan Ecol A/F (Sigma Shinto Coatings Company Ltd, Amagasaki, Japan). Copper-pigmented acrylic resin: Ecoflex SPC 600 (Nippon Paint Company Ltd, Osaka, Japan). Silver-pigmented silicone resin: AF Biosuper HG (Usagida Chemical Industries Ltd, Kobe, Japan). Silver-pigmented polyurethane resin: Ureall RIM PD-30 (Kawakami Paint MFG Company Ltd, Amagasaki, Japan). (From Matsui et al. 2012, Biofouling, 18:140, reproduced with permission from Taylor & Francis, License number 3422200658786)

of mussel abundances among the 64 survey points was similar during both surveys ($R=0.77$, $p<0.001$). Densities were lowest in the branched section of the pipe, probably due to high water velocity and turbulence, whereas highest values were recorded on the inner side of curved sections, where flow velocities were lowest (Matsui et al. 2002). Differences in pipe wall materials between the straight (mortar lining) and curved (cast iron) sections were concluded to have no effect on colonization by golden mussels.

Impacts on Irrigation Facilities and Case Studies in Japan

Publications on the impacts of *L. fortunei* are often reported in local languages and are not readily accessible to the international community. The following section will review some salient examples of *L. fortunei*-related fouling problems in various Japanese facilities.

A water treatment company in the Yodo River area (Osaka, Japan) reported the frequency of operational problems associated with macrofouling by *L. fortunei.* The first incident occurred in October 1994 (Nakanishi and Kukai 1997). Figure 10 illustrates the number of fouling-related problems recorded at this plant throughout the year between April 1995 and March 1999. Problems tend to increase in the summer season, from July to September, when the reproductive activity of the mussel is highest (see Chapter "Reproductive Output and Seasonality of *Limnoperna fortunei*" in this volume). Approximately 40% of the problems were due to pipe-clogging. *Limnoperna fortunei* grows inside large diameter water intake pipes (diameter: 1800 mm, line length: 1.8 km, flow rate: 0.6 m/s); occasionally mussel beds with dead and live mussels detach in large clumps from the pipe walls are carried into the grit chambers where they sediment. These shells can be removed with the aid of screens, but the volumes to be disposed of are so high (up to 50 m^3) that the company opted for increasing periodic pipe-cleaning operations in the summer season (Nakanishi and Kukai 1997). The river water intake facility is equipped with a water quality monitor whose pipes (polyvinyl chloride, 15–50 mm in diameter, 2.5 m in length) periodically clog up with mussel shells. Shell fragments also enter other pipes and pumps, and this interferes with water-flow and the operation of monitoring devices.

The company investigated *L. fortunei*-related problems and made the following observations. Fouling is not constant but varies between sites and between years. *Limnoperna fortunei* attaches preferably to pipe walls and does not form multi-layered beds. Fouling is highest during summer. From autumn to spring, the shell length increases at a rate of roughly 1.0 mm/month, whereas in spring to autumn, the shell length increases ~2.0 mm/month. One year old individuals are ~18 mm. Most individuals die after reaching ~18 mm; the ones that survive continue to grow to about 30 mm, at which time (~2 years) they also die. Shells detach from the walls a few months after death, and therefore no additional efforts are needed to dislodge dead specimens. This company's report, based on 6 years of observations,

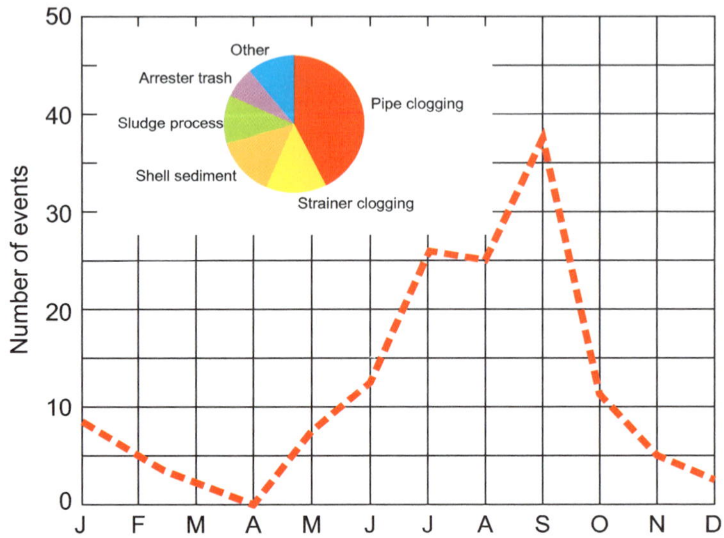

Fig. 10 Frequency of *L. fortunei*-related fouling problems reported by the Hanshin Water Supply Authority. (Modified from Nakanishi and Kukai 1997)

concluded that fouling by *L. fortunei* is unlikely to cause major problems at this facility, but monitoring of the problem was still deemed necessary.

Concluding Remarks

This review of on-going studies on the biochemical aspects of fouling by *L. fortunei* shows that presently our knowledge is restricted to only two glue proteins: Lffp-1 and Lffp-2. The underwater adhesion mechanism of byssus formation is still only a working hypothesis, where most processes are derived from information gained from marine mussels. Differences in molecular adhesion mechanisms between marine and freshwater mussels involve protein aggregation, presumably as a response to dissimilar environmental adaptation processes.

In both marine and freshwater mussels, the essential biochemical component involved in adhesion mechanisms is DOPA. This allows us to propose antifouling surface designs that utilize low SFE surfaces. This knowledge will help to develop antifouling paints that target attachment of *L. fortunei*. Silicone-based and epoxy-based antifouling paints are being subjected to field tests, and several of them have been found to be effective for a year, but their antifouling activity decreases significantly over time, in particular due to the development of biofilms.

There is a trade-off between the use of antifouling coatings with and without biocidal effects. Alternative fouling control strategies must be investigated, including methods that take into account the reproductive biology of the golden mussel, predator-based strategies, and others. In 2013, the Ministry of Agriculture, Forestry and Fisheries of Japan published a manual that reviews actions for the mitigation of *L. fortunei* fouling. This guide stressed the importance of: (1) real-time information on the development of mussel colonies at new sites, (2) early stage control before explosive population increases occur, and (3) enhancement of tools leading to a better understanding and prediction of the spread of *L. fortunei* larvae throughout the network of agricultural and irrigation waterways.

In Japan, fouling-related problems tend to affect residential and farming areas more than large facilities like those of water treatment plants (Nakano et al. 2011; Nakano et al. 2012). Thus, the development of safer and more effective methods for controlling fouling by the golden mussel should also target those settings where resources and potential methods are more limited. The biological, biochemical, and molecular studies of adhesion by *L. fortunei* will hopefully contribute toward the development of mitigation technologies useful for a wide spectrum of users.

Acknowledgements The authors thank Dr. Isamu Sakaguchi and Takuya Kobayashi, CRIEPI, for supplying the *L. fortunei* foot samples, which were used in the molecular cloning experiments. This work was supported by Grants-in-Aid No. 26288101, and partly by the Japanese Association for Marine Biology (JAMBIO) No 23–73.

References

Anderson KE, Waite JH (2001) Biochemical characterization of a byssal protein from *Dreissena bugensis* (Andrusov). Biofouling 18:37–45

Boltovskoy D, Correa N, Cataldo D, Sylvester F (2006) Dispersion and ecological impact of the invasive freshwater bivalve *Limnoperna fortunei* in the Río de la Plata watershed and beyond. Biol Invasions 8:947–963

Brown CH (1950) Quinone tanning in the animal kingdom. Nature 165:275

Inoue K, Takeuchi Y, Miki D, Odo S (1995) Mussel adhesive plaque protein gene is a novel member of epidermal growth factor-like gene family. J Biol Chem 270:6698–6701

Kamino K (2008) Underwater adhesive of marine organisms as the vital link between biological science and material science. Mar Biotechnol 10:111–121

Matsui Y, Nagaya K, Yuasa A, Naruto H, Yamamoto H, Ohkawa K, Magara Y (2001) Attachment strength of *Limnoperna fortunei* on substrates, and their surface properties. Biofouling 17:29–39

Matsui Y, Nagaya K, Funahashi G, Goto Y, Yuasa A, Yamamoto H, Ohkawa K, Magara Y (2002) Effectiveness of antifouling coatings and water flow in controlling attachment of the nuisance mussel *Limnoperna fortunei*. Biofouling 18:137–148

Nagaya K, Matsui Y, Ohira H, Yuasa A, Yamamoto H, Ohkawa K, Magara Y (2001) Attachment strength of an adhesive nuisance mussel, *Limnoperna fortunei*, against water flow. Biofouling 17:263–274

Nakanishi M, Kukai S (1997) Methods and case studies on extermination of Asian freshwater mussel *Limnoperna fortunei* at a water treatment facility. J Water Waste 39:1017–1020 [In Japanese]

Nakano D, Kobayashi T, Endo N, Sakaguchi I (2011) Growth rate and settlement of *Limnoperna fortunei* in a temperate reservoir. J Molluscan Stud 77:142–148

Nakano D, Kobayashi T, Sakaguchi I (2012) Changes in the drift and the settlement of the freshwater mussel *Limnoperna fortunei* along a headrace channel. Water Sc Technol 65:426–433

Ohkawa K, Nishida A, Honma R, Matsui Y, Nagaya K, Yuasa A, Yamamoto H (1999a) Studies on fouling by the freshwater mussel *Limnoperna fortunei* and the antifouling effects of low energy surfaces. Biofouling 13:337–350

Ohkawa K, Nishida A, Ichimiya K, Matsui Y, Nagaya K, Yuasa A, Yamamoto H (1999b) Purification and characterization of a Dopa-containing protein from the foot of the Asian freshwater mussel, *Limnoperna fortunei*. Biofouling 14:181–188

Ohkawa K, Ichimiya K, Nishida A, Yamamoto H (2001) Synthesis and surface chemical properties of adhesive protein of the Asian freshwater mussel, *Limnoperna fortunei*. Macromol Biosci 1:376–386

Ohkawa K, Nishida A, Yamamoto H, Waite JH (2004) A glycosylated byssal precursor protein from the green mussel *Perna viridis* with modified Dopa side-chains. Biofouling 20:101–115

Ohkawa K, Nagai T, Nishida A, Yamomoto H (2009) Purification of DOPA-containing foot proteins from green mussel, *Perna viridis*, and adhesive properties of synthetic model copolypeptides. J Adhesion 85:770–791

Oliveira MD, Takeda AM, Fernandes de Barros L, Savio Barbosa D, Kawakami de Resende E (2006) Invasion by *Limnoperna fortunei* (Dunker, 1857) (Bivalvia, Mytilidae) of the Pantanal wetland, Brazil. Biol Invasions 8:97–104

Paz MA, Fluckiger R, Boak A, Kagan HM, Gallop PM (1991) Specific detection of quinoproteins by redox-cycling staining. J Biol Chem 266:689–692

Rzepecki LM, Waite JH (1993a) The byssus of the zebra mussel, *Dreissena polymorpha*. I: morphology and in situ protein processing during maturation. Mol Mar Biol Biotechnol 2:255–266

Rzepecki LM, Waite JH (1993b) The byssus of the zebra mussel, *Dreissena polymorpha*. II: Structure and polymorphism of byssal polyphenolic protein families. Mol Mar Biol Biotechnol 2:267–279

Rzepecki LM, Chin SS, Waite JH, Lavin MF (1991) Molecular diversity of marine glues: polyphenolic proteins from five mussel species. Mol Mar Biol Biotechnol 1:78–88

Rzepecki LM, Hansen KM, Waite JH (1992) Characterization of a cystine-rich polyphenolic protein family from the blue mussel *Mytilus edulis* L. Biol Bull 183:123–137

Stewart RJ, Ransom TC, Hlady V (2011) Natural underwater adhesives. J Polym Sci Part B: Polym Phys 49:757–771

Stewart RJ, Weaver JC, Morse DE, Waite JH (2004) The tube cement of *Phragmatopoma californica* : a solid foam. J Exp Biol 207:4727–4734

Taylor SW, Waite JH, Ross MM, Shabanowitz J, Hunt DF (1994) trans-2,3-cis-3,4-dihydroxyproline, a new naturally occurring amino acid, is the sixth residue in the tandemly repeated consensus decapeptides of an adhesive protein from *Mytilus edulis*. J Am Chem Soc 116:10803–10804

Waite JH (1983) Evidence for a repeating 3,4-dihydroxyphenylalanine- and hydroxyproline-containing decapeptide in the adhesive protein of the mussel, *Mytilus edulis* L. J Biol Chem 285:2911–2915

Waite JH (1990) The phylogeny and chemical diversity of quinone-tanned glues and varnishes. Comp Biochem Physiolo Part B: Biochem & Mol Biol 97:19–29

Waite JH, Andersen SO (1980) 3,4-Dihydroxyphenylalanine (DOPA) and sclerotization of periostracum in *Mytilus edulis* L. Biol Bull 158:164–173

Waite JH, Qin XX (2001) Polyphosphoprotein from the adhesive pads of *Mytilus edulis*. Biochemistry 40:2887–2893

Waite JH, Housley TJ, Tanzer ML (1985) Peptide repeats in a mussel glue protein: theme and variations. Biochemistry 24:5010–5014

Waite JH, Hansen DC, Little KT (1989) The glue protein of ribbed mussels (*Geukensia demissa*): a natural adhesive with some features of collagen. J Comp Physiol B 159:517–525

Waite JH, Jensen RA, Morse DE (1992) Cement precursor proteins of the reef-building polychaete *Phragmatopoma californica* (Fewkes). Biochemistry 31:5733–5738

Waite JH, Qin XX, Coyne KJ (1998) The peculiar collagens of mussel byssus. Matrix Biol 17:93–106
Xu W, Faisal M (2009) Development of a cDNA microarray of zebra mussel (*Dreissena polymorpha*) foot and its use in understanding the early stage of underwater adhesion. Gene 436:71–80
Yamamoto H, Ogawa T, Ohara S, Nishida A, Hirata Y, Tatehata H, Ohkawa K (1996) Adhesion and wettability of marine adhesive proteins in aqueous systems. J Adhesion 59:17–24
Zhao H, Sun C, Stewart RJ, Waite JH (2005) Cement proteins of the tube-building polychaete *Phragmatopoma californica*. J Biol Chem 280:42938–42944
Zhao H, Sagert J, Hwang DS, Waite JH (2009) Glycosylated hydroxytryptophan in a mussel adhesive protein from *Perna viridis*. J Biol Chem 284:23344–23352

Chemical Strategies for the Control of the Golden Mussel (*Limnoperna fortunei*) in Industrial Facilities

Renata Claudi and Marcia Divina de Oliveira

Abstract There are a number of available chemical strategies for the control of macrofouling by bivalves within piping systems such as those that carry raw water into water treatment plants, cooling water to vital areas of power plants and industrial facilities, and fire protection water in any industry. A number of these strategies have been tested for the control of golden mussels and some of the methods tested hold promise for industrial use. Treatment strategies generally fall into proactive and reactive treatments. Proactive treatments will prevent settlement of mussels or prevent the development of adults. Reactive treatments will allow the settlement and growth of adult mussels, periodically removing settled adults. The chemicals used for both strategies are divided into nonoxidizing and oxidizing chemicals. The actual choice of chemicals and the mode of application will depend on several factors such as regulatory approval, economic viability, and preference of the individual user.

Keywords *Limnoperna fortunei* · Golden mussel · Biological invasion · Biofouling control · Impact · Industrial plants · Power plants · Irrigation facilites · Chemical treatment · Biocides · Molluscicides

Introduction

The golden mussel, a macrofouling bivalve, was introduced into Argentina from Asia around 1990 (Pastorino et al. 1993) and within a decade spread to four other South American countries (see Chapter "Colonization and Spread of *Limnoperna fortunei* in South America" in this volume). Before the widespread invasion of the golden mussels in South America, the invasion of the zebra mussel

R. Claudi (✉)
RNT Consulting Inc., Picton, ON, Canada
e-mail: rnt@idirect.com

M. D. de Oliveira
Embrapa Pantanal, Rua 21 de Setembro, Corumbá, MS 1880, Brazil
e-mail: marcia.divina@embrapa.br

D. Boltovskoy (ed.), *Limnoperna fortunei,* Invading Nature - Springer Series in Invasion Ecology 10, DOI 10.1007/978-3-319-13494-9_23

(a morphologically similar bivalve) into North American waters demonstrated the vulnerability of industrial raw water systems to macrofouling species and the need for control of infestation.

The golden mussel (Mytilidae) and the zebra and quagga mussels (Dreissenidae) share many characteristics. Both the dreissenids and the golden mussel cause serious problems for industry because they possess byssal threads with which they hold on to the substrate enabling them to settle in cooling water pipes. Their free-living larvae are carried by raw water and can gain access and settle in most industrial raw water systems. These two characteristics allow both the dreissenids and the golden mussel to foul all structures and surfaces exposed to raw water. Mussel settlement and growth inside cooling water piping can decrease flow and cause numerous maintenance problems for most industrial facilities.

Different industrial facilities face different problems depending on their materials of construction, cooling water system configuration, and the way they use raw water. A vulnerability assessment can pinpoint the areas of a facility most likely to suffer from infestation and allow management to focus control efforts on those areas.

In this chapter, we shall review available information on chemical strategies for the control of golden mussels within piping systems such as those which carry raw water into water treatment plants, cooling water to vital areas of power plants, and fire protection water in any industry. The methods and chemicals included in this review are restricted to those which have been tested with *L. fortunei* and which hold promise for industrial use. The actual number of chemicals that could be potentially used is far larger (Sprecher and Getsinger 2000; Mackie and Claudi 2010; Rajagopal et al. 2012).

Control Strategies for Internal Piping Systems

The approach and the materials used for the control of golden mussels are very similar to those used for the control of dreissenid mussels in Europe and North America. However, it is important to note that in some instances there are differences between the response of the dreissenids and the response of the golden mussel to some strategies. The golden mussel exhibits a wider tolerance of some ecological parameters than the dreissenids (Karatayev et al. 2010) (see Chapter "Parallels and Contrasts Between *Limnoperna fortunei* and Species of *Dreissena*" in this volume). Golden mussels often thrive under extremely adverse environmental conditions (e.g., high pollution levels, low oxygen concentrations, very low calcium levels, low pH; Karatayev et al. 2007), an indication that it is a highly tolerant species. For example, under anoxic conditions at 25 °C, 100 % mortality of *Dreissena* is achieved in only 4 days (Matthews and McMahon 1994). For *L. fortunei* 13 days are required for 100 % mortality under similar conditions (Perepelizin and Boltovskoy 2011).

Due to wide tolerances of environmental conditions by the golden mussels, individual chemical strategies that have been used successfully for the control of dreissenids need to be retested on the golden mussel to verify their effectiveness.

Proactive versus Reactive Treatments

Prior to discussing individual control strategies, the concept of proactive versus reactive treatment must be considered. Depending on the vulnerability of the individual facility or system, the operators must decide if they will treat in a manner which will prevent settlement of mussels (proactive treatment), or if they will allow settlement to occur and periodically remove settled adults (reactive treatment). The more vulnerable the system and the larger the population density of the mussels, the more likely it is that the operators will choose preventative treatment.

The choice of proactive versus reactive treatment will dictate both the choice of treatment strategy and the method of application. For example, the use of small-pore self-cleaning filters to remove veligers and continuous irradiation of the raw water flowing within a pipe with UV lamps to disable the veligers are both considered proactive treatments. Continuous pH adjustment and low-level continuous dosing with oxidizing and nonoxidizing chemicals to prevent settlement (without necessarily causing mortality of ready to settle veligers) are also proactive treatment strategies.

In contrast, periodic application of oxidizing or nonoxidizing chemicals, mechanical removal and application of hot water to remove adults are examples of reactive treatment.

Proactive Application of Chemicals for the Control of L. fortunei

Nonoxidizing Chemicals

pH Adjustment

Salto de Cashias is a hydroelectric plant in Brazil that belongs to the COPEL company. The facility has been coping with golden mussels for a number of years. The population density of the mussels has been reported to be as high as 150,000 ind./m^2.

The initial system installed for control of mussels in 2000 was that of pH adjustment. Sodium hydroxide was added to the service water stream to achieve a pH of approximately 9. This pH appeared to eliminate mussels from the service water system. In addition, biofilm was removed from service water piping. The removal of the biofilm did expose some underdeposit corrosion in the piping system, including some pinhole leaks (Calazans and Fernandes 2012). The control of settlement is thought to have been achieved by adjustment of the pH to outside of the preferred range of the golden mussel rather than by the direct toxicity of sodium hydroxide.

Before considering pH adjustment as a mitigation strategy, it is important to test pH adjustment in the raw water. In water with high alkalinity or high calcium content, increasing the pH can cause precipitation of calcium carbonate. This would be an undesirable side effect and the use of pH adjustment would not be practical under those circumstances.

Montresor et al. (2013) used concentrations of NaOH from 40 to 800 mg/L, which resulted in a pH range of 11.24–13.04. The authors recorded a LC_{50} (the concentration capable of killing 50 % of the individuals exposed) in 96 h exposures to 88.51 mg/L (pH ~ 11.5). This is comparable to the response of zebra mussels observed by Claudi et al. (2012). Zebra mussels exposed to very high pH levels (i.e., pH 10, 11, and 12) caused by addition of NaOH experienced 90 % mortality after 120 h at pH 12. Results from both studies suggest that very high pH could be used as an end of season treatment for elimination of adult mussels, as well as a preventative strategy using a modest pH adjustment for settlement prevention.

To continuously adjust the pH at Salto de Cashias had a cost of approximately US$ 200/day. In 2009, plant management decided to switch to the use of the MXD-100 product (see below). One of the primary reasons was worker safety when handling large volumes of sodium hydroxide.

MXD-100

MXD-100 is a product of the Brazilian Company Maxclean Ambiental e Química. It is considered an antifouling and antimicrobial agent. The composition of this product appears fairly complex. According to the manufacturer, it contains plant-derived tannins, isothiazolone, EDTA (ethylenediaminetetraacetic acid), active cationic surfactants, and nonionic glycols. Given the complexity of the product, it is hard to speculate on the mode of action. However, when Pereyra et al. (2011) tested the toxicity of three plant-derived tannin preparations to *L. fortunei*, the LC_{50} values ranged from 138.53 to 1273.73 mg/L, depending on the size of the specimens (larvae or 13 and 19 mm adults) and on the chemical compound tested (ECOTECs-UA, ECOTECs-L and ECOTECs-MC) (Table 1). The observed tannin toxicity was far below the toxicity documented for MXD-100 by Montresor et al. (2013), which suggests that the importance of tannins in the formulation of this compound is marginal. Montresor et al. (2013) tested the toxicity of several concentrations of MXD-100 (0.05, 0.5, 1, 10, 100, and 500 mg/L) on adult *L. fortunei* at ambient water temperature of 23 to 27 °C. The authors found LC_{50} concentrations of 45.49 mg/L after 48 h, 13.69 mg/L after 72 h, and 11.1 mg/L after 96 h (Table 1). When Salto de Cashias switched to the use of MXD-100, the product was injected in the same location as was used for sodium hydroxide but it was added three times each day for 11 min at a time. The amount injected is approximately 0.8 ppm of MXD, and the cost is similar to the sodium hydroxide treatment (L. C. Montresor pers. comm.). In Brazil, the treatment with MXD-100 is usually applied over a period of 90 days, three times a day for 10 min each, at a concentration range of 1–7 mg/L (as registered by the Brazilian Environmental Institute–IBAMA–under the number 4722/11-10) (Mata et al. 2013; Montresor et al. 2013). The three times per day injection appears to eliminate freshly settled individuals and is therefore a preventative strategy.

Oxidizing Chemicals

An oxidizing chemical is the element or compound in a chemical reaction that accepts an electron from another species. Because the oxidizing chemical is gaining electrons, it is said to have been reduced. The reactant element or compound in the chemical reaction is oxidized by having its electrons taken away by the oxidizing agent. This oxidation-reduction chemical reaction is often referred to as a redox reaction.

Ozone

The hydroelectric plant Itaipu (Brazil-Paraguay, upper Paraná River) has tested an ozone addition system in the service water of Unit 10. The ozone was injected continuously into the service water system resulting in a concentration of 0.15–0.20 mg/L. Even at this low concentration there appeared to be good control of mussels downstream of the injection. Furthermore, ozone appeared to provide excellent control of biofilm (Rothe 2007).

Chlorine Dioxide

Chlorine dioxide has been used for the control of *L. fortunei* by a municipal water system in Southern Brazil by Nalco (Nalco Inc. 2013). The water utility used Purate™ chlorine dioxide technology as the primary oxidant starting in 2007. The objective was to prevent the attachment of the *L. fortunei* veligers in the incoming water pipeline. Chlorine dioxide was injected at the point of water withdrawal in the river and again at the intake to the treatment plant. A continuous dose of approximately 0.8 ppm of chlorine dioxide at the intake prevented settlement in the pipeline. The addition of chlorine dioxide was also effective for color and odor removal caused by Cyanobacteria. According to the study, by using chlorine dioxide, this municipality saved more than 80 % of the expenses previously used to manually remove golden mussels. In 2006, the expenses for golden mussel control were US$ 260,000. In 2008, the expense was less than US$ 50,000.

Chlorine

Chlorine as chlorine gas or either sodium or calcium hypochlorite has been widely used for prevention of macrofouling and water disinfection. Morton (1976) reported on the efficacy of chlorine for control of *L. fortunei* in Hong Kong's raw water supply system. Continuous application of 0.5 mg chlorine/L was found to be sufficient to prevent infestation of the system without adversely affecting sand filters or leaving a taste in the water (Table 1). The prevention of settlement by chlorine has also

Table 1 Summary of results of chemical control methods assayed with *Limnoperna fortunei*. Toxicant [commercial name]: (1) 2,5′ -dichloro-4′ -nitrosalicylanilide (active ingredient: 70 %) [Bayluscide WP70]; (2) Chlorine dioxide; (3) Diallyl dimethyl ammonium chloride polymer (active ingredient: 40 %) [Veligon TL-M]; (4) Didecyl dimethyl ammonium chloride (active ingredient: 50 %) [H130M]; (5) N-alkil dimethylbenzyl ammonium chloride (active ingredient: 50 %) [Clam-Trol CT-2]; (6) N-alkil dimethylbenzyl ammonium chloride (active ingredient: 50 %) [Spectrus CT1300]; (7) Poly(oxyethylene (dimethyliminio) ethylene (dimethyliminio) ethylene dichloride) [BULAB 6002]; (8) Poly(oxyethylene (dimethyliminio) ethylene (dimethyliminio) ethylene dichloride) [BULAB 6002]; (9) Poly-diallyldimethyl ammonium chloride (microencapsulated); (10) Potassium chloride; (11) Potassium chloride (microencapsulated); (12) Quaternary ammonium+tannin extracts [MXD-100]; (13) *Schinopsis balansae* tannins at 70 % [ECOTEC-L]; (14) *Schinopsis balansae* tannins at 74 % [ECOTEC-UA]; (15) *Schinopsis balansae* tannins at 86.5 % [ECOTEC-MC]; (16) Sodium chloride; (17) Sodium dichloroisocyanurate; (18) Sodium hydroxide; (19) Sodium hypochlorite; (20) Total ammonia; (21) Trichloroisocyanuric acid; (22) Unionized ammonia (NH_3-N); (23) Potassium permanganate; (24) Copper sulfate

Toxicant	LC_{50} or % dead	Exposure [postexposure] times (h)	No. of concentrations tested (range, ppm)	Stage (size in mm)	Temp. (ºC)	Test conditions	Reference
1	1.0 ppm	48 [24–264]	7 (0.25-8)	A (15–25)	15	Static	Cataldo et al. (2003)
1	0.8 ppm	48 [24–264]	7 (0.25-8)	A (15–25)	20	Static	Cataldo et al. (2003)
1	0.3 ppm	48 [24–264]	7 (0.25-8)	A (15–25)	25	Static	Cataldo et al. (2003)
2	427.6 ppm	48 [48]	14 (1-800)	A (15–25)	25	Static	Calazans et al. (2013)
3	3.88 ppm	720 [0]	5 (2–50)	A	20–22	Static	Boltovskoy and Cataldo (2003)
3	815.04 ppm	264 [0]	5 (0.5–10)	L	20–22	Static	Boltovskoy and Cataldo (2003)
4	0.56 ppm	720 [0]	5 (0.5–10)	A	20–22	Static	Boltovskoy and Cataldo (2003)
4	1.03 ppm	264 [0]	5 (0.5–10)	L	20–22	Static	Boltovskoy and Cataldo (2003)
4	2.9 ppm	48 [24–264]	8 (0.5–30)	A (15–25)	15	Static	Cataldo et al. (2003)
4	1.7 ppm	48 [24–264]	8 (0.5–30)	A (15–25)	20	Static	Cataldo et al. (2003)
4	0.8 ppm	48 [24–264]	8 (0.5–30)	A (15–25)	25	Static	Cataldo et al. (2003)
5	2.43 ppm	36 [252]	5 (1–3)	A	20–22	Static	Boltovskoy and Cataldo (2003)

Table 1 (continued)

Toxicant	LC_{50} or % dead	Exposure [postexposure] times (h)	No. of concentrations tested (range, ppm)	Stage (size in mm)	Temp. (°C)	Test conditions	Reference
5	0.98 ppm	36 [480]	5 (1–3)	A	30	Static	Boltovskoy and Cataldo (2003)
5	1.28 ppm	36 [480]	5 (1–3)	A	25	Static	Boltovskoy and Cataldo (2003)
5	2.43 ppm	36 [480]	5 (1–3)	A	20	Static	Boltovskoy and Cataldo (2003)
5	0.88 ppm	48 [480]	5 (1–3)	A	30	Static	Boltovskoy and Cataldo (2003)
5	1.38 ppm	48 [480]	5 (1–3)	A	25	Static	Boltovskoy and Cataldo (2003)
5	2.52 ppm	48 [480]	5 (1–3)	A	20	Static	Boltovskoy and Cataldo (2003)
5	0.90 ppm	720 [0]	5 (0.5–10)	A	20–22	Static	Boltovskoy and Cataldo (2003)
5	0.71 ppm	264 [0]	5 (0.5–10)	L	20–22	Static	Boltovskoy and Cataldo (2003)
5	34.9 ppm	48 [24–264]	5 (1–30)	A (15–25)	15	Static	Cataldo et al. (2003)
5	1.3 ppm	48 [24–264]	5 (1–30)	A (15–25)	20	Static	Cataldo et al. (2003)
5	1.2 ppm	48 [24–264]	5 (1–30)	A (15–25)	25	Static	Cataldo et al. (2003)
6	% dead: 41.75	12 [120]	1 (2.3)	A (5–35)	22.5–23.5	FT	Boltovskoy and Cataldo (2003)
6	% dead: 41.45	24 [120]	1 (2.3)	A (5–35)	22.5–23.5	FT	Boltovskoy and Cataldo (2003)
6	% dead: 62.15	36 [120]	1 (2.3)	A (5–35)	22.5–23.5	FT	Boltovskoy and Cataldo (2003)

Table 1 (continued)

Toxicant	LC_{50} or % dead	Exposure [postexposure] times (h)	No. of concentrations tested (range, ppm)	Stage (size in mm)	Temp. (ºC)	Test conditions	Reference
6	% dead: 92.85	48 [120]	1 (2.3)	A (5–35)	22.5–23.5	FT	Boltovskoy and Cataldo (2003)
6	% dead: 92.05	72 [120]	1 (2.3)	A (5–35)	22.5–23.5	FT	Boltovskoy and Cataldo (2003)
6	% dead: 63	24 [168]	1 (2.5)	A (6–43)	24–25	FT	Boltovskoy et al. (2005)
6	% dead: 94	48 [168]	1 (2.5)	A (6–43)	24–25	FT	Boltovskoy et al. (2005)
6	% dead: 99	72 [168]	1 (2.5)	A (6–43)	24–25	FT	Boltovskoy et al. (2005)
7	7.185 ppm	24 [0]	5 (5–75.5)	L	ND	Static	Darrigran et al. (2001)
8	0.88 ppm	720 [0]	4 (0.5–20)	A	20–22	Static	Boltovskoy and Cataldo (2003)
8	1.51 ppm	264 [0]	4 (0.5–10)	L	20–22	Static	Boltovskoy and Cataldo (2003)
8	% dead: 77–100%	116–168 [0]	3 (8–20)	A	24	Static	Darrigran and Damborenea (2001)
9	1313.3 ppm	6 [48]	1 (90)	A (15–25)	25	FT	Calazans et al. (2013)
9	270.9 ppm	48 [48]	11 (12–1000)	A (15–25)	25	Static	Calazans et al. (2013)
10	1439.0 ppm	48 [48]	8 (10–10000)	A (15–25)	25	Static	Calazans et al. (2013)
11	8303.1 ppm	6 [48]	4 (90–1000)	A (15–25)	25	FT	Calazans et al. (2013)
11	2536.9 ppm	48 [48]	8 (12–6000)	A (15–25)	25	Static	Calazans et al. (2013)
12	45.49 ppm	48 [0]	6 (0.05–500)	A (21–26)	23–27	Static	Montresor et al. (2013)

Table 1 (continued)

Toxicant	LC_{50} or % dead	Exposure [postexposure] times (h)	No. of concentrations tested (range, ppm)	Stage (size in mm)	Temp. (ºC)	Test conditions	Reference
12	13.69 ppm	72 [0]	6 (0.05–500)	A (21–26)	23–27	Static	Montresor et al. (2013)
12	11.10 ppm	96 [0]	6 (0.05–500)	A (21–26)	23–27	Static	Montresor et al. (2013)
12[a]	% dead: 99	8760 [0]	1 (1)	A	18–26	FT	Netto (2011)
13	138.54 ppm	24 [0]	ND	L	21.8–23.7	Static	Pereyra et al. (2011)
14	160.21 ppm	24 [0]	ND	L	21.8–23.7	Static	Pereyra et al. (2011)
15	983.27 ppm	168 [0]	ND	A (13)	21.8–23.7	Static	Pereyra et al. (2011)
15	309.92 ppm	168 [0]	ND	A (13)	21.8–23.7	Static	Pereyra et al. (2011)
15	160.1 ppm	168 [0]	ND	A (13)	21.8–23.7	Static	Pereyra et al. (2011)
15	1273.73 ppm	168 [0]	ND	A (19)	21.8–23.7	Static	Pereyra et al. (2011)
15	442.14 ppm	168 [0]	ND	A (19)	21.8–23.7	Static	Pereyra et al. (2011)
15	283.4 ppm	168 [0]	ND	A (19)	21.8–23.7	Static	Pereyra et al. (2011)
15	138.53 ppm	24 [0]	ND	L	21.8–23.7	Static	Pereyra et al. (2011)
16	% dead: 90	240 [0]	1 (2000)	A	15–22	Static	Angonesi et al. (2008)
16	% dead: 92	240 [0]	1 (4000)	A	15–22	Static	Angonesi et al. (2008)
16	% dead: 100	240 [0]	1 (6000)	A	15–22	Static	Angonesi et al. (2008)
16	% dead: 100	240 [0]	1 (8000)	A	15–22	Static	Angonesi et al. (2008)
16	% dead: 100	240 [0]	1 (12000)	A	15–22	Static	Angonesi et al. (2008)
16	8336.7 ppm	48 [48]	8 (1000–20000)	A (15–25)	25	Static	Calazans et al. (2013)
17	376.0 ppm	48 [48]	9 (1–2000)	A (15–25)	25	Static	Calazans et al. (2013)
17[a]	% dead: 86	8760 [0]	1 (1)	A	18–26	FT	Netto (2011)
18	344.95 ppm	48 [0]	7 (40–800)	A (21–26)	23–27	Static	Montresor et al. (2013)

Table 1 (continued)

Toxicant	LC_{50} or % dead	Exposure [postexposure] times (h)	No. of concentrations tested (range, ppm)	Stage (size in mm)	Temp. (°C)	Test conditions	Reference
18	113.14 ppm	72 [0]	7 (40–800)	A (21–26)	23–27	Static	Montresor et al. (2013)
18	88.51 ppm	96 [0]	7 (40–800)	A (21–26)	23–27	Static	Montresor et al. (2013)
18[a]	% dead: 99	8760 [0]	ND	A	18–26	FT	Netto (2011)
19	% dead: 2	24 [168]	1 (0.5)	A (6–43)	24–25	FT	Boltovskoy et al. (2005)
19	% dead: 1	48 [168]	1 (0.5)	A (6–43)	24–25	FT	Boltovskoy et al. (2005)
19	% dead: 0.2	72 [168]	1 (0.5)	A (6–43)	24–25	FT	Boltovskoy et al. (2005)
19	663.6 ppm	48 [48]	4 (10–1000)	A (15–25)	25	Static	Calazans et al. (2013)
19	% dead: 100	720–2160 [0]	1 (1)	A	ND	FT	Cepero (2003)
19	300 ppm: 6 d	144 [0]	9 (0.2–400)	A	~20	FT	Morton et al. (1976)
19	400 ppm: 6 d	144 [0]	9 (0.2–400)	A	~20	FT	Morton et al. (1976)
19	200 ppm: 6.5 d	156 [0]	9 (0.2–400)	A	~20	FT	Morton et al. (1976)
19	1 ppm: 15.3 d	367.2 [0]	9 (0.2–400)	A	~20	FT	Morton et al. (1976)
19	0.8 ppm: 23 d	552 [0]	9 (0.2–400)	A	~20	FT	Morton et al. (1976)
20	46.54 ppm	24 [0]	5 (5–80)	A (21–26)	23–27	Static	Montresor et al. (2013)
20	19.84 ppm	48 [0]	5 (5–80)	A (21–26)	23–27	Static	Montresor et al. (2013)
20	14.29 ppm	72 [0]	5 (5–80)	A (21–26)	23–27	Static	Montresor et al. (2013)

Table 1 (continued)

Toxicant	LC_{50} or % dead	Exposure [postexposure] times (h)	No. of concentrations tested (range, ppm)	Stage (size in mm)	Temp. (°C)	Test conditions	Reference
20	11.53 ppm	96 [0]	5 (5–80)	A (21–26)	23–27	Static	Montresor et al. (2013)
21	368.2 ppm	48 [48]	6 (10–2000)	A (15–25)	25	Static	Calazans et al. (2013)
22	0.58 ppm	24 [0]	5 (5–80)	A (21–26)	23–27	Static	Montresor et al. (2013)
22	0.35 ppm	48 [0]	5 (5–80)	A (21–26)	23–27	Static	Montresor et al. (2013)
22	0.29 ppm	72 [0]	5 (5–80)	A (21–26)	23–27	Static	Montresor et al. (2013)
22	0.25 ppm	96 [0]	5 (5–80)	A (21–26)	23–27	Static	Montresor et al. (2013)
23	% dead: 100	720–2160 [0]	1 (1)	A	ND	FT	Cepero (2003)
24	% dead: 100	720–2160 [0]	1 (1 ppm of Cu2+)	A	ND	FT	Cepero (2003)
24	% dead: 100	2160 [0]	3 (0.25–1 ppm of Cu2+)	A	ND	FT	Cepero (2003)

ND no data, *A* adults, *L* larvae, *FT* flow through

[a] Testing was carried out under plant conditions, comparing fouling developing on clean test vouchers enclosed in a steel pipe through which untreated (control) and treated water was circulated. Animals exposed to pH 9 for 24 h/day during 1 year. Mortality assessed by comparison with simultaneous control

been documented by numerous authors for dreissenids (Mackie and Claudi 2010). The mechanism appears to be the closing of the valves by veligers in the presence of an oxidant, thus preventing attachment to the substrate. Semicontinuous application of chlorine has been used with varying success. In adults, the application has to be such as to prevent them from opening their shells and recovering between chlorination events. Chlorination regime of 15 min on and 45 min off has been used successfully by several facilities in Canada for a number of years. Twice per day treatment of 1 h at 2 ppm has been shown to eliminate new settlement that has occurred in the last 12 h. However, this regime will not eliminate adult mussels (Mackie and Claudi 2010).

Reactive Application of Chemicals for Control of *L. fortunei*

Proprietary Nonoxidizing Chemicals

Nonoxidizing chemicals are generally used as reactive treatment to eliminate established adult population. This is primarily due to the fact that many of the proprietary nonoxidizing molluscicides must be detoxified prior to their release into open water environment. The proprietary chemicals also tend to be costly, making continuous application for settlement prevention not feasible. It is therefore their effect on adults that is of greatest interest. The effect of nonoxidizing chemicals on *L. fortunei* has been reported to be quite different from the effect some of these chemicals have on dreissenids.

Clam-Trol CT-2/Spectrus CT1300

The primary active ingredients of this product [50 % n-alkyl (C12-50 %, C14-40 %, and C16-10 %) dimethylbenzyl ammonium chloride] are cationic surfactants of the alkyldimethyl-benzyl ammonium chloride (ADBAC) family. There were three formulations of the product: CT-1, CT-2, and CT-3. The proportions of the various components varied in each formulation. Clam-Trol CT-2 was the product most tested on the golden mussel and it has been renamed relatively recently to Spectrus CT1300.

Boltovskoy and Cataldo (2003) tested the effect of Clam-Trol CT-2 at different concentrations (0.5–10 ppm). At the lowest concentration of 0.5 ppm no mortalities were observed. At 1 ppm LC_{50} was reached in 192 h, at 2 ppm it was 120 h, at 5 ppm it was 96 h, and at 10 ppm it was 72 h (Table 1). Interestingly, total mortality for 2 ppm, 5 ppm, and 10 ppm concentrations was reached at the same time: 192 h (Table 1). By comparison, the North American experience for dreissenids is that 2–5 ppm applied for 6–24 h will result in 100 % adult mortality (McMahon 2008).

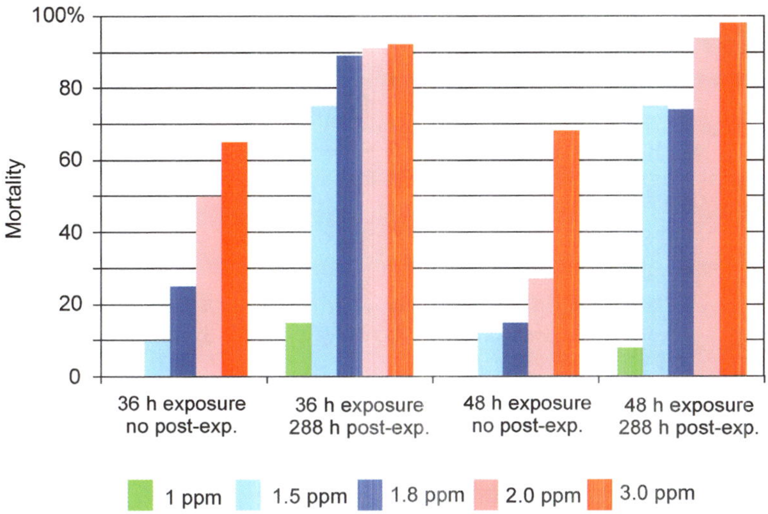

Fig. 1 Clam-Trol CT-2 exposure for 36 h followed by 288 h recovery. (Based on data from Cataldo et al. 2003)

Boltovskoy and Cataldo (2003) tested the effect of short-term exposure on adult *L. fortunei* followed by a period of recovery. They exposed the adult mussels to five concentrations (1, 1.5, 1.8, 2.0, and 3 ppm) for 36 and 48 h. These exposures were followed by a 12-day recovery during which the mussels were observed. After the recovery period, the authors observed significant postexposure mortality at all concentrations above 1 ppm (Fig. 1). This finding of postexposure mortality has significantly reduced the amount of product that needs to be used for reactive treatments in industrial facilities.

These postexposure mortality results are similar to those recorded by Cataldo et al. (2003). The authors report that 48 h exposure to concentration of 2.5 mg/L (ambient water temperature of 20 and 25 °C) resulted in 80–90 % mortalities after 3 days in recovery. It is interesting to note that at higher concentrations of the chemical, the authors noted that mussels in the test closed their shells and ceased filtering.

In a field experiment at the Itaipu Hydroelectric Power Plant in October 2002, the effect of Spectrus CT1300 (produced by GE Betz Inc.), with identical composition as Clam-Trol CT-2, was tested on *L. fortunei* in a continuous flow environment using bioboxes (Boltovskoy and Cataldo 2003). Adult mussels of various size classes were exposed to concentrations between 2.2 and 2.4 ppm of the product. The ambient water temperature was 22.5–23.5 °C. Mussels were exposed for periods of 12, 24, 36, 48, 60, and 72 h. After the exposure period mussels were moved to the control biobox for a period of 5 days. Mortalities in the controls were less that 5 %. The mortalities in the treated groups increased with exposure time, and ranged from 42 % mortality at 12 h to 90 % at 72 h. The mortality rate began to decrease after 48 h (Table 1). There were no significant differences in mortality between different size classes.

Boltovskoy et al. (2005) tested Spectrus CT1300 once again on adult *L. fortunei* at Embalse Rio Tercero Nuclear Power Plant (Argentina). Chlorine as sodium hypochlorite was also included in the study. The tests were carried out using flow-through bioboxes. The ambient temperature was 24–25 °C. Spectrus CT1300 was tested with a nominal concentration of 2.5 ppm (1.25 ppm active ingredient). Exposure times were 24, 48, and 72 h and postexposure (recovery) time was between 4 and 7 days. Chlorine was added for 4.5 h per day, and the nominal concentration was 1.5 ppm at the injection point and 0.5 ppm in the biobox. The flow rate to the bioboxes was set at 100 L per minute.

After 72 h, there was zero mortality in the control and only 1.2 % mortality in the biobox treated with chlorine. In the biobox treated with Spectrus CT1300, mortalities increased with time: 63 % of the mussels were killed in 24 h, 94 % in 48 h and 99 % in 72 h (Table 1). No significant differences were found between the mortality of different size classes of adult mussels indicating that the product works similarly across age classes. These results are comparable to those obtained by the authors at Itaipu in 2002 (Boltovskoy and Cataldo 2003).

The lack of difference in response by various size classes is in contrast to findings by Waller et al. (1993) working with dreissenids. These authors found a significant difference between the amount of chemical required to cause an LC_{50} in 48 h in 20–25-mm-long dreissenid mussels (0.738 mg/L of active ingredient) and 5–10-mm-long dreissenids (0.29 mg/L).

McMahon et al. (1994) found that Clam-Trol CT-1 had rapid zebra mussel toxicity at relatively low use rates (1.0–2.0 mg/L for 6–24 h).

Currently, the nuclear power plant Embalse de Río Tercero is using Spectrus at 2.5 ppm 2–3 times a year for 2 days. The 2-day treatment eliminates all settled adult golden mussels.

BULAB 6002

BULAB 6002 (Poly [oxyethylene (dimethyliminio) ethylene (dimethyliminio) ethylene dichloride]) is a liquid cationic polyquaternary ammonium compound used for the control of algae in swimming pools and as a microbicide for the control of microorganisms in commercial and industrial water systems. It also considered an effective molluscicide (Waller et al. 1993). Darrigran and Damborenea (2001) tested the effectiveness of BULAB 6002 on both juvenile and adult golden mussels. Cumulative mortality was assessed in six experiments conducted with different size classes of mussels and different concentrations of the product (8, 12, and 20 mg/L of active substance). At 24 °C, the results show that 20 mg/L resulted in 100 % mortality of adult golden mussels in 144 h (Table 1). Boltovskoy and Cataldo (2003) tested several concentrations of BULAB 6002 (2, 5, 10, and 20 ppm) on adult golden mussels. At ambient temperature of 20–22 °C LC_{50} was reached in 408 h at a concentration of 2 ppm of BULAB 6002, and 288 h for concentration of 20 ppm. Total mortality was reached after 672 h for the 2 ppm concentration and 552 h for the 20 ppm concentration. Interestingly, there was only 24 h difference between reaching total mortality at 10 ppm versus 20 ppm (Table 1). In comparison, during trials with dreissenids, adult mortality at 0.5 ppm was reached in 826 h, at 2 ppm the time required was 313 h, and at 8 ppm total adult mortality was reached in 197 h (McMahon et al. 1993).

H-130 (Didecyl Dimethyl Ammonium Chloride)

This compound is a nonoxidizing liquid containing a solution of polyquaternary alkyl ammonium registered in North America for use as a molluscicide in industrial once-through freshwater cooling water systems. Because of its need for proper deactivation prior to discharge, in North America, it is sold only as part of a complete Calgon mollusk treatment application service, and is to be used only with supervision from a Calgon representative.

Boltovskoy and Cataldo (2003) tested the effect of several concentrations of H130 (0.5, 1, 2, 5, and 10 ppm) on adult golden mussels. At 0.5 ppm there was less than 50% mortality after 30 days. At 1 ppm LC_{50} mortality was reached in 96 h, while concentrations of 2,5 and 10 ppm reached LC_{50} in 48 h and total mortality in 120 h.

Cataldo et al. (2003) tested several concentrations of H130 (2.5, 5, 10, 20, and 30 mg/L) at 15, 20, and 25 °C ambient temperature using a 48 h exposure followed by a recovery period in clean water. At 15 °C, none of the tested concentrations achieved 100% mortality in postexposure recovery experiments. At 20 °C, only doses >10 mg/L achieved 100% mortality. At 25 °C, all concentrations down to 2.5 mg/L were 100% effective in 1 week or less in causing 100% mortality. Once again, during treatments with the higher concentrations of the chemical, authors noted that the mussels in the test chamber closed their shells and ceased filtering.

For comparison, in dreissenids, 1 ppm for 24 h causes 100% mortality in adults (McMahon 2008).

Bayluscide

Bayluscide (dichloro-2'nitro-4' salicylanilide) was originally developed as a molluscicide to eliminate snails. It is not considered to be persistent in the environment; it breaks down in natural water and sediment systems through hydrolysis, photolysis, and microbial degradation (Dawson 2003).

Cataldo et al. (2003) tested several concentrations of Bayluscide (0.25, 0.5, 1, 2, 4, 6, and 8 mg/L) at 15, 20 and 25 °C using a 48 h exposure followed by a recovery period in clean water. At 15 °C, 48 h exposure at concentrations of 4 mg/L caused greater than 80% mortality. Concentrations between 1 and 2 mg/L killed 60–70%, and concentrations of 0.5 mg/L eliminated <20% of adults. All concentrations yielded postexposure mortalities only for the first 2–3 days during recovery. At 20 °C the final mortalities were generally higher and occurred faster than with comparable concentrations at 15 °C with similar postexposure mortalities for up to 5 days. At 25 °C, the adult mussel response to concentrations between 0.5 and 8 mg/L was almost identical and >90 % of the adult mussels were eliminated after 4 days during recovery. At 0.25 mg/L, mortality rate peaked to 30 % (Table 1).

By comparison, a concentration of 0.05 mg/L has been reported to cause 70 % mortality in 24 h in dreissenids, and 0.1 mg/L achieved 100 % mortality in the same time period (Waller et al. 1993).

Veligon [Poly (Dimethyl Diallyl Ammonium Chloride)]

High-charge cationic coagulant dimethyl diallyl ammonium chloride (DMDAAC) compounds have been used in water treatment plants as flocculants and clarification aids. There are various Veligon formulations of these compounds and they differ in their molecular weight and cationic charge density. Flocculation binds up the veligers in the floc and there is also reported impact on adult dreissenids. In North America, it is approved for use in potable water.

Boltovskoy and Cataldo (2003) tested the effect of several concentrations of Veligon TL-M (2, 5, 10, 20 and 50 ppm) on adult golden mussels. At a concentration of 2 ppm there was no mortality during the 48 h exposure or after 30 days in recovery. For concentrations of 5, 10, 20, and 50 ppm the following mortalities were noted after 48 h: 30, 50, 75, and 80 %. However, all the concentrations tested resulted in 100 % delayed mortality with 11 and 30 days (Table 1).

By comparison, median reported LC_{50} at 96 h for adult zebra mussels is between 1.5 and 3.0 mg/L (Blanck et al. 1996).

Nonproprietary Nonoxidizing Chemicals

There are numerous other nonoxidizing chemicals that impact both adult and larval golden mussels. The challenge is to select those that are economical and have the least possible negative impact on the environment.

Copper Sulfate

The toxicity of copper to marine life has been recognized for centuries, and hence the use of sheets of copper on bows of sailing ships to prevent hull fouling. Copper-based antifouling paints are widely used on marine and freshwater watercraft. Mollusks are particularly sensitive to the presence of copper in the environment. Elevated levels of copper can result in such diverse effects as decreased growth rate, reproductive impairment, enzyme inhibition, reductions or alterations in protein synthesis, and disruptions of ATP synthesis and Ca^{2+} homeostasis (Clayton et al. 2000). The toxicity of copper in freshwater systems is greatly influenced by the total

Table 2 Effect of copper-based algaecides on dreissenid mussel adults. Average percent mortality after 96 h of exposure, and mortality following maximum recovery time allowed. (After Claudi et al. 2014)

	Low concentration				High concentration			
	Zebra		Quagga		Zebra		Quagga	
Algaecide	96 h	Max	96 h	Max	96 h	Max	96 h	Max
Copper sulfate (July)	0.8%	6.7%	–	–	25.0%	69.9%	–	–
Copper sulfate (Nov)	10.4%	14.2%	65.0%	92.4%	36.3%	52.1%	96.5%	99.4%
Natrix™ (July)	16.2%	62.2%	–	–	63.0%	97.2%	–	–
Natrix™ (Nov)	55.3%	64.4%	93.0%	99.6%	70.7%	84.6%	100%	–
Captain™	26.0%	43.5%	94.1%	–	72.5%	86.7%	94.1%	–
EarthTec®	100%	–	97.1%	–	100%	–	99.3%	–

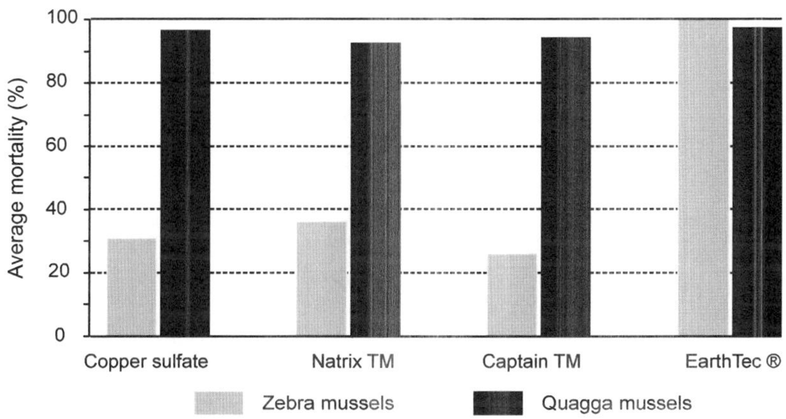

Fig. 2 Effect of copper-based algaecides (0.5 mg/L copper equivalent) on dreissenid mussel adults. (After Claudi et al. 2014)

hardness of water. Claudi et al. (2014) found four copper-based algaecides to be a viable tool for managing adult zebra and quagga mussel infestations, particularly in water bodies which require the use of these chemicals for the control of algae or aquatic plants (Table 2). Figure 2 shows the observed mortality in zebra and quagga mussels after 96 h exposure to 0.5 mg/L copper concentration of the four algaecides tested.

Soares et al. (2009) evaluated copper sulfate toxicity to golden mussels. The aim was to establish LC_{50} and LC_{95} doses which could be used for adult mussel control. Adult *L. fortunei* were exposed to 1.24, 2.33, 3.88, 5.43, 7.76, 10.08, 13.19, and 15.50 mg/L copper concentrations through analogous copper sulfate solutions ($CuSO_4\ 5H_2O$). Test organisms (1.8–2 mm in length) were acclimatized

(20±1 °C, 24 h) in aquaria with air pumps in river water prior to exposure to the copper solution for 48 h. The average, minimum, and maximum LC_{50} and LC_{95} values were 2.16 (1.70 and 2.65), and 4.86 (3.97 and 6.47) mg/L, respectively.

Salinity

Salinity tolerance of the golden mussel was investigated by Angonesi et al. (2008). Ninety percent of the individuals of *L. fortunei* survived for at least 10 days after they were placed in salinity of 2‰. When exposed to salinity of 4 and 6‰, the adult mussels exhibited high tolerance only in the first 96 and 72 h, respectively, with a sharp decline thereafter. Similar salinity tolerance can be found in *L. fortunei* in the Asian and South American estuarine regions, such as the Changjiang River (China) (Huang et al. 1981), Río de la Plata estuary (Argentina) (Darrigran and Damborenea 2006), and in the Patos Lagoon (Brazil) (Capítoli and Bemvenuti 2004). In the studies above, *L. fortunei* survival differs from the findings by Deaton et al. (1989), where 80% of the animals survived in a salinity of 6.8‰ for time periods of 3 weeks or more, and less than a week in a salinity of 13.6‰. Angonesi et al. (2008) found *L. fortunei* specimens could only tolerate a salinity of 2‰ for a period of up to 10 days. For this same time period, salinities starting at 4‰ were fatal for at least 80% of the organisms. The higher tolerance found by Deaton et al. (1989) through salinity and osmotic regulation experiments (hemolymph osmotic and ionic composition and tissue amino acid content) may be due to the species' ecological adaptation in Asia, the species' origin (Deaton et al. 1989).

In agreement with the results of Angonesi et al. (2008), *L. fortunei* was found to be restricted to salinities below 2–3‰ in several South American brackish waterbodies (Capítoli and Bemvenuti 2004; Brugnoli et al. 2005; Darrigran and Mansur 2006; Darrigran et al. 2011). However, in 2004 and 2009 *L. fortunei* was recorded in the Río de la Plata estuary in the vicinity of Montevideo (Uruguay) (Giberto and Sardiña 2009), in an area periodically influenced by salinities in excess of 20‰ (Sylvester et al. 2013). This suggests that while the mussel is not tolerant of continuous salinities above 2 or 3‰, it is able to tolerate relatively high intermittent salinity exposure. Sylvester et al. (2013) evaluated the tolerance of *L. fortunei* to intermittent higher salinity by testing mussel mortality in 30-day experiments using both constant and fluctuating salinities at different temperatures in the laboratory. Test conditions simulated different seasons of the year and locations with increasing influence of marine waters in the Río de la Plata estuary. Significant mortality (31% after 30 days) was observed at a constant salinity of 2‰, increasing to 45 and 57% at 5 and 10‰, respectively. In contrast, considerably greater tolerances were observed when conditions in the experimental chamber fluctuated between saltwater and freshwater. No significant mortality was observed in mussels exposed to a salinity cycle with abrupt salinity changes ranging from 1 to 23‰ (mean 2.68‰) over a month. Tolerance was not affected by different temperatures. As mussels were observed to close their shells in higher salinity regimes (Boltovskoy pers. comm.), the tolerance of *L. fortunei* to short term salinity changes is likely due to the ability of the adults to avoid noxious conditions.

Calazans et al. (2013) observed LC_{50} of sodium chloride to be 8.3 ppt for a 48 h exposure. This suggests that high salinity could be used to eliminate settled adult golden mussels on freshwater ship hulls or ballast water tanks. By taking ships in need of clearing golden mussel infestation to very saline parts of the estuary for more than 2 days could help eliminate any fouling of the hull and of the cooling water systems. Filling ballast water tanks with saline waters could have the same effect thereby limiting further expansion of the geographic range of the mussel through ballast water introduction.

Ammonium Chloride

Montresor et al. (2013) tested the effect of unionized ammonia (TA-N) on adult *L. fortunei* using a solution of ammonium chloride. The concentrations of NH_3-N tested were 0.14, 0.21, 0.31, 0.50, and 0.72 mg/L at temperatures between 23 and 27 °C. The authors note that the current legal limits for concentrations of TA-N in Brazil are 5.6 mg/L TA-N at pH between 7.5 and 8.0. Given the high limits for TA-N in Brazil, ammonium chloride could be considered a viable treatment for elimination of adult mussels in this country.

Biobullets

Microencapsulated poison has been created in England to combat *D. polymorpha* in Britain. Calazans et al. (2013) tested microencapsulated chemicals, along with the traditional dissolved chlorine and potassium chloride (KCl), for the control of *L. fortunei*. The "biobullets" tested were a commercial blend of microencapsulated KCl and quaternary ammonia. The encapsulation coating was made of a mixture of binder starch, oil, and wax. The amount of microencapsulated KCL required to cause 50 % mortality was ten times lower than for the dissolved form of the same chemical. The same study demonstrated similar effects for other microencapsulated substances. Since the amount of chemical released into the environment in microencapsulated form is substantially lower, Biobullets may be a more environmentally friendly alternative to deliver chemical treatment to adult mussels. At this point, the technology may still require some improvement. In a study commissioned by the Spanish government, the use of Biobullets was not rated as the best available technique due to the need to improve the stability of the microparticles and standardizing the Biobullet size to enhance particle retention by the target molluscs (Calazans pers. comm.).

Oxidizing Chemicals

Chlorine

Chlorine as chlorine gas, liquid sodium hypochlorite, calcium hypochlorite pellets, or pellets of sodium dichloroisocyanurate are used by a large majority of the plants in South America that use chemical control. Porto Primavera CESP (Brazil) uses sodium dichloroisocyanurate; Bariri Ibitinga AES Tietê (Brazil) use chlorine gas; Rosana, Taquaruçu, Canoas I and Canoas II of Duke Energy (Brazil) use chlorine gas, Itaipu (Brazil-Paraguay) uses chlorine gas and sodium hypochlorite; Tucuruí Eletronorte (Brazil) uses granular chlorine; Central Puerto and Termoeléctrica General Belgrano (Argentina) use sodium hypochlorite. Whatever the source, chlorine is consistently toxic at approximately the same concentration of Total Residual Chlorine (TRC) or Free Available Chlorine (FAC).

Numerous studies exist detailing the toxicity of chlorine to adults of *L. fortunei*. Cataldo et al. (2003) tested sodium hypochlorite concentrations (as free available chlorine) of 1, 5, 10, 25, 50, and 100 mg/L at three different exposure temperatures (15, 20, and 25 °C). The authors report that at 15 °C, there were no mortalities at any concentration of chlorine for 2 weeks. After that, except for the lowest dose of 1 mg/L, all concentrations resulted in 100 % mortality in 2–4 weeks. At 1 mg/L, there was only 30 % mortality at the end of the experiment on day 47. At 20 °C, chlorine concentrations between 5 and 100 mg/L resulted in 100 % mortality after about 1 month of exposure; at a dose of 1 mg/L 50 % mortality was observed at the end of the experiment. At 25 °C, the effects of chlorine were almost identical throughout the entire range of concentrations tested; 100 mg/L chlorine required 11 days, and 1 and 5 mg/L chlorine required 17 days for total mortality. The calculated LC_{50} values at 15 °C were 25 days at 93.2 mg/L, 30 days at 51.7 mg/L, 35 days at 27.2 mg/L, 40 days at 14.0 mg/L, and 45 days at 2.1 mg/L (Table 1). At 20 °C the calculated LD50 values were 20 days at 3.3 mg/L and 25 days at 1.2 mg/L. At 25 °C, the calculated LD50 value was 10 days at 5.5 mg/L (Table 1).

Adult mussels recognize chlorine as a noxious substance and keep their shell closed as long as possible before they are forced to take in water for respiration. The length of time they are able to keep the shells closed depends on the ambient water temperature. The warmer the water, the shorter is the time of shell closure. This explains why no mortality was observed at 15 °C for 2 weeks. Higher doses of chlorine do not translate into significantly lower time to mortality as reflected by the LC_{50} values: at 15 °C 25 days were needed at 93.2 mg/L, and 45 days at 2.1 mg/L (Cataldo et al. 2003). These results generally align with earlier studies on *L. fortunei* (Morton et al. 1976).

CESPI (Companhia Energetica de São Paulo) reports good results when treating cooling water systems for 2 h/day by addition of sodium hypochlorite at a concentration of 1.5 ppm residual in the system. This type of strategy has been shown to eliminate freshly settled juveniles of *Dreissena*. The freshly settled juveniles have relatively fragile shells that can be oxidized, exposing the juvenile to chlorine attack. Adults are not affected as their shells are robust and can be kept closed during short periods of chlorination.

Sodium Dichloroisocyanurate

This compound is mainly used as a disinfectant, biocide, industrial deodorant, and detergent. It is found in some modern water purification tablets and filters. In these applications, it is a source of slow release of chlorine in low concentrations at a relatively constant rate. CESPI has successfully tested the use of sodium dichloroisocyanurate as an alternative source of chlorine for golden mussel control. The advantage is easier storage and no formation of trihalomethanes. On the negative side, the product is effervescent and forms microbubbles that can negatively impact pump performance and must be eliminated prior to the addition of the dissolved product. The strategy is to maintain 1.5 ppm residual of chlorine in the pipe for 2 h during the breeding season and to cut back to 0.5 ppm residual for 2 h in the winter.

Concluding Remarks

The advantage of chemical control is the ability of such a strategy to protect the entire system, from the point of chemical injection to the point of water discharge. It can be applied continuously or semicontinuously to eliminate settlement or it can be applied periodically to eliminate adults. Chemical addition is quick to implement with generally modest capital expenditure. However, in many parts of North America and Europe, chemical controls for macrofouling are more and more difficult to implement due to strict regulatory limits on the presence of chemicals in the discharge water.

Many chemicals will cause mortality in golden mussels; however, worker safety, cost, and protection of the environment, protection of materials of construction and ambient water quality all have to be evaluated when contemplating a chemical protection strategy. As requirements and conditions differ between facilities, no one method or chemical will be suitable at every location. The same is true of proactive versus reactive treatments. Facilities coping with massive infestation levels are more likely to require proactive or frequent periodic treatment to keep sensitive components from failing. During the periodic treatments, it is essential to apply the chemical continuously until mortality of all adults has been reached. Any break in chemical application will allow the adults to recover and the overall length of treatment will increase.

The timing of periodic or end of season treatments needs to take ambient water temperature into consideration. Most, if not all chemicals will cause mortality more quickly and at lower concentration in warm water (20–30 °C) then in temperature below 15 °C (Fig. 3). To minimize treatment time and lower the cost of chemicals required, warm water treatments are recommended.

Finally, from the above review of the chemical methods of control we must conclude that the adults of *L. fortunei* appear to require higher doses or longer treatment periods than adult dreissenids. This observation is likely the result of the higher tolerance *L. fortunei* has for a variety of environmental conditions and polluted water

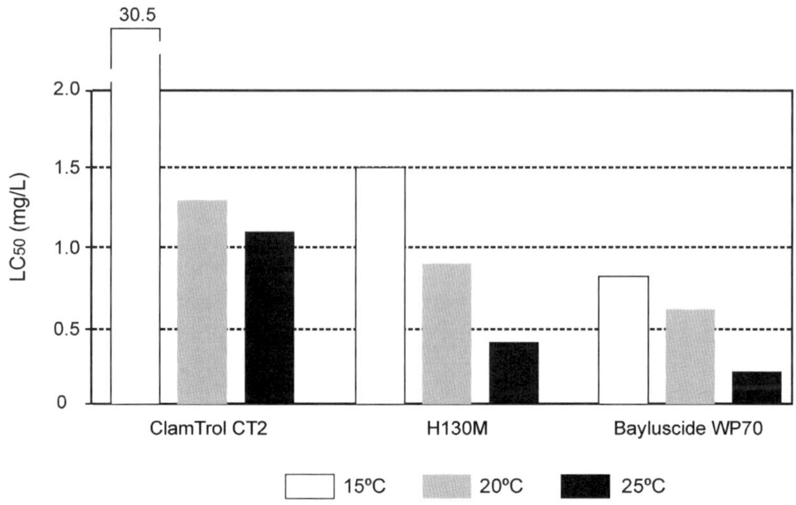

Fig. 3 LC_{50} values for three different molluscicides at 15, 20 and 25 °C. In all cases exposure times are 2 days; concentrations are based on percent active ingredient. (Based on data from Cataldo et al. 2003)

(Villar et al. 1999; Belaich et al. 2006; Boltovskoy et al. 2006; Young et al. 2014). It is, however, important to note that the impact of oxidizing chemicals on veligers or freshly settled juveniles is identical between the two species. In proactive treatments preventing settlement, the veligers do detect the oxidant as a noxious substance, close their shells and do not attach. In freshly settled juveniles, the impact of the oxidizing chemical is due to an attack on the fragile shell provided the oxidant dose is high enough for a long period of time.

References

Angonesi LG, da Rosa NG, Bemvenuti CE (2008) Tolerance to salinities shocks of the invasive mussel *Limnoperna fortunei* under experimental conditions. Iheringia (Série Zoologia) 98:66–69

Belaich M, Oliver C, Pilloff M, Porta A (2006) Evaluation of a biomarker of Cd(II) exposure on *Limnoperna fortunei*. Environ Pollut 144:280–288

Blanck CA, Mead DF, Adams DJ (1996) Effective control of zebra mussels using a high molecular weight polymer. In: sixth international zebra mussel and other aquatic nuisance species conference, Dearborn

Boltovskoy D, Cataldo D (2003) Estudios de evaluación, monitoreo y análisis de medidas de control de incrustaciones de moluscos bivalvos (*Limnoperna fortunei*) en las instalaciones de la central hidroeléctrica Itaipu. Infome final de trabajos. Central Hidroelectrica Itaipú. CHI, pp 1–281

Boltovskoy D, Cataldo D, Hechem M, Matsubara A (2005) Evaluación de la colonización y control de *Limnoperna fortunei* en la Central Nuclear Embalse de Río Tercero, Córdoba. Informe final. Central Nuclear Embalse. Embalse, pp 1–46

Boltovskoy D, Correa N, Cataldo D, Sylvester F (2006) Dispersion and ecological impact of the invasive freshwater bivalve *Limnoperna fortunei* in the Río de la Plata watershed and beyond. Biol Invasions 8:947–963

Brugnoli E, Clemente J, Boccardi L, Borthagaray A, Scarabino F (2005) Golden mussel *Limnoperna fortunei* (Bivalvia: Mytilidae) distribution in the main hydrographical basins of Uruguay: update and predictions. Anais da Academia Brasileira de Ciências 77:235–244

Calazans SHC, Fernandes FC (2012) Microencapsulados. In: Mansur MCD, Santos CP, Pereira D, Padula Paz IC, Leite Zurita ML, Raya Rodriguez MT, Vilar Nehrke M, Aydos Bergonci PE (eds) Moluscos límnicos invasores no Brasil. Biologia, prevenção, controle, Redes Editora, Porto Alegre, pp 299–302

Calazans SHC, Americo JA, Fernandes FD, Aldridge DC, Rebelo MD (2013) Assessment of toxicity of dissolved and microencapsulated biocides for control of the golden mussel *Limnoperna fortunei*. Mar Environ Res 91:104–108

Capítoli RR, Bemvenuti CE (2004) Distribuição do mexilhão dourado *Limnoperna fortunei* (Dunker 1857) na área estuarina da Lagoa dos Patos e Canal São Gonçalo. In: VI Simposio de Ecossistemas Brasileiros, Sao Jose dos Campos

Cataldo D, Boltovskoy D, Pose M (2003) Toxicity of chlorine and three non-oxidizing molluscicides to the invasive pest mussel *Limnoperna fortunei*. J Am Water Works Assoc 95:66–78

Cepero E (2003) Control de *Limnoperna fortunei*. Infestación de plantas potabilizadoras de agua. In: 13 Congreso Argentino de Saneamiento y Medio Ambiente, Buenos Aires

Claudi R, Graves A, Taraborelli AC, Prescott R, Mastitsky S (2012) Impact of pH on survival and settlement of dreissenid mussels. Aquat Invasions 7:21–28

Claudi R, Prescott TH, Mastisky S, Coffey H (2014) Efficacy of copper based algaecides for control of Quagga and Zebra Mussels. Report for California department of water resources, aquatic nuisance species program, pp 1–58

Clayton ME, Steinmann R, Fent K (2000) Different expression patterns of heat shock proteins hsp 60 and hsp 70 in zebra mussels (*Dreissena polymorpha*) exposed to copper and tributyltin. Aquat Toxicol 47:213–226

Darrigran G, Damborenea C (2001) Concentraciones letales de un biocida para adultos del molusco invasor *Limnoperna fortunei* (Mytilidae). In: Seminario Internacional - Gestión Ambiental e Hidroelectricidad, Salto Grande

Darrigran GA, Mansur MCD (2006) Distribución, abundancia y dispersión. In: Darrigran GA, Damborenea C (eds) Bioinvasión del mejillón dorado en el continente americano. Editorial Universitaria de La Plata, La Plata, pp 93–110

Darrigran GA, Damborenea C (eds) (2006) Bioinvasión del mejillón dorado en el continente americano. Editorial Universitaria de La Plata, La Plata, pp 1–221

Darrigran GA, Damborenea C, Drago EC, Ezcurra de Drago I, Paira A (2011) Environmental factors restrict the invasion process of *Limnoperna fortunei* (Mytilidae) in the Neotropical region: a case study from the Andean tributaries. Ann Limnol Int J Limnol 47:221–229

Darrigran GA, Maroñas ME, Colautti DC (2001) Primeras estimaciones de concentraciones letales de un biocida para el molusco invasor *Limnoperna fortunei* (Mytilidae). In: Seminario Internacional - Gestión Ambiental e Hidroelectricidad, Salto Grande (Argentina-Uruguay)

Dawson VK (2003) Environmental fate and effects of the lampricide Bayluscide: a review. J Gt Lakes Res 29:475–492

Deaton LE, Derby JGS, Subhedar N, Greenberg MJ (1989) Osmoregulation and salinity tolerance in two species of bivalve mollusc: *Limnoperna fortunei* and *Mytilopsis leucophaeta*. J Exp Mar Biol Ecol 133:67–79

Giberto DA, Sardiña P (2009) *Mytella charruana* d'orbigny 1842 y *Limnoperna fortunei* (Dunker, 1857) (Bivalvia: Mytilidae) en la zona mixohalina del Río de La Plata: ¿bancos residuales o futuras poblaciones locales? In: VII Jornadas Nacionales de Ciencias del Mar, Bahía Blanca

Huang Z, Li C, Zhang L, Li F, Zheng C (1981) The distribution of fouling organisms in Changjiang River estuary. Oceanologia et Limnologia Sinica 12:531–537

Karatayev AY, Boltovskoy D, Padilla DK, Burlakova LE (2007) The invasive bivalves *Dreissena polymorpha* and *Limnoperna fortunei*: parallels, contrasts, potential spread and invasion impacts. J Shellfish Res 26:205–213

Karatayev AY, Burlakova LE, Karatayev VA, Boltovskoy D (2010) *Limnoperna fortunei* versus *Dreissena polymorpha*: Population densities and benthic community impacts of two invasive freshwater bivalves. J Shellfish Res 29:975–984

Mackie GL, Claudi R (2010) Monitoring and control of macrofouling mollusks in fresh water systems. CRC, Boca Raton, pp 1–508

Mata FAR, Dias C, Mader O (2013) Utilização do remediador MXD-100 para controle de incrustação em sistemas de resfriamento de usinas hidrelétricas, ocasionada por *Limnoperna fortunei* (Dunker, 1857). In: 1 Congreso Argentino de Malacología, La Plata

Matthews MA, McMahon RF (1994) The survival of Zebra Mussels (*Dreissena polymorpha*) and Asian Clams (*Corbicula fluminea*) under extreme hypoxia. In: 4th international zebra mussel conference, Madison

McMahon RF (2008) Nonoxidizing biocides, thermal shocking, dessication, oxygen deprivation, carbon dioxide, and low frequency agitation. In: Quagga Mussel conference, Central Utah Water Conservancy District, Orem

McMahon RF, Shipman BN, Long DP (1993) Laboratory efficacies of nonoxidizing molluscicides on the zebra mussel (*Dreissena polymorpha*) and the Asian clam (*Corbicula fluminea*). In: Nalepa TF, Schloesser D (eds) Zebra mussels: biology, impacts, and control, Lewis, Boca Raton, pp 575–598

McMahon RF, Ussery TA, Clarke M (1994) Review of zebra mussel control methods (Technical Note ZMR-2-14). U.S. Army Corps of Engineers Waterways Experiment Station. Vicksburg, pp 1–12

Montresor LC, Miranda-Filho KC, Paglia A, Luz DM, Araujo JM, Silva MJS, Gerhard L, Martinez CB, Vidigal TH (2013) Short-term toxicity of ammonia, sodium hydroxide and a commercial biocide to golden mussel *Limnoperna fortunei* (Dunker, 1857). Ecotoxicol Environ Saf 92:150–154

Morton B, Au CS, Lam WW (1976) Control of *Limnoperna fortunei*. J Inst Water Eng Sci 30:147–156

Nalco Inc. (2013) Drinking water golden mussel control. Case study—drinking water CH-1568. pp 1–2

Netto OSM (2011) Controle da incrustação de organismos invasores em materiais de sistemas de resfriamento de usinas hidrelétricas. MSc Thesis, Universidade Federal de Curitiba (Brazil), pp 1–113

Pastorino G, Darrigran GA, Martín SM, Lunaschi L (1993) *Limnoperna fortunei* (Dunker, 1857) (Mytilidae), nuevo bivalvo invasor en aguas del Río de la Plata. Neotropica 39:101–102

Perepelizin PV, Boltovskoy D (2011) Resistance of the invasive pest mussel *Limnoperna fortunei* to anoxia. J Am Water Works Assoc 103:79–85

Pereyra PJ, Rossini GB, Darrigran GA (2011) Toxicity of three commercial tannins to the nuisance invasive species *Limnoperna fortunei* (Dunker, 1857): implications for control. Fresenius Environ Bull 20:1432–1437

Rajagopal S, Jenner HA, Venugopalan VP (eds) (2012) Operational and environmental consequences of large industrial cooling water systems. Springer, New York, pp 1–477

Rothe M (2007) Control of golden mussel by ozonation of cooling water in power plants. In: 15th international conference on aquatic invasive species, Nijmegen

Soares MF, Pereira D, Santos CP, Mansur MCD, Pires M, Breintenbach JO, Grespan C (2009) Toxicidade do sulfato de cobre ao mexilhão dourado, *Limnoperna fortunei* (Dunker, 1857), em água bruta. J Braz Soc Ecotoxicol 4:37–48

Sprecher SL, Getsinger KD (2000) Zebra mussel chemical control guide. Erdc/El Tr-00-1, U.S. Army Engineer Research and Development Center, Vicksburg, pp 1–89

Sylvester F, Cataldo D, Notaro C, Boltovskoy D (2013) Fluctuating salinity improves survival of the invasive freshwater golden mussel at high salinity: implications for the introduction of aquatic species through estuarine ports. Biol Invasions 15:1355–1366

Villar C, Stripeikis J, D'Huicque L, Tudino M, Troccoli O, Bonetto C (1999) Cd, Cu and Zn concentrations in sediments and the invasive bivalves *Limnoperna fortunei* and *Corbicula fluminea* at the Rio de la Plata basin, Argentina. Hydrobiologia 416:41–49

Waller DL, Rach JJ, Cope GW, Marking LL, Fisher SW, Dabrowska H (1993) Toxicity of candidate molluscicides to zebra mussels (*Dreissena polymorpha*) and selected nontarget organisms. J Gt Lakes Res 19:695–702

Young SS, Yang HN, Huang DJ, Liu SM, Huang YH, Chiang CT, Liu JW (2014) Using benthic macroinvertebrate and fish communities as bioindicators of the Tanshui River Basin around the greater Taipei area—multivariate analysis of spatial variation related to levels of water pollution. Int J Environ Res Public Health 11:7116–7143

Control of *Limnoperna fortunei* Fouling by Thermal Treatments

Pablo V. Perepelizin and Demetrio Boltovskoy

Abstract Thermal treatments for the control of mussel fouling can be applied in two different ways: (1) Water temperature can be gradually increased at a specific rate until target organisms die off (acute upper lethal temperature), or (2) animals can be exposed to a constant (high) temperature for periods long enough to achieve 100% mortality (chronic upper thermal limit). Exposure of *L. fortunei* to temperature increase rates of 1 °C per 5, 15, or 30 min yields 100% mortality after 1.8–15.3 h, at temperatures between 43.6 and 50.2 °C. In chronic treatments, at 34–36 °C total mortality takes 25.0–644.3 h, whereas at 38–43 °C all mussels die after 0.7–17.5 h.

Keywords *Limnoperna fortunei* · Golden mussel · Biological invasion · Biofouling control · Impact · Industrial plants · Power plants · Thermal treatment · Upper lethal temperature

Introduction

All organisms are highly sensitive to ambient temperature. In both freshwater and marine habitats, temperature is usually the most important abiotic factor which determines vital activities, including reproduction, growth, and survival. Animals and plants have thermal ranges where they successfully thrive. Metabolic activity increases with temperature up to a certain point, but shortly after the optimum temperature is exceeded, metabolic activity is affected adversely. Thus, the optimum temperature is normally much closer to the upper thermal limit than to the lower

P. V. Perepelizin (✉) · D. Boltovskoy
Museo Argentino de Ciencias Naturales 'Bernardino Rivadavia'-CONICET,
Av. Angel Gallardo 470, 1405 Buenos Aires, Argentina
e-mail: perepelizin@macn.gov.ar

D. Boltovskoy
Instituto de Ecología, Genética y Evolución de Buenos Aires (IEGEBA), Facultad de Ciencias Exactas y Naturales, Universidad de Buenos Aires-CONICET,
Ciudad Universitaria, 1428 Buenos Aires, Argentina
e-mail: demetrio@ege.fcen.uba.ar

D. Boltovskoy (ed.), *Limnoperna fortunei,* Invading Nature - Springer Series in Invasion Ecology 10, DOI 10.1007/978-3-319-13494-9_24

one (Mordukhai-Boltovskoi 1975). This trait has allowed for the use of heated water to eliminate fouling organisms from industrial and power plant cooling systems (Graham et al. 1975; Claudi and Mackie 1994; Rajagopal et al. 1995; Maddox 1998), as well as to mitigate hull-fouling-related transport of introduced species (Beyer et al. 2011). Thermal shock is often considered as "one of the most efficient, environmentally sound and cost effective methods" of controlling introduced fouling organisms (O'Neill and MacNeill 1991).

Abundant field data on the distribution of *Limnoperna fortunei* indicate that the mussel has an extremely wide thermal range, spanning from ca. 0 to 35 °C (Karatayev et al. 2007; see Chapter "Parallels and contrasts between *Limnoperna fortunei* and species of *Dreissena*" in this volume), but surveys on its tolerance to extreme temperatures applicable to the design of control strategies are few. Montalto and Marchese (2003) carried out tests of the combined effects of pH (5 and 10) and temperature (5 and 35 °C) on survival of the mussel. Their experimental design was not aimed at the development of control measures, and therefore the results are of limited practical use. Oliveira et al. (2010) performed experiments on the thermal tolerance of *L. fortunei* to low (rather than high) temperatures. Rolla and Mota (2010) reported data on survival rates at 10–40 °C, but the lack of methodological details (acclimation temperatures, number of experimental specimens, controls, mussel size, etc.) and inadequate observation intervals (24 h) make these data of little value.

Thermal treatments can be applied in two different ways. Water temperature can be gradually increased at a specific rate until target organisms die off; in this procedure, the temperature necessary for killing a given proportion of the fouling population is referred to as the acute upper lethal temperature. Results are expressed as lethal temperatures (e.g., the temperature required to kill 50 or 100 % of the animals) (McMahon and Ussery 1995; Perepelizin and Boltovskoy 2011b). An alternative method consists in exposing organisms to a constant (high) temperature for periods long enough to achieve 100 % mortality, known as chronic upper thermal limit (McMahon et al. 1995; Perepelizin and Boltovskoy 2011a). In both cases, initial or acclimation temperatures may affect results significantly (McMahon et al. 1995).

The only studies aimed specifically at the control of *L. fortunei* in industrial installations by means of thermal shock are those of Perepelizin (2011), subsequently published by Perepelizin and Boltovskoy (2011a, b), where the authors assessed the acute upper lethal temperature and the chronic upper lethal temperature of the mussel.

Acute Upper Lethal Temperature

Perepelizin and Boltovskoy (2011b) exposed small (6 mm) and large (20 mm) mussels to different temperature increase rates (1 °C per 5, 15, or 30 min, which are within the range of operationally feasible values at industrial raw water cooling systems; McMahon and Ussery 1995), starting from three initial temperatures: 12, 23, and 28 °C. Experiments were monitored until all animals (total n: 1112) were dead. Under these experimental conditions, 100 % mortality occurred between 1.8

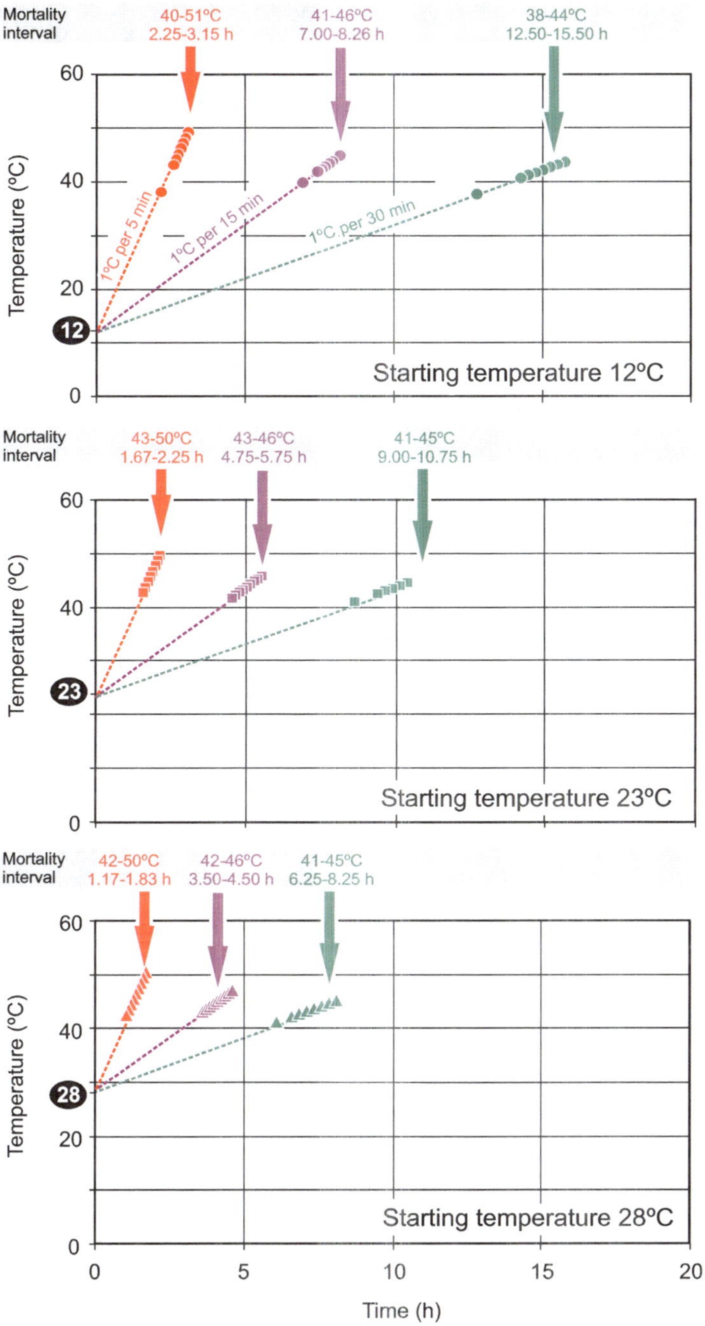

Fig. 1 Temperature and time of 100 % *Limnoperna fortunei* mortality exposed to increasing temperature at three different heating rates, starting from three different initial temperatures (pooled data for small and large specimens). *Dotted lines* indicate temperature increase rates; *symbols* denote death of one or more individuals. (Based on data from Perepelizin and Boltovskoy 2011b)

and 15.3 h, at temperatures between 43.6 and 50.2 °C (Fig. 1). The results of this work showed that faster rates of temperature increase take less time to kill mussels (around 2–3 h), but the temperature at which 100 % mortality occurs is several degrees higher than those at slower heating rates (48 to > 50 °C; Fig. 1). The lowest temperature at which all mussels died was around 43–44 °C, attained at the slowest rates of temperature increase (1 °C per 30 min) after 8–15 h. Neither the mussel size nor the starting temperature was associated with either temperature or time to yield 100 % mortality. Longer survival at fast temperature increase rates is most probably due to a delay in the biological response of the animals to the harmful thermal conditions.

Chronic Upper Lethal Temperature

In this survey (Perepelizin and Boltovskoy 2011a), experiments were performed with small (7 mm) and large (21 mm) animals (overall n: 1700) acclimated at two different temperatures characteristic of summer (28 °C) and winter (12 °C) conditions in the Río de la Plata estuary. Exposure temperatures (selected on the basis of the ambient temperatures where *L. fortunei* is known to thrive, and the operationally feasible conditions in industry) were 34, 36 (exposures of up to 4 weeks), 38, 40, 41, 42, and 43 °C (exposures of up to 24 h). Temperature was raised from initial acclimation levels to each treatment value at a rate of 0.1 °C/min (the rate generally feasible at power plants) and maintained constant until 100 % mortality in all (3) replicates.

These experiments showed that at temperatures ≥ 38 °C 100 % mortality takes between < 1 h (at 43 °C) and around 16 h (at 38 °C) (Fig. 2), regardless of the initial temperature or mussel size. At 34–36 °C the time needed to kill all animals varied between 25 and 644 h, but acclimation temperatures affected mortality rates significantly, with considerably higher survival times in the summer (acclimation temperature: 28 °C) than in the winter (acclimation temperature: 12 °C) (Fig. 2a).

Concluding Remarks

Thermal shock has some important advantages over other methods of control, especially in subtropical and tropical areas, where ambient water temperatures are already closer to the upper tolerance limit of the species (Mackie and Claudi 2010). It can be cost-effective, especially when increase in the water temperature is attained through thermal backwash or recirculation of thermal discharge, and it does not flush toxic substances into the environment.

On the other hand, it also has limitations and drawbacks. Government regulations often limit the temperature of discharge water, which may require additional installations for reducing effluent temperature. In order to use excess heat generated by the plant, industrial installations must be provided with additional water circuits,

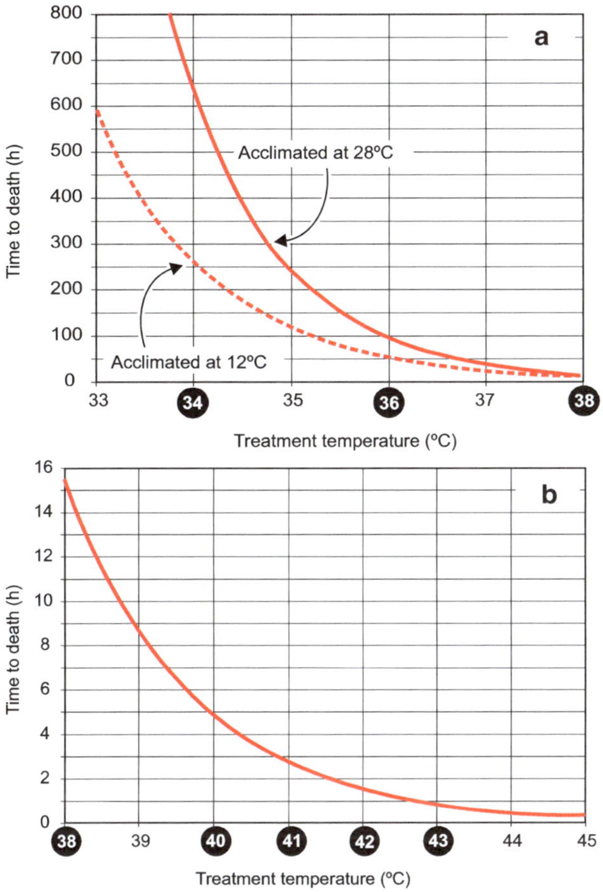

Fig. 2 Exposure times necessary to kill 100 % of the *Limnoperna fortunei* specimens at different temperatures (actual experimental temperatures are enclosed in *black circles*). (Modified from Perepelizin and Boltovskoy 2011a)

or retrofitted in order to be able to use this method. These modifications are costly and not always possible. Even when this method is applicable, it usually involves plant shutoffs or operation at reduced power level during the period of treatment (Mackie and Claudi 2010), which may significantly raise costs, especially considering that, for any given treatment temperature, the total operational time required is higher than those derived from laboratory experiments because the latter do not include the time necessary to attain the target temperature. Lead times vary depending on ambient water temperature and operational possibilities (Whitehouse et al. 1985; Perepelizin and Boltovskoy 2011a). Injecting steam or hot water generated ad hoc for this purpose (Miller et al. 1992) involves an additional expenditure. The method is nonselective, thus killing all or most of the organisms present in the cooling water, both fouling ones and those that pass through without causing harm.

Despite these shortcomings, thermal shock has been used successfully both in Europe and in North America for controlling zebra mussel fouling (Mackie and Claudi 2010).

Acknowledgements This work was partially financed by grants from the University of Buenos Aires, Argentina (UBA X-020 and 20020100100035) and from the Agencia Nacional de Promoción Científica y Tecnológica, Argentina (PICT 2007 1968) to DB.

References

Beyer J, Moy P, De Stasio B (2011) Acute upper thermal limits of three aquatic invasive invertebrates: hot water treatment to prevent upstream transport of invasive species. Environ Manage 47:67–76

Claudi R, Mackie GL (1994) Controlling zebra mussels at water treatment plant intakes. CRS Press, Boca Raton, pp 1–227

Graham JW, Moncreiff RW, Benson PH, Stock JN (1975) Heat treatment for the control of marine fouling at coastal electric generating stations. In: Ocean 75 Conference, San Diego (USA)

Karatayev AY, Boltovskoy D, Padilla DK, Burlakova LE (2007) The invasive bivalves *Dreissena polymorpha* and *Limnoperna fortunei* : parallels, contrasts, potential spread and invasion impacts. J Shellfish Res 26:205–213

Mackie GL, Claudi R (2010) Monitoring and control of macrofouling mollusks in fresh water systems. CRC Press, Boca Raton, pp 1–508

Maddox T (1998) The thermal treatment methodology for zebra mussel elimination. In: 8th International Zebra Mussel and Aquatic Nuisance Species Conference, Sacramento (USA)

McMahon RF, Ussery TA (1995) Thermal tolerance of zebra mussels (*Dreissena polymorpha*) relative to rate of temperature increase and acclimation temperature. US Army Engineer Waterways Experimental Station (Vicksburg, MS), Technical Report: EL-95-10, pp 1–19

McMahon RF, Matthews M, Ussery TA, Chase R, Clarke M (1995) Studies of heat tolerance of zebra mussels: effects of temperature, acclimation and chronic exposure to lethal temperatures. US Army Engineer Waterways Experimental Station (Vicksburg, MS), Technical Report EL-95-9, pp 1–25

Miller AC, Payne BS, Neilson F, McMahon RF (1992) Control strategies for zebra mussel infestations at public facilities. US Army Engineer Waterways Experimental Station (Vicksburg, MS), Technical Report EL-92-25, pp 1–24

Montalto L, Marchese M (2003) *Limnoperna fortunei* (Dunker, 1857) (*Bivalvia, Mytilidae*) tolerance to temperature and pH in experimental conditions. Neotropica 49:26–34

Mordukhai-Boltovskoi FD (1975) Teplovye elektrostantzii i zhizn vodoemov [Thermal power stations and life in waterbodies]. Priroda 1:57–66 [In Russian]

O'Neill CR, MacNeill DB (1991) The zebra mussel (*Dreissena polymorpha*): an unwelcome North American invader. Coastal Resources Fact Sheet. Sea Grant-Coastal Resources, Cornell Cooperative Extension, State University of New York (USA), pp 1–12

Oliveira MD, Hamilton SK, Jacobi CM (2010) Forecasting the expansion of the invasive golden mussel *Limnoperna fortunei* in Brazilian and North American rivers based on its occurrence in the Paraguay River and Pantanal wetland of Brazil. Aquat Invasions 5:59–73

Perepelizin PV (2011) Alternativas ambientalmente inocuas (no químicas) para el control del mejillón incrustante *Limnoperna fortunei* en instalaciones industriales. PhD Thesis, Universidad de Buenos Aires (Argentina), pp 1–126

Perepelizin PV, Boltovskoy D (2011a) Hot water treatment (chronic upper lethal temperature) mitigates biofouling by the invasive Asian mussel *Limnoperna fortunei* in industrial installations. Environ Sci Technol 45:7868–7873

Perepelizin PV, Boltovskoy D (2011b) Thermal tolerance of *Limnoperna fortunei* to gradual temperature increase and its applications for biofouling control. Biofouling 27:667–674

Rajagopal S, Nair KVK, Azariah A (1995) Response of brown mussel, *Perna indica*, to elevated temperatures in relation to power plant biofouling control. J Therm Biol 20:461–467

Rolla ME, Mota HR (2010) Response of a major Brazilian utility to the golden mussel invasion. In: Mackie GL, Claudi R (eds) Monitoring and control of macrofouling mollusks in fresh water systems. CRC Press, Boca Raton, pp 396–403

Whitehouse JE, Khalanski M, Saroglia MG, Jenner HA (1985) The control of biofouling in marine and estuarine power stations: a collaborative research working group report for use by station designers and station managers. Central Electricity Generating Board, NW Region 191-9-85, CEGB (England), EdF (France), ENEL (Italy), KEMA (The Netherlands), pp 1–48

Control of *Limnoperna fortunei* Fouling by Oxygen Deprivation

Pablo V. Perepelizin and Demetrio Boltovskoy

Abstract The only detailed survey aimed at assessing the efficacy of oxygen deprivation for controlling fouling by the golden mussel indicates that, at dissolved oxygen levels <0.16 mg/L, total mortality is achieved after 10–12 days (at 27 °C) to 21–29 days (at 20 °C). At 20 °C (but not at 27 °C), small (7 mm) mussels are significantly less tolerant than large (20 mm) individuals. Oxygen deprivation may be a viable alternative for the control of mussel fouling in industrial installations.

Keywords *Limnoperna fortunei* · Golden mussel · Biological invasion · Biofouling control · Impact · Industrial plants · Power plants · Oxygen deprivation · Anoxia

Experimental assessment of the tolerance of *Limnoperna fortunei* to oxygen deprivation was carried out by Perepelizin and Boltovskoy (2011). They exposed small (7 mm) and large (20 mm) mussels to anoxic conditions (<0.16 mg O_2/L, or 1.8–2.2 % saturation) at 20 and 27 °C, for periods ranging between 8 and 31 days, until 100 % mortality in all (3) replicates (with 43 mussels each) occurred.

At 20 °C, small mussels started dying on day 3, with the last of 137 specimens enduring anoxia for 22 d (Fig. 1). Average time to 100 % mortality in all replicates was 20.7 d (LT_{50}: 9.5 d). Large mussels were significantly more resistant: individual

P. V. Perepelizin (✉) · D. Boltovskoy
Museo Argentino de Ciencias Naturales 'Bernardino Rivadavia'-CONICET, Av. Angel Gallardo 470, 1405 Buenos Aires, Argentina
e-mail: perepelizin@macn.gov.ar

D. Boltovskoy
Instituto de Ecología, Genética y Evolución de Buenos Aires (IEGEBA), Facultad de Ciencias Exactas y Naturales, Universidad de Buenos Aires-CONICET, Ciudad Universitaria, 1428 Buenos Aires, Argentina
e-mail: demetrio@ege.fcen.uba.ar

D. Boltovskoy (ed.), *Limnoperna fortunei,* Invading Nature - Springer Series in Invasion Ecology 10, DOI 10.1007/978-3-319-13494-9_25

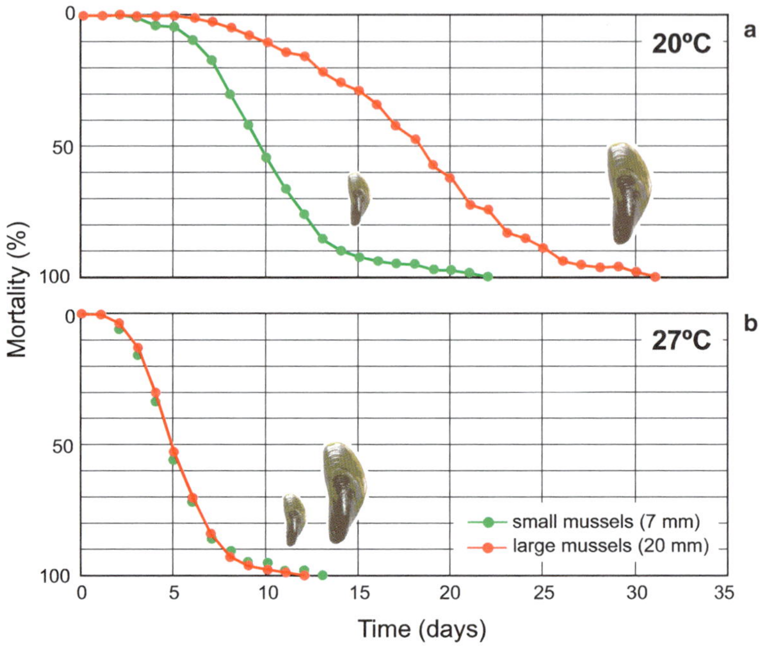

Fig. 1 Mortality rates of small (7±2 mm) and large (20±3 mm) *Limnoperna fortunei* under anoxia. Each *curve* represents the mean of three replicates. (Modified from Perepelizin and Boltovskoy 2011, from *Journal AWWA* 103(3) by permission. Copyright © 2011 the American Water Works Association)

survival times were 6–31 d, and average time to 100% mortality in all replicates was 29.3 d (LT_{50}: 18.0 d) (Fig. 1).

In contrast with experiments at 20 °C, large and small animals behaved almost identically at 27 °C. Individual survival times were between 1 and 13 days, and mean time to 100% mortality was 10.2 (large; LT_{50}: 4.7 d) to 11.6 (small; LT_{50}: 4.7 d) days (Fig. 1).

The distribution of *L. fortunei* in South America, where it often thrives under extremely adverse environmental conditions (e.g., high pollution levels, low oxygen concentrations, very low calcium levels, low pH; Karatayev et al. 2007) indicates that it is a highly tolerant species. For example, under anoxic conditions at 25 °C, 100% mortality of *Dreissena polymorpha* is achieved in only 4 days (Matthews and McMahon 1994) (as opposed to ca. 13 days for *L. fortunei*). Although the golden mussel can withstand dissolved oxygen levels as low as 0.5 mg/L for extended periods (Karatayev et al. 2007), its ability to survive under extreme hypoxia is limited (Liu et al. 2006). In raw reservoir or river water pipelines, fouling by *L. fortunei* has been observed to decrease with distance from the intake in association with decreasing dissolved oxygen concentrations (Ye et al. 2011). Ample differences in mussel densities in two closely spaced Japanese reservoirs were ascribed to the lack of dissolved oxygen in one of them (Nakano et al. 2010).

In the upper Paraguay River, seasonally changing water levels are responsible for extensive flooding of vegetated lowlands. During inundation, aquatic plants colonize the area, but die shortly thereafter during the dry season, when water levels subside. In the subsequent flood pulse, the dead vegetation is leached and decomposed, strongly depressing O_2 levels and pH values, and raising CO_2. These conditions, known locally as "dequada," which usually last for several weeks, have been found to produce massive mortalities of *L. fortunei* (Oliveira et al. 2010).

Thus, oxygen deprivation may constitute an adequate solution to *L. fortunei* fouling. It is environmentally innocuous, economical, and does not involve hazardous substances or operations. As shown by the information reviewed above, as well as by data on other fouling species (Sprung 1995; Johnson and McMahon 1998), it is particularly efficacious at high water temperatures, such as those prevailing for at least some months of the year in most water bodies invaded by this mussel around the world.

Oxygen depletion can be achieved simply by sealing off pipes and precluding water circulation, or with the aid of oxygen scavengers such as sodium metabisulfite or hydrogen sulfide gas (Smithson 1986; O'Neill 1995). Several water treatment plants in Wisconsin, provided with dual intakes, add sodium bisulfite and seal off one of the intakes for 4–10 weeks achieving 100% mortality of infesting zebra mussels (Mackie and Claudi 2010). Small water treatment facilities drawing water from the reservoir Embalse de Río Tercero (Córdoba Province, Argentina) experienced intake pipe clogging problems due to growth of *L. fortunei* (Anonymous 2006). Some of them are provided with dual intakes, which allows for temporary capping of a fouled one while using the other until infesting mussels die. Backflushing the sealed-off intake at the end of the inactive period is important for clearing the pipe of dead, loose, and weakly adhering mussels.

However, as with most other control methods, anoxia also has drawbacks. Sealing off an intake for an extended period of time requires that the plant is provided with dual intake pipes, which is not always the case, and retrofitting in order to provide this alternative may be costly and complicated. This obviously also applies to sections other than intake pipes, where duplication of components may be even more costly and problematic. At high temperatures, the response of *L. fortunei* to the lack of oxygen is fairly fast (around 2 weeks for 100% mortality at 27 °C; see Fig. 1), but at lower temperatures response times increase significantly. Depending on plant design and operating conditions, the possibilities of shutting off an important section for weeks or months may be limited. In addition, lack of oxygen frequently enhances the abundance of sulfate-reducing bacteria, which are responsible for microbially induced corrosion (Mackie and Claudi 2010).

Anoxia is a nonselective method, killing practically all organisms, but since it normally involves treating a limited volume of water, its environmental impact is negligible.

Acknowledgements This work was partially financed by grants from the University of Buenos Aires, Argentina (UBA X-020 and 20020100100035) and from the Agencia Nacional de Promoción Científica y Tecnológica, Argentina (PICT 2007 1968) to DB.

References

Anonymous (2006) El mejillón dorado (*Limnoperna fortunei*). Los serios problemas que origina en la planta potabilizadora. NotiCoop (Boletín Informativo de la Cooperativa de Obras y Servicios Públicos Limitada de Villa del Dique, Provincia de Córdoba, Argentina) 42:22–23

Johnson PD, McMahon RF (1998) Effects of temperature and chronic hypoxia on survivorship of the zebra mussel (*Dreissena polymorpha*) and Asian clam (*Corbicula fluminea*). Can J Fisheries Aquat Sci 55:1564–1572

Karatayev AY, Boltovskoy D, Padilla DK, Burlakova LE (2007) The invasive bivalves *Dreissena polymorpha* and *Limnoperna fortunei* : parallels, contrasts, potential spread and invasion impacts. J Shellfish Res 26:205–213

Liu L, You Z, Luo F, Zhang J, Ouyang S (2006) Killing and removing *Limnoperna fortunei* from raw water pipe by isolating oxygen. Water Wastewater Eng 22:40–42, 47

Mackie GL, Claudi R (2010) Monitoring and control of macrofouling mollusks in fresh water systems. CRC Press, Boca Raton, pp 1–508

Matthews MA, McMahon RF (1994) The survival of Zebra Mussels (*Dreissena polymorpha*) and Asian Clams (*Corbicula fluminea*) under extreme hypoxia. In: 4th International Zebra Mussel Conference, Madison (USA)

Nakano D, Kobayashi T, Sakaguchi I (2010) Differences in larval dynamics of golden mussel *Limnoperna fortunei* between dam reservoirs with and without an aeration system. Landsc Ecol Eng 6:53–60

Oliveira MD, Hamilton SK, Calheiros DF, Jacobi CM (2010) Oxygen depletion events control the invasive golden mussel (*Limnoperna fortunei*) in a tropical floodplain. Wetlands 30:705–716

O'Neill CR (1995) Control of zebra mussels in residential water systems. New York Sea Grant Institute, New York, pp 2–10

Perepelizin PV, Boltovskoy D (2011) Resistance of the invasive pest mussel *Limnoperna fortunei* to anoxia. J Am Water Works Assoc 103:79–85

Smithson JA (1986) Development of a *Corbicula* control treatment at the Baldwin power station. Am Malacol Bull 2:63–67 (Special Edition)

Sprung M (1995) Physiological energetics of the zebra mussel *Dreissena polymorpha* in lakes III. Metabolism and net growth efficiency. Hydrobiologia 304:147–158

Ye BM, Cao XW, Xu M, Wang ZY, Lin CC (2011) Study of *Limnoperna fortunei* invasion in long distance water transmission project. Water Wastewater Eng 37:99–102 [In Chinese]

Control of *Limnoperna fortunei* Fouling by Desiccation

Luciana Montalto

Abstract When exposed to air, *Limnoperna fortunei* can survive between 3 and >10 days, larger individuals being more resilient than small ones. Desiccation can constitute an innocuous, cost-effective nonchemical control strategy for freshwater fouling mussels, but in order to be applicable, fouled components must be taken off-line for variable periods of time, which may involve the need for parallel backup components unless production is stopped. Survival in air is important in waterbodies with significant fluctuations in water levels, and survival in air can also facilitate overland dispersion of the golden mussel.

Keywords *Limnoperna fortunei* · Golden mussel · Biological invasion · Biofouling control · Impact · Industrial plants · Power plants · Emersion · Desiccation

Introduction

Knowledge of the ecological requirements of *Limnoperna fortunei* has increased significantly after it started spreading outside of its native range. Information on the tolerance of this invasive mussel to different physical and chemical stressors allows prediction of its potential distribution and provides a basis for developing preventive or corrective measures to control its fouling of man-made structures.

Studies on the resistance of the golden mussel to desiccation are important because they (1) provide valuable information for assessing the potential of various dispersal vectors, in particular overland transport (e.g., waterfowl and other aquatic and semi-aquatic vertebrates, trailered boats, fishing gear, net cages for fish farming, etc.), (2) help assess the impact of variable water levels on the survival and

L. Montalto (✉)
Instituto Nacional de Limnología (INALI, Consejo Nacional de Investigaciones Científicas y Técnicas–Universidad Nacional del Litoral), Facultad de Humanidades y Ciencias (Universidad Nacional del Litoral) 3000, Santa Fe, Argentina
e-mail: lmontalto@inali.unl.edu.ar

D. Boltovskoy (ed.), *Limnoperna fortunei,* Invading Nature - Springer Series in Invasion Ecology 10, DOI 10.1007/978-3-319-13494-9_26

population growth of this mussel, and (3) can be used in biofouling control strategies in industrial installations (see "Impacts of *Limnoperna fortunei* on Man-made Structures and Control Strategies: General Overview" in this volume).

Tidal emersion and air exposure tolerance of marine bivalves have been extensively documented in the literature (Griffiths 1981; Griffiths and Buffenstein 1981; Widdows and Shick 1985; Hicks and McMahon 2003; among others). Marine species typically respond to tidal emersion by closing their valves, reducing metabolic rates, and respiring anaerobically, thus minimizing evaporative water loss (Widdows et al. 1979; McMahon 1988; Hicks and McMahon 2003). For freshwater bivalves, several studies have been made on *Dreissena* species indicating that desiccation enhances evaporative water loss (tissue water and water from the mantle cavity), resorption of energy stores because of starvation and stress, depletion of energy reserves, toxic buildup of anaerobic metabolites, and acidosis (Byrne et al. 1990; McMahon and Paine 1992; Ricciardi et al. 1995; Tucker et al. 1997; Paukstis et al. 1999).

This chapter reviews the results of desiccation experiments carried out with *L. fortunei* focusing on its tolerance to air exposure as a function of temperature, relative humidity, and mussel size Iwasaki 1997; Montalto and Ezcurra de Drago 2003; Darrigran et al. 2004).

Desiccation Tolerance Studies with *L. fortunei*

Iwasaki (1997) and Darrigran et al. (2004) carried out studies of air exposure tolerance of *L. fortunei* under laboratory conditions, whereas Montalto and Ezcurra de Drago (2003) performed both laboratory and outdoor experiments. The results of these tests (Table 1; Fig. 1; 2) clearly indicate that smaller mussels (< 10 mm) are less tolerant to prolonged emersion, but mortality rates of medium-sized (~ 10–20 mm) and large specimens (15 to > 20 mm) do not seem to differ (Fig. 1). The effects of humidity on mortality rates are less clear. Small mussels seem to stay alive for longer periods of time when humidity is high, whereas medium and large mussels die more rapidly at humidity levels ~ 100 % than at ~ 70 % (Fig. 1). Faster mortality rates at higher humidity levels, especially at high temperatures, may be due to the inability to cool tissues through evaporation in excessively moist air (Mackie and Claudi 2010). However, in the experiments of Darrigran et al. (2004), higher humidity seemed to increase survival times.

Data listed in Table 1 are illustrated in Fig. 2, stressing the extremely high variability between the results of different studies. These differences are most likely in part due to the fact that laboratory protocols differed between surveys. For example, Iwasaki (1997) used isolated individuals, whereas Montalto and Ezcurra de Drago (2003) and Darrigran et al. (2004) used either clusters of 20 individuals or druses. The gregarious mode of life of *L. fortunei* favors the retention of water between the valves upon emersion, and thus likely attenuates the effects of air exposure. Temperature also differed between experiments, although relationships between air temperature and time to total mortality do not show a consistent trend in these results.

Table 1 Summary of studies investigating the tolerance of *Limnoperna fortunei* to desiccation. (Based on data from: (1) Iwasaki (1997); (2) Montalto and Ezcurra de Drago (2003); (3) Darrigran et al. (2004))

Size class (mm)	Number of experimental mussels per replicate	Replicates	Air temperature range throughout the experiment (°C)	Relative humidity range throughout the experiment (%)	Time to 100% mortality (h)	Observations	Ref.
<10	ND	0	26–30	72–81	120	L, TM: 135[a]	1
10–15	ND	0	26–30	72–81	216	L, TM: 135[a]	1
15–20	ND	0	26–30	72–81	240	L, TM: 135[a]	1
>20	ND	0	26–30	72–81	240	L, TM: 135[a]	1
<6	20	3–5	9.1–16.5	63.4–78.4	72	L, TM: 360–600	2
6–15	20	3–5	9.1–16.5	63.4–78.4	192	L,TM: 960–1600	2
15–27	20	3–5	9.1–16.5	63.4–78.4	276	L, TM: 1380–2300	2
<6	20	3–5	14.8–16.6	64.8–92.7	72	O, TM: 360–600	2
6–15	20	3–5	14.8–16.6	64.8–92.7	96	O, TM: 480–800	2
15–27	20	3–5	14.8–16.6	64.8–92.7	108	O, TM: 540–900	2
<10	ND	4	24.6–25.4	49–63	72–96	L, TM: ND	3
10–20	ND	4	24.6–25.4	49–63	72–96	L, TM: ND	3
>20	ND	4	24.6–25.4	49–63	96–120	L, TM: ND	3
<10	ND	2	24.6–25.4	100	144	L, TM: ND	3
10–20	ND	2	24.6–25.4	100	144	L, TM: ND	3
>20	ND	2	24.6–25.4	100	144–168	L, TM: ND	3

L laboratory experiments, *O* outdoors experiments, *TM* total number of mussels, *ND* no data, *a* refers to all the four size classes used

In order to mimic natural conditions more closely, Montalto and Ezcurra de Drago (2003) repeated outdoors the trials carried out in the laboratory. The outcome of this experiment showed that survival times outdoors were lower, especially for mid-sized and large mussels (Fig. 2). This result, ascribed to the effects of direct sunlight and different temperature and humidity settings (see Table 1), suggests that laboratory tests may overestimate actual mussel survival times in nature. On the other hand, survival in pipes and other fouled components of industrial and power plants are probably more similar to (or even higher than) those observed in laboratory settings.

In summary, these surveys have shown that when exposed to air, small (<10 mm) *L. fortunei* specimens can survive around 3–7/8 days, whereas large (>20 mm) ani-

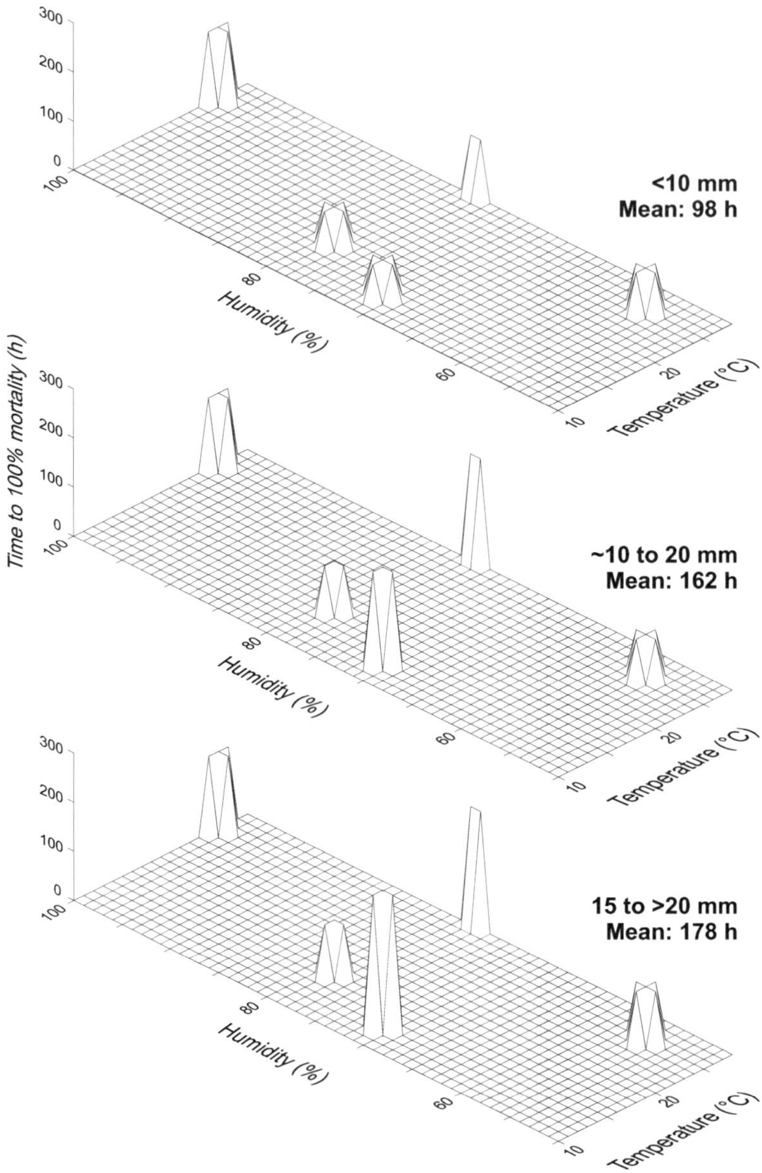

Fig. 1 Survival of *Limnoperna fortunei* exposed to air as a function of mussel size, temperature, and humidity. Values for identical settings are averaged. Based on data from Iwasaki (1997); Montalto and Ezcurra de Drago (2003); Darrigran et al. (2004)

mals survive up to 11–12 days. These results generally align with those obtained for *Dreissena polymorpha*. Ricciardi et al. (1995) concluded that at 10–15 °C and high relative humidity (95 %) zebra mussels survive ~10 days. At higher temperatures (20 °C), however, survival times drop noticeably (5–6 days). Large specimens

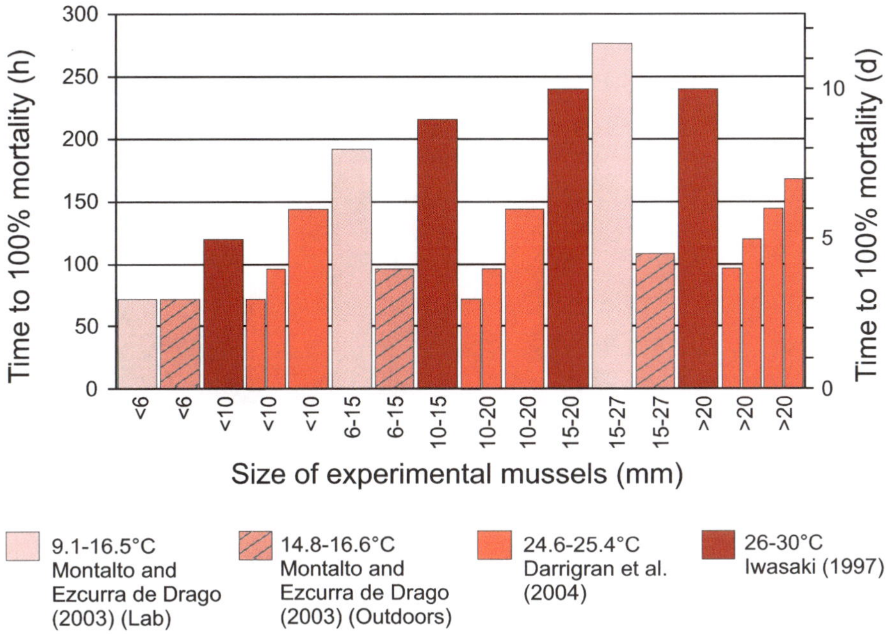

Fig. 2 Survival of *Limnoperna fortunei* exposed to air as a function of mussel size. *Narrow bars* are paired to denote the range of values reported. Based on data listed in Table 1)

were also found to survive desiccation for longer periods of time than small ones, presumably because they are less susceptible to evaporative water loss and the toxic effects of acidic products which occur in anaerobic metabolism (Ricciardi et al. 1995; Tucker et al. 1997; Paukstis et al. 1999).

Desiccation as a Control Strategy

Desiccation can constitute an innocuous, cost-effective nonchemical control strategy for freshwater fouling mussels (McMahon et al. 1993; Mackie and Claudi 2010). However, in order to be applicable, fouled components must be taken off-line for variable periods of time, which may involve the need for parallel backup components unless production is stopped. Manual cleaning may be required after desiccation to clear away dead mussels and shells. Laboratory tests indicate that ~4 to 12 days may be needed to kill all the mussels in a fouled system. Treatment times will probably be considerably shorter during the summer than at colder times of the year, and times may be further shortened by injecting heated air into the drained components (Ricciardi et al. 1995).

Fig. 3 Massive mortality of *Limnoperna fortunei* after exposure of colonized tree trunks in the Middle Paraná River

Tolerance to Desiccation: Importance as a Deterrent to Population Growth and Geographic Expansion

Previous studies have confirmed that periods of low water levels can cause extensive mortalities in *L. fortunei* populations. Air exposure periods as short as 4 days can result in massive mussel kills. This mechanism is likely very important in the large South American floodplain rivers, which are characterized by very marked flood–drought pulses where water levels rise and fall several meters over huge areas. Reservoirs also undergo significant fluctuations in their water levels, exposing large areas of mussel-colonized substrata, usually for periods much longer than mussel survival times. Such events have been recorded repeatedly in several Argentine reservoirs, including Salto Grande and Embalse de Río Tercero (see Figs. 5 and 14 in Chapter "*Limnoperna fortunei* Colonies: Structure, Distribution and Dynamics" in this volume).

Populations of the mussel are often restricted to—or much denser in—shallow areas along the shore, where hard substrata are more widespread, than in deep areas, usually covered by thick layers of loose sediments (mud and silt) (Boltovskoy et al. 2009; see "*Limnoperna fortunei* Colonies: Structure, Distribution and Dynamics" in this volume), and this makes them particularly vulnerable to water-level changes (Montalto and Ezcurra de Drago 2003, Spaccesi and Rodrigues Capitulo 2012, Spaccesi 2013; Fig. 3). On the other hand, in estuaries subjected to tidal water-level changes, like the Río de la Plata estuary, emersion periods are not long enough to produce massive kills, even when wind conditions significantly modify tidal water-level changes.

Tolerance to desiccation is also important for the overland transport of mussels attached to trailered boat hulls, fishing equipment, net cages for fish farming, etc. Colonization of the inland waterbodies of the Mar Chiquita endorheic basin in Argentina is thought to have occurred via overland transport by a leisure boat trailered from somewhere on the Paraná River (see "Colonization and Spread of *Limnoperna fortunei* in South America" in this volume).

Acknowledgments This chapter was partially supported by grants PICT N° 1965 and SECTeI 2010-042-13 Res.117/13.

References

Boltovskoy D, Karatayev A, Burlakova L, Cataldo D, Karatayev V, Sylvester F, Mariñelarena A (2009) Significant ecosystem-wide effects of the swiftly spreading invasive freshwater bivalve *Limnoperna fortunei*. Hydrobiologia 636:271–284

Byrne RA, Gnaiger E, McMahon RF, Dietz TH (1990) Behavioral and metabolic responses to emersion and subsequent reimmersion in the freshwater bivalve, Corbicula fluminea. Biological Bulletin of the Marine Biological Laboratory, Woods Hole, 17:251–259

Darrigran GA, Maroñas ME, Colautti DC (2004) Air exposure as a control mechanism for the golden mussel, *Limnopema fortunei*, (Bivalvia: Mytilidae). J Freshw Ecol 19:461–464

Griffiths RJ (1981) Aerial exposure and energy balance in littoral and sublittoral *Choromytilus meridionalis* (Kr.) (Bivalvia). J Exp Mar Biol Ecol 52:231–241

Griffiths RJ, Buffenstein R (1981) Aerial exposure and energy input in the bivalve *Choromytilus meridionalis* (Kr.). J Exp Mar Biol Ecol 52:219–229

Hicks DW, McMahon RF (2003) Temperature and relative humidity effects on water loss and emersion tolerance of *Perna perna* (L.) (Bivalvia: Mytilidae) from the Gulf of Mexico. BulletinMar Sci 72:135–150

Iwasaki K (1997) Climbing behaviour and tolerance to aerial exposure of a freshwater mussel, *Limnoperna fortunei*. Venus (Jpn J Malacol) 56:15–25

Mackie GL, Claudi R (2010) Monitoring and control of macrofouling mollusks in fresh water systems. CRC Press, Boca Raton, pp 1–508

McMahon RF (1988) Respiratory response to periodic emergence in intertidal mollusks. Am Zool 28:97–114

McMahon RF, Paine BS (1992) Effects of temperature and relative humidity on desiccation resistance in zebra mussel (*Dreissena polymorpha*): is aerial exposure a viable control option? In: *Dreissena polymorpha* Information Review, Special Conference Issue

McMahon RF, Ussery TA, Clarke M (1993) Use of emersion as a Zebra mussel control method. US Army Corps of Engineers, Waterways Experiment Station (Vicksburg, MS), Report EL-93-1. Vicksburg (USA), pp 1–27

Montalto L, Ezcurra de Drago I (2003) Tolerance to desiccation of an invasive mussel, *Limnoperna fortunei* (Dunker, 1857) (Bivalvia, Mytilidae), under experimental conditions. Hydrobiologia 498:161–167

Paukstis GL, Tucker JK, Bronikowski AM, Jansen FJ (1999) Survivorship of aerially-exposed zebra mussels (*Dreissena polymorpha*) under laboratory conditions. J Freshw Ecol 14:511–517

Ricciardi A, Serrouya R, Whoriskey FG (1995) Aerial exposure tolerance of zebra and quagga mussels (Bivalvia: Dreissenidae): implications for overland dispersal. Can J Fish Aquat Sci 52:470–477

Spaccesi FG (2013) Abundance, recruitment, and shell growth of the exotic mussel *Limnoperna fortunei* in the Río de la Plata (Argentina). Zool Stud 52:1–11

Spaccesi FG, Rodrigues Capitulo A (2012) Benthic communities on hard substrates covered by *Limnoperna fortunei* Dunker (Bivalvia, Mytilidae) at an estuarine beach (Río de la Plata, Argentina). J Limnol 71:144–153

Tucker JK, Theiling CH, Janzen FJ, Paukstis GL (1997) Sensitivity to aerial exposure: potencial of system-wide drawdowns to manage zebra mussels in the Mississippi River. Regul Rivers: Res Manag 13:479–487

Widdows J, Shick JM (1985) Physiological responses of *Mytilus edulis* and *Cardium edule* to aerial exposure. Mar Biol 85:217–232

Widdows J, Bayne BL, Livingstone DR, Newell RIE, Donkin P (1979) Physiological and biochemical responses of bivalve mollusks to exposure to air. Comp Biochem Physiol 62:301–308

Alternative Strategies for Control of Golden Mussel (*Limnoperna fortunei*) in Industrial Facilities

Renata Claudi and Marcia Divina de Oliveira

Abstract There are numerous chemically based control strategies for macrofouling by bivalves within piping systems such as those that carry raw water into water treatment plants, cooling water to vital areas of power plants and industrial facilities, and fire protection water in any industry. As the use of chemicals is continuously scrutinized by regulators in many countries, alternative control strategies for macrofouling control are increasing in popularity. Nonchemical controls such as filtration, ultraviolet light, electric currents, acoustics, manipulation of water velocity, magnetic fields, turbulence, and sacrificial substrates have been tested and in some cases are being utilized for macrofouling control in industrial settings. In some cases, these alternative strategies are limited in their applicability by either plant design or quality of the raw water, but they are generally seen as more environmentally friendly as they do not result in release of chemicals when the cooling water is returned to the environment.

Keywords *Limnoperna fortunei* · Golden mussel · Biological invasion · Biofouling control · Impact · Industrial plants · Power plants · Irrigation facilities · Filtration · Ultraviolet light · Electric currents · Acoustic treatment · Magnetic fields · Water velocity · Turbulence

R. Claudi (✉)
RNT Consulting Inc., Picton, ON, Canada
e-mail: rnt@idirect.com

M. D. de Oliveira
Embrapa Pantanal, Rua 21 de Setembro, 1880, Corumbá, MS, Brazil
e-mail: marcia.divina@embrapa.br

D. Boltovskoy (ed.), *Limnoperna fortunei,* Invading Nature - Springer Series in Invasion Ecology 10, DOI 10.1007/978-3-319-13494-9_27

Introduction

Nonchemical macrofouling control strategies for industry are generally favored by regulators in many parts of the world, since such techniques do not result in release of chemicals when the cooling water is returned to the environment. A subset of the nonchemical strategies potentially useful to prevent macrofouling by invasive bivalves is described below. Other nonchemical strategies such as thermal treatment, anoxia, desiccation, use of fouling-resistant surfaces, and antifouling coatings are discussed in other chapters of this volume.

Filtration

Filtration has been used as a nonchemical control strategy in numerous industrial systems in North America to prevent dreissenid settlement in locations downstream of the filter. In many cases, the filters had been installed to eliminate silt and debris from sensitive areas. The elimination of ready-to-settle veligers had become a bonus.

When correctly designed, this technology is capable of removing all stages of macrofouling organisms and protecting all downstream systems and components from infestation. Filtration can be carried out using media filters (e.g., sand filters) or mechanical self-cleaning filter units. In both cases, it is important to verify that the selected filter is capable of removing all particles in the size range of ready-to-settle *Limnoperna fortunei* veligers (~200 µm).

Many filters are very good at removing all or most particles from relatively small streams or raw water, but most filters are not able to process large volumes of water efficiently, especially if the turbidity of the raw water is high. Most of the successful filters have an automatic backwash system that can be used while the filter is in service. A filter capable of filtering 45 m^3/min was tested by Ontario Hydro at the Nanticoke Station on Lake Erie. This filter was equipped with a 40-µm nominal mesh and was generally called upon to filter 23 m^3/min. The filter performed well under different silt-loading scenarios, but the backwash frequency increased with increasing silt presence in the raw water. Approximately, 15–20 % of the water filtered was lost in backwash. After having been in service for an entire mussel-breeding season, no mussel settlement was observed past the filter.

Filters that use stainless steel, square weave mesh, and automatic backwash seem to have the best balance between particle removal efficiency and volume of water filtered. Slot-based filtration units have difficulty in removing veligers due to the shape and flexibility of the veliger shell.

Usually, a pressure of at least 2.5 kg/cm^2 (35 psi) is required in the piping system to be protected by a filter. This pressure is needed to operate the backwash systems effectively.

The use of filters offers an additional advantage to removal of ready-to-settle veligers. In many cases, the filter will also remove inorganic material such as silt that causes operational problems in many cooling water systems. The impact turbidity has on self-cleaning filters will depend on the size of the particles present in the water. In some cases, the particulates are smaller than the screen pore of the filter and will have very little impact on filter performance. In other locations, the particulates will be large enough to collect on the screen. Knowing the particulate size distribution in the raw water is helpful when considering filtration. Generally, self-cleaning filters will use between 3 and 5 % of the water they filter to backwash the screens. In very turbid water, the performance of most filtration systems may be compromised and self-cleaning filters may have to backwash more frequently, which leads to diminished efficiency. If large amounts of organic debris, such as aquatic weeds, are present in the raw water source, it is advisable to consider a two-step filtration. The first step is to remove the coarse debris with a self-cleaning strainer, and only then remove the veligers with a self-cleaning small-pore filter.

The use of self-cleaning, high-efficiency filters for elimination of organisms in the ship's ballast water has resulted in a number of excellent mechanical self-cleaning filters being available. Claudi et al. (2014) tested the efficiency of one such filter equipped with either 57- or 120-μm mesh. Both mesh sizes allowed the passage of particles somewhat greater than the manufacturer's stated pore size. This suggests that pore size should be smaller than the smallest-size veliger to be eliminated from the raw water. In the case of *L. fortunei,* a pore size of 150 μm is recommended to eliminate all ready-to-settle veligers. If some settlement is acceptable, the pore size can be increased.

Ultraviolet Light

The term ultraviolet (UV) is applied to that portion of the electromagnetic spectrum between visible light and X-rays, typically between 190 and 400 nm. This region is commonly subdivided into UVA, UVB and UVC, where UVA corresponds to the longer wavelength (lower-energy regime), through to UVC that corresponds to the shorter wavelength (higher energy) end of the UV spectrum.

In UV irradiation studies, the dose (= intensity × exposure time) and nature of the light source are important factors. Medium-pressure mercury lamps have proven to be a convenient source of UV radiation since they provide reasonable intensity over the range 240–310 nm.

Several studies carried out in the 1990s have shown that flow-through UV systems have the ability to prevent attachment of dreissenid veligers to downstream surfaces (Chalker-Scott et al. 1993; Lewis and Whitby 1993; Chalker-Scott et al. 1994; Evans et al. 1995; Lewis and Whitby 1996; Lewis and Cairns 1998). The available body of evidence suggests that medium-pressure lamps with UV wavelengths between 200 and 400 nm would inhibit downstream settlement of dreissenid veligers if the veligers were exposed to a radiation dose of approximately 100 mW s/cm^2 (=100 mJ/cm^2).

In 1999, Ontario Power Generation (then called Ontario Hydro) embarked on a full-size UV pilot installation to test the efficacy of UV under field conditions in an open, concrete channel. The flow treated was 760 L/s. The computed UV dose delivered to each particle passing through the UV system was between 70 and 100 mW s/cm^2. The system was operational for one breeding season of the mussels. Despite numerous outages, there was an 85 % reduction in settlement downstream of the UV system when compared to control chambers upstream (Pickles 2000).

Hoover Dam, located on the Lower Colorado River (USA), installed an Aquafine medium-pressure UV system in late 2010 to protect a relatively small cooling water circuit (3300 L/min) on Unit A1. Monitoring of the system performance (carried out in 2012 from May to November) confirmed that no settlement occurred downstream of the UV system which was delivering a dose of approximately 100 mW s/cm^2 (Claudi and Prescott 2013).

In the spring of 2013, Davis Dam (on the Colorado River, USA) installed a full-sized medium-pressure UV system from Atlantium Technologies to protect all of the cooling water (total flow of 13500 L/min) on power-generating Unit 3. Part of the purpose for the installation was to find the minimum UV dose that would prevent at least 95 % of possible settlement. Claudi et al. (2014) tested various doses of the medium-pressure UV (Fig. 1) and found good control of veliger settlement at all dose levels tested. These results may only be applicable to the Atlantium UV system tested. This system uses a quartz reflective liner in the reaction vessel that may aid the effectiveness of this system for veliger control. Comparative studies using commonly available UV systems have not been done.

Working with *L. fortunei* veligers and using low-pressure UV lamp, Santos et al. (2012a) found that the dose to inactivate 50 % of the larvae in flows ranging from 1400 to 4200 L/h was 324 mW s/cm^2, whereas for 100 % mortality the dose needed

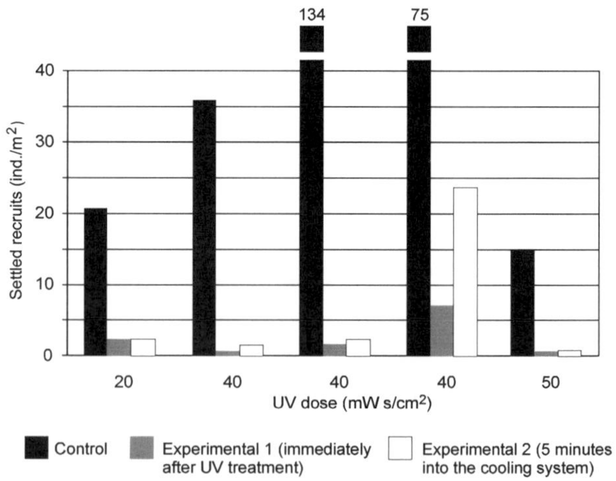

Fig. 1 Results of 30 to 35 day experiments on the effects of different doses from medium pressure UV system on the settling of *Dreissena polymorpha* larvae in the cooling water circuit of Davis Dam (Colorado River, USA) (Based on data from Claudi et al. 2014)

was 781 mW s/cm^2. However, it must be noted that experimental conditions in this study were poorly defined: The absorbance of the ambient water used was not assessed, experimental doses were calculated on the basis of the lamp manufacturer's specifications rather than measured, and delayed mortality was not established.

Perepelizin and Boltovskoy (2014), working in laboratory settings in a closed system with a low-pressure lamp, found that doses of 149 mW s/cm^2 at 23 °C achieved 100 % mortality of veligers, while at 25.8 °C for 100 % mortality a dose of only 103 mW s/cm^2 was required. Immediately after exposure, larvae were alive but had reduced mobility. The proportion of active larvae increased after 24 h, but fell again at 48 and 72 h to levels similar to those immediately after exposure. The highest mortality rates were always recorded at the last observation, 72 h after exposure. These results indicate that the larvae of *L. fortunei* are highly sensitive to UVC but the mortality is not immediate, developing over a period of 72 h.

Oliveira et al. (unpublished) tested the efficacy of a medium-pressure UV system from Atlantium Technologies that was installed in a cooling water system of a hydroelectric plant on the Paraná River. The UV transmissibility (UVT) was measured as 82, and the flow in the system was approximately 100 m^3/h. In this experiment, a dose of 80 mW s/cm^2 was sufficient to eliminate 70–80 % of *L. fortunei* veligers and prevent settlement downstream of the UV system.

The above studies suggest that UV irradiation has great potential as a control strategy in raw water systems provided the raw water transmissibility allows for the delivery of an adequate UV dose. In most South American rivers colonized by the golden mussel, this is a major limitation because suspended solid loads are very high (typically around 160 mg/L, absorbance: 0.255 for 254 nm UV; Depetris and Kempe 1993; Perepelizin and Boltovskoy 2014). In these conditions, Perepelizin and Boltovskoy (2014) estimated that in order to deliver 103 mW s/cm^2 (the 100 % mortality dose at 25.8 °C) throughout a 10-cm deep water column, the UV source must yield 40,700 mW s/cm^2. Reduction of flow rates and/or the thickness of the water layer irradiated could improve effectiveness, but these extremely high turbidity levels would still make UV treatments economically nonviable, unless the raw water is filtered prior to irradiation. However, considering that UV technology continues to improve all the time, and that new UV systems can now be used for milk sterilization, this technology has great promise.

Electric Currents

Electric current has been tested in flowing streams of water to shock and disable incoming veligers. Electricity has also been tested as impressed currents creating an electrical field on or just above a substrate to prevent the settlement of macrofoulers. In experiments aimed at veliger destruction, veligers, post veligers, and juveniles were killed when passing through a strong electrostatic field. Exposure to 100 V/cm continuous alternating current (AC) field for 0.25 s seems to result in permanent physiological damage to veligers (McKay 1991).

The literature suggests that successful mitigation requires either a very high electrical intensity and short exposure, or a longer exposure time at lower intensity to achieve the desired effect. Lange et al. (1993) tested the effects of low current fields, using both AC and direct current (DC). Using AC, at field strengths of up to 17 V/cm and veliger exposure of 0.1 s, there was no observed decrease in mussel settlement after they passed through the field. The DC voltages tested appeared to enhance rather than discourage settlement. Schoenbach et al. (1997) found that increased voltage at a reduced pulse rate was more effective than lower voltage at a higher pulse rate. Smythe and Lange (1999), using an electric pulse of 6000–8000 V/cm at 40 Hz, determined that increasing the pulse frequency showed a corresponding increase in veliger mortality and decrease in settlement activity.

Smythe and Dardoe (1999), in reviewing the relevant literature, found that pulse rate, amplitude, duration, and amplitude shape were all important factors affecting the success of a system. Smythe and Miller (2003) found that when using a current of 100–110 A, approximately 45 pulses produced a maximum mortality rate of 40 % in planktonic veligers.

Studies performed on *L. fortunei* pediveligers found that pulsed electric fields of >40 V/cm at 25 pulses/s and >60 V/cm at 6 pulses/s were effective in stunning veligers and that the time required for larval recovery increased proportionately with increasing voltage (Satuito et al. 2000).

The effect of DC on the swimming behavior of *L. fortunei* veligers was examined by Katsuyama et al. (2005). They constructed an experimental apparatus, which had two channels for water flow. One was used as control and the other for applying DC to swimming larvae passing through the channel (Fig. 2). This antifouling strategy aims to stun the swimming larva of *L. fortunei* and thereby prevent settlement. There were two different arrangements of electrodes for generating pulse in the channel (Fig. 2). Pulse voltages of 5–7 kV were examined for their ability to stun larvae of *L. fortunei*. The water flow rates inside the channels varied (0.2, 0.1, 0.025 m/s), and the pulse frequencies examined were 2, 4, and 6 pulses per second.

Under configuration 1 (Fig. 2a), the authors found that a pulse voltage of 5 kV and an electric stimulus (i.e., the recorded total pulse discharge time multiplied by the pulse voltage) between 800 and 2000 resulted in percentages of stunned larvae of 20 and 100 %, respectively. With a pulsed voltage of 7 kV, ~80 % of the larvae were stunned at an electric stimulus of 400, whereas a stimulus above 3000 stunned 100 % of the larvae with none being able to swim after 20 min. Configuration 2 (Fig. 2b) resulted in 75 % of the larvae being stunned at an electric stimulus of 270, which proved the most effective in the experiment.

Additional laboratory experiments performed by Katsuyama et al. (2005) showed that under static conditions 100 % of the larvae were immobilized immediately at 0.8 kV with 200 pulses/s. Recovery was observed after approximately 15–20 min with configuration 1 (at 5 and 7 kV) and 2 (7 kV). These results suggested that, in order to suppress larval recovery, stimuli around 2000–3000 are required, resulting in mortality of most veligers. However, since the retention time of cooling water in the system is usually around 20 min, lower electrical stimuli may be sufficient to prevent their attachment and allow them to be flushed out.

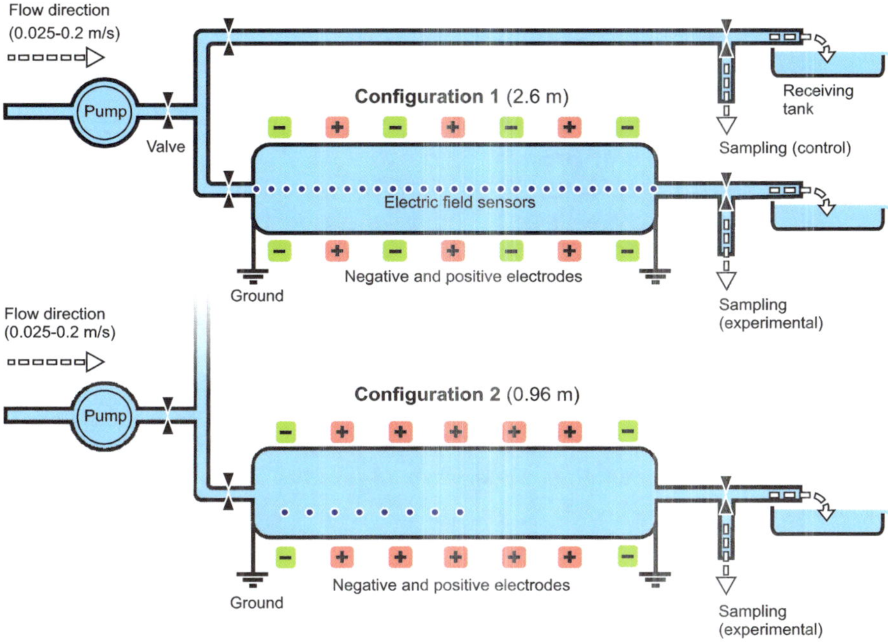

Fig. 2 Configuration of electrodes for DC current experiments to control of *Limnoperna fortunei* attachment carried out by Katsuyama et al. DC direct current (2005; Modified after Katsuyama et al. 2005)

Field studies suggest that electric current densities commonly used for protection of steel surfaces against corrosion in freshwater environments deter zebra mussel settlement. Significant reduction in settlement was accomplished at 65–97 mA/m^2 as compared to the control group (Lewis and Pawson 1993).

Tests conducted at a power plant on the Great Lakes (Canada–USA) using anode mesh panels in the pump wells showed a reduction of settled mussels versus control using a current density of less than 108 mA/m^2. Unfortunately, the experiment was discontinued before long-term studies could be completed.

Electrodes installed at the Nanticoke Thermal Generating Station (Ontario, Canada) on both concrete and steel panels, when using a voltage of 20 V and currents of 1.52 and 1.88 A, respectively, prevented mussel settlement on either electrified surface, while the control panels had mussel densities of 2877 ind./m^2 (Fears and Mackie 1997). In a similar study, complete prevention of settlement was achieved on both steel and wood surfaces utilizing a voltage of 3.15 V/cm (Fears and Mackie 1995).

The overall conclusion at this time is that the voltages necessary, the required length of exposure, and the amount of power needed make this technique of control impractical for most industrial applications. We are not aware of any commercial systems currently using this technology.

Acoustics

Although the use of sound as a deterrent to mussel settlement is frequently mentioned, the research performed in this area often produced controversial results. The earliest recorded work is by Breitig (1965), who worked with dreissenids. He used high-frequency ultrasound at 22 and 800 kHz, with sound intensities ranging from 2 to 5 W/cm. In this experiment, he determined that mortality was achieved through the effects of cavitation alone, and that it could cause veliger destruction (22 kHz) or perforation of the soft body (800 kHz). Differences in mortality rates between the two frequencies were negligible; at both frequencies, 70 % mortality was recorded after 3 s, 90 % after 1 min, and 100 % after 3 min.

Later macrofouling control tests have been performed using sound frequencies between 20 and 60 Hz, and also at frequencies greater than 20 kHz. Results were often inconclusive or impractical, possibly because of the difficulty of effectively measuring, installing, and reliably operating some of the sound-generating devices proposed.

Empire State Electric Energy Research Corporation (ESEERCO 1992) describes how acoustic energy in the range of 39–41 kHz fragments veligers "within a few seconds" in flowing water. Attached adults are killed in 19–24 h. Donskoy and Ludyanskiy (1995) found that low-frequency (200 Hz) sound waves combined with sound waves of pressure intensity 315 Pa produced vibration that could prevent settlement. The combined vibratory effect was found to kill veligers, achieving a near-100 % mortality rate using a sound frequency of < 200 Hz and a sound level of 177 dB. This combination also prevented translocating juveniles and adults from settling on surfaces being bombarded with the treatment.

Santos et al. (2012b) tested the effect of ultrasound in the laboratory. They found that, at a dose of 44 kWs/L, high frequencies (> 100 kHz) killed up to 30 % of *L. fortunei* larvae, whereas 100 % mortality was achieved at 20 kHz. Low frequencies (20 kHz) yielded total veliger mortalities at doses > 10 kWs/L.

Although acoustic deterrents appear to be feasible, the reproducibility of some of the experiments is questionable and at this time there appear to be no commercial devices on the market aimed at the control of either dreissenids or golden mussels.

Water Velocity

When the speed of flow in a raw water system continuously exceeds 1.5 m/s, there is minimal, if any dreissenid veliger settlement, and in some power plants, clusters of mussels have been observed to detach (Jenner et al. 1998). Dense settlements of zebra mussels have been observed in areas where water velocity ranges from 10 to 50 cm/s, but mussels are essentially absent in areas where water velocity exceeds 1 m/s (Jenner et al. 1998).

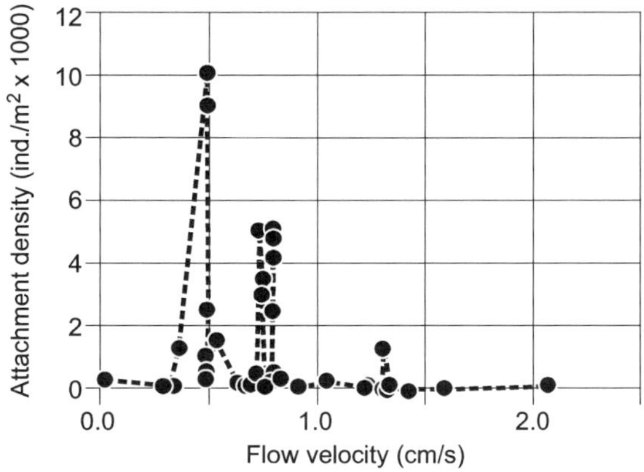

Fig. 3 Correlation between attachment density of larvae of *Limnoperna fortunei* and flow velocity along a tunnel. (After Xu et al. 2012)

Correlation between settlement and flow speed has also been observed for golden mussels. Matsui et al. (2002) concluded that the threshold water velocity values below which golden mussel attachment increased significantly were around 1.2–1.3 m/s. These results generally align with those of Xu et al. (2012), who found that the highest golden mussel densities occurred within 0–1 km from the pipeline entrance and in flow velocities of 0.4–0.9 m/s (Fig. 3). To remove attached adult mussels from the substrate, water velocities in excess of 2.2 m/s were required.

Very few systems are designed for fast flow rates, and it would involve a major expense to redesign them even if it were possible. In fact, intake structures are frequently designed to maintain as slow a rate of flow as possible to prevent entrainment of fish. However, avoiding taking a system temporarily out of service during the mussel-breeding season, whenever possible, is a sound strategy. Temporary shutdown of hydroelectric units usually means their penstocks remain filled with water. During the shutdown, any veligers present in the water will be able to attach to the penstock walls, an area which usually is free of mussels due to flow. After restart, these mussels may be able to stay attached and continue to grow, especially if flow interruptions are a daily occurrence.

Magnetic Fields

Several studies have been performed to determine whether a standing magnetic field can perceptibly affect mussel settlement. Although no solid conclusions about the mechanism responsible for any magnetic effects can be confirmed at this time, the dominant theories maintain that either the magnetism causes soft tissue damage directly, especially to the gills (Barnes et al. 1998), or that elemental uptake and availability of certain minerals (such as magnesium and calcium) may be impaired. The studies performed in this area have yielded very mixed results.

Smythe et al. (1997) found no significant mortality or settlement inhibition after treating mussels with water that passed through a 3000-Gs (Gauss) field. After 99 days, the treated mussels were not significantly different from the mussels in the control group in terms of mortality, shell length, or calcium–magnesium content in soft tissues. In contrast, Ryan (1998) concluded that zebra mussels exposed to an induced electromagnetic field in a closed system died within 7–15 days. Loss of calcium by the mussels was thought to be the cause of mortality, but the experimental details were not well documented.

Magnetic field is considered by Companhia Riograndense de Saneamento (CORSAN; Brazil; Dengo and Carraro 2013) as one of the effective technologies to control golden mussels in a water intake. In 2005, CORSAN installed a commercial magnetic device on a small water intake. After 3 months of operation, it appeared that there was a decrease in the infestation of the pipe, while an increase of pump efficiency was noted (Ratkiewicz 2006).

Turbulence

Turbulence in water is initiated when flow is destabilized, while moving along a surface that has sufficient roughness to cause formation of unstable patterns of eddies. This may occur in a pipe or in an open channel. Some properties of turbulence are of interest when considering particles within the fluid. First, turbulence is diffusive. It rapidly increases mixing in the fluid so that all particles in all areas of the fluid from the pipe wall to the pipe center are exposed to eddies. Next, turbulence is dissipative. This means that eddies formed are initially large and then shed into progressively smaller lengths until becoming small enough that viscous forces convert them into heat. In an aquatic environment, turbulent mixing affects a number of biological processes (Peters and Marrase 2000). It may act beneficially by increasing contact opportunities for planktonic organisms and their food source, and increasing contact between gametes in events of external fertilization. Conversely, it may hinder these processes, carrying elements away from one another or causing physical damage to organisms and gametes.

Dreissenid veligers, like other planktonic organisms, can be affected by turbulent environments. Research suggests that there is a rapid increase in mortality when the size of the veliger is greater than the size of the smallest eddy in turbulent flow. The time of exposure is also an important factor. Horvath and Lamberti (1999) observed an exponential reduction in veliger survival with distance downstream in Christiana Creek (Michigan, USA). They suggested that veliger mortality may have been a result of prolonged exposure to turbulence which may be a key factor in limiting mussel distribution. In a laboratory study where dreissenid veligers were exposed to low and high turbulence for 1, 24, and 48 h, Horvath and Crane (2010) observed a significant increase in mortality after 48 h for both the low- and high-turbulent environments. To date, laboratory tests have not established the relationship of combination of turbulence intensity with contact time versus veliger mortality levels.

The exact mechanism by which turbulence affects dreissenid veligers is unknown; however, it is believed to be related to inertial and shear forces present in the turbulent flow. Inertial forces dominate in large-scale turbulent eddies. These large eddies transfer kinetic energy and give rise to smaller eddies until they are small enough to be influenced by viscous forces which dissipate the kinetic energy of the eddies. The point where dominance by inertial forces transitions to dominance by viscous forces is referred to as the Kolmogorov microscale. The Kolmogorov length scale, which is the length of the smallest eddy in a turbulent fluid, is found using:

$$\eta = \left(\frac{\upsilon^3}{\varepsilon} \right)^{\frac{1}{4}},$$

where υ is the kinematic viscosity of the fluid in m^2/s and ε is the rate of energy dissipation in m^2/s^3. Particles in a fluid that are larger than η will be influenced by turbulent shear forces, whereas particles that are smaller than η will be moved along with the flow.

In an effort to explain the effects of turbulence on dreissenid veligers, Rehmann et al. (2003) defined d^* as the ratio of the size of a veliger, d_{veliger}, to the Kolmogorov length scale, η, or:

$$d^* = \frac{d_{\text{veliger}}}{\eta}.$$

They concluded for the ratio where $d^* > 0.9$, veligers are large enough relative to the smallest turbulent eddies that they are significantly affected by turbulent forces. According to Jessopp (2007), bivalve veligers are thin shelled and therefore prone to damage when colliding with suspended particles or the substrate. He observed signs of shell damage in veligers collected downstream from a turbulent stretch of water in Lough Hyne Marine Nature Reserve (Ireland). Jessopp (2007) was unable to discern, however, if the damage was from water transport or sample collection. Horvath and Lamberti (1999) frequently found empty, unbroken veliger shells in samples from Christiana Creek, suggesting the veligers were pulled open. Veligers may be subject to high shear stress in turbulent environments that may pull apart their shells when feeding or actively swimming (Jessopp 2007).

Rehmann et al. (2003) observed zebra mussel veligers being killed in the turbulent flow created by an aerating pump when the eddy scale was comparable with veliger size. Working with *L. fortunei,* Xu et al. (2013) reported that exposure to turbulent flow created by forcing water through perforated plates for more than 5 min resulted in greater than 80 % mortality of veligers. Mortality increased to 100 % when the exposure time was greater than 10 min.

Sacrificial Substrate as Means of Decreasing Downstream Settlement

The use of preferred substrates to attract settling molluscs is a technique used in the aquaculture of marine mussels worldwide. The same concept was described by Smit et al. (1993) when devising a "biological filter" which consisted of a fine-mesh net which had been colonized by dreissenids, rolled up, and used as filter to remediate the water from a polluted stream before it reached a small inland lake. More recently, Nakazato (2004) and Bodamer and Bossenbroek (2008) suggested that water flowing through heavily vegetated wetlands will lose most of the ready-to-settle veligers due to attachment to aquatic plants.

Hanging easy to remove fine mesh cloth in front of areas to be protected from settlement could be an alternative to having to mechanically clean those areas on a regular basis. The issue is not efficacy, veligers will readily settle on such substrates, but efficiency in having to deploy and remove large lengths of the sacrificial substrate. Xu (2013) suggested bamboo and cloth to attract the veliger settlement and then removing the colonized substrate prior to the mussels reaching reproductive age.

References

Barnes DK, Alix MS, Camp JW (1998) The effects of magnetic water treatment on Zebra mussels in laboratory and industrial settings. In: 8th International Zebra Mussel and Aquatic Nuisance Species Conference, Sacramento (USA)

Bodamer BL, Bossenbroek JM (2008) Wetlands as barriers: effects of vegetated waterways on downstream dispersal of zebra mussels. Freshw Biol 53:2051–2060

Breitig G (1965) The use of ultrasound in the eradication of larvae. PhD Thesis, Mathematisch-Naturwissenschaftliche Fakultat, University Greiswald, pp 1–169

Chalker-Scott L, Carnevale R, Smith K, Scott J (1993) Comparison of acute and chronic mid-range Ultraviolet radiation (280–320 nm) effect on adult zebra mussels. In: Third International Zebra Mussel Conference, Toronto (Canada)

Chalker-Scott L, Scott J, Titus J, Scalia J (1994) Influence of wide-range ultraviolet radiation upon behavior and mortality of *Dreissena polymorpha* . In: 4th International Zebra Mussel Conference, Madison (USA)

Claudi R, Prescott TH (2013) Evaluation of UV Technology at Hoover Dam as means of eliminating downstream settlement of dreissenid mussel veligers. Bureau of Reclamation, Task Order Number: R12PD80180, IDIQ Contract: R10PC80264, pp 1–22

Claudi R, Prescott TH, Coffey H (2014) Atlantium Technologies medium pressure UV dose required for minimizing downstream settlement of quagga mussel veligers. Report for Bureau of Reclamation, IDIQ Contract R13PD80500, pp 1–29

Dengo RR, Carraro I (2013) Controle preventivo do mexilhão dourado em sistemas de abastecimento de água da CORSAN. Trabalho prévio de seleção de tecnologias de controle. CORSAN Report. Porto Alegre (Brazil), pp 1–38

Depetris PJ, Kempe S (1993) Carbon dynamics and sources in the Paraná River. Limnol Oceanogr 38:382–395

Donskoy DM, Ludyanskiy ML (1995) Low frequency sound as a control measure for zebra mussel fouling. In: 5th International Zebra Mussel and other Aquatic Nuisance Organisms Conference, Toronto (Canada)
ESEERCO (Empire State Electric Energy Research Corporation) (1992) Evaluation of ultrasound for zebra mussel mitigation. Research Report EP 91-17, pp 1–27
Evans DW, Lewis D, Whitby E, McKenna G (1995) UV light for zebra mussel control. In: US Army Corps of Engineers Zebra Mussel Workshop, New Orleans (USA)
Fears CD, Mackie GL (1995) Efficacy of low level electric current (AC) for controlling quagga mussels in the Welland Canal. In: 5th International Zebra Mussel and other Aquatic Nuisance Organisms Conference, Toronto (Canada)
Fears CD, Mackie GL (1997) Use of low level electric current (AC) to prevent settlement of zebra/quagga mussels on concrete and steel panels at Nanticoke TGS. In: D'Itri FM (ed) Zebra mussels and aquatic nuisance species. Lewis Publishers, Boca Raton, pp 407–416
Horvath TG, Crane L (2010) Hydrodynamic forces affect larval zebra mussel (*Dreissena polymorpha*) mortality in a laboratory setting. Aquat Invasions 5:379–385
Horvath TG, Lamberti GA (1999) Mortality of zebra mussel, *Dreissena polymorpha*, veligers during downstream transport. Freshw Biol 42:69–76
Jenner HA, Whitehouse JA, Taylor CJL, Khalanski M (1998) Cooling water management in European power stations. Biology and control of fouling. Electricité de France, Paris, pp 1–225
Jessopp MJ (2007) The quick and the dead: larval mortality due to turbulent tidal transport. J Mar Biol Assoc U K 87:675–680
Katsuyama I, Satuito CG, Maeda T, Oonishi M, Kumagai T (2005) The effect of DC-pulse electric stimulus on the swimming behavior of larvae of the freshwater mussel *Limnoperna fortunei* in flowing water within a pipe. Sess Org 2:1–5 [In Japanese]
Lange CL, Smythe GA, Doyle JF, Sawyko PM (1993) Application of low voltage electric fields to deter attachment of zebra mussels to structures. In: Third International Zebra Mussel Conference, Toronto (Canada)
Lewis DP, Cairns W (1998) Ultraviolet control for Zebra Mussel—the evolution of a non chemical alternative. In: 8th International Zebra Mussel and Aquatic Nuisance Species Conference, Sacramento (USA)
Lewis DP, Pawson RL (1993) Use of cathodic protection as a means of control and removal of zebra mussels. In: Third International Zebra Mussel Conference, Toronto (Canada)
Lewis DP, Whitby GE (1993) Use of ultraviolet radiation for zebra mussel control. In: Third International Zebra Mussel Conference, Toronto (Canada)
Lewis DP, Whitby GE (1996) Control of Zebra Mussel with medium and low pressure ultraviolet radiation: optimization study. Aquatic Sciences, Project M9295 pp 1–38
Matsui Y, Nagaya K, Funahashi G, Goto Y, Yuasa A, Yamamoto H, Ohkawa K, Magara Y (2002) Effectiveness of antifouling coatings and water flow in controlling attachment of the nuisance mussel *Limnoperna fortunei* . Biofouling 18:137–148
McKay S (1991) Control of zebra mussels by electroshocking. Ontario Hydro Report 91-171 K, pp 1–41
Nakazato H (2004) Method of decreasing a sessile bivalve population. The United States Patent and Trademark Office (Pub. No.: 2004/0234629 A1)
Perepelizin PV, Boltovskoy D (2014) Effects of 254 nm UV irradiation on the mobility and survival of larvae of the fouling invasive mussel *Limnoperna fortunei*. Biofouling 30:197–202
Peters F, Marrase C (2000) Effects of turbulence on plankton: an overview of experimental evidence and some theoretical considerations. Mar Ecol Prog Ser 205:291–306
Pickles SB (2000) Use of ultraviolet radiation for zebra mussel control at Ontario Power Generation. In: 10th International Aquatic Nuisance Species and Zebra Mussels Conference, Toronto (Canada)
Ratkiewicz MLA (2006) Relatório da experiência piloto dos equipamentos da LGM-DANSUL para controle da infestação por mexilhões dourados na EBAB e incrustações na adutora de água bruta da US de Barra do Ribeiro. DEPEDET/SUMOP/CORSAN. Porto Alegre (Bazil), pp 1–18

Rehmann CR, Stoeckel JA, Schneider DW (2003) Effect of turbulence on the mortality of zebra mussel veligers. Can J Zool 81:1063–1069

Ryan MF (1998) Extremely low frequency electromagnetism: an effective nonchemical method for control of zebra mussel infestation. In: 8th International Zebra Mussel and Aquatic Nuisance Species Conference, Sacramento (USA)

Santos CP, Raya Rodriguez MT, Mansur MCD, Vilar Nehrke M, Leite Zurita ML (2012a) Controle de *Limnoperna fortunei* com aplicação de radiação ultravioleta. In: Mansur MCD, Santos CP, Pereira D, Padula PIC, Leite ZML, Raya RMT, Vilar NM, Aydos BPE (eds) Moluscos límnicos invasores no Brasil. Biologia, prevenção, controle. Redes Editora, Porto Alegre, pp 335–338

Santos CP, Vicenzi J, Berutti FA, Mansur MCD, Perez Bergmann C, Raya Rodriguez MT, Vilar Nehrke M, Leite Zurita ML (2012b) Controle de bivalves com a utilização do ultrassom. In: Mansur MCD, Santos CP, Pereira D, Padula PIC, Leite ZML, Raya RMT, Vilar NM, Aydos BPE (eds) Moluscos límnicos invasores no Brasil. Biologia, prevenção, controle. Redes Editora, Porto Alegre, pp 339–341

Satuito CG, Katuyama I, Den S (2000) Effect of electric fields on the planktonic-staged larvae of fouling organisms. In: 10th International Aquatic Nuisance Species and Zebra Mussels Conference, Toronto (Canada)

Schoenbach KH, Abou-Ghazala A, Alden RW, Turner R, Fox TJ (1997) Biofouling prevention with pulsed electric fields. In: 7th International Zebra Mussel and Aquatic Nuisance Species Conference, New Orleans (USA)

Smit H, bij de Vaate A, Reeders HH, van Nes EH, Noordhuis R (1993) Colonization, ecology, and positive aspects of zebra mussels (*Dreissena polymorpha*) in The Netherlands. In: Nalepa TF, Schloesser D (eds) Zebra mussels: biology, impacts, and control. Lewis Publishers, Boca Raton, pp 55–77

Smythe AG, Dardoe EA (1999) Overview of electrical technologies for controlling dreissenids with emphasis on pulsed-power systems. U.S. Army Engineer and Research and Development Center (Vicksburg, USA). Zebra Mussel Technical Notes Collection (ZMR-3-22), pp 1–6

Smythe AG, Lange CL (1999) Application of pulse-power electric fields as a control for settlement and planktonic-stage zebra mussels. In: 9th International Zebra Mussel and Aquatic Nuisance Species Conference, Duluth (USA)

Smythe AG, Miller AC (2003) Pulse Power: a possible alternative to chemical treatment for zebra mussel control. Summary of 2000 field studies. U.S. Army Engineer and Research and Development Center, Vicksburg, USA. ANSRP Technical Note Collection (ERDC TN-ANSRP-03-2), pp 1–11

Smythe AG, Lange CL, Sawyko PM (1997) Evaluation of an in-line magnetic device as a control method for adult and settlement-stage zebra mussels. In: 7th International Zebra Mussel and Aquatic Nuisance Species Conference, New Orleans (USA)

Xu M (2013) Experimental study of bio-fouling control of *Limnoperna fortunei* in water transfer tunnels. In: 35 International Association of Hydraulic Engineering & Research Congress, Chengdu (China)

Xu M, Cao XW, Wang ZY, Wang XZ (2012) Attachment characteristics of golden mussels (*Limnoperna fortunei*) in water transport projects. J Tsinghua Univ [Qinghua Daxue Xuebao] 52:170–176 [In Chinese]

Xu M, Wang ZY, Lin CC, Pan BZ, Zhao N (2013) Experimental study of invasion and biofouling of freshwater mussel *Limnoperna fortunei*. Int J Geosciences 4:1–7

Printed by Printforce, the Netherlands